Advances in Intelligent Systems and Computing

1459

Series Editor

Janusz Kacprzyk, *Systems Research Institute, Polish Academy of Sciences, Warsaw, Poland*

Advisory Editors

Nikhil R. Pal, *Indian Statistical Institute, Kolkata, India*
Rafael Bello Perez, *Faculty of Mathematics, Physics and Computing, Universidad Central de Las Villas, Santa Clara, Cuba*
Emilio S. Corchado, *University of Salamanca, Salamanca, Spain*
Hani Hagras, *School of Computer Science and Electronic Engineering, University of Essex, Colchester, UK*
László T. Kóczy, *Department of Automation, Széchenyi István University, Gyor, Hungary*
Vladik Kreinovich, *Department of Computer Science, University of Texas at El Paso, El Paso, USA*
Chin-Teng Lin, *Department of Electrical Engineering, National Chiao Tung University, Hsinchu, Taiwan*
Jie Lu, *Faculty of Engineering and Information Technology, University of Technology Sydney, Sydney, Australia*
Patricia Melin, *Graduate Program of Computer Science, Tijuana Institute of Technology, Tijuana, Mexico*
Nadia Nedjah, *Department of Electronics Engineering, University of Rio de Janeiro, Rio de Janeiro, Brazil*
Ngoc Thanh Nguyen , *Faculty of Computer Science and Management, Wrocław University of Technology, Wrocław, Poland*
Jun Wang, *Department of Mechanical and Automation Engineering, The Chinese University of Hong Kong, Shatin, Hong Kong*

The series "Advances in Intelligent Systems and Computing" contains publications on theory, applications, and design methods of Intelligent Systems and Intelligent Computing. Virtually all disciplines such as engineering, natural sciences, computer and information science, ICT, economics, business, e-commerce, environment, healthcare, life science are covered. The list of topics spans all the areas of modern intelligent systems and computing such as: computational intelligence, soft computing including neural networks, fuzzy systems, evolutionary computing and the fusion of these paradigms, social intelligence, ambient intelligence, computational neuroscience, artificial life, virtual worlds and society, cognitive science and systems, Perception and Vision, DNA and immune based systems, self-organizing and adaptive systems, e-Learning and teaching, human-centered and human-centric computing, recommender systems, intelligent control, robotics and mechatronics including human-machine teaming, knowledge-based paradigms, learning paradigms, machine ethics, intelligent data analysis, knowledge management, intelligent agents, intelligent decision making and support, intelligent network security, trust management, interactive entertainment, Web intelligence and multimedia.

The publications within "Advances in Intelligent Systems and Computing" are primarily proceedings of important conferences, symposia and congresses. They cover significant recent developments in the field, both of a foundational and applicable character. An important characteristic feature of the series is the short publication time and world-wide distribution. This permits a rapid and broad dissemination of research results.

Indexed by DBLP, INSPEC, WTI Frankfurt eG, zbMATH, Japanese Science and Technology Agency (JST).

All books published in the series are submitted for consideration in Web of Science.

For proposals from Asia please contact Aninda Bose (aninda.bose@springer.com).

Daniel H. de la Iglesia · Juan F. de Paz Santana ·
Alfonso J. López Rivero
Editors

New Trends in Disruptive Technologies, Tech Ethics, and Artificial Intelligence

The DITTET 2024 Collection

Editors
Daniel H. de la Iglesia
Faculty of Science
University of Salamanca
Salamanca, Spain

Juan F. de Paz Santana
Faculty of Science
University of Salamanca
Salamanca, Spain

Alfonso J. López Rivero
Faculty of Informatics
Pontifical University of Salamanca
Salamanca, Spain

ISSN 2194-5357 ISSN 2194-5365 (electronic)
Advances in Intelligent Systems and Computing
ISBN 978-3-031-66634-6 ISBN 978-3-031-66635-3 (eBook)
https://doi.org/10.1007/978-3-031-66635-3

© The Editor(s) (if applicable) and The Author(s), under exclusive license to Springer Nature Switzerland AG 2024

This work is subject to copyright. All rights are solely and exclusively licensed by the Publisher, whether the whole or part of the material is concerned, specifically the rights of translation, reprinting, reuse of illustrations, recitation, broadcasting, reproduction on microfilms or in any other physical way, and transmission or information storage and retrieval, electronic adaptation, computer software, or by similar or dissimilar methodology now known or hereafter developed.
The use of general descriptive names, registered names, trademarks, service marks, etc. in this publication does not imply, even in the absence of a specific statement, that such names are exempt from the relevant protective laws and regulations and therefore free for general use.
The publisher, the authors and the editors are safe to assume that the advice and information in this book are believed to be true and accurate at the date of publication. Neither the publisher nor the authors or the editors give a warranty, expressed or implied, with respect to the material contained herein or for any errors or omissions that may have been made. The publisher remains neutral with regard to jurisdictional claims in published maps and institutional affiliations.

This Springer imprint is published by the registered company Springer Nature Switzerland AG
The registered company address is: Gewerbestrasse 11, 6330 Cham, Switzerland

If disposing of this product, please recycle the paper.

Preface

In recent years, we have witnessed significant advances in technologies such as artificial intelligence, big data, the Internet of Things, and bioinformatics. These advancements have made it evident that there is a need for a thorough review of current ethical patterns. One of the research fields that is rapidly growing and has a broad future is technology ethics or tech ethics. Until a few years ago, this type of research was a small part and did not involve many technology researchers. Currently, due to the proliferation of new applications of artificial intelligence, numerous initiatives, declarations, principles, guides, and analyses focused on measuring the social impact of these systems and the development of more ethical technology have flourished.

The International Conference on "Disruptive Technologies, Tech Ethics, and Artificial Intelligence" (DITTET 2024) provides a forum to present and discuss the latest scientific and technical advances and their implications in the field of ethics. It also offers an opportunity for experts to present their latest research on disruptive technologies, promoting knowledge transfer. This conference provides a unique platform for bringing together experts from various fields, academia, and professionals to exchange their experiences in the development and deployment of disruptive technologies, artificial intelligence, and their ethical problems.

The aim of DITTET is to gather researchers and developers from industry, humanities, and academia to report on the latest scientific advancements and the application of artificial intelligence, as well as its ethical implications in diverse fields such as climate change, politics, economy, or security in today's world.

Each paper submitted to DITTET underwent a rigorous peer review process by three members of each workshop's international committee. Of the proposals received from over 130 authors, 42 were selected for presentation at the conference.

We would like to express our deepest gratitude to all the contributing authors, as well as to the members of the Program Committees and the Organizing Committee, for their hard work and invaluable contributions. Their dedication and efforts were instrumental to the success of the DITTET 2024 event. Thank you for your support; DITTET 2024 would not have been possible without your contribution.

May 2024

Daniel H. de la Iglesia
Juan F. de Paz Santana
Alfonso J. López Rivero

Organization

Program Committee

Daniel H. de la Iglesia	University of Salamanca, Spain
Juan F. De Paz Santana	University of Salamanca, Spain
Alfonso J. López Rivero	Pontifical University of Salamanca, Spain
Javier Bajo Pérez	Polytechnic University of Madrid, Spain
Gabriel Villarrubia	University of Salamanca, Spain

Scientific Committee

Benjamín Sahelices	University of Valladolid, Spain
Valderi Reis Quietinho Leithardt	Instituto Politécnico de Portalegre, Portugal
Cristian Iván Pinzón	Universidad Tecnológica de Panamá, Panama
María N. Moreno García	University of Salamanca, Spain
Shridhar Devamane	APS College of Engineering, India
Filipe Caldeira	Polytechnic Institute of Vise, Portugal
Vivian F. López Batista	University of Salamanca, Spain
Rui Duarte	Polytechnic Institute of Vise, Portugal
Luiz Faria	Institute of Engineering - Polytechnic of Porto, Portugal
María Navarro Cáceres	University of Salamanca, Spain
Celestino Goncalves	Instituto Politecnico da Guarda, Portugal
Gustavo Rivadera	Universidad Católica de Salta, Argentina
Tiancheng Li	Northwestern Polytechnical University, China
Roberto Berjón	Pontifical University of Salamanca
Vidal Alonso Secades	Pontifical University of Salamanca
M. Encarnación Beato	Pontifical University of Salamanca
Jose Neves	University of Minho, Portugal
Igor Kotenko	St. Petersburg Federal Research Center of the Russian Academy of Sciences, Russia
Sylvie Ratté	École de technologie supérieure, Canada
Juan Pavón	Universidad Complutense de Madrid, Spain
Alejandro Rodriguez	Universidad Politécnica de Madrid, Spain
Rosalia Laza	University of Vigo, Spain
Antonio Marques	Instituto Politécnico do Porto, Portugal
Manuel Martín-Merino Acera	Pontifical University of Salamanca

Manuel José Ruiz García | ProFuturo Foundation - Telefónica, Spain
Armando Pinho | University of Aveiro, Portugal
Satish Gajawada | Indian Institute of Technology Roorkee Alumnus, India
Taha Duri | College of Design, American University in the Emirates, United Arab Emirates
Lauri Tuovinen | University of Oulu, Finland
Anouar Abtoy | National School of Applied Sciences of Tetouan (ENSATe), Abdelmalek Essaâdi University, Morocco
Anne Marte Gardenier | Eindhoven University of Technology, Nederlands
Gualtiero Volpe | InfoMus-DIST-University of Genoa, Italy
Luís Moniz Pereira | NOVA Laboratory for Computer Science and Informatics (NOVA LINCS), Universidade Nova de Lisboa
Chun-Hua Tsai | University of Omaha, United States
Syuan-Yi Chen | Department of Electrical Engineering, National Taiwan Normal University, Taiwan
Hwang-Cheng Wang | National Ilan University, Taiwan
Juddy Yinet Morales Peña | Universidad Distrital Francisco José de Caldas, Colombia
Gianluca Schiavo | University of Trento, Italy
Ulrik Franke | RISE Research Institutes of Sweden & KTH Royal Institute of Technology, Sweden
Francesco Zirilli | Sapienza Universita Roma, Italy
Marcus Düwel | Technical University of Darmstadt, Germany
Francesc Fusté Forné | Universitat de Girona/Lincoln University, Spain
Dr. Khaled Almakadmeh | The Hashemite University, Jordan
Mohammad Abido | King Fahd University of Petroleum & Minerals (KFUPM), Saudi Arabia
Beth Kewell | University of Exeter Business School, England
Dariusz Jacek Jakóbczak | Technical University of Koszalin, Poland
Dr. Mohammad Siraj | Universiti Teknologi Malaysia and King Saud University, Saudi Arabia
Nolen Gertz | University of Twente, Netherlands
Ikvinderpal Singh | Trai Shatabdi GGS Khalsa College, India
Grigorios Beligiannis | University of Patras, Greece
Arnold Aribowo | Pelita Harapan University, Indonesia
Pujianto Yugopuspito | Pelita Harapan University, Indonesia
Amel Ourici | University Badji Mokhtar Annaba, Algeria
Giovanna Sanchez Nieminen | VTT Technical Research Centre of Finland, Finland
Narendra Nath Joshi | IBM Research, United States

Souad Taleb Zouggar	Es senia University Oran, Algeria
Smain Femmam	Université de Haute-Alsace, France
K.-L. Du	Concordia University, Canada
Bui Trung Thanh	Hung Yen University of Technology and Education, Vietnam
Albert Bakhtizin	Central Economics and Mathematics Institute of Russian Academy of Sciences, Russia
Siddhartha Bhattacharyya	VSB Technical University of Ostrava, Czech Republic
Ruhaidah Samsudin	University of Technology Malaysia, Malaysia
Matthew Dennis	Eindhoven University of Technology, Nederlands
João Calado	ISEL - Instituto Superior de Engenharia de Lisboa, Portugal
Prabhat Mahanti	University of New Brunswick, Canada
Sanjay Goel	The State University of New York at Albany, United States
Thomas M. Powers	University of Delaware, United States
Solomiia Fedushko	Lviv Polytechnic National University, Ukraine
Ridda Laouar	Tebessa University, Algeria
Jenifer Sunrise Winter	University of Hawaii at Manoa, United States
Alessio Ishizaka	University of Portsmouth, England
Ong Pauline	Universiti Tun Hussein Onn Malaysia, Malaysia
Hamed Taherdoost	University Technology Malaysia, Malaysia
Hamidreza Rokhsati	Sapienza University of Rome, Italy
Mohamed Waleed Fakhr	Arab Academy for Science, Technology & Maritime Transport, Egypt
Sanna Lehtinen	University of Helsinki/Aalto University, Finland
Altino Sampaio	Instituto Politécnico do Porto, Portugal
Ficco Massimo	Second University of Naples, Italy
Davide Carneiro	Polytechnic Institute of Porto, Portugal
Guillermina Nievas	Universidad Católica de Salta, Argentina
Beatriz Parra	Universidad Católica de Salta, Argentina
Fernando Rivera	Universidad Católica de Salta, Argentina
Giovanny Mauricio Tarazona	Universidad Distrital Francisco José de Caldas, Colombia
Montserrat Mateos Sánchez	Pontifical University of Salamanca
Ana María Fermoso García	Pontifical University of Salamanca
André Fabiano De Moraes	Instituto Federal Catarinense, Brasil

Organising Committee

Fernando Lobato Alejano	Pontifical University of Salamanca
Diego M. Jiménez-Bravo	University of Salamanca
André Sales Mendes	University of Salamanca
Luis Augusto Silva	University of Salamanca
Álvaro Lozano Murciego	University of Salamanca
Héctor Sánchez San Blas	University of Salamanca
Jorge Zakour	Pontifical University of Salamanca
Carlos Chinchilla Corbacho	Pontifical University of Salamanca
Sergio García	University of Salamanca

DITTET 2024 Organizers

DITTET 2024 Collaborates

Contents

Track on Artificial Intelligence

Integrating AI Techniques for Enhanced Management and Operation
of Electric Vehicle Charging Stations 3
   *Filipe Cardoso, José Rosado, Marco Silva, Pedro Martins, Paulo Váz,
   José Silva, and Maryam Abbasi*

Ensemble Learning Models for Wind Power Forecasting 15
   *Samara Deon, José Donizetti de Lima, Geremi Gilson Dranka,
   Matheus Henrique Dal Molin Ribeiro, Julio Cesar Santos dos Anjos,
   Juan Francisco de Paz Santana, and Valderi Reis Quietinho Leithardt*

Predictive Model for Estimating Body Weight Based on Artificial
Intelligence: An Integrated Approach to Pre-processing and Evaluation 28
   Diana M. Figueiredo, Rui P. Duarte, and Carlos A. Cunha

Requirements on and Procurement of Explainable
Algorithms—A Systematic Review of the Literature 40
   *Ulrik Franke, Celine Helgesson Hallström, Henrik Artman,
   and Jacob Dexe*

Mapping Changes in Procurement Management for Project Implementation
in the Food Sector: An Applied Review to the Principles PMBOK®Guide
v.7 with DotProject+ 53
   *Jhony Reinheimer Magalhães, André Fabiano de Moraes,
   and Luis Augusto Silva*

Plant Disease Classification Through Image Representations
with Embeddings 62
   *Arturo Álvarez-Sánchez, Diego M. Jiménez-Bravo, Luís Augusto Silva,
   Álvaro Lozano Murciego, and André Sales Mendes*

Explaining Social Recommendations Using Large Language Models 73
   Md. Ashaduzzaman, Thi Nguyen, and Chun-Hua Tsai

Utilizing Retrieval-Augmented Large Language Models for Pregnancy
Nutrition Advice 85
   *Taranum Bano, Jagadeesh Vadapalli, Bishwa Karki,
   Melissa K. Thoene, Matt VanOrmer, Ann L. Anderson Berry,
   and Chun-Hua Tsai*

Vineyard Leaf Disease Prediction: Bridging the Gap Between Predictive
Accuracy and Interpretability 97
   *Noor E. Mobeen, Sarang Shaikh, Livinus Obiora Nweke,
Mohamed Abomhara, Sule Yildirim Yayilgan, and Muhammad Fahad*

Application of Unmanned Aerial Vehicles for Autonomous Fire Detection 109
   *José Silva, David Sousa, Paulo Vaz, Pedro Martins,
and Alfonso López-Rivero*

Track on Disruptive Technologies

Smart City Air Quality Monitoring: A Mobile Application for Intelligent
Cities 123
   *Pedro Martins, Diogo Silva, João Pinto, José Varanda, Paulo Váz,
José Silva, and Maryam Abbasi*

On the Use of Message Brokers for Real-Time Monitoring Systems 133
   Manuel Lopes, Luciano Correia, João Henriques, and Filipe Caldeira

Incorporating Electric Vehicles in Strategic Management or Value Creation
Initiatives with a Focus on Sustainability? 148
   *Sónia Gouveia, Daniel H. de la Iglesia, José Luís Abrantes,
Alfonso J. López Rivero, Elisabete Silva, Eduardo Gouveia,
and Vasco Santos*

A Framework for Monitoring Pollution Levels in Smart Cities 159
   *Diogo Silva, João Pinto, José Varanda, João Henriques,
Filipe Caldeira, and Cristina Wanzeller*

Value Creation and Strategic Management in the Era of Digital
Transformation: A Bibliometric Analysis and Systematic Literature Review ... 171
   *Sónia Gouveia, Daniel H. de la Iglesia, José Luís Abrantes,
and Alfonso J. López Rivero*

A Framework for Wood Moisture Control in Industrial Environment 180
   *Ricardo Cláudio, Francisco Soares, Jorge Leitão, João Henriques,
Filipe Caldeira, and Cristina Wanzeller*

EgiCool - IoT Solution for Datacenter Environmental Control and Energy
Consumption Monitoring 194
   *Vicente Gonçalves, José Daniel, Celestino Gonçalves, Filipe Caetano,
and Clara Silveira*

Development of an Autonomous Device for People Detection 207
  José Silva, Gabriel Raperger, Paulo Vaz, Pedro Martins,
  and Alfonso López-Rivero

PetWatcher – Ubiquitous Device Proximity Location System 219
  Laura Fernandim, Celestino Gonçalves, Filipe Caetano,
  and Clara Silveira

Integration of a Mobile Robot in the ROS Environment. Analysis
of the Implementation in Different Versions 231
  Sergio García González, Vidal Moreno Rodilla,
  Francisco Javier Blanco Rodríguez, Belen Curto Diego,
  Héctor Sánchez San Blas, André Filipe Sales Mendes,
  and Gabriel Villarrubia González

The Biomimicry Database: An Integrated Platform to Enhancing
Knowledge Sharing in Biomimetics 244
  Vagner Bom Jesus, Clara Silveira, and Carlos Carreto

A Monitoring Framework for Smart Building Facilities Management 256
  Eduardo Pina, José Ramos, João Henriques, Filipe Caldeira,
  and Cristina Wanzeller

Airsense – Low-Cost Indoor Air Quality Monitoring Wireless System 269
  Rafael Marques, Celestino Gonçalves, and Rui Ferreira

An Audit Framework for Civil Construction Safety Management
and Supervision 281
  Luciano Correia, Manuel Lopes, Filipe Caldeira, and João Henriques

Track on Technological Ethics

Bytes and Battles: Pathways to De-escalation in the Cyber Conflict Arena 295
  Lambèr Royakkers

A Positive Perspective to Redesign the Online Public Sphere:
A Deliberative Democracy Approach 307
  Roxanne van der Puil

A Comparative Analysis of Model Alignment Regarding AI Ethics
Principles 319
  Guilherme Palumbo, Davide Carneiro, and Victor Alves

The Fictional Pact in Human-Machine Interaction. Notes on Empathy
and Violence Towards Robots 331
  Daniel Blanco Parra and Lucía Martín-Gómez

Moral Asymmetries in LLMs 346
  Nadiya Slobodenyuk

Against Skepticism and Deterioration: Art and the Construction of Peace 355
  Taha Duri

Track on Doctoral Consortium

A Comparison of DoDAF, TOGAF, and FEAF: Architectural Frameworks
for Effective Systems Design 363
  Bernardo Gaudêncio, João Ferraz, Pedro Martins, Paulo Váz,
  José Silva, Maryam Abbasi, and Filipe Cardoso

From Waste to Wealth: Circular Economy Approaches for Recycled EV
Batteries in Energy Storage 372
  Alejandro H. de la Iglesia, Carlos Chinchilla Corbacho,
  Jorge Zakour Dib, and Fernando Lobato Alejano

An IoT-Based Framework for Smart Homes 380
  André Bastos, Carlos Silva, Luís Pais, João Henriques, Filipe Caldeira,
  and Cristina Wanzeller

An IoT Framework for Improved Vineyards Treatment in Grape Farming 389
  Henrique Jorge, Manuel Aidos, Rodrigo Cristovam, João Henriques,
  Filipe Caldeira, and Cristina Wanzeller

Strengthening the Role of Citizens in Governing Disruptive Technologies:
The Case of Dutch Volunteer Hackers 399
  Anne Marte Gardenier

Consumer Behaviour in the AI Era 410
  Ana Ribeiro, Alfonso Rivero, and José Luís Abrantes

Evaluation of the Effect of Side Information on LLM Rankers
for Recommender Systems 416
  Adrián Valera Román, Álvaro Lozano Murciego,
  and María N. Moreno-García

A Monitoring Framework to Assess Air Quality on Car Parks 423
  Tiago Almeida, Pedro Monteiro, João Henriques, Filipe Caldeira,
  and Cristina Wanzeller

Disruptive Technologies Applied to Digital Education

Gamification, PBL Methodology and Neural Network to Boost
and Improve Academic Performance 431
   Filipe William C. Almeida, André Fabiano de Moraes,
   Rafael de Moura Speroni, and Luis Augusto Silva

Wrap Your Mind Around Education: Applying Hugging Face to a Chatbot
with AI 444
   Dafne Itzel Rojas González, Josue Aaron Soriano Rivero,
   and Jatziri Hernandez Hernandez

An Explainable Clustering Methodology for the Categorization
of Teachers in Digital Learning Platforms Based on Their Performance 455
   Emma Pérez García, María Rosa Hortelano Díaz,
   and Eva Martín Rodríguez

Educational Platform for Inclusive Learning with Deep Camera Integration
and Serious Games 464
   Héctor Sánchez San Blas, Rocío Galache Iglesias,
   Enrique Maya-Cámara, Blanca García-Riaza,
   Ana Paula Couceiro Figueira, Josué Prieto-Prieto,
   and André Sales Mendes

Author Index 475

Track on Artificial Intelligence

Integrating AI Techniques for Enhanced Management and Operation of Electric Vehicle Charging Stations

Filipe Cardoso[1], José Rosado[3], Marco Silva[3], Pedro Martins[2(✉)], Paulo Váz[2], José Silva[2], and Maryam Abbasi[3]

[1] Polytechnic Institute of Santarém, Santarém Higher School of Management and Technology, Porto, Portugal
filipe.cardoso@esg.ipsantarem.pt
[2] CISeD - Digital Services Research Center, Polytechnic of Viseu, Viseu, Portugal
{pedromom,paulovaz,jsilva}@estgv.ipv.pt
[3] Applied Research Institute, Coimbra Polytechnic, Coimbra, Portugal
{jfr,msilva}@isec.pt, maryam.abbasi@ipc.pt

Abstract. Managing Electric Vehicle (EV) charging stations poses challenges due to limited contracted power. To address this, we introduce the Intelligent Electric Vehicle Charging Controller (IEVCC) prototype. IEVCC optimizes power utilization without additional costs and can operate independently or in a network. A designated manager allocates power among chargers through load balancing and prioritization.

Our system incorporates predictive charging, using AI to anticipate users' needs based on historical data and preferences, ensuring availability. Additionally, AI-driven load balancing efficiently distributes power among chargers, preventing overload.

Dynamic pricing strategies incentivize efficient charging by adjusting tariffs based on demand and available power, promoting sustainability

In conclusion, the IEVCC offers a holistic solution for managing EV charging stations. By optimizing power usage, integrating predictive charging, load balancing, and dynamic pricing, it enhances efficiency and ensures prudent resource utilization in the transition towards sustainable transportation.

Keywords: Electric Vehicle (EV) · Charging Stations · Intelligent Charging Controller · Predictive Charging · Dynamic Pricing

1 Introduction

The surge in Electric Vehicle (EV) sales worldwide, driven by factors such as low Total Cost of Ownership (TCO), environmental concerns, and government incentives, underscores the increasing adoption of EVs [1,2]. Despite temporary setbacks due to the COVID-19 pandemic, EV sales witnessed a remarkable 108% increase last year compared to 2020 [3], with projections indicating continued growth albeit at a slightly slower pace [4–6]. These trends align with legislative initiatives like the European Commission's target of achieving climate neutrality by 2050 and a 55% reduction in greenhouse gas emissions by 2030 [2]. The urgency to reduce fossil fuel dependency has made the transition to EVs imperative, as evidenced by major automakers' commitments to become fully electric by 2030–2035 [13].

However, despite the increasing popularity of EVs, challenges persist, particularly regarding access to private charging facilities and the availability and affordability of public charging stations. Limited access to convenient charging infrastructure and inadequate residential electric infrastructure pose significant barriers. The standard 16A Schuko plug commonly found in residential buildings often fails to meet the charging demands of modern EVs, which require higher charging capacities.

To address these challenges, this paper proposes an Intelligent Electric Vehicle Charger Controller that utilizes AI techniques to enhance EV charging station operation and management. The system integrates Predictive Charging, Load Balancing, and Dynamic Pricing to optimize charging schedules, maximize power utilization efficiency, and incentivize sustainable charging practices. Experimental evaluations demonstrate the effectiveness of this approach in enhancing charging infrastructure efficiency, availability, and sustainability.

In conclusion, the integration of AI-driven techniques offers a promising solution to the challenges faced in managing EV charging stations. By addressing limitations in charging infrastructure and promoting sustainable practices, this research contributes to the advancement of EV technology and supports the transition to a greener transportation system.

2   AI driven Electric Vehicle Charging Controller

In order to support the AI-driven Predictive Charging, Load Balancing, and Dynamic Pricing strategies, the hardware (Fig. 1) of the Intelligent Electric Vehicle Charger Controller (IEVCC) system is designed to accommodate these functionalities (Fig. 2).

Fig. 1. IEVCC board

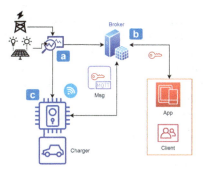

Fig. 2. Diagram of the Standalone Version.

Figure 2 shows a diagram of the system, which is composed by the:

- Consumption monitor hardware (a);
- Consumption monitor system - broker (b);
- Intelligent Electric Vehicle Charging Controller (c).

Firstly, the hardware incorporates advanced processing capabilities to enable predictive charging. The AI algorithms responsible for predicting the charging behavior of EV owners based on historical data and user preferences require computational power for analysis and decision-making. The hardware is equipped with a powerful processor or microcontroller, such as the expressif ESP32 board, which can handle complex calculations and perform real-time predictions. Using the processing capabilities of the hardware, the system can optimize charging schedules and ensure that charging stations are available when they are most needed.

Second, the hardware is designed to facilitate load balancing among multiple electric vehicle chargers. Load balancing techniques involve intelligent distribution of the available power among chargers within a building or parking lot. The hardware integrates communication interfaces and protocols to enable efficient data exchange between the chargers and the central controller. This allows the controller to monitor the power usage of individual chargers and dynamically allocate power resources based on demand and availability. By coordinating the charging activities of multiple EVs, the hardware ensures that the electrical infrastructure is not overloaded and maximizes the utilization of the available power.

Lastly, the hardware supports the implementation of dynamic pricing strategies. This involves the ability to adjust charging prices based on various factors, such as demand, time of day, and availability of renewable energy. The hardware is designed to facilitate communication with external systems or networks that provide information related to pricing and energy availability. When integrated with these systems, the hardware can receive real-time data and make pricing decisions accordingly. The hardware includes the necessary interfaces, such as Wi-Fi or cellular connectivity, to establish communication and receive pricing information. This enables the system to incentivize users to charge their vehicles during off-peak hours or when renewable energy sources are abundant, effectively managing the demand on the charging infrastructure.

Overall, the hardware of the IEVCC system is carefully designed to support AI-driven predictive charging, load balancing, and dynamic pricing functionalities. Its processing capabilities, communication interfaces, and connectivity options enable the system to optimize charging schedules, efficiently distribute power, and implement dynamic pricing strategies. By leveraging the hardware capabilities, the IEVCC system enhances the operation and management of EV charging stations, contributing to the availability, efficiency, and sustainability of the charging infrastructure.

2.1 Mesh Version

In the mesh version of the Intelligent Electric Vehicle Charger Controller (IEVCC) system, the integration of AI-driven Predictive Charging, Load Balancing, and Dynamic Pricing further enhances its functionality and addresses the challenges specific to condominium environments.

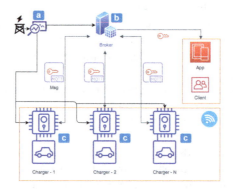

Fig. 3. Diagram of the Mesh Version.

To support Predictive Charging in the Mesh version (Fig. 3), AI algorithms are used to predict the charging behavior of EV owners residing in the condominium. By analyzing historical data and user preferences, the system can anticipate the charging patterns of individual users, even with multiple users having different needs or rights. The hardware includes a consumption monitor (a) that captures the charging data and communicates it to the central Broker/Manager (b). The AI algorithms running on the Broker/Manager analyze the data to optimize charging schedules, ensuring the availability of charging stations when they are most needed. This integration enables efficient charging management for multiple users within the condominium, addressing the unique infrastructure conditioning factors, such as cable thermal limits or the maximum available power from the energy provider.

Load Balancing in the Mesh version becomes more challenging due to the presence of multiple users with different charging needs. The hardware incorporates the Intelligent Electric Vehicle Charging Controller with energy meter (c), which facilitates load balancing among the chargers in the condominium. The central Broker/Manager receives information about the power usage and demands from individual chargers and dynamically allocates power resources based on the available power and user profiles. By utilizing AI-based load balancing techniques, the system ensures that the electrical infrastructure is not overloaded and maximizes power utilization among charging stations.

Dynamic Pricing is another essential aspect integrated into the Mesh version of the IEVCC system. Each user is provided with an NFC card that activates the charger and identifies them on the system. The card is linked to the user's account, which defines their access rights, such as the maximum power they can charge and priority over other users. Based on the user profile, the number of users charging at any given moment, and the available power, the system dynamically adjusts the charging rates and prioritizes charging based on predefined criteria. The chargers communicate all information about the charging sessions, including usage time and used energy, to the Broker/Manager. This information is then used to apply dynamic pricing strategies and charging fees. Through AI algorithms, the system optimizes pricing based on factors such as demand, time of day, and available power capacity, incentivizing users to charge their vehicles during periods of low demand and high renewable energy availability.

By integrating the AI-driven Predictive Charging, Load Balancing, and Dynamic Pricing functionalities, the Mesh version of the IEVCC system ensures efficient utilization of charging stations in condominium environments. It addresses the challenges of multiple users, varying infrastructure factors, and the need to make charging decisions based on user profiles and available power resources. With the hardware components, including the consumption monitor, Broker/Manager, and Intelligent Electric Vehicle Charging Controller, the system enables effective management of charging sessions, adjustments to charging rates, and accurate fee calculations based on AI-driven algorithms.

2.2 Consumption Monitor Hardware

The consumption monitor hardware (Fig. 4) is pivotal in overseeing power consumption within the house or building. Utilizing a PZEM-004T V3.0 board, based on the Vango V98xx IC, it measures critical parameters like Voltage, Current, and Active Power at regular intervals [11]. These values are then transmitted to the consumption monitor system, stored in a database for user access, providing insights into electricity costs associated with various home devices.

To augment system capabilities, artificial intelligence (AI) techniques are integrated into the consumption monitor system. AI algorithms analyze collected consumption data, predicting future energy usage patterns. This AI-driven Predictive Charging optimizes schedules, ensuring charging stations are available when needed, by anticipating charging patterns and adjusting schedules accordingly [14].

Fig. 4. Consumption Energy Monitor

Additionally, current consumption monitor data is communicated to a broker using the MQTTS protocol [12]. This facilitates the Intelligent Electric Vehicle Charging Controller (IEVCC) to not only adjust charging current based on contracted power information but also utilize AI-based Load Balancing techniques. These techniques intelligently distribute available power among multiple EV chargers, preventing overload and maximizing utilization [15].

Furthermore, the system integrates AI-driven Dynamic Pricing strategies, adjusting charging prices based on factors like demand and available power capacity [16]. This incentivizes users to charge during off-peak hours or when renewable energy is abundant, promoting efficiency and sustainability while managing infrastructure demand.

Moreover, the system supports renewable energy utilization, directing power from sources like solar panels directly to chargers. This AI-driven integration empowers users to monitor and optimize energy consumption, maximize renewable energy utilization, and make informed charging decisions.

3 Results

In this study, comprehensive tests were conducted in real-world parking spots connected directly to household power. Residents continued daily activities without disruption while the Intelligent Electric Vehicle Charging Controller (IEVCC), integrated with AI, autonomously adjusted charging power. This AI-driven adaptation ensured efficient charging operations, seamlessly accommodating electric vehicle needs alongside household activities.

3.1 Standalone Version

This paper examines how AI integration optimizes electric vehicle charging dynamics, focusing on the inaugural version of the standalone Intelligent Electric Vehicle Charging Controller (IEVCC). Through real-world charging sessions, the study demonstrates AI's role in adaptive power management, addressing household power constraints and responding promptly to energy demand fluctuations.

Our experiment utilized a 2015 Nissan Leaf 24 kWh with a user-contracted power of 5.75 kVA, allowing a maximum current of 25 A. The dwelling featured an intelligent meter, surpassing conventional overcurrent residual current breaker (RCBO) responsiveness. To ensure compatibility with the intelligent meter, an AI-driven buffer of 3 A was ingeniously implemented. This buffer serves as an intelligent safeguard, preventing household current from exceeding the meter's defined threshold and mitigating the risk of interruptions.

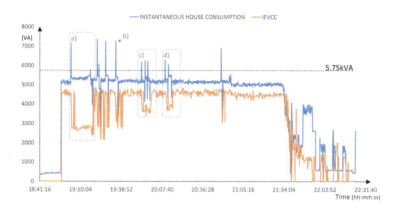

Fig. 5. Charging session data

The IEVCC, empowered by AI algorithms, orchestrates a dynamic dance between the charging requirements of the electric vehicle and the instantaneous power consumption of the home. Through continuous monitoring and analysis, the IEVCC effortlessly adapts the charging current to maintain a harmonious balance, ensuring that the sum of the current drawn by household appliances and the vehicle remains below the pre-defined maximum value. This seamless coordination is visible in Fig. 5, where the charging session unfolds.

Analyzing the details of Fig. 6, moments marked as a, b, c, and d reveal occasional peaks in power consumption. These spikes occur when household appliances activate, momentarily exceeding the maximum threshold. However, the AI-driven consumption monitor system, operating at a 10-s interval, detects these changes in real-time. Upon identification of a peak, the system promptly communicates with the IEVCC, triggering a rapid response. Leveraging its AI capabilities, the IEVCC swiftly adjusts the charging current available to the electric vehicle, bringing consumption back within defined limits.

In cases where available power cannot support household consumption and charging requirements simultaneously, the AI-powered IEVCC intelligently intervenes. It may temporarily suspend the charging process until the power supply becomes adequate to resume seamlessly. This AI-driven decision-making ensures optimal power utilization, mitigates the risk of power overload, and safeguards overall electrical system stability.

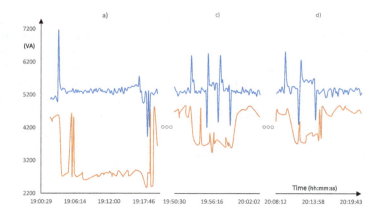

Fig. 6. Charging session data details of Fig. 5

The integration of AI in the standalone IEVCC not only enables efficient power management but also heralds a future of intelligent charging. Through machine learning algorithms, the system learns from past charging patterns, user preferences, and energy demand forecasts to dynamically optimize charging schedules. This AI-driven Predictive Charging maximizes charging station availability during peak demand periods, enhancing user convenience and alleviating strain on charging infrastructure.

In conclusion, our real-world experiments demonstrate the tangible benefits of integrating AI technologies into the standalone IEVCC. The seamless coordination between the intelligent meter, consumption monitor system, and AI-driven charging controller ensures optimal charging experiences while maintaining harmonious power balance within the household. This research lays the groundwork for intelligent, efficient, and sustainable charging systems, advancing us towards a future powered by smart energy management.

3.2 Optimizing Multiple EV Charging Scenarios

In the quest for efficient electric vehicle (EV) charging infrastructure, load balancing techniques integrated with AI algorithms have emerged as a promising solution. In this phase of our research, we conducted experimental simulations to evaluate the performance of load balancing in a multi-charger scenario. The simulation environment consisted of three chargers operating alongside the charging manager, showcasing the adaptability and intelligence of the system.

The ability to dynamically adjust the charging power based on global consumption patterns remains a key feature of our AI-driven load balancing approach. Unlike traditional systems, our solution does not require aggressive update intervals of less than 10 s, enabling smoother operation with minimal computational overhead.

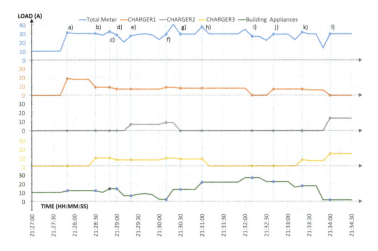

Fig. 7. Mesh Charger Network session sample

Figure 7 provides a snapshot of a sample interval illustrating load balancing dynamics among three connected chargers. Changes in overall consumption, unrelated to EV charging, trigger swift adjustments in the allocation of available energy to each charger, ensuring that the maximum load threshold (30 A) is not surpassed. Several situations depicted in Fig. 7 are described below, shedding light on the system's operation:

Charger 1: Initially requested to charge at 19.9 A, Charger 1 adjusted to 19 A due to limitations in integer values. As household appliance consumption increased to 12 A, only 18 A remained available for charging, prompting a reduction in charging current to 18 A in the next cycle.

Charger 3: Charger 3 registered and requested to start charging, with available power equally divided between both chargers, allowing them to charge at 9 A each.

Consumption Limit Exceeded: Building consumption exceeded contracted power, necessitating a reduction in available power for charging.

Reduced Charging Power: Following the previous power limit violation, charging power for both chargers was further reduced.

Charger 2: Charger 2, having registered earlier, initiated the charging process as available power allowed all chargers to operate simultaneously.

Increased Available Power: A reduction in building consumption increased available power for each charger, allowing them to increase charging current to 9 A.

Standby Mode: Rising building consumption led to insufficient power for all chargers to operate at the minimum power level of 6 A, placing charger 2 in standby mode.

Single Charger Operation: With rising building consumption, only one charger could operate, with charger 1 continuing charging.

Power Shortage: Increased building consumption led to insufficient power for both chargers, resulting in charger 1 being instructed to stop charging until power became available again.

Resumption of Charging: A subsequent reduction in building consumption allowed charger 1 to resume charging.

Charger 2 Rejoins: Another reduction in building consumption enabled charger 2 to resume charging, with charging current adjusted from 7 A to 6 A in the next time step as building consumption slightly increased.

Charging Progression: Charger 1 completed its charging session, coinciding with a decrease in building consumption, leading charger 3 to increase charging power and charger 2 to resume its charging session.

These simulation scenarios demonstrate the efficiency and adaptability of our AI-driven load balancing system in a multi-charger environment. Dynamic allocation of available power ensures optimal charging experiences, minimizing power limitations, and maximizing charging resource utilization. This research underscores the potential of AI in load balancing and its pivotal role in shaping a smarter and more sustainable future for EV charging infrastructure.

4 Conclusion

In this groundbreaking study, we unveil the Intelligent Electric Vehicle Charging Controller (IEVCC) as a game-changing solution propelled by Artificial Intelligence (AI) to transform the landscape of electric vehicle (EV) charging. The IEVCC not only dynamically optimizes charging power but also taps into AI capabilities to deliver unparalleled benefits.

With its AI-driven intelligence, the IEVCC seamlessly adjusts charging power for one or multiple electric vehicles based on a comprehensive range of conditions, ensuring that contracted power limits are never exceeded and averting potential disruptions like RCBO tripping. Through continual real-time data monitoring and analysis, the IEVCC fine-tunes the charging process for maximum efficiency and performance.

The IEVCC offers a spectrum of operation modes, blending AI automation with manual control options, granting users flexibility while upholding AI-driven safeguards. This amalgamation of AI intelligence and user adaptability sets the IEVCC apart as a truly versatile and user-centric charging solution.

One of the standout advantages of the IEVCC lies in its adaptive nature, effortlessly responding to changing conditions and external factors. By dynamically redistributing available power among chargers to achieve optimal load balance without surpassing pre-defined maximum load limits, the IEVCC ensures efficient energy utilization while preventing overloading scenarios.

Moreover, the IEVCC presents substantial cost-saving opportunities by intelligently managing the charging process, negating the necessity for users to increase their contracted power and optimizing charging efficiency to reduce charging times and minimize energy wastage.

Operating within a mesh network, the IEVCC efficiently shares resources and manages loads, offering particular value in settings with limited energy supply or within public parking buildings.

Integration with renewable energy sources further enhances sustainability, channeling solar energy toward EV charging and diminishing reliance on non-renewable sources, thereby aligning electric mobility with green energy practices.

Through AI-driven data analytics, the IEVCC not only provides crucial insights into user behavior and system performance but also empowers manufacturers and service providers to make data-informed decisions and refine future iterations.

In conclusion, the Intelligent Electric Vehicle Charging Controller (IEVCC) heralds a new era in EV charging, blending AI for dynamic power adjustment, load balancing, cost savings, mesh networking, renewable energy integration, and data-driven insights. This innovative approach accelerates the global transition toward sustainable transportation systems.

Acknowledgments. "This work is funded by National Funds through the FCT - Foundation for Science and Technology, I.P., within the scope of the project Ref. UIDB/05583/2020. Furthermore, we would like to thank the Research Center in Digital Services (CISeD) and the Instituto Politécnico de Viseu for their support."

Maryam Abbasi thanks the national funding by FCT - Foundation for Science and Technology, Pi.I., through the institutional scientific employment program-contract (CEECINST/00077/2021).

The authors would like to thank INESC Coimbra for their support. This work is partially funded by National Funds through the FCT - Foundation for Science and Technology, I.P., within the scope of the projects UIDB/00308/2020.

References

1. Liikennevirta Oy (Ltd.): The global electric vehicle market in 2021 - virta. Virta.global (2019). https://www.virta.global/global-electric-vehicle-market. Accessed 11 Apr 2021
2. International Energy Agency: Global EV Outlook 2021 (2021). https://www.iea.org/reports/global-ev-outlook-2021. Accessed 06 May 2021
3. Liikennevirta Oy (Ltd.): The global electric vehicle market in 2021 - virta. Virta.global (2019). https://www.virta.global/global-electric-vehicle-market. Accessed 11 Mar 2021
4. Muratori, M., et al.: The rise of electric vehicles-2020 status and future expectations. Prog. Energy **3**, 022002 (2021). https://doi.org/10.1088/2516-1083/abe0ad
5. Wu, M., Chen, W.: Forecast of electric vehicle sales in the world and china based on PCA-GRNN. Sustainability. **14** (2022). https://www.mdpi.com/2071-1050/14/4/2206
6. Graham, J., Brungard, E.: Consumer adoption of plug-in electric vehicles in selected countries. Future Trans. **1**, 303–325 (2021). https://www.mdpi.com/2673-7590/1/2/18

7. Alghamdi, T.G., Said, D., Mouftah, H.T.: Decentralized game-theoretic scheme for D-EVSE based on renewable energy in smart cities: a realistic scenario. IEEE Access **8**, 48274–48284 (2020)
8. SAE International: J1772: SAE Electric Vehicle and Plug in Hybrid Electric Vehicle Conductive Charge Coupler. SAE International, Warrendale (2016)
9. Stegen Electronics Smart EVSE. https://www.smartevse.nl/. Accessed 10 May 2022
10. Expressif - Technical Document. https://www.espressif.com/en/products/socs. Accessed 2 July 2022
11. Vango Technologies, Inc.: V98XX Datasheet. http://www.vangotech.com/uploadpic/164299406070.pdf. Accessed 12 July 2022
12. MQTT - The Standard for IoT Messaging (2019). https://docs.oasis-open.org/mqtt/mqtt/v5.0/mqtt-v5.0.pdf. Accessed 10 Apr 2021
13. International Energy Agency: Global EV Outlook 2021: accelerating ambitions despite the pandemic. OECD (2021). https://iea.blob.core.windows.net/assets/ed5f4484-f556-4110-8c5c-4ede8bcba637/GlobalEVOutlook2021.pdf
14. Helmus, J., Hoed, R.: Unraveling user type characteristics: towards a taxonomy for charging infrastructure. World Electr. Veh. J. **7**, 589–604 (2015)
15. Xiuj, Y.: Charging optimization of massive electric vehicles in distribution network. Electr. Power Autom. Equip. **35**, 31–36 (2015)
16. Almaghrebi, A., Aljuheshi, F., Rafaie, M., James, K., Alahmad, M.: Data-driven charging demand prediction at public charging stations using supervised machine learning regression methods. Energies **13**, 4231 (2020)

Ensemble Learning Models for Wind Power Forecasting

Samara Deon[1], José Donizetti de Lima[1], Geremi Gilson Dranka[1], Matheus Henrique Dal Molin Ribeiro[1], Julio Cesar Santos dos Anjos[2,3], Juan Francisco de Paz Santana[4], and Valderi Reis Quietinho Leithardt[5,6,7(✉)]

[1] Industrial and Systems Engineering Graduate Program, Federal University of Technology - Parana, Pato Branco, Brazil
{donizetti,geremidranka,mribeiro}@utfpr.edu.br
[2] Federal University of Ceará, Campi Itapaje, Fortaleza, CE 62600-000, Brazil
[3] Graduate Program in Teleinformatics Engineering (PPGETI/UFC), Center of Technology, Campus of Pici, Fortaleza, CE 60455-970, Brazil
jcsanjos@ufc.br
[4] Expert Systems and Applications Laboratory, University of Salamanca, 37700 Salamanca, Spain
fcofds@usal.es
[5] Lisbon School of Engineering (ISEL), Polytechnic University of Lisbon (IPL), 1549-020 Lisbon, Portugal
valderi.leithardt@isel.pt
[6] Center of Technology and Systems (UNINOVA-CTS) and Associated Lab of Intelligent Systems (LASI), 2829-516 Caparica, Portugal
[7] FIT-ISEL, 1959-007 Lisboa, Portugal

Abstract. Wind power, a clean and sustainable energy source, has experienced substantial growth in Brazil's energy capacity over recent decades. Accurate wind power forecasting is crucial for effectively harnessing wind energy and ensuring the reliable operation of power systems. However, due to the unique characteristics of wind power generation time series, developing statistical models for forecasting can be a challenging endeavor. This paper introduces a comprehensive approach by proposing forty-two ensemble learning models designed for forecasting wind power time series. These ensembles are created by combining various machine learning models and utilizing different aggregation methods, which incorporate various statistical measures such as the arithmetic average, harmonic average, median, and weighted average, with weights determined through metrics like mean absolute percentage error (MAPE), mean absolute error, and root mean squared error, was employed to evaluate its performance in forecasting wind power time series. This evaluation was conducted at time intervals of 10, 30, 60, and 120 min for two wind farms situated in Bahia, Brazil. The findings suggest that the ensemble method, which combines forecasts from individual models using weighted averages based on MAPE-derived weights, proved to be effective achieving the lowest percentage error in 87.5% of the evaluated cases. Conversely, the ensemble utilizing the harmonic average exhibited a higher error rate compared to the alternatives in 75% of the cases.

Keywords: Wind power generation · time series forecasting · ensembles

1 Introduction

Wind energy is one of the main renewable energy sources in the Brazilian power sector, representing 11% of the energy matrix in the country [1]. Furthermore, wind power has considerable economic and environmental advantages compared to other power sources. Within this context, wind power forecasting is essential to the decision-making process regarding the exploitation of the wind energy potential but also for power system operation and planning. Thus, the development of forecasting systems is essential [2].

Due to the characteristics of external factors of wind power production, such as climate concerns and features geographies of wind farms, developing statistical models in the context of time series forecasting becomes challenging. Using artificial intelligence (AI) techniques can be appealing to solve this problem. AI is a field of computer science capable of recognizing tasks [3], analyzing data [4,5], time series evaluation, and making decisions by an algorithm defined by specialists [6]. In this context, smart grids and the internet of things have been also widely explored in this field. In this sense, the ensemble learning method combines several machine learning models, aiming to overcome the weakness of using a single model [7].

Based on the success of ensemble learning, this paper aims to evaluate individual techniques' performance and their combinations for wind power time series forecasting. The ensembles generated are combinations of the individual techniques through the average, weight average, harmonic average, and median. We will compare the individual models with the ensembles generated in this research to point to the more successful approach.

The contributions of this study can be summarized as:

- Different time series forecast ensembles are developed, applied, and evaluated to forecast the wind power generation of a wind farm;
- A comparative study between the individual techniques and the ensemble generated regarding wind power time series accuracy is performed;
- A comparison and analysis of wind power time series forecast obtained through the combined models developed using the performance indicators will be conducted.

The subsequent sections of this paper are structured in the following manner: Sect. 2 details the dataset used in this paper, Sect. 3 presents the methods and the procedures adopted, respectively. Finally, Sect. 4 presents the results, followed by the main research findings and conclusions in Sect. 5.

2 Problem Description and Laboratory Analysis

The dataset used in this paper is from Ribeiro et al. [8]. The dataset adopted includes two-time series consisting of values of wind power generation (in kilowatt-hour, kWh) from two wind farms (WDF1 and WDF2), both located in Bahia, Brazil, in January 2021. WDF1 is located in São Sebastião do Alto, has 12 turbines and an installed capacity of 19.20 megawatt-hours (MW). WDF2 is located in Rio Verde and it has 19 turbines, each with a generation capacity of 1.68 MWh. The duration of the time series reached 4,322 samples for January 2021, collected every 10 min. The complete data represent the entire series of wind power generation data, the training set expresses the first 70% of complete data, and the test set comprises the remaining 30% of the whole data.

3 Methodology

This section details the main aspects regarding machine learning approaches and the combination of forecasting methods for ensemble generation.

3.1 Machine Learning

Machine learning (ML) is currently widely used in various fields, such as power systems [9], medicine, internet search engines, banking systems for fraud detection, security systems for data, industry [10,11], aerospace sector [12], and power systems [13–16]. Among the utilities of machine learning, we can also mention the prediction of time series [17].

Machine learning learns through repeated observations and its main objective is to develop a system that learns from a database [18], in this case, the time series. In the end, it generates a prediction [19], classification [20,21], or detection [22]. In the context of time series forecasting, machine learning has been widely used in several fields: wind speed forecast [23,24], fault prediction [25], wind speed forecast, and wind power forecast [26].

3.2 Combination of Forecasting Techniques

The combination of different forecasting techniques (using ensemble learning methods) has been studied by several authors such as in [27,28], and [29]. De Lima et al. [30] cite the following arguments for adopting forecast combinations: (i) there is no perfect forecasting technique; (ii) the variability in the accuracy of a technique applied across different periods, forecasting horizons, and datasets; (iii) the sensitivity to various factors of individual techniques, and (iv) the combination of techniques can generate a better result compared to individual techniques.

In the forecast methods combination, individual time series forecasts are generated from the data obtained, each one based on different forecasting models [31]. Then, all these individual forecasts are combined to obtain the time series' final forecast. Subsequently, the main aspects related to the combination of prediction methods will be presented as described in [30]. The combination of forecasting techniques is given by:

$$F_c(t) = \sum_{i=1}^{k} w_i F_i(t) \; with \; \sum_{i=1}^{k} w_i \leq 1 \; and \; 0 \leq w_i \leq 1, \qquad (1)$$

in which $F_c(t)$ is the forecast result at time t, k is the number of forecast techniques combined, $F_i(t)$ is the result of the i-th forecast, and w_i is the weight assigned to the i-th forecast.

Among the existing forms of combinations, there is the combination by arithmetic average, in which all weights are equals [30], and is expressed by:

$$F_c(t) = \frac{1}{k} \sum_{i=1}^{k} F_i(t), \qquad (2)$$

Furthermore, these same authors [30] present the combination by the harmonic average, which is given by:

$$F_c(t) = \frac{k}{\sum_{i=1}^{k} \frac{1}{F_i(t)}}, \qquad (3)$$

In addition, the median forecasts of the individual methodologies employed during the study might be considered as a combination. Another approach pointed out by [30] is the combination of predictions in which the weights are defined according to the performance of each individual technique. In this case, the technique that presents the best performance receives greater weight. The performance indicators MAPE, RMSE, and MAE were used to evaluate the effectiveness of the models developed in this research in terms of out-of-sample forecasting (for the test set).

In the development of this study, the data adopted are those used in [8]. After obtaining the data, ensembles were generated by combining two, three, four, and five forecast base models such as support vector regression (SVR), extreme learning machines (ELM), GP, ridge regression (RR), and random forests (RF) by arithmetic average, harmonic average, and weighted average, the last one generated three different forecasts. In addition, the median central tendency measure was adopted, as it eliminates the bias of extreme values. In the context of weighted average combination, weights were obtained according to Eqs. (4), (5) and (6). In the same way, for combinations through arithmetic and harmonic average, Eqs. (2) and (3) were used.

The choice of base models for each ensemble was based on the model with the lowest RMSE error presented by [8], taking into account the forecast horizon. For example, to generate an ensemble from two base models for the 10-minute forecast horizon, the two base models with the lowest RMSE error in the 10-minute forecast horizon were chosen. Based on this criterion, forty-two ensembles were created.

Once the forecasts of the combined models have been made, the MAPE, RMSE, and MSE metrics were computed to analyze the accuracy of the predictions. Finally, the forecasts developed were compared with each other and with the forecasts obtained in the reference work to evaluate which model presents the best performance: the smallest error compared to the metrics indicated at the beginning of this paragraph.

The ensembles generated in this study were named as "XCombYz", in which X indicates the number of base models used, Y indicates which were the base models used according to Table 1, and z points which combination was used. For example, the "2Comb1c" model is the ensemble generated from the SVR and ELM models combined through the harmonic average.

Table 1. Indices of the base models and combination used.

Index (Y)	Base models	Index (z)	Combination
1	SVR - ELM	a	Average
2	SVR - RR	b	Median
3	SVR - ELM - GP	c	Harmonic
4	SVR - ELM - RR	d	RMSE weighted
5	SVR - RR - GP	e	MAE weighted
6	SVR - ELM - GP - RR	f	MAPE weighted

4 Experiments and Discussion

This section describes the main results achieved by the proposed data modeling. Table 2 illustrates the performance measures for the ensemble learning models that achieved the lowest, second, and most significant forecasting error. Also, the remaining ensembles are omitted. For WDF1 data, the wind power forecast (with 10 min horizon) with the lowest error according to the RMSE metric was the one obtained by the 2Comb1d model (SVR-ELM combined through weighted average with RMSE), presenting an error of 122.59 kWh and performance 0.13% better than the second model with the lowest error (2Comb1a, SVR-ELM combined through the arithmetic average) and 97.46% better than the model that showed the highest error (2Comb1c, SVR-ELM combined via harmonic average).

Still on WDF1, for 30 min horizon, the 3Comb4f model (SVR-ELM-RR combined through the weighted average with MAPE) obtained a lower error than the others (320.73 kWh) according to the MAE metric, presenting a performance 0.51% better than the second model with the lowest error in this category (3Comb4d, SVR-ELM-RR combined via weighted average with RMSE) and 69.27% better than the model with the highest error (2Comb2f, SVR-RR combined through weighted average with MAPE).

For the forecast horizon of 60 min, the 4Comb6f model (SVR-ELM-GP-RR combined through the weighted average with MAPE) obtained the lowest error according to the MAPE metric, presenting an error of 63.75% and performance 0.48% better than the second model with the lowest error (4Comb6d, SVR-ELM-GP-RR combined through weighted average with RMSE) and 75.55% better than the model with the highest error (4Comb6c, SVR -ELM-GP-RR combined through harmonic average).

For the forecast horizon of 120 min, on WDF1, 4Comb6f model (SVR-ELM-GP-RR combined through the weighted average with MAPE) obtained the lowest error according to the RMSE metric, presenting error of 1198.06 kWh and

Table 2. RMSE, MAE, and MAPE metrics of generated forecasts.

WDF	Forecast horizon	Criterion (kWh)	Lowest error	Second lowest error	Larger error
1			**2Comb1d**	**2Comb1a**	**2Comb1c**
	10 min	RMSE	122.59	122.75	4,827.61
		MAE	78.36	78.42	755.19
		MAPE	10.35%	10.45%	1,187.29%
			3Comb4f	**3Comb4d**	**2Comb2f**
	30 min	RMSE	452.43	454.84	1,448.97
		MAE	320.73	322.38	1,043.56
		MAPE	24.43%	24.45%	101.13%
			4Comb6f	**4Comb6d**	**4Comb6c**
	60 min	RMSE	873.76	878.67	1,560.29
		MAE	630.83	634.32	899.24
		MAPE	63.75%	64.06%	260.71%
			4Comb6f	**4Comb6d**	**5Comb7c**
	120 min	RMSE	1,198.06	1,204.74	6,753.03
		MAE	843.97	848.72	1,509.53
		MAPE	114.75%	115.20%	246.78%
2			**4Comb6f**	**4Comb6d**	**5Comb7c**
	10 min	RMSE	1106.24	1175.55	12617.35
		MAE	786.13	835.71	1476.07
		MAPE	31.88%	34.21%	121.10%
			3Comb4f	**3Comb4e**	**2Comb2c**
	30 min	RMSE	848.38	950.97	2803.64
		MAE	601.11	675.64	2018.85
		MAPE	23.45%	27.76%	100.77%
			4Comb6f	**4Comb6e**	**5Comb7c**
	60 min	RMSE	1661.41	1745.17	3819.08
		MAE	1199.15	1261.10	1912.32
		MAPE	62.47%	66.56%	190.60%
			4Comb6f	**4Comb6e**	**4Comb6c**
	120 min	RMSE	2279.03	2390.51	5097.02
		MAE	1604.47	1686.46	2600.67
		MAPE	112.95%	118.70%	492.05%

The nomenclature of the models is shown in bold in this table.

performance 0.55% better than second model with lowest error (4Comb6d, SVR-ELM-GP-RR combined through weighted average with RMSE), 82.26% better than the model with the highest error (5Comb7c, SVR-ELM-GP-RR combined through the harmonic average), and 54.55% better than the second highest error model (4Comb6c, SVR-ELM-GP-RR-RF combined via harmonic average). The errors obtained according to the RMSE, MAE and MAPE metrics of the first two forecasts with the lowest error and the forecasts with the highest error are shown in Table 2.

For the WDF2 data, the forecast with a 10-min horizon that obtained the result with the lowest error according to the RMSE metric was the one obtained by the 4Comb6f model (SVR-ELM-GP-RR combined through weighted average with MAPE), presenting an error of 1106.24 kWh and performance 4.83% better than the second model with the lowest error (4Comb6e, SVR-ELM-GP-RR combined through weighted average with MAE) and 52.44% better than the model with the highest error (3Comb5b, SVR-RR-GP combined through the median).

Still in WDF2, for the forecast horizon of 30 min, 3Comb4f model (SVR-ELM-RR combined through weighted average with MAPE) obtained a lower error than the others (601.11 kWh) according to MAE metric, presenting a performance 11.03% better than the second model with the lowest error in this category (3Comb4e, SVR-ELM-RR combined by MAE-weighted average) and 70.23% better than the model with the highest error (2Comb2c, SVR-RR combined by harmonic mean).

For the forecast horizon of 60 min, 4Comb6f model (SVR-ELM-GP-RR, combined through weighted average with MAPE) obtained the lowest error according to RMSE metric, presenting an error of 1661.41 kWh and performance 4,80% better than the second model with the lowest error (4Comb6e, SVR-ELM-GP-RR, combined using MAE-weighted average) and 56.50% better than the model with the highest error (5Comb7c, SVR-ELM- GP-RR-RF, combined through harmonic mean).

Continuing in WDF2, for the forecast horizon of 120 min, 4Comb6f model (SVR-ELM-GP-RR, combined through weighted average with MAPE) obtained the lowest error according to MAPE metric, presenting an error of 112.95% and performance 4.85% better than the second model with the lowest error (4Comb6e, SVR-ELM-GP-RR combined through weighted average with MAE), 77.05% better than the model with the highest error (4Comb6c, SVR- ELM-GP-RR combined through harmonic mean), and 67.30% better than the second highest error model (5Comb7c, SVR-ELM-GP-RR-RF combined through harmonic mean).

Therefore, for WDF1, the models with the best performance in terms of lowest error were 2Comb1d (SVR-ELM, combined using the weighted average with RMSE), 3Comb4f (SVR-ELM-RR, combined using the weighted average with MAPE), 4Comb6f (SVR-ELM-GP-RR, combined via MAPE-weighted average) and 4Comb6f (SVR-ELM-GP-RR, combined via MAPE-weighted average) for forecast horizons 10, 30, 60 and 120 min, and according to the metrics RMSE, MAE, MAPE and RMSE, respectively.

While for WDF2 the ensembles with the best performance were 4Comb6f (SVR-ELM-GP-RR, combined using the weighted average with MAPE), 3Comb4f (SVR-ELM-RR, combined using the weighted average with MAPE), 4Comb6f (SVR-ELM-GP-RR, combined via MAPE-weighted average) and 4Comb6f (SVR-ELM-GP-RR, combined via MAPE-weighted average), for forecast horizons of 10, 30, 60 and 120 min and in relation to the metrics RMSE, MAE, RMSE and MAPE, respectively. Table 3 summarizes what is described in this paragraph.

Table 3. Best-performed models in terms of the lowest error in each category.

WDF	Forecast Horizon (min)	Criterion	Model	Combination
1	10	RMSE	SVR-ELM	weighted average with RMSE
	30	MAE	SVR-RR-ELM	weighted average with MAPE
	60	MAPE	SVR-RR-GP-ELM	weighted average with MAPE
	120	RMSE	SVR-RR-GP-ELM	weighted average with MAPE
2	10	RMSE	SVR-RR-GP-ELM	weighted average with MAPE
	30	MAE	SVR-RR-ELM	weighted average with MAPE
	60	RMSE	SVR-RR-GP-ELM	weighted average with MAPE
	120	MAPE	SVR-RR-GP-ELM	weighted average with MAPE

In 87.5% of cases, the combination employing the weighted average MAPE was deemed the most effective, while, in 75% of scenarios, the combination utilizing the harmonic mean exhibited the highest error compared to other methods.

4.1 Comparison with Individual Techniques

Among the individual models used in this research (SVR, ELM, RR, GP and RF), those that presented the best performance in terms of the lowest error in each of the categories were chosen, that is, for the horizons of 10 min, 30 min, 60 min and 120 min, from WDF1 and WDF2. Designated models from each category were compared with the best performing combinations in terms of lowest error obtained in this study.

For WDF1, the individual model that obtained the lowest error in terms of RMSE for the 10 min forecast horizon was the SVR, presenting an error of 1,191.70 kWh. Thus, the 2Comb1d model (SVR-ELM combined through weighted average with RMSE) performed 89.71% better, with an error of 122.59 kWh in terms of RMSE. For the 30 min forecast horizon, the individual model that obtained the lowest error in terms of MAE was the SVR, presenting an error of 1,030.43 kWh. Therefore, the 3Comb4f model (SVR-ELM-RR combined through the weighted average with MAPE) performed 68.87% better, with an error of 320.73 kWh in the MAE metric.

Still in WDF1, the individual model that obtained the lowest error in terms of MAPE for the 60 min forecast horizon was the GP, presenting an error of 18.76%. It performed 70.57% better than the 4Comb6f model (SVR-ELM-GP-RR combined through weighted average with MAPE), which presented an error of 63.75% in terms of MAPE. Finally, for the 120 min forecast horizon, the individual SVR model obtained the lowest error in terms of RMSE among the individual models mentioned at the beginning of this section, showing an error of 2,430.56 kWh. Thus, the 4Comb6f model (SVR-ELM-GP-RR combined using the weighted average with MAPE) performed 50.71% better, with an error of 1,198.06 kWh in terms of the RMSE metric.

For WDF2, the individual model that obtained the lowest error in terms of RMSE for 10 min forecast horizon among the individual models mentioned at the beginning of this section was SVR, presenting an error of 2,303.97 kWh. Therefore, the 4Comb6f model (SVR-ELM-GP-RR combined through weighted average with MAPE) performed 51.99% better, with an error of 1,106.24 kWh in terms of RMSE. For the 30 min forecast horizon, the SVR model obtained the lowest error in terms of MAE among the individual models mentioned at the beginning of this section, presenting an error of 1,992.21 kWh. Therefore, the 3Comb4f model (SVR-ELM-RR combined through weighted average with MAPE) performed 69.83% better, with an error of 601.11 kWh in terms of the MAE metric.

Still in WDF2, the model that obtained the lowest error for the 60 min forecast horizon in terms of RMSE was the SVR, with an error value of 3,455.34 kWh. Therefore, the 4Comb6f model (SVR-ELM-GP-RR combined through weighted average with MAPE) performed 51.92% better, with an error of 1,661.41 kWh in terms of the RMSE metric. On the other hand, the model that obtained the lowest error in terms of MAPE for the 120 min forecast horizon among the individual models mentioned at the beginning of this section was GP, exhibiting 22.12% error. It performed 80.42% better than the 4Comb6f model (SVR-ELM-GP-RR combined through weighted average with MAPE), which presented an error of 112.95% in terms of MAPE.

However, it can be seen that the combinations developed in this study have a lower error concerning the RMSE and MAE metrics compared to the individual models. Nevertheless, the individual models compared above have a lower error in terms of the MAPE metric compared to the combinations adopted here.

Figure 1 compares the time series of observed and predicted values with a forecast horizon of 60 and 120 min, respectively. Both predictions were obtained using the 4Comb6f ensemble (SVR-ELM-GP-RR combined through weighted average with MAPE). The values are very close, but they are far apart at the maximum and minimum points. Similar results were observed for WDF2.

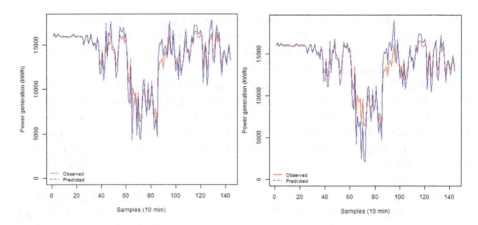

Fig. 1. Observed and predicted values WDF1: ensemble 4Comb6f with forecast horizon 60 min and 120 min, respectively.

5 Conclusion

In the present paper, forty-two ensemble learning models were proposed for forecasting wind power. Ensemble models were created by amalgamating various base models through methods like arithmetic average, harmonic average, median, and weighted average, with weight assignments based on RMSE, MAE, and MAPE metrics.

The ensembles were applied to generate the wind power time series forecast for forecast horizons of 10, 30, 60, and 120 min for two wind farms WDF1 and WDF2, located in Bahia, Brazil. For WDF1, the best-performing models were 2Comb1d (SVR-ELM, weighted average combined with RMSE), 3Comb4f (SVR-ELM-RR, weighted average combined with MAPE), 4Comb6f (SVR-ELM-GP-RR, weighted average combined with MAPE) and 4Comb6f (SVR-ELM-GP-RR, weighted average combined with MAPE) for forecast horizons 10, 30, 60, and 120 min, respectively.

And for WDF2, the ensembles with the best performance were 4Comb6f (SVR-ELM-GP-RR, weighted average combined with MAPE), 3Comb4f (SVR-ELM-RR, weighted average combined with MAPE), 4Comb6f (SVR -ELM-GP-RR, weighted average combined with MAPE) and 4Comb6f (SVR-ELM-GP-RR, weighted average combined with MAPE), for forecast horizons of 10, 30, 60 and 120 min, respectively.

Furthermore, the combination utilizing the MAPE-weighted average demonstrated superior performance by achieving the lowest error in 87.5% of cases, while the combination employing the harmonic average resulted in a higher error compared to other methods in 75% of the instances. Furthermore, the 4Comb6f ensemble (SVR-ELM-GP-RR, combined through weighted average with MAPE) was considered the best performance regarding the smallest error in 62.5% of the situations.

Although the results presented in this study are promising in terms of forecast accuracy, it is possible to identify the following opportunities for future studies: (i) develop a systematic literature review regarding the application of combination methods in time series forecasting; (ii) studying the applicability of the models adopted in this study in databases of the same wind farm, in January of previous years, would be interesting to verify whether the effectiveness of this method remains for that month; and (iii) carry out a study applying these models to the dataset used by [30].

Acknowledgments. The authors would like to acknowledge the Portuguese FCT program, Center of Technology and Systems (CTS) UIDB-00066-2020-UIDP-00066-2020, for partially funding the research. This work was supported by the Junta De Castilla y León-Consejería De Economía Y Empleo: System for simulation and training in advanced techniques for the occupational risk prevention through the design of hybrid-reality environments with ref. J118.

References

1. da Silva, R.G., Moreno, S.R., Ribeiro, M.H.D.M., Larcher, J.H.K., Mariani, V.C., dos Santos Coelho, L.: Multi-step short-term wind speed forecasting based on multi-stage decomposition coupled with stacking-ensemble learning approach. Int. J. Electr. Power Energy Syst. **143**, 108504 (2022). https://doi.org/10.1016/j.ijepes.2022.108504
2. Stefenon, S.F., Seman, L.O., Aquino, L.S., dos Santos Coelho, L.: Wavelet-Seq2Seq-LSTM with attention for time series forecasting of level of dams in hydroelectric power plants. Energy **274**, 127350 (2023). https://doi.org/10.1016/j.energy.2023.127350
3. Yamasaki, M., Freire, R.Z., Seman, L.O., Stefenon, S.F., Mariani, V.C., dos Santos Coelho, L.: Optimized hybrid ensemble learning approaches applied to very short-term load forecasting. Int. J. Electr. Power Energy Syst. **155**, 109579 (2024). https://doi.org/10.1016/j.ijepes.2023.109579
4. Starke, L., Hoppe, A.F., Sartori, A., Stefenon, S.F., Santana, J.F.D.P., Leithardt, V.R.Q.: Interference recommendation for the pump sizing process in progressive cavity pumps using graph neural networks. Sci. Rep. **13**(1), 16884 (2023). https://doi.org/10.1038/s41598-023-43972-4
5. Corso, M.P., Stefenon, S.F., Singh, G., Matsuo, M.V., Perez, F.L., Leithardt, V.R.Q.: Evaluation of visible contamination on power grid insulators using convolutional neural networks. Electr. Eng. **105**, 3881–3894 (2023). https://doi.org/10.1007/s00202-023-01915-2
6. Westarb, G., Stefenon, S.F., Hoppe, A.F., Sartori, A., Klaar, A.C.R., Leithardt, V.R.Q.: Complex graph neural networks for medication interaction verification. J. Intell. Fuzzy Syst. **44**(6), 10383–10395 (2023). https://doi.org/10.3233/JIFS-223656
7. Seman, L.O., Stefenon, S.F., Mariani, V.C., dos Santos Coelho, L.: Ensemble learning methods using the Hodrick-Prescott filter for fault forecasting in insulators of the electrical power grids. Int. J. Electr. Power Energy Syst. **152**, 109269 (2023). https://doi.org/10.1016/j.ijepes.2023.109269

8. Ribeiro, M.H.D.M., da Silva, R.G., Moreno, S.R., Mariani, V.C., dos Santos Coelho, L.: Efficient bootstrap stacking ensemble learning model applied to wind power generation forecasting. Int. J. Electr. Power Energy Syst. **136**, 107712 (2022)
9. da Silva, E.C., Finardi, E.C., Stefenon, S.F.: Enhancing hydroelectric inflow prediction in the Brazilian power system: a comparative analysis of machine learning models and hyperparameter optimization for decision support. Electr. Power Syst. Res. **230**, 110275 (2024). https://doi.org/10.1016/j.epsr.2024.110275
10. Stefenon, S.F., et al.: Electric field evaluation using the finite element method and proxy models for the design of stator slots in a permanent magnet synchronous motor. Electronics **9**(11), 1975 (2020). https://doi.org/10.3390/electronics9111975
11. Borré, A., Seman, L.O., Camponogara, E., Stefenon, S.F., Mariani, V.C., Coelho, L.S.: Machine fault detection using a hybrid CNN-LSTM attention-based model. Sensors **23**(9), 4512 (2023). https://doi.org/10.3390/s23094512
12. dos Santos, G.H., Seman, L.O., Bezerra, E.A., Leithardt, V.R.Q., Mendes, A.S., Stefenon, S.F.: Static attitude determination using convolutional neural networks. Sensors **21**(19), 6419 (2021). https://doi.org/10.3390/s21196419
13. Singh, G., Stefenon, S.F., Yow, K.-C.: Interpretable visual transmission lines inspections using pseudo-prototypical part network. Mach. Vis. Appl. **34**(3), 41 (2023). https://doi.org/10.1007/s00138-023-01390-6
14. Stefenon, S.F., Furtado Neto, C.S., Coelho, T.S., Nied, A., Yamaguchi, C.K., Yow, K.-C.: Particle swarm optimization for design of insulators of distribution power system based on finite element method. Electr. Eng. **104**, 615–622 (2022). https://doi.org/10.1007/s00202-021-01332-3
15. Stefenon, S.F., Americo, J.P., Meyer, L.H., Grebogi, R.B., Nied, A.: Analysis of the electric field in porcelain pin-type insulators via finite elements software. IEEE Lat. Am. Trans. **16**(10), 2505–2512 (2018). https://doi.org/10.1109/TLA.2018.8795129
16. Ribeiro, M.H.D.M., et al.: Variational mode decomposition and bagging extreme learning machine with multi-objective optimization for wind power forecasting. Appl. Intell. **54**, 3119–3134 (2024). https://doi.org/10.1007/s10489-024-05331-2
17. Klaar, A.C.R., Stefenon, S.F., Seman, L.O., Mariani, V.C., Coelho, L.S.: Optimized EWT-Seq2Seq-LSTM with attention mechanism to insulators fault prediction. Sensors **23**(6), 3202 (2023). https://doi.org/10.3390/s23063202
18. Surek, G.A.S., Seman, L.O., Stefenon, S.F., Mariani, V.C., Coelho, L.S.: Video-based human activity recognition using deep learning approaches. Sensors **23**(14), 6384 (2023). https://doi.org/10.3390/s23146384
19. Stefenon, S.F., Seman, L.O., da Silva, L.S.A., Mariani, V.C., dos Santos Coelho, L.: Hypertuned temporal fusion transformer for multi-horizon time series forecasting of dam level in hydroelectric power plants. Int. J. Electr. Power Energy Syst. **157**, 109876 (2024). https://doi.org/10.1016/j.ijepes.2024.109876
20. Stefenon, S.F., Singh, G., Souza, B.J., Freire, R.Z., Yow, K.-C.: Optimized hybrid YOLOu-Quasi-ProtoPNet for insulators classification. IET Gener. Transm. Distrib. **17**(15), 3501–3511 (2023). https://doi.org/10.1049/gtd2.12886
21. Stefenon, S.F., Seman, L.O., Klaar, A.C.R., Ovejero, R.G., Leithardt, V.R.Q.: Hypertuned-YOLO for interpretable distribution power grid fault location based on EigenCAM. Ain Shams Eng. J. **15**(6), 102722 (2024). https://doi.org/10.1016/j.asej.2024.102722
22. Glasenapp, L.A., Hoppe, A.F., Wisintainer, M.A., Sartori, A., Stefenon, S.F.: OCR applied for identification of vehicles with irregular documentation using IoT. Electronics **12**(5), 1083 (2023). https://doi.org/10.3390/electronics12051083

23. Moreno, S.R., Mariani, V.C., dos Santos Coelho, L.: Hybrid multi-stage decomposition with parametric model applied to wind speed forecasting in Brazilian northeast. Renew. Energy **164**, 1508–1526 (2021)
24. Moreno, S.R., Seman, L.O., Stefenon, S.F., dos Santos Coelho, L., Mariani, V.C.: Enhancing wind speed forecasting through synergy of machine learning, singular spectral analysis, and variational mode decomposition. Energy **292**, 130493 (2024). https://doi.org/10.1016/j.energy.2024.130493
25. Stefenon, S.F., Seman, L.O., Sopelsa Neto, N.F., Meyer, L.H., Mariani, V.C., dos Santos Coelho, L.: Group method of data handling using Christiano-Fitzgerald random walk filter for insulator fault prediction. Sensors **23**(13), 6118 (2023). https://doi.org/10.3390/s23136118
26. Wang, Y., Zou, R., Liu, F., Zhang, L., Liu, Q.: A review of wind speed and wind power forecasting with deep neural networks. Appl. Energy **304**, 117766 (2021)
27. Du, S., et al.: A systematic data-driven approach for production forecasting of coalbed methane incorporating deep learning and ensemble learning adapted to complex production patterns. Energy **263**, 126121 (2023)
28. Wang, Z., Gao, R., Wang, P., Chen, H.: A new perspective on air quality index time series forecasting: a ternary interval decomposition ensemble learning paradigm. Technol. Forecast. Soc. Change **191**, 122504 (2023)
29. Brahma, B., Wadhvani, R.: A residual ensemble learning approach for solar irradiance forecasting. Multimedia Tools Appl. **82**(21), 33087–33109 (2023)
30. de Lima, J.D., Oliveira, G.A., Trentin, M.G., Batistus, D.R., Pozza, C.B.: A study of the performance of individual techniques and their combinations to forecast urban water demand. Espacios **37**(22), 5 (2016)
31. Sauer, J., Mariani, V.C., dos Santos Coelho, L., Ribeiro, M.H.D.M., Rampazzo, M.: Extreme gradient boosting model based on improved Jaya optimizer applied to forecasting energy consumption in residential buildings. Evol. Syst. **13**, 577–588 (2022). https://doi.org/10.1007/s12530-021-09404-2

Predictive Model for Estimating Body Weight Based on Artificial Intelligence: An Integrated Approach to Pre-processing and Evaluation

Diana M. Figueiredo[1], Rui P. Duarte[1,2(✉)], and Carlos A. Cunha[1,2]

[1] Polytechnic Institute of Viseu, Viseu, Portugal
estgv17365@alunos.estgv.ipv.pt, {pduarte,cacunha}@estgv.ipv.pt
[2] CISeD - Research Center in Digital Services, Viseu, Portugal

Abstract. Body weight is much more than just a number on a scale. This value can indicate various diseases, as both excess and insufficient weight have implications for an individual's health. Excess weight is associated with heart disease, obesity, diabetes, high blood pressure, and respiratory disorders, among others. Meanwhile, extreme underweight is associated with problems such as nutritional deficiency, weakened immune system, osteoporosis, and hormonal imbalances. Due to these issues, there is a need to monitor and analyse body changes to adopt a diet and lifestyle balanced with individual needs. The weight control process is complicated and depends on various factors. This paper aims to develop a machine-learning model to predict future weight based on dietary records, physical exercise, and basal metabolic rate to demonstrate three days' impact on future weight. Results of the model's performance show that the coefficient of determination yielded a value of 0.75, which is considered good for this metric. The mean square and absolute errors demonstrate that the model could learn patterns in the data without significant overfitting.

Keywords: machine learning · body weight · basal metabolic rate · harris-Benedict · neural networks

1 Introduction

Human body weight is widely used in health and nutrition as an indicator of overall health and well-being [1]. Monitoring changes in body weight reduces the risk of diseases associated with unhealthy eating habits [2]. Therefore, understanding the differences in weight gain trajectories and the ability to predict future weight has significant implications for promoting healthy living and developing personalised intervention strategies to prevent lifestyle-related diseases. This study proposes the research and development of a machine learning (ML) model to analyse differences in human body weight and predict future weight, aiming to facilitate understanding of the impact of diet and physical exercise on changes in body mass. Moreover, individuals often interpret body weight as a number on the scale, encompassing much more than mere numerical value. This indicator is important in human life, influencing the aesthetic appearance and impacting physical health [3].

Obesity is one of the diseases that has become a global public health concern. According to the World Health Organisation, it is estimated that in 2022, over 1.9 billion adults worldwide will be overweight, with over 650 million classified as obese [4]. These alarming numbers reflect changes in dietary patterns, sedentary lifestyles, and socioeconomic factors contributing to poor nutrient intake and consumption of processed foods leading to weight gain [5]. The relationship between obesity and cardiovascular diseases is undeniable. Studies have shown that obesity is associated with a significant increase in the risk of developing hypertension, diabetes, respiratory diseases, heart diseases, among others [6]. According to the World Heart Federation, approximately 2.8 million deaths per year are attributed to obesity, underscoring the urgency of preventive approaches [7].

Predictive systems are believed to be a solution for preventing diseases associated with weight gain, thus controlling this factor that greatly impacts human health [8]. In recent years, several studies have been developed regarding predictive systems that forecast the impact of lifestyle on health, body weight, or the risk of contracting diseases such as obesity, hypertension, or other health issues [9]. For future weight prediction, advantages can be found in using historical data more effectively. ML models and their sub-fields have proven proficient in analysing large sets of longitudinal data, identifying patterns and correlations that would escape conventional analyses [10]. This retrospective analysis allows for increasingly accurate predictions of changes in body weight. Another crucial advancement is the incorporation of relevant variables in building predictive models. In contrast to traditional approaches that rely on isolated factors, ML models can integrate a variety of variables, from historical physical activity data to dietary habits [11]. This more comprehensive approach allows for a more precise and personalised analysis, considering the uniqueness of each individual. Therefore, this paper aims to understand the reasons for low adherence to meal plans and develop strategies that encourage individuals to maintain their meal plans by developing ML models to predict future weight based on the intake of calories during a period, thus promoting a healthy and active lifestyle. This raises the following research questions:

- *Q1*: Is it feasible to develop predictive models capable of accurately and immediately estimating the short-term consequences of calorie intake and physical exercise on body weight?
- *Q2*: What data pre-processing techniques should be considered to improve the quality and effectiveness of predictive models seeking to estimate the immediate consequences of calorie intake and physical exercise on body weight, including factors such as normalisation, handling missing data, and selecting relevant features?

– *Q3*: What is the performance of the body weight prediction model considering calorie intake and physical exercise?

To reach these goals, the work is organised in the following manner. This section introduces the topic to gain a general understanding of the proposed theme. Section 2 conducts a literature review, addressing the importance of body weight in human life, factors influencing weight changes, the importance of monitoring weight changes, and methods for calculating ideal weight. The bibliographic search was conducted with an emphasis on related works. Section 3 outlines the methodology used for model development. This section describes the research process, including the sample, data collection methods, techniques used in the study, justification for the chosen methods, and potential limitations. Section 4 details the model implementation process, presenting an analysis of the results. Finally, Sect. 5 presents the main conclusions, focusing on the limitations and challenges inherent in the work and prospects for future work.

2 Literature Review

In recent years, the significant increase in overweight and obesity has posed considerable challenges to physical health, particularly concerning the cardiovascular and metabolic systems [12]. Individuals with excess weight often face an additional burden on the heart due to increased blood volume and vascular resistance, elevating the risk of hypertension and heart diseases [13]. Body weight is influenced by various factors, including genetic aspects, lifestyle, eating behaviours, physical activity, and mental and emotional health [14]. Genetic factors are crucial in regulating body weight, from energy metabolism efficiency to appetite and satiety regulation. They also influence body fat distribution, contributing to variation in body shapes among individuals [15].

On the other hand, lifestyle, including eating behaviours and physical activity levels, plays a significant role in weight variations. Food availability, exercise practices, stress levels, and sleep quality can affect energy balance and body weight. Currently, the predominant lifestyle is more sedentary, with a preference for fast and high-saturated-fat meals, increasing the risk of chronic diseases such as obesity, diabetes, and cardiovascular problems. Furthermore, eating behaviours and levels of physical activity are fundamental to body weight regulation. A balanced diet, consisting of adequate macro-nutrients and micro-nutrients, is essential for physical and mental development and the prevention of chronic diseases. Additionally, regular physical activity promotes calorie expenditure, muscle development, and metabolic efficiency [16].

The relationship between mental health and weight is complex and bidirectional. Conditions such as anxiety, depression, and stress can lead to inadequate eating behaviours, while changes in body weight can affect mental health. Issues like emotional eating are common and can exacerbate challenges related to weight and mental health [17]. Moreover, the stigma associated with being overweight can have negative impacts on both physical and mental health, increasing the risk of problems like depression, anxiety, and social isolation [18]. It is important to recognise that strict diets can lead to psychological problems, such as binge eating, if not accompanied by qualified professionals. Binge eating is common in people with restrictive diets and can result in feelings of loss of control and compulsive eating [19].

Medicine and nutrition have developed several metrics to determine whether a weight is healthy. Among them, the tools most commonly used by professionals in the field are Body Mass Index (BMI) and Basal Metabolic Rate (BMR) [20].

BMI is a common measure in nutrition used to assess a person's relative weight relative to height. BMI is calculated by dividing weight (in kilograms) by the square of height (in meters). The equation for this calculation is represented by Eq. 1.

$$BMI = \frac{Weight(kg)}{Height(m)^2} \tag{1}$$

After the calculation, BMI is interpreted based on classifications established by the World Health Organisation and other health institutions. These classifications divide BMI into ranges indicating different weight states. For example, a BMI below 18.5 is considered underweight, while values between 18.5 and 24.9 indicate normal weight. Individuals with a BMI between 25 and 29.99 are classified as overweight, and those with a BMI equal to or greater than 30 are considered obese [21].

BMR is vital for determining the body's minimum energy requirements at rest. Nutritionists rely on BMR calculations to tailor nutrition plans, factoring in weight management goals. BMR reflects energy needs for essential bodily functions like breathing and circulation. Personalised caloric recommendations depend on understanding BMR, which considers individual traits such as age and gender. The Harris-Benedict equation [22] calculates BMR based on weight, height, age, and gender. Values may differ by gender, emphasising the equation's specificity, as shown in Eq. 2.

$$\text{BMR} = \begin{cases} \textit{If gender is male}: \\ 88.362 + (13.397 \times \text{weight}) + (4.799 \times \text{height}) - (5.677 \times \text{age}) \\ \textit{If gender is female}: \\ 447.593 + (9.247 \times \text{weight}) + (3.098 \times \text{height}) - (4.330 \times \text{age}) \end{cases} \tag{2}$$

Although the Harris-Benedict equation is widely used [23], it is important to note that different versions and adaptations of the formula exist [24].

In [25], a Multi-Layer Perceptron (MLP) technique was employed to predict heart disease risk. The neural network, trained using the back-propagation algorithm, utilised 13 clinical features as input to predict the presence or absence of heart disease in patients. With a dataset of 303 records, including various clinical attributes, the model achieved a maximum accuracy of 89%. This reliable model assists doctors in accurate disease prediction and diagnosis, ensuring individual safety and health.

Related to the prediction of obesity, in [26], a machine learning-based approach is used. Over 1100 data samples from individuals with and without obesity were collected. Nine ML algorithms were applied, with Logistic Regression (LR) achieving the highest accuracy of 97.09% among all classifiers. In [27], a Deep Learning (DL) solution was proposed for personalised diet recommendations based on individual health conditions. DL and ML algorithms were applied to a medical dataset of 30 patients with 13 distinct characteristics and 1000 food products. The LSTM model demonstrated superior performance with precision, recall, and F1 scores of up to 98% and 99% for allowed classes and 89% and 73% for disallowed classes, respectively. In [28], the K-means algorithm was employed to group children in Indonesia based on their nutritional status. Four clusters were identified, indicating varying levels of malnutrition or poor nutritional conditions. [29] developed a hybrid food recommendation system using graph clustering and DL-based approaches. The hybrid system outperformed methods solely based on user or food content, with an average performance increase of 11.8%. Additionally, including the user's trust network and considering the time factor further improved the system's effectiveness. In [30], a system using rule-based reasoning and genetic algorithms for dietary recommendations was proposed. DL algorithms showed superior performance compared to ML algorithms, leveraging individual activity data and disease history for personalized recommendations.

3 Methodology

This section delineates the methodology adopted for the development of the ML model. The creation of ML models adheres to a method that enables the description of the process followed for model creation and its respective evaluation. It is based on the CRoss-Industry Standard Process for Data Mining (CRISP-DM) methodology [31].

3.1 Data Understanding

The dataset used for model creation ("2018 calorie, exercise, and weight changes") is publicly available on the Kaggle repository [32]. The dataset comprises 14 columns, as detailed in Table 1.

This dataset corresponds to the record of an individual who initiated a weight loss diet in early January, following the Christmas period. This record includes the following information: daily weight, approximate calorie intake, calories expended, and some information about the foods consumed, such as whether the individual ate five doughnuts, drank a glass of wine, or had a protein-rich meal, as well as whether they engaged in a run and the type of physical exercise practised.

Table 1. Attributes of the dataset

Date	Date of observations
Stone, Pounds, Ounces	Observed weight
Weight_oz	Total weight in ounces
Calories	Approximate calories consumed on that date
cals_per_oz	calories for that day divided by weight in ounces
five_donuts	Day where main meal was bag of five jam donuts, 1 = TRUE, 0 = FALSE
walk	Day included at least one brisk walk of over 20 min, 1 = TRUE, 0 = FALSE
run	Day included at one short run of 2.5 miles, 1 = TRUE, 0 = FALSE
wine	Day with at least one large glass of wine when not accompanied by other fluids 1 = TRUE, 0 = FALSE
prot	Day with a considered high protein diet, 1 = TRUE, 0 = FALSE
weight	Day with weight-based exercise; includes rest days as part of period, 1 = TRUE, 0 = FALSE
change	Change in weight in oz, calculated from following day's observed weight

3.2 Data Preparation

Data preparation is the stage where the necessary data for model development is gathered and transformed. The chosen data adequately addresses the model's needs by providing records of daily calorie intake, body weight, and changes in weight over time. After analysing the dataset, three columns were selected: *weight_oz*, *calories*, and *change*. The data preparation and analysis were performed using the Python programming language, utilising the Pandas library. Table 2 represents the data after processing.

Table 2. Table with pre-processed data and BMR calculation.

#	Ingested Calories (IC)	Spent Calories (SC)	Weight (W)	Basal Metabolic Rate (BMR)
0	1950	−30	2726	37340.1
1	2600	8	2696	36938.2
2	2500	0	2704	37045.3
3	1850	−40	2704	37045.3
4	2900	14	2664	36509.5
5	3600	14	2678	36697
6	2400	−2	2692	36884.6
7	3100	6	2690	36857.8
8	2200	−8	2696	36938.2
9	1800	−40	2688	36831
10	2300	−18	2648	36295.1
11	3000	12	2630	36054
12	4000	34	2642	36214.7
13	2800	−12	2676	36670.2
...	...	...	...	...

To calculate the BMR for each record in the dataset, the Harris-Benedict function is applied (Eq. 2) by adding a new column with this calculation. It uses a the height, weight, age, and gender to compute the BMR, as presented in Table 2.

After data aggregation, a function was developed to calculate future weight based on weight measurements, BMR, and calorie intake and expenditure over the past three days. This function is structured in the following stages. First, from the current record of spent calories (SC), ingested calories (IC), weight (W), and BMR, new columns representing the values from the columns of days t-n are created, where t is the current temporal reference of the record and n is the number of days. This column can be observed in Table 3. Next, a new column named *FutureWeight* is introduced, representing the calculation of the future weight prediction based on the values of columns $t-1$, $t-2$, and $t-3$. Finally, this results in a larger and more detailed dataset, facilitating the understanding of weight changes over time. This new dataset is presented in Table 3.

Table 3. Calculation of future weight taking into account the last three days.

#	$IC-t_1$	$SC-t_1$	$W-t_1$	$BMR-t_1$	$IC-t_2$	$SC-t_2$	$W-t_2$	$BMR-t_2$	$IC-t_3$	$SC-t_3$	$W-t_3$	$BMR-t_3$	Future Weight
0	2600	8	2696	36938.2	2500	0	2704	37045.3	1850	−40	2704	37045.3	2706.05
1	2500	0	2704	37045.3	1850	−40	2704	37045.3	2900	14	2664	36509.5	2675.21
2	1850	0	2704	37045.3	2900	14	2664	36509.5	3600	14	2678	366697.0	2680.11
3	2900	14	2664	36509.5	3600	14	2678	36697.0	2400	−2	2692	36884.6	2678.65
4	3600	14	2678	36697.0	2400	−2	2692	36884.6	3100	6	2690	36857.8	2638.05
5	2400	−2	2692	36884.6	3100	6	2690	36857.8	2200	−8	2696	36938.2	2655.99
6	3100	6	2690	36857.8	2200	−8	2696	36938.2	1800	−40	2688	36831.0	2671.81
7	2200	−8	2696	36938.2	1800	−40	2688	36831.0	2300	−18	2648	36295.1	2671.81
8	1800	−40	2688	36831.0	2300	−18	2648	36295.1	3000	12	2630	36054.0	2675.58
9	2300	−18	2648	36295.1	3000	12	2630	36054.0	4000	34	2642	36214.7	2661.51
10	3000	12	2630	36054.0	4000	34	2642	36214.7	2800	−12	2676	36670.2	2620.10
11	4000	34	2642	36214.7	2800	−12	2676	36670.2	2300	−2	2664	36509.5	2604.06
12	2800	−12	2676	36670.2	2300	−2	2664	36509.5	2400	−4	2662	36482.7	2620.52
13	2300	−2	2664	36509.5	2400	−4	2662	36482.7	2100	−10	2658	36429.1	2656.53
14	2400	−4	2662	36482.7	2100	−10	2658	36429.1	2450	10	2648	36295.1	2644.13
15	2100	−10	2658	36429.1	2450	10	2648	36295.1	3400	22	2658	36429.1	2639.35
16	2450	10	2648	36295.1	3400	22	2658	36429.1	3600	14	2680	36723.8	2631.13
17	3400	22	2658	36429.1	3600	14	2680	36723.8	4200	16	2694	36911.4	2616.15
18	3600	14	2680	36273.8	4200	16	2694	36911.4	2000	−18	2710	37125.7	2630.03
19	4200	16	2694	36911.4	2000	−18	2710	37125.7	2150	−20	2692	36884.6	2656.08
20	2000	−18	2710	37125.7	2150	−20	2692	36884.6	2300	−26	2672	36616.6	2675.39
...	...	...	...	...	...	...	...	...	...	...	...	...	

At the end of this process, the final dataset is obtained, and it is possible to proceed to the phase of creating two datasets, which will be used for training and testing the model. These two datasets are called the train split and the test split. The train split is used to train the model and accounts for 70% of the data volume, while the test split represents the remaining 30%, used to predict the target and evaluate the model's performance.

3.3 Modelling and Evaluation

The neural network comprises three layers: input, intermediate, and output. The input layer, with twenty neurons and ReLU activation, receives and distributes information to the intermediate layers via weighted connections. The intermediate layer, with five neurons and ReLU activation, facilitates problem-solving by introducing nonlinearities. In the output layer, one neuron predicts future weight without a specific activation function, given the regression nature of the problem. Hyperparameters include 2000 epochs, 20 neurons in the input layer, 5 in the intermediate layer, and 1 in the output layer, employing ReLU activation and MSE and MAE loss functions, with RMSprop as the optimizer.

Training initiates by feeding input variables into the network, with test data aiding model validation. *Y_pred* stores predicted future weight, evaluated against *Y_test*, the test weight. Evaluation metrics like MAE, MSE, and R^2 gauge model accuracy and generalisation, comparing predicted and actual data to assess prediction error accurately.

4 Results

The primary objective of this work is to create a predictive model for a person's weight in the short term based on their behaviour in the days preceding the prediction. This behaviour includes calorie intake, calories burned at rest (associated with BMR), and calories expended during physical activity. For evaluation, the loss function was utilised to measure the model's performance at each iteration, reflecting how much the predicted outcome deviates from the actual outcome. During the training phase, the loss function involves calculating the MAE. After training, the MSE is computed. Analysing the two plots depicted in Fig. 1, it can be observed that the generalisation error of the models is reduced. The model demonstrates consistent performance on both training and testing data, mitigating overfitting concerns, particularly evident with increased training epochs. The congruent behaviour of the curves signifies the model's accuracy in predicting outcomes across datasets.

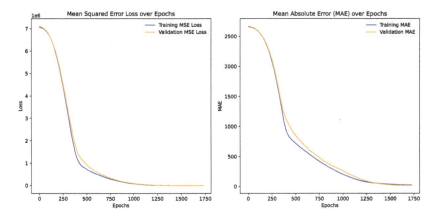

Fig. 1. Calculation of MSE and MAE for each epoch.

Accuracy assessment employs the R^2 metric, ranging from 0 to 1, where 0 indicates no data variability accounted for and 1 denotes a perfect fit. The obtained result for this project's model was 0.75, indicating a favourable fit. Evaluation metric analysis reveals minor prediction errors from the data's linear patterns. Figure 2 depicts the comparison between predicted and actual values, illustrating the model's close approximation to reality.

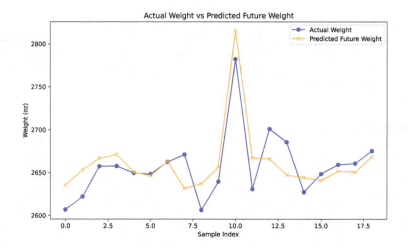

Fig. 2. Comparison between actual and predicted weight.

The experimental findings address the research questions outlined in Sect. 1. Concerning *Q1*, the model accurately predicts the short-term impact of lifestyle on weight. Using such a model enables individuals to assess and anticipate the effects of their dietary and exercise habits on weight management, aiding decision-making in health and nutrition.

Regarding *Q2*, reprocessing techniques like feature selection and data handling significantly improved predictive model quality. These steps, along with the introduction of new attributes like BMR, enhanced model effectiveness. By incorporating individual physiological aspects and past behavior through temporal shifting, the model captured short-term metabolic variations.

Finally, for *Q3*, selecting a neural network architecture and optimising hyperparameters minimised prediction errors. Adapting the model to the dataset's size ensured maximum performance. Evaluation revealed a strong generalisation capability (with $R^2 = 0.75$), supported by consistent performance across various training epochs, as depicted in Fig. 1.

5 Conclusions

This paper explores the feasibility and utility of developing predictive models that estimate the consequences of calorie intake and exercise on body weight. The results presented above indicate that such models have the potential to accurately predict the impact of lifestyle on weight in the short term. In addition to analysing the impact of lifestyle on weight, it is essential to consider the importance of body weight in an individual's overall health. Excess weight and obesity are associated with a range of health problems, as discussed earlier. Therefore, understanding and monitoring body weight is crucial for preventing and treating these conditions.

To enhance the model's accuracy, it would be beneficial to incorporate contextual variables that influence weight, such as the level of physical activity, diet quality, sleep habits, and environmental factors. Incorporating these variables can lead to a more comprehensive and personalised understanding of weight gain or loss patterns, making the models more robust and applicable to a broader range of scenarios. Furthermore, integrating biometric data, such as body composition and hormonal markers, and using more advanced technology, such as devices that monitor physiological data in real-time, can be explored to provide a rich source of information that healthcare professionals can analyse.

Acknowledgements. This work is funded by National Funds through the FCT – Foundation for Science and Technology, I.P., within the scope of the project Ref. UIDB/05583/2020. Furthermore, we would like to thank the Research Centre in Digital Services (CISeD), under the internal project PIDI/CISeD/2022/007 and the Polytechnic of Viseu for their support.

References

1. Svendsen, M.T., et al.: Associations of health literacy with socioeconomic position, health risk behavior, and health status: a large national population-based survey among Danish adults. BMC Public Health **20**(1), 1–12 (2020)
2. Blüher, M.: Metabolically healthy obesity. Endocr. Rev. **41**(3), bnaa004 (2020)
3. Abhari, S., et al.: A systematic review of nutrition recommendation systems: with focus on technical aspects. J. Biomed. Phys. Eng. **9**(6), 591–602 (2019)
4. Horta, B.L., Rollins, N., Dias, M.S., Garcez, V., Pérez-Escamilla, R.: Systematic review and meta-analysis of breastfeeding and later overweight or obesity expands on previous study for world health organization. Acta Paediatr. **112**(1), 34–41 (2023)
5. Buoncristiano, M., et al.: Socioeconomic inequalities in overweight and obesity among 6-to 9-year-old children in 24 countries from the World Health Organization European region. Obes. Rev. **22**, e13213 (2021)
6. Powell-Wiley, T.M., et al.: Obesity and cardiovascular disease: a scientific statement from the American Heart Association. Circulation **143**(21), e984–e1010 (2021)
7. Lopez-Jimenez, F., et al.: Obesity and cardiovascular disease: mechanistic insights and management strategies. A joint position paper by the World Heart Federation and World Obesity Federation. Eur. J. Prev. Cardiol. **29**(17), 2218–2237 (2022)
8. Safaei, M., Sundararajan, E.A., Driss, M., Boulila, W., Shapi'i, A.: A systematic literature review on obesity: understanding the causes & consequences of obesity and reviewing various machine learning approaches used to predict obesity. Comput. Biol. Med. **136**, 104754 (2021)
9. Zheng, Z., Ruggiero, K.: Using machine learning to predict obesity in high school students. In: 2017 IEEE International Conference on Bioinformatics and Biomedicine (BIBM), pp. 2132–2138 (2017)
10. Alkhalaf, M., Yu, P., Shen, J., Deng, C.: A review of the application of machine learning in adult obesity studies. Appl. Comput. Intell. **2**(1), 32–48 (2022)

11. Zhao, H., et al.: Predicting the risk of hypertension based on several easy-to-collect risk factors: a machine learning method. Front. Public Health **9**, 619429 (2021)
12. Sun, M., Jia, W., Chen, G., Hou, M., Chen, J., Mao, Z.H.: Improved wearable devices for dietary assessment using a new camera system. Sensors **22**(20), 8006 (2022)
13. Haththotuwa, R.N., Wijeyaratne, C.N., Senarath, U.: Worldwide epidemic of obesity. In: Obesity and Obstetrics, pp. 3–8. Elsevier (2020)
14. Pellegrini, M., et al.: Effects of time-restricted feeding on body weight and metabolism. A systematic review and meta-analysis. Rev. Endocr. Metab. Disord. **21**, 17–33 (2020)
15. Hill, J.O., Pagliassotti, M.J., Peters, J.C.: Nongenetic determinants of obesity and body fat topography. In: The Genetics of Obesity, pp. 35–48. CRC Press (2020)
16. Zhao, F.Y., et al.: Effects of a 12-week exercise-based intervention on weight management in overweight or obese breast cancer survivors: a randomized controlled trial. Support. Care Cancer **32**(2), 98 (2024)
17. Van Strien, T.: Causes of emotional eating and matched treatment of obesity. Curr. Diab. Rep. **18**, 1–8 (2018)
18. Alimoradi, Z., Golboni, F., Griffiths, M.D., Broström, A., Lin, C.Y., Pakpour, A.H.: Weight-related stigma and psychological distress: a systematic review and meta-analysis. Clin. Nutr. **39**(7), 2001–2013 (2020)
19. de Witt Huberts, J.C., Evers, C., de Ridder, D.T.: Double trouble: restrained eaters do not eat less and feel worse. Psychol. Health **28**(6), 686–700 (2013)
20. Anthanont, P., Jensen, M.D.: Does basal metabolic rate predict weight gain? Am. J. Clin. Nutr. **104**(4), 959–963 (2016)
21. Khanna, D., Peltzer, C., Kahar, P., Parmar, M.S.: Body mass index (BMI): a screening tool analysis. Cureus **14**(2), e22119 (2022)
22. Roza, A.M., Shizgal, H.M.: The Harris Benedict equation reevaluated: resting energy requirements and the body cell mass. Am. J. Clin. Nutr. **40**(1), 168–182 (1984)
23. Jagim, A.R., et al.: Accuracy of resting metabolic rate prediction equations in athletes. J. Strength Conditioning Res. **32**(7), 1875–1881 (2018)
24. Bendavid, I., et al.: The centenary of the Harris-Benedict equations: how to assess energy requirements best? Recommendations from the ESPEN expert group. Clin. Nutr. **40**(3), 690–701 (2021)
25. Sonawane, J.S., Patil, D.R.: Prediction of heart disease using multilayer perceptron neural network. In: International Conference on Information Communication and Embedded Systems (ICICES 2014), pp. 1–6. IEEE (2014)
26. Ferdowsy, F., Rahi, K.S.A., Jabiullah, M.I., Habib, M.T.: A machine learning approach for obesity risk prediction. Curr. Res. Behav. Sci. **2**, 100053 (2021)
27. Iwendi, C., Khan, S., Anajemba, J.H., Bashir, A.K., Noor, F.: Realizing an efficient IoMT-assisted patient diet recommendation system through machine learning model. IEEE Access **8**, 28462–28474 (2020)
28. Nagari, S.S., Inayati, L.: Implementation of clustering using k-means method to determine nutritional status. J. Biometrika dan Kependud **9**(1), 62 (2020)
29. Rostami, M., Oussalah, M., Farrahi, V.: A novel time-aware food recommender-system based on deep learning and graph clustering. IEEE Access **10**, 52508–52524 (2022)
30. Raut, M., Prabhu, K., Fatehpuria, R., Bangar, S., Sahu, S.: A personalized diet recommendation system using fuzzy ontology. Int. J. Eng. Sci. Invention **7**(3), 51–55 (2018)

31. Chapman, P., et al.: The CRISP-DM user guide. In: 4th CRISP-DM SIG Workshop in Brussels in March, vol. 1999 (1999)
32. CHRISBOW: 2018 calorie, exercise and weight changes (2018). https://www.kaggle.com/datasets/chrisbow/2018-calorie-exercise-and-weight-changes/data

Requirements on and Procurement of Explainable Algorithms—A Systematic Review of the Literature

Ulrik Franke[1,2]([✉])[iD], Celine Helgesson Hallström[1], Henrik Artman[1][iD], and Jacob Dexe[1,2][iD]

[1] KTH Royal Institute of Technology, 100 44 Stockholm, Sweden
{ulrikf,celhal,artman,jacode}@kth.se
[2] RISE Research Institutes of Sweden, P.O. Box 1263, 164 29 Kista, Sweden

Abstract. Artificial intelligence is making progress, enabling automation of tasks previously the privilege of humans. This brings many benefits but also entails challenges, in particular with respect to 'black box' machine learning algorithms. Therefore, questions of transparency and explainability in these systems receive much attention. However, most organizations do not build their software from scratch, but rather procure it from others. Thus, it becomes imperative to consider not only requirements on but also procurement of explainable algorithms and decision support systems. This article offers a first systematic literature review of this area. Following construction of appropriate search queries, 503 unique items from Scopus, ACM Digital Library, and IEEE Xplore were screened for relevance. 37 items remained in the final analysis. An overview and a synthesis of the literature is offered, and it is concluded that more research is needed, in particular on procurement, human-computer interaction aspects, and different purposes of explainability.

Keywords: Requirements · Procurement · Explainable Artificial Intelligence (XAI) · Transparency · Explainability

1 Introduction

Artificial intelligence (AI) and digitalization are reshaping the world. Software can now do many things, such as reading and writing texts or identifying objects in images, which used to be the exclusive privilege of humans. This brings many benefits, but also new challenges, in particular with respect to 'black box' algorithms which have been shown to sometimes disadvantage, e.g., poorer people and those from minorities [42]. As a result, questions of transparency and explainability have received much attention, with calls for more transparency [20], reviews of explainability techniques [24], and practical developer tools.

However, most organizations do not build their own software. Instead, they procure (partial) solutions from others [4]. Consider a highly specialized company which creates a code-component for natural language processing. This code is then used by another company and integrated within a commercial-off-the-shelf (COTS) product for customer interaction, which is then procured by yet another company wanting to employ it on its web site, followed by a project where the COTS product is customized to the requirements of the procuring company, before it starts to interact with the customers, whose input refines the resulting chatbot. This chatbot will be 'a black box', not only because it includes opaque technology (e.g. deep neural networks) but also because no single person will be in a good position to understand and explain its behavior. To 'open' this black box, technical explainability solutions are necessary, but not sufficient.

In complex processes such as this one, the end-product is the result of both *development* and *procurement*, where the latter sets boundary conditions on the former. Recall that both processes occur not only within the narrow context envisioning the system-in-use using requirements engineering and user-centered design expertise—user involvement in both processes is a central success factor!—but also within a broader context of laws, culture, computational infrastructure and resources, to name a few. However, requirements engineering techniques focus on development, and offers little support for procurement.

This is worrying, because with increasing use of AI, the stakes are high. Returning to the chatbot, much can go wrong. For example, chatbots easily pick up and internalize, e.g., racist or sexist language which it encounters. Such problems plague application areas. To alleviate them, organizations must master both requirements and procurement; they need to understand not only code and architecture, but also supply, demand, and markets.

It is against this background that this article offers a systematic review of the literature on requirements on and procurement of explainable algorithms and decision support systems, with the following precise research question: *What support can the extant literature give to procurement or requirements engineering efforts to ensure explainable algorithms and decision support systems?*

2 Related Work

The research question posed in the previous section is located at the intersection of several research areas. One set of questions concerns technical feasibility and are addressed within the field of Explainable Artificial Intelligence (XAI) [24]. Such methods are obviously very useful, but equally obviously, they only address a very small part of the problem. In this literature, users are often very idealized.

Another set of questions concerns how technical solutions should be made to match the goals of organizations and users. In the of non-functional requirements paradigm [23] from requirements engineering, explainability is often seen as yet another non-functional requirement, closely related to usability, and its procurement [40]. Ahmad [1] presents a project plan on human-centric requirements engineering for AI including a literature review, survey and validation.

Yet another set of questions concerns what, normatively, *should* be aimed for. Most agree that transparency is a *prima facie* good, but sometimes even more important goods take precedence. For example, Holm [28] argues that black box methods are useful in two cases; (i) when the cost of being wrong is much lower than the value of being right, and (ii) when it produces the best results. London [39] argues that Holm's second case holds in many medical contexts, where actually curing someone is more important than understanding how.

One more set of questions concerns what is required by laws, such as the European General Data Protection Regulation (GDPR). All these sets of questions are not—and should not be—distinct. Ideally, the corresponding perspectives should *all* be included in the development and procurement of future information systems. Still, the literature rarely covers all of them at the same time, as reflected in field-specific literature reviews. This article offers a broader view.

3 Method

3.1 Search Queries

To cover the field described in Sect. 1, the authors designed database queries together with a professional university librarian. To capture three key features of the relevant literature—it must be about *explainability* (1) *requirements* (2) posed on *algorithms* (3)—each of these three concepts was expanded with synonyms or related terms (such as *transparency* or *explicability* for 1, *procurement* for 2, and *decision support* or *AI* for 3).

This preliminary search query was refined iteratively, with the librarian working both off-line and in an interactive meeting where different search strategies were tried. Based on these trials, the requirements parts of the conjunction was expanded to include more terms from the requirements engineering field, such as *non functional requirement* and *software specification*. Similarly, the selection of databases was refined. By defining the search query in this iterative and interactive way, the results became more robust and relevant.

After the follow-up meeting, the librarian further refined the search query based and conducted the final search on December 22, 2021, in Scopus (400 items), ACM Digital Library (81 items), and IEEE Xplore (144 items). The database interfaces differ somewhat, but the Scopus query is representative:

```
TITLE-ABS-KEY ( transparen* OR explainab* OR explicab* OR accountab* OR intelligib*
OR understandab* OR interpretab* )
AND
TITLE-ABS-KEY ( "requirement* engineering" OR purchas* OR "software quality" OR procurement
OR "legal requirement*" OR "non functional requirement*" OR "software requirement*" OR
"software
specification*" OR "system requirement*" OR "system specification*" OR "requirement
specification*" )
AND
TITLE-ABS-KEY ( "decision support" OR "artificial intelligence" OR algorithm* OR "machine
learning" )
```

Removing duplicates, 503 unique items remained.

3.2 Relevance Screening

The next step was to screen the 503 items for relevance. Developing the queries, it had been observed that they gave many false positives. It is easy to see why: instead of explainability requirements posed on algorithms, false positives include, e.g., (i) algorithms for explainable requirements, (ii) explainable algorithms for managing requirements, or (iii) algorithms explaining procurement, etc.

Thus, the 503 items were independently assessed for relevance by two of the authors, based on title and abstract only. Full texts were not read at this time. The two authors found 66 and 87, respectively, relevant items. To avoid discarding any relevant items, the union of the two sets—98 items—was retained and the remaining 405 items were discarded.

3.3 Content Evaluation and Analysis

To read and evaluate the 98 remaining items, these were divided equally between the four authors, who evaluated them according to the protocol in Fig. 1. Consistency was maintained through regular meetings to interpret the protocol.

1. Is the item, upon closer inspection, still relevant to the research question?
2. Does the item have an emphasis on requirements or procurement?
3. Does the item have a particular focus on transparency or explainability, or is this just one among many quality attributes?
4. Does the item focus on supporting a particular role (e.g., developer, clerk, customer, lawyer)? If so, which one?
5. Does the item address transparency or explainability of AI black boxes or broader decision support?
6. Where in a system life-cycle process does the item fit? For example, is it equally applicable to newly developed systems and maintenance of existing ones?
7. Does the item include an empirical component, and if so, what kind? For example, user experiment, case study of procurement, text analysis of requirements, experiment with code?
8. Does the item address transparency about what kind of input data is used by a system? For example, transparency about whether a system uses only the data entered by the user, or also data from other sources.

Fig. 1. Items were evaluated with respect to the questions in this evaluation protocol.

First in the protocol was a more profound relevance assessment using the inclusion (IC) and exclusion (EC) criteria in Fig. 2. These IC and EC are the precise operationalization of the first question (relevance) in the evaluation protocol in Fig. 1. Following closer inspection and application of these criteria, 37 items remained in the final analysis. The evaluation protocol as such was drafted based on the research questions, refined during the relevance screening and finalized before the content evaluation and analysis described in this section.

IC 1: The item should give meaningful practical support to procurement or requirements engineering efforts beyond a single community.
EC 1: The item should not be too abstract, e.g., only articulating or discussing ethical standards without supporting practitioners in upholding them.
EC 2: The item should not be too narrow, e.g., only supporting transparency or explainability efforts for very particular cases.
EC 3: The item should not (only) be a dictionary or a standard.
EC 4: The item should not be (only) an abstract describing, e.g., a panel or a tutorial at a conference, so that the entire material is not actually accessible in the literature.

Fig. 2. Items were included or excluded with respect to these criteria.

4 Results

Table 1 offers an overview of the 37 items, based on the protocol in Fig. 1. A few quantitative observations are evident. First, the literature is relatively recent, with the oldest items being from 2017 and 2018. This reflects the timeliness of questions about explainability requirements. Second (column 2) most items are about requirements rather than procurement (though the categories are not fully distinct). Third (column 3) considering explainability, transparency, or interpretability to be non-functional requirements (just like, e.g., performance, interoperability, or availability) is common. Fourth (column 4) there is considerably more literature supporting technical roles such as requirements or software engineers compared to non-technical roles such as purchasers. Fifth (column 5) much of the literature is concerned with AI, but some more generic articles address broader decision support systems (DSS) matters. Sixth (column 6, mirroring column 4) there is much support for requirements elicitation or development, though some articles address the entire life cycle. Seventh (column 7) much of the work is still conceptually rather than empirically driven. Eighth (column 8) transparency about input data used is sometimes addressed in the literature.

From a qualitative point of view, the literature contains many different approaches to explainability requirements. In the following, the "pathetic dot theory" first proposed in a paper by Lessig [36], and later developed at book length [37], is used to structure the material. The theory describes how any *thing* is regulated by four forces: Architecture, Law, Norms and Markets. Architecture, such as speed bumps, walls, or website interfaces, regulates by limiting the technical possibilities. Law regulates by punishing with force. Norms also regulate by punishing behaviors, but with social means such as exclusion or angry glances. Finally, markets regulate through the price mechanism.

4.1 Architecture

Architecture is the most straightforward of Lessig's four forces. Building systems to predictably conform to user needs and specifications is the rationale of software engineering at large and requirements of engineering in particular. Thus, much of the literature broadly belongs to architecture in Lessig's sense.

The goal of the software engineering discipline is to make building software an engineering practice, not a craft that can only be learned from experience.

Table 1. An overview of the items. The numbered columns correspond to the questions in Fig. 1: 2. Requirements or procurement, 3. Special quality attributes, 4. Supports particular role, 5. Narrow AI or broader DSS, 6. Life cycle, 7. Empirical component, 8. Transparent input data. Abbreviations: NFR = Non Functional Requirement, QA = Quality Assurance, DL = Deep Learning.

Reference	2	3	4	5	6	7	8
IEEE [32]	Req.	Transp.		Broader	Entire process		
Ahmad [1]	Req.	One of many			Dev.		
Alison Paprica et al. [2]	Req.	Transp.	Domain pract.				
Andrus et al. [3]	Req.	One of many					
Barclay and Abramson [4]	Req.	Transp.	Several		Dev. & Maint.		Partly
Bibal et al. [5]	Req.	Expl.		Both	Req.& dev.		
Brkan and Bonnet [7]	Req.	Expl.		Both	Req.& dev.		
Calo [8]	Proc.		Policymakers	AI	Early stage		
Chazette et al. [10]	Req.		Dev.				
Curcin et al. [11]	Req.	Transp., expl., traceability	Req. eng.	Broader DSS	Req.		Yes
Cysneiros and do Prado Leite [12]	Req.	One of many					
Cysneiros et al. [13]	Req.	Transp.	Req. eng.	AI	Req.	Lit. rev.	
de Cerqueira et al. [9]	Req.	Broad AI ethics	Req. eng., dev.	AI (ML, DL)	Req.	Lit. rev.	
Dor and Coglianese [16]	Proc.	Transp.	Purchaser, policymaker	AI	Proc.		
Drobotowicz et al. [17]	Req.	Transp. (trust)	Purchaser, req. eng.	AI	Req.	Interviews & design workshop	Partly
Fagbola and Thakur [18]	Req.	Interpret. & Expl.	Req. eng., dev.	AI	Req., dev.	Lit. rev.	
Felderer and Ramler [19]	Req.	Interpret.	Software eng., QA eng., req. eng.	AI, black boxes	QA (mainly) & req.		
Habibullah and Horkoff [25]	Req.	Many NFRs	Req. eng.	AI/ML	Req.	Interviews	Yes
Hamon et al. [26]	Req.	Expl.		AI	Evaluation	Case study	Yes
Hepenstal et al. [27]	Req.	Transp. & Expl.		AI	Dev.	Interviews	
Hong and Fong [29]	Req.	Transp. & Interpret.	Software eng.	Black box	Dev. & audit		Yes
Hussain et al. [30]	Proc.	One of many	Software eng.		Dev.	Interviews & case studies	
Hutchinson et al. [31]	Req.	Transp. & Qual.	Data set eng.		Yes		Yes
Kuwajima and Ishikawa [34]	Req.	Expl.	Req. eng.	AI (ML)	Req.	Document study	Yes
Langer et al. [35]	Req.	Expl.	Auditing	Both	Auditing		
Liu et al. [38]	Req.	Expl.	Several	AI (DL)	Entire process	Yes	Yes
London [39]	Both	Expl.		Black box	Dev. & implementation		Yes
Martin et al. [41]	Req.	Expl.	Req. eng., dev.	Black box	Dev. & evaluation	User experiments	Yes
Nguyen et al. [43]	Req.	Aspects of Expl.	Req. eng.	Both	Req.		
Sadeghi et al. [45]	Req.	Expl.	Dev.?	Broader DSS			
Schoonderwoerd et al. [46]	Req.	Expl.	Req. eng., dev.	Both	Entire process	User studies	Yes
Serrano and do Prado Leite [47]	Req.	Transp.	Req. eng.	Broader DSS	Req. elicitation	Observation study	
Simmler et al. [48]	Proc.	One of many	Several	Both		Lit. rev.	
van Otterlo and Atzmueller [44]	Req.	Expl.	Several	AI			
Villamizar et al. [51]	Req.	One of many	Req. eng.	AI (ML)	Entire process	Lit. rev.	As data quality
Vogelsang and Borg [52]	Req.	One of many	Req. eng.	AI (ML)	Entire req. eng. process	Interviews	Yes
Vojíř and Kliegr [53]	Req.	Expl.	Researchers	AI	Req. elicitation	User studies	

Quality should not depend on who wrote the code, just as the tensile strength of a bolt or the impedance of an electric circuit should not depend on who worked in the factory. Therefore, software engineering employs methods such as elicitation of functional and non-functional requirements beforehand, use of best practice patterns and architecture models when software is built, and testing afterwards.

Now, such practices are challenged by modern AI 'black boxes', especially machine learning (ML), and new non-functional requirements such as explainability is the most researched response [25,51]. A paradigmatic change from coding to training also entails other challenges, such as articulating functional requirements [52], finding test cases, and handling changing environments [19,38].

Broadly speaking, AI and ML make *quality assurance*—systems reliably and predictably doing what they should—more difficult. Partial solutions to this quality assurance problem include adapting existing practices and methodologies such as the ISO 25000 series [34], or to encapsulate ML components within symbolic reasoning or ensemble decision-making [29] to control irregularities.

With respect to explainability within a software (or requirements) engineering perspective, different aspects of explainability are relevant to different stakeholders in different contexts [10]. Thus, methods for involving relevant stakeholders become important [43,47], including experimental approaches [53]. The elicitation process can also be guided by taxonomies [45], templates [11], or knowledge catalogs [10], so that it is not unnecessarily limited by individual experience, but makes full use of the existing state-of-the-art.

4.2 Law

Legal requirements drive greater explainability in automated decision-making. In particular, Article 15 in the General Data Protection Regulation (GDPR) mandates a legal right to "meaningful information about the logic involved" in "automated decision-making" based on the personal data of the individual [22]. Though the exact meaning of this right is still subject to debate [6,15], the majority of the items found are of European origin and interpret the GDPR as granting a right to explanations of algorithmic decisions.

More precisely, the requirements and procurement literature tries to translate legal rights into (technical) methods for different types of explanations [5,7,26]. Some difficulties arise from a lack of a consensus legal definition of AI [8]— though this may change with the EU AI Act [50]. Therefore, as a complement to regulations, other frameworks and policies can also offer guidance on explainable AI [44]. Some work connects non-functional requirements with particular legal settings, such as audits [27] or criminal justice [48].

4.3 Norms

Software affects humans both directly (as users) and indirectly (as part of datasets). Since automated systems sometimes exhibit bias [42], norms and human values are of utmost importance when procuring and designing systems

[21]. The items found include preliminary guidelines based on literature reviews [2,18], interviews [2,3,30] and deductive reasoning [32,35]. However, guidelines are still quite open-ended, and many research questions remain [18].

Hussain et al. [30] empirically investigate human values in the design of systems. Through interviews with 31 software engineers in two organizations, they discover two different approaches; one focusing on non-functional requirements such as privacy, security and accessibility, the other focusing on broader social justice and equality. Alison Paprica et al. [2] offer another example of empirical work related to values, where literature and interviews are used to develop practical guidelines for data trust. Results include the importance of accountable governing bodies, comprehensive data management, training of data users, and stakeholder engagement. Andrus et al. [3] also address normative aspects of data use, viz. procuring and acquiring sensitive demographic data. Langer et al. [35] outline a visionary multi-disciplinary perspective on explainability which include technical, psychological, legal and ethical considerations, much in line with the broad scope of this review.

4.4 Markets

Out of the four forces, markets—impacting the design of technical artifacts through prices, supply and demand—are the one least represented in the literature. None of the 37 items can straightforwardly be classified as belonging only there. However, a few combine markets with one of the other forces.

4.5 Combinations of the Forces

Since all four forces are always present in Lessig's model, the most interesting literature items are the ones explicitly acknowledging such combinations. The most important combination is architecture and norms. It is widely believed that if ethical norms were incorporated into technical artifacts, significant progress would be made (though this can be questioned: see [54]). Thus, Cysneiros and do Prado Leite [12] bridge social and design sciences by incorporating criteria for socially responsible software from the former into the non-functional requirements of the latter. Similarly, Barclay and Abramson [4] identify the roles and responsibilities needed to improve the development and deployment of AI systems in software engineering practice, and Schoonderwoerd et al. [46] offer design patterns to facilitate human-understandable explanations. Another combination of architecture and norms is that establishing best practice methods or ethical models requires new modes of working and new social norms for architecture to be improved, see, e.g., [35,43,47].

Combining architecture with markets, Cysneiros et al. [13] argue that to increase market demand for self-driving cars—for customers to travel with them and insurers to insure them—more transparent operations are needed, illustrating the need for transparency to become a proper non-functional requirement.

5 Discussion

Today, procurement of transparency and explainability is in the state usability was some 20 years ago. Research is still in its infancy, but advancing rapidly. More precisely, five observations can be made:

First, the dominant paradigm is non-functional requirements from requirements engineering; a community clearly at work to address AI.

Second, however, there is much less literature explicitly addressing procurement (see Table 1)—a shame, since most organizations procure rather than build software. If the procuring organization cannot grasp how the supplier can be expected to work it is difficult to create appropriate requests for tender.

Third, there is a lack of literature exploring human-computer interaction aspects of explainable systems. There are psychology insights [33] to harness, but since systems change as we use them, as the vendors update them, and as the underpinning data changes, good explanations are a moving target, even if user groups are meticulously specified. A further complication is the context of use, which may make the user misunderstand or ignore [49] explanations given.

Fourth, the literature does not sufficiently address the fact that explainability can serve many different purposes, and that depending on the purpose, different kinds of explainability are required. Even if transparency is seen as valuable, companies may struggle to make use of it [14].

Fifth, market characteristics such as the business cycles, with booms and busts, matter. We are now in an AI boom, with plans to automate all kinds of things with ML algorithms learning from the data available in modern society. In some cases this is reasonable, in some cases this is overconfidence and hype. Booms are always followed by busts, but appropriate transparency and explainability may help separating some of the wheat from the chaff.

6 Conclusion

As AI and in particular ML is increasingly used, concerns over lack of transparency and explainability mount. How can we understand what 'black box' algorithms do? However, since most organizations do not build their AI software from scratch, but rather procure it from others, transparency and explainability must be addressed not only from the perspective of the software engineers *building* systems, but also from the perspective of those *procuring* them. In terms of Lessig's forces, organizations need to learn to master not only architecture, but also markets. Procurement and requirements engineering must go hand in hand.

Acknowledgments. The authors thank librarian Magdalena Svanberg as well as Thomas Olsson who read and commented a draft of the paper. This research was partially supported by the Swedish Competition Authority, grant no 456/2021.

Disclosure of Interests. The authors have no competing interests to declare.

References

1. Ahmad, K.: Human-centric requirements engineering for artificial intelligence software systems. In: 29th International Requirements Engineering Conference (RE), pp. 468–473. IEEE (2021). https://doi.org/10.1109/RE51729.2021.00070
2. Alison Paprica, P., et al.: Essential requirements for establishing and operating data trusts. Int. J. Popul. Data Sci. **5**(1) (2020). https://doi.org/10.23889/IJPDS.V5I1.1353
3. Andrus, M., Spitzer, E., Brown, J., Xiang, A.: What we can't measure, we can't understand. In: ACM Conference on Fairness, Accountability, and Transparency, pp. 249–260 (2021). https://doi.org/10.1145/3442188.3445888
4. Barclay, I., Abramson, W.: Identifying roles, requirements and responsibilities in trustworthy AI systems. In: Adjunct Proceedings of UbiComp/ISWC, pp. 264–271 (2021). https://doi.org/10.1145/3460418.3479344
5. Bibal, A., Lognoul, M., de Streel, A., Frénay, B.: Legal requirements on explainability in machine learning. Artif. Intell. Law **29**(2), 149–169 (2021). https://doi.org/10.1007/s10506-020-09270-4
6. Bottis, M., Panagopoulou-Koutnatzi, F., Michailaki, A., Nikita, M.: The right to access information under the GDPR. Int. J. Technol. Policy Law **3**(2), 131–142 (2019). https://doi.org/10.1504/IJTPL.2019.104950
7. Brkan, M., Bonnet, G.: Legal and technical feasibility of the GDPR's quest for explanation of algorithmic decisions. Eur. J. Risk Regul. **11**(1), 18–50 (2020). https://doi.org/10.1017/err.2020.10
8. Calo, R.: Artificial intelligence policy: a primer and roadmap. Univ. Bologna Law Rev. **3**(2), 180–218 (2018). https://doi.org/10.6092/issn.2531-6133/8670
9. de Cerqueira, J.A.S., Althoff, L.S., de Almeida, P.S., Canedo, E.D.: Ethical perspectives in AI. In: HICSS-54, pp. 5240–5249. AIS (2020). https://hdl.handle.net/10125/71257
10. Chazette, L., Brunotte, W., Speith, T.: Exploring explainability. In: 29th International Requirements Engineering Conference (RE), pp. 197–208 (2021). https://doi.org/10.1109/RE51729.2021.00025
11. Curcin, V., Fairweather, E., Danger, R., Corrigan, D.: Templates as a method for implementing data provenance in decision support systems. J. Biomed. Inform. **65**, 1–21 (2017). https://doi.org/10.1016/j.jbi.2016.10.022
12. Cysneiros, L.M., do Prado Leite, J.C.S.: Non-functional requirements orienting the development of socially responsible software. In: Nurcan, S., Reinhartz-Berger, I., Soffer, P., Zdravkovic, J. (eds.) BPMDS & EMMSAD 2020. LNBIP, pp. 335–342. Springer, Cham (2020). https://doi.org/10.1007/978-3-030-49418-6_23
13. Cysneiros, L.M., Raffi, M., do Prado Leite, J.C.S.: Software transparency as a key requirement for self-driving cars. In: 26th International Requirements Engineering Conference (RE), pp. 382–387. IEEE (2018). https://doi.org/10.1109/RE.2018.00-21
14. Dexe, J., Franke, U., Rad, A.: Transparency and insurance professionals. Geneva Pap. Risk Insur. Issues Pract. **46**, 547–572 (2021). https://doi.org/10.1057/s41288-021-00207-9
15. Dexe, J., et al.: Explaining automated decision-making–a multinational study of the GDPR right to meaningful information. Geneva Pap. Risk Insur. Issues Pract. **47**, 669–697 (2022). https://doi.org/10.1057/s41288-022-00271-9
16. Dor, L.M.B., Coglianese, C.: Procurement as AI governance. IEEE Trans. Technol. Soc. **2**(4), 192–199 (2021). https://doi.org/10.1109/TTS.2021.3111764

17. Drobotowicz, K., Kauppinen, M., Kujala, S.: Trustworthy AI Services in the Public Sector: What Are Citizens Saying About It? In: Dalpiaz, F., Spoletini, P. (eds.) REFSQ 2021. LNCS, vol. 12685, pp. 99–115. Springer, Cham (2021). https://doi.org/10.1007/978-3-030-73128-1_7
18. Fagbola, T.M., Thakur, S.C.: Towards the development of artificial intelligence-based systems. In: ICIIBMS, pp. 200–204. IEEE (2019). https://doi.org/10.1109/ICIIBMS46890.2019.8991505
19. Felderer, M., Ramler, R.: Quality assurance for AI-based systems: overview and challenges (introduction to interactive session). In: Winkler, D., Biffl, S., Mendez, D., Wimmer, M., Bergsmann, J. (eds.) SWQD 2021. LNBIP, vol. 404, pp. 33–42. Springer, Cham (2021). https://doi.org/10.1007/978-3-030-65854-0_3
20. Fleischmann, K.R., Wallace, W.A.: A covenant with transparency. Commun. ACM **48**(5), 93–97 (2005). https://doi.org/10.1145/1060710.1060715
21. Friedman, B., Kahn, P.H., Borning, A., Huldtgren, A.: Value sensitive design and information systems. In: Doorn, N., Schuurbiers, D., van de Poel, I., Gorman, M.E. (eds.) Early Engagement and New Technologies: Opening Up the Laboratory. POET, vol. 16, pp. 55–95. Springer, Cham (2013). https://doi.org/10.1007/978-94-007-7844-3_4
22. GDPR: General data protection regulation. Off. J. EU (OJ) L **119**, 4.5, pp. 1–88. (2016). http://data.europa.eu/eli/reg/2016/679/oj
23. Glinz, M.: On non-functional requirements. In: 15th IEEE International Requirements Engineering Conference (RE 2007), pp. 21–26. IEEE (2007). https://doi.org/10.1109/RE.2007.45
24. Guidotti, R., Monreale, A., Ruggieri, S., Turini, F., Giannotti, F., Pedreschi, D.: A survey of methods for explaining black box models. ACM Comput. Surv. (CSUR) **51**(5), 1–42 (2018). https://doi.org/10.1145/3236009
25. Habibullah, K.M., Horkoff, J.: Non-functional requirements for machine learning. In: 29th International Requirements Engineering Conference (RE), pp. 13–23. IEEE (2021). https://doi.org/10.1109/RE51729.2021.00009
26. Hamon, R., Junklewitz, H., Malgieri, G., Hert, P.D., Beslay, L., Sanchez, I.: Impossible explanations? In: ACM Conference on Fairness, Accountability, and Transparency, pp. 549–559 (2021). https://doi.org/10.1145/3442188.3445917
27. Hepenstal, S., Zhang, L., Kodagoda, N., William Wong, B.L.: What are you thinking? Explaining conversational agent responses for criminal investigations. In: ExSS-ATEC'2, vol. 2582. CEUR-WS (2020)
28. Holm, E.A.: In defense of the black box. Science **364**(6435), 26–27 (2019). https://doi.org/10.1126/science.aax0162
29. Hong, G.Y., Fong, A.C.M.: Multi-prong framework toward quality-assured AI decision making. In: IC3I, pp. 106–110. IEEE (2019). https://doi.org/10.1109/IC3I46837.2019.9055640
30. Hussain, W., et al.: Human values in software engineering. IEEE Trans. Softw. Eng. (2020). https://doi.org/10.1109/TSE.2020.3038802
31. Hutchinson, B., et al.: Towards accountability for machine learning datasets. In: ACM Conference on Fairness, Accountability, and Transparency, pp. 560–575 (2021). https://doi.org/10.1145/3442188.3445918
32. IEEE: Standard Model Process for Addressing Ethical Concerns During System Design. Std 7000-2021 (2021). https://doi.org/10.1109/IEEESTD.2021.9536679
33. Keil, F.C.: Explanation and understanding. Annu. Rev. Psychol. **57**, 227–254 (2006). https://doi.org/10.1146/annurev.psych.57.102904.190100

34. Kuwajima, H., Ishikawa, F.: Adapting SQuaRE for quality assessment of artificial intelligence systems. In: ISSREW, pp. 13–18. IEEE (2019). https://doi.org/10.1109/ISSREW.2019.00035
35. Langer, M., Baum, K., Hartmann, K., Hessel, S., Speith, T., Wahl, J.: Explainability auditing for intelligent systems. In: International Requirements Engineering Conference Workshops (REW), pp. 164–168. IEEE (2021). https://doi.org/10.1109/REW53955.2021.00030
36. Lessig, L.: The new Chicago school. J. Leg. Stud. **27**(S2), 661–691 (1998). https://doi.org/10.1086/468039
37. Lessig, L.: Code: Version 2.0. Basic Books (2006)
38. Liu, Y., Ma, L., Zhao, J.: Secure deep learning engineering: a road towards quality assurance of intelligent systems. In: Ait-Ameur, Y., Qin, S. (eds.) ICFEM 2019. LNCS, vol. 11852, pp. 3–15. Springer, Cham (2019). https://doi.org/10.1007/978-3-030-32409-4_1
39. London, A.J.: Artificial intelligence and black-box medical decisions. Hastings Cent. Rep. **49**(1), 15–21 (2019). https://doi.org/10.1002/hast.973
40. Markensten, E., Artman, H.: Procuring a usable system using unemployed personas. In: Proceedings of the Third Nordic Conference on Human-Computer Interaction, pp. 13–22 (2004). https://doi.org/10.1145/1028014.1028017
41. Martin, K., Liret, A., Wiratunga, N., Owusu, G., Kern, M.: Developing a catalogue of explainability methods to support expert and non-expert users. In: Bramer, M., Petridis, M. (eds.) SGAI 2019. LNAI, vol. 11927, pp. 309–324. Springer, Cham (2019). https://doi.org/10.1007/978-3-030-34885-4_24
42. Nature: More accountability for big-data algorithms. Nature **537**(7621), 449 (2016). https://doi.org/10.1038/537449a
43. Nguyen, M.L., Phung, T., Ly, D.H., Truong, H.L.: Holistic explainability requirements for end-to-end machine learning in IoT cloud systems. In: International Requirements Engineering Conference Workshops (REW), pp. 188–194. IEEE Computer Society (2021). https://doi.org/10.1109/REW53955.2021.00034
44. van Otterlo, M., Atzmueller, M.: On requirements and design criteria for explainability in legal AI. In: XAILA Workshop at JURIX. CEUR-WS (2019)
45. Sadeghi, M., Klös, V., Vogelsang, A.: Cases for explainable software systems. In: International Requirements Engineering Conference Workshops (REW), pp. 181–187. IEEE (2021). https://doi.org/10.1109/REW53955.2021.00033
46. Schoonderwoerd, T.A.J., Jorritsma, W., Neerincx, M.A., van den Bosch, K.: Human-centered XAI. Int. J. Hum. Comput. Stud. **154** (2021). https://doi.org/10.1016/j.ijhcs.2021.102684
47. Serrano, M., do Prado Leite, J.C.S.: Capturing transparency-related requirements patterns through argumentation. In: RePa, pp. 32–41 (2011). https://doi.org/10.1109/RePa.2011.6046723
48. Simmler, M., Canova, G., Schedler, K.: Smart criminal justice. Int. Rev. Adm. Sci. (2021). https://doi.org/10.1177/00208523211039740
49. Steinfeld, N.: "I agree to the terms and conditions": (how) do users read privacy policies online? An eye-tracking experiment. Comput. Hum. Behav. **55**, 992–1000 (2016). https://doi.org/10.1016/j.chb.2015.09.038
50. Veale, M., Zuiderveen Borgesius, F.: Demystifying the Draft EU Artificial Intelligence Act–analysing the good, the bad, and the unclear elements of the proposed approach. Comput. Law Rev. Int. **22**(4), 97–112 (2021). https://doi.org/10.9785/cri-2021-220402

51. Villamizar, H., Escovedo, T., Kalinowski, M.: Requirements engineering for machine learning: a systematic mapping study. In: SEAA, pp. 29–36. IEEE (2021). https://doi.org/10.1109/SEAA53835.2021.00013
52. Vogelsang, A., Borg, M.: Requirements engineering for machine learning: perspectives from data scientists. In: International Requirements Engineering Conference Workshops (REW), pp. 245–251. IEEE (2019). https://doi.org/10.1109/REW.2019.00050
53. Vojíř, S., Kliegr, T.: Editable machine learning models? A rule-based framework for user studies of explainability. Adv. Data Anal. Classif. **14**(4), 785–799 (2020). https://doi.org/10.1007/s11634-020-00419-2
54. Wong, P.H.: Democratizing algorithmic fairness. Philos. Technol. **33**, 225–244 (2020). https://doi.org/10.1007/s13347-019-00355-w

Mapping Changes in Procurement Management for Project Implementation in the Food Sector: An Applied Review to the Principles PMBOK®Guide v.7 with DotProject+

Jhony Reinheimer Magalhães[1](✉) ⓘ, André Fabiano de Moraes[1] ⓘ, and Luis Augusto Silva[2] ⓘ

[1] IT - Information Systems, IFC - Camboriú, Camboriú, SC, Brazil
jhonymagalhaes2610@gmail.com, andre.moraes@ifc.edu.br
[2] Departamento de Informática y Automática, USAL - Salamanca, Salamanca, Spain
luisaugustos@usal.es

Abstract. With the increasing demand from organizations for effective project management methods, considering the continuous evolution of existing methods in the market, this study aims to provide an overview of the changes that occurred in the transition from PMBOK®Guide v.6 to PMBOK®Guide v.7, with special attention to the knowledge area applied to procurement management. The main objective is to identify the favorable and unfavorable points of the implemented changes. To carry out this analysis, a derivation of the open-source software DotProject+ was developed, mainly due to its prior alignment with PMBOK®Guide v.6 and flexibility with the principles outlined in PMBOK®Guide v.7. Through this research, we intend to contribute to a deeper understanding of the changes introduced in the seventh edition of the PMBOK®Guide, with a specific focus on procurement management, and provide valuable insights for professionals and researchers interested in implementing projects in the food industry.

Keywords: Business Model · Principles and Domains · Agile Methods

1 Introduction

Project is the development of a temporary effort aimed at producing a unique product, service, or result [1]. It is understood that project management plays a fundamental role in integrating and coordinating diverse resources to achieve goals and deliver value to stakeholders [1, 2]. In this context, the PMBOK (Project Management Body of Knowledge) Guide [3, 4], and [5] emerges as a vital resource for professionals and organizations worldwide, offering a set of best practices and guidelines for effective project management.

According to Dinsmore & Cabanis-Brewin (2014), the PMBOK® Guide, in its first edition, aimed to identify and describe the knowledge and practices generally accepted, that is, those applicable to most projects most of the time, and about which there is widespread consensus on their value and usefulness. Version 6 represented a milestone by consolidating concepts, processes, and knowledge areas, offering a structured and recognized approach. Its acceptance reflects confidence in its guidelines to enhance project management. With the release of version 7, there was a response to emerging demands in the field. The adoption of version 7 is an evolving process, with organizations and professionals evaluating the proposed changes and their impact. As version 7 is implemented and explored, its impact and acceptance are expected to increase, further consolidating the role of the PMBOK Guide as a fundamental resource in project management [6–8], and [9].

2 Comparative Study - Guide Pmbok V6 x Pmbok V7

2.1 Version 6 PMBOK Guide

The version 6 of the PMBOK® Guide, released in 2017, represented a significant evolution in the structure and content of the guide. It established a solid foundation for project management by organizing the process into five groups: Initiating, Planning, Executing, "Monitoring and Controlling", and Closing. According to study [6], each process group was outlined to address the key phases of the project lifecycle, providing detailed guidance on the activities required at each stage. Additionally, version 6 introduced ten knowledge areas, covering essential aspects of project management such as scope, schedule, cost, quality, resources, communication, risks, procurement [10], and stakeholders. Although widely recognized and adopted as a standard reference in the field of project management, version 6 of the PMBOK Guide also received criticism for its predominantly traditional approach and its emphasis on processes and detailed documentation.

2.2 Guide PMBOK Version 7

The version 7 of the PMBOK Guide [1] marked a significant shift in project management approach. This version adopted a more holistic and flexible perspective, reorganizing the guide around principles, processes, and project management activities. Instead of focusing exclusively on sequential processes, version 7 recognized the need for adaptability in an increasingly dynamic and complex project environment. This was reflected in the integration of traditional and agile methodologies, encouraging a hybrid mindset that allows project professionals to select and adapt approaches according to the specific project needs. Additionally, version 7 placed renewed emphasis on delivering value and results as per [1] and [6], prioritizing impact over bureaucratic process execution. As shown in Fig. 1, the reduced emphasis on prescriptive documentation and the adoption of a more accessible and practical language were also distinctive features of version 7, making it more relevant in a variety of project contexts.

In summary, the main differences between versions 6 and 7 of the PMBOK Guide reflect a fundamental change in project management approach. While version 6 established a solid foundation with its traditional structure and focus on sequential processes and detailed documentation, version 7 adopted a more flexible and adaptable perspective.

Furthermore, the integration of traditional and agile methodologies, the prioritization of delivering value and results, and the reduction of emphasis on prescriptive documentation are distinctive aspects of version 7, making it more relevant and applicable in a variety of contemporary project contexts. This evolution demonstrates a continuous commitment to providing updated and practical guidance for professionals and organizations involved in project management, empowering them to tackle challenges and seize opportunities in an ever-changing world.

3 Knowledge Area in Procurement Management

The Procurement Knowledge Area [12], as described in version 6 of the PMBOK Guide, plays an essential role in project management, focusing on the acquisition of goods and services necessary for project execution. This knowledge area covers everything from procurement planning to contract closure. During the procurement planning process, supplier selection criteria, contracting strategies, and types of contracts to be used are defined. Version 6 of the PMBOK Guide provides detailed and practical guidelines for each phase of the procurement lifecycle, aiming to ensure efficiency and effectiveness in acquiring resources necessary for project success.

In version 7, the Procurement Knowledge Area has been substantially revised to align with updated project management principles and approaches. This knowledge area continues to play a crucial role, providing guidance for the acquisition of goods and services necessary for successful project execution. However, version 7 emphasizes a more flexible and adaptable approach to procurement, recognizing the importance of considering a variety of factors, including project complexity, risks involved, and specific stakeholder needs. This new approach allows for greater customization and adaptation of procurement practices according to the project context. Additionally, version 7 incorporates agile and hybrid concepts into procurement management, encouraging a more collaborative and interactive approach to supplier selection and management.

4 Detecting Favorable and Unfavorable Aspects

Both methodologies are widely used in practice, but the choice between them depends on the specific objectives and the intended approach for managing the project in question. Therefore, it is crucial to carefully consider these aspects when making decisions related to project management. Version 7 introduced a more flexible and adaptable perspective compared to the prescriptive and sequential approach of version 6. This flexibility allows project professionals greater customization and adjustment of procurement practices according to the specific needs of each project and the environment in which they operate. Additionally, the integration of agile and hybrid methodologies [13] enables a faster response to changes in project requirements and greater adaptation to dynamic and complex project environments. Prioritizing the delivery of value and results represents a significant change, redirecting the focus of project professionals to customer needs and project objectives. The reduction in emphasis on prescriptive documentation promotes a more pragmatic and results-oriented approach to procurement management.

The transition to version 7 of the PMBOK Guide presents challenges and important considerations compared to its predecessor, version 6. Firstly, there is a potential increase in complexity of implementation due to the introduction of more flexible and adaptable approaches. This requires a deeper understanding and skills in the application of new methodologies, especially for professionals less familiar with agile or less structured approaches. Additionally, the reduction in process prescription may lead to greater reliance on professional judgment and experience of project professionals, increasing the risk of errors or suboptimal decisions. This highlights the importance of additional investments in training and capacity building to ensure that teams are prepared to implement the recommended practices in the new version.

5 Results

With the aim of obtaining relevant data in the food sector, a joint monitoring was conducted with the managers of four restaurants. The objective was to understand the main difficulties faced in the procurement of products and services, in order to formulate proposals for process improvement, using agile methodologies [4, 11, 13–16, 19] and [20] as a basis. The following processes were defined for the procurement sector:

5.1 Counting and Ordering of Products

Inventory counting occurs weekly and is conducted by the stockkeeper. This activity is crucial to facilitate the preparation of purchase orders by buyers [10]. After the counting is completed, buyers prepare purchase orders, which are shared for review and validation. Subsequently, this data is recorded in a collaborative Google Sheets spreadsheet, providing transparency and access to all involved in the process. However, one of the main obstacles faced in this process is the accuracy and timeliness of orders. This results in delivery delays and hampers buyers' ability to negotiate effectively with suppliers.

5.2 Receipt of Products

During the receiving stage, a thorough inspection is conducted between the received items, the corresponding invoice, and the purchase order, aiming to ensure all quality criteria are met. The stock controller is responsible for recording this information in the Purchasing Spreadsheet, attaching the invoice to the drive, and identifying the corresponding folder with the company name and receipt date for the finance department to make the payment. Subsequently, the products are properly stored in their designated locations. One of the challenges encountered in this process is rework in delivering information, which requires data entry in the Purchasing Spreadsheet and subsequent delivery to the finance.

5.3 Filling Out Minimum and Ideal Stock

The ideal inventory quantity is measured by the buyer. Where the inventory is divided into 3 Levels: Level 1 - Off-peak season where we work with reduced inventory to

Mapping Changes in Procurement Management for Project Implementation 57

avoid waste. Level 2 - Intermediate season, inventory is still reduced, but prepared with a surplus in case of atypical movement, which in some weeks generates unnecessary waste. Level 3 - Made to meet high demand.

5.4 Food Handling

In recent years, the factor that most concerns people's daily lives is food. Therefore, it is essential to have healthy and safe food [2]. As shown in Fig. 1, data has revealed a lack of care in handling products before they are sent to the production department in the restaurant. It is important to implement food safety in the midst of processes.

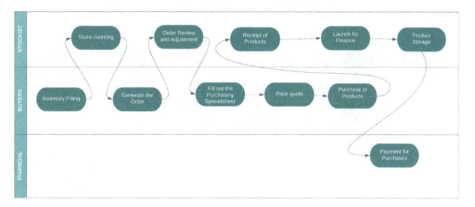

Fig. 1. Flowchart of initially identified processes.

5.5 Unified Database for the Purchasing Sector

A spreadsheet was created to structure the unified database (Relational Model) for the purchasing department, as shown in Fig. 2, containing all relevant product information. Disparity in data usage among departmental collaborators was observed, resulting in ineffective communication between them. Therefore, a restructuring of duties began, starting with processes related to Inventory Levels and Product Orders. In this reorganization, the stockkeeper was assigned the responsibility of determining stock quantities and placing orders, aiming to increase accountability, reduce rework, and promote efficiency in delivery. Additionally, a goal was established to encourage purchases within established limits, aiming to improve management.

As a result of this restructuring, the processes previously under the buyer's responsibility have been optimized, focusing on price quoting with suppliers and maintaining product quality. These changes, as shown in Fig. 3, have significantly contributed to improving efficiency, agility, and communication among departments compared to previous procedures, with the aim of increasing the company's profitability. The integration between spreadsheets in Google Sheets provided an additional modification in the receiving process. Currently, only the warehouse professional is responsible for filling out the receipt spreadsheet, whose summarized data is automatically forwarded to the finance department.

1	2	3	4	5	6	7	8	9	10	11
SITUAÇÃO ATIVO / INATIVO	PRODUTO	PRODUTO DESCRIÇÃO COMPLETA	EMB. UNT.	QTD. EMB. UNT.	UND. EMB. UNT.	UND. EMB. COMPRAS	QTD. EMB. COMPRAS	UND. MED. ESTOQ.	GRUPO	CATEGORIA
TIVO	ALCATRA MIOLO	ALCATRA MIOLO - KG	PCT		KG			UND	ALIMENTOS	CARNES E FRUTOS DC
NATIVO	BACON FATIADO	BACON FATIADO - 1KG	PCT	1	KG			KG	ALIMENTOS	CARNES E FRUTOS DC
NATIVO	BALLOTINE FRANGO (C/10)	BALLOTINE FRANGO (C/10) - 70G	PÇ	70	G	PCT	10	PCT	ALIMENTOS	CARNES E FRUTOS DC
NATIVO	BALLOTINE SALMÃO (C/10)	BALLOTINE SALMÃO (C/10) - 70G	PÇ	70	G	PCT	10	PCT	ALIMENTOS	CARNES E FRUTOS DC
TIVO	ABACAXI	ABACAXI	CX	10	KG			KG	ALIMENTOS	HORTIFRUTI
NATIVO	CAMARÃO CINZA 61/70 DESC	CAMARÃO CINZA 61/70 DESCASCADO E EVISCERADO - 5KG	PCT	5	KG			KG	ALIMENTOS	CARNES E FRUTOS DC
NATIVO	CAMARÃO SEM CASCA CONG	CAMARÃO SEM CASCA CONGELADO - KG	PCT		KG			KG	ALIMENTOS	CARNES E FRUTOS DC
TIVO	ABACAXI EM CALDA RODELAS	ABACAXI EM CALDA RODELAS - 400G	LT	400	G	UND	1	UND	ALIMENTOS	FONDUE DOCE
TIVO	ABÓBORA CABOTIA	ABÓBORA CABOTIA	KG		KG			KG	ALIMENTOS	HORTIFRUTI
TIVO	ABÓBORA MORANGA	ABÓBORA MORANGA	KG		KG			KG	ALIMENTOS	HORTIFRUTI
TIVO	ABOBRINHA	ABOBRINHA	KG		KG			KG	ALIMENTOS	HORTIFRUTI
TIVO	ABRIDOR DE GARRAFA	ABRIDOR DE GARRAFA - UND	UND		UND			UND	UTENSILIOS	BAR
NATIVO	COSTELA SUINA TEMPERADA	COSTELA SUINA TEMPERADA - KG	KG		KG			KG	ALIMENTOS	CARNES E FRUTOS DC
NATIVO	COXÃO MOLE COM CAPA	COXÃO MOLE COM CAPA - KG	PCT		KG			KG	ALIMENTOS	CARNES E FRUTOS DC

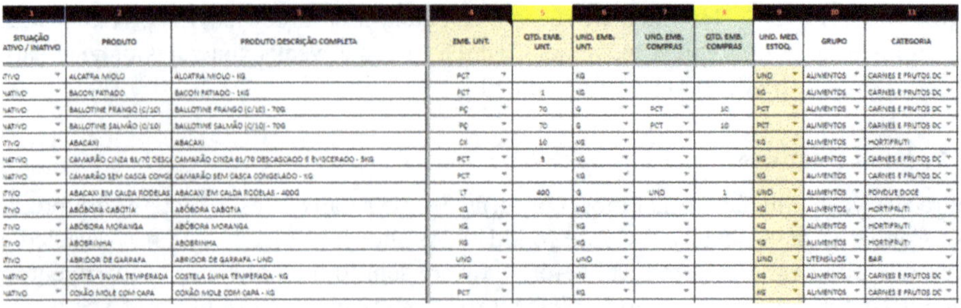

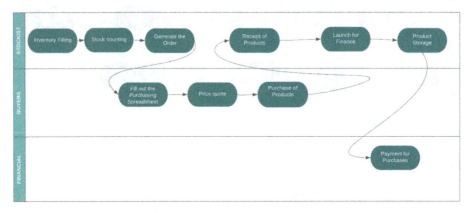

Fig. 2. Unified database for the purchasing department

Fig. 3. Flowchart of processes after changes.

6 Discussion

The current scenario lacks a free software platform adapted to version 7 of the PMBOK Guide. In this context, DotProject+ [18] emerges as an open-source solution for project management, as shown in Figs. 4 and 5. This software offers a range of comprehensive features, including planning, tracking, team collaboration, time and resource tracking, reporting and analysis, customization, and extensibility. One of the advantages of DotProject+ is its compatibility with Docker, which facilitates its installation and deployment in different hosting environments. Additionally, there are already versions that have been adapted to the PMBOK Guide in its version 6.

However, in DotProject+, procurement management features an important functionality, which is the inclusion of costs in the project, divided into human and non-human resources, and distributed over time. That is, with the enhancement of the functionality, the cost performance domain and the specification of the responsible party were added, as illustrated in Fig. 5.

These changes allowed for a more effective identification of the purpose of the cost and the assignment of responsibility related to each cost, thus facilitating the financial management of the project. With adaptation to version 7, when completed, it will be

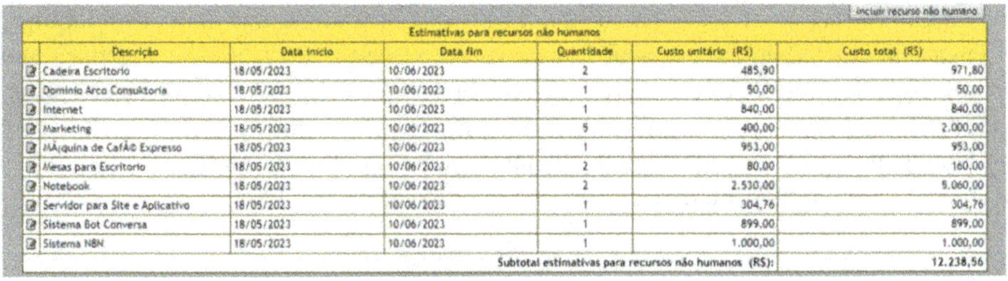

Estimativas para recursos não humanos					
Descrição	Data início	Data fim	Quantidade	Custo unitário (R$)	Custo total (R$)
Cadeira Escritorio	18/05/2023	10/06/2023	2	485,90	971,80
Dominio Arco Consuktoria	18/05/2023	10/06/2023	1	50,00	50,00
Internet	18/05/2023	10/06/2023	1	840,00	840,00
Marketing	18/05/2023	10/06/2023	5	400,00	2.000,00
MÃ¡quina de CafÃ© Expresso	18/05/2023	10/06/2023	1	953,00	953,00
Mesas para Escritorio	18/05/2023	10/06/2023	2	80,00	160,00
Notebook	18/05/2023	10/06/2023	2	2.530,00	5.060,00
Servidor para Site e Aplicativo	18/05/2023	10/06/2023	1	304,76	304,76
Sistema Bot Conversa	18/05/2023	10/06/2023	1	899,00	899,00
Sistema NBN	18/05/2023	10/06/2023	1	1.000,00	1.000,00
Subtotal estimativas para recursos não humanos (R$):					12.238,56

Fig. 4. Functionality of Non-Human Cost Recording in DotProject+

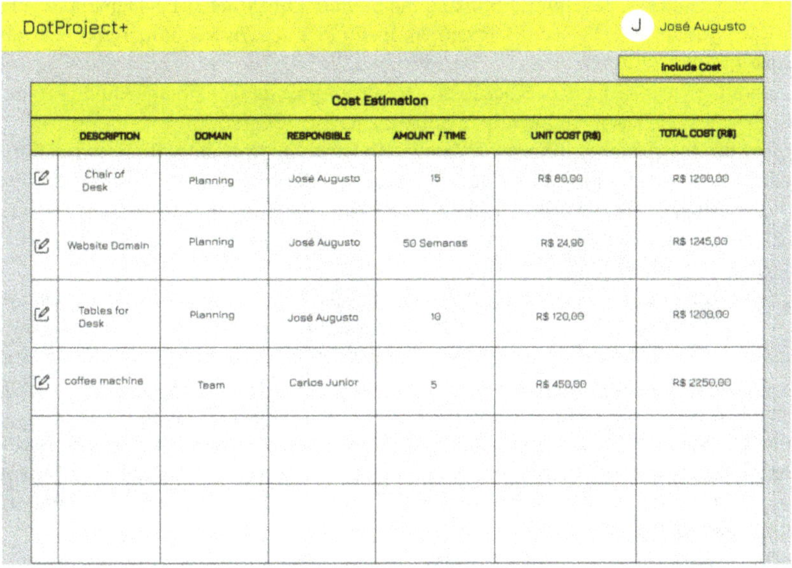

Cost Estimation					
DESCRIPTION	DOMAIN	RESPONSIBLE	AMOUNT / TIME	UNIT COST (R$)	TOTAL COST (R$)
Chair of Desk	Planning	José Augusto	15	R$ 80,00	R$ 1200,00
Website Domain	Planning	José Augusto	50 Semanas	R$ 24,90	R$ 1245,00
Tables for Desk	Planning	José Augusto	10	R$ 120,00	R$ 1200,00
coffee machine	Team	Carlos Junior	5	R$ 450,00	R$ 2250,00

Fig. 5. Prototyping of the Non-Human Cost Registration Functionality.

available in the GitHub repository - https://github.com/KingDarkZeira/Dot_Projetc_Plus_PMBOK_7.

7 Conclusion

Due to the accumulation of tasks and the high cost of dynamically controlling changes in projects focused on the food industry, it was possible to understand the procedures and provide a contribution through this research, which proposes the identification of favorable and unfavorable aspects of implementing the project management methodology from the PMBOK®Guide v.7, assisting in a better understanding of the impact of the changes introduced, specifically focusing on intelligent process management for acquisitions.

Moreover, through this work, its adoption may also imply challenges that require careful consideration and planning by the organizations and professionals involved. It is essential to recognize these challenges and invest in training and skills development to ensure a smooth and successful transition to the new recommended practices and approaches. Version 7 reflects a recognition of the need for adaptability and agility, making it more relevant and applicable in a variety of contemporary project contexts. By embracing these changes and adapting to new practices, the team can seize opportunities in an ever-changing world, driving success and delivering results.

References

1. Project Management Institute: A Guide to the Project Management Body of Knowledge (PMBOK®Guide)—Seventh Edition and the Standard of Project Management (2021)
2. Newtown Square, PA: Project Management Institute. https://www.pmi.org/pmbok-guide-standards/foundational/pmbok. Accessed 14 Mar 2023
3. Cardozo, L.F.: Um Guia para Seleção de Métricas Ágeis de Gerenciamento de Projetos para Organizações de Desenvolvimento de Software. CC-UFSC (2020)
4. Bernardino, M., Abner, L., Pieva, L., Araujo, T., Dias, M., Machado, R.: Empirical studies of an educational tool for project management based on PMBOK using problem-based learning. In: Proceedings of the XXXVI Brazilian Symposium on Software Engineering (SBES 2022), pp. 168–177. Association for Computing Machinery, New York (2022). https://doi.org/10.1145/3555228.3555260
5. Iglesias-Cuzcano, R.A., Quiroz-Rodriguez, L.E., Quiroz-Rodriguez, J.C., Diaz-Garay, B.H., Vasquez-Rivas-Plata, R.: The business model for a superfood company based on lean startup techniques, digital marketing, and PMBOK guidelines. In: Proceedings of the 8th International Conference on Industrial and Business Engineering, pp. 321–328. Association for Computing Machinery, New York (2022). https://doi.org/10.1145/3568834.3568855
6. Imran, R., Soomro, T.R.: Mapping of agile processes into project management knowledge areas and processes. In: International Conference on Business Analytics for Technology and Security (ICBATS), Dubai, United Arab Emirates, pp. 1–12 (2022). https://doi.org/10.1109/ICBATS54253.2022.9759013
7. Lonchina, A.E., Eroshkin, A.V.: Contemporary standards in IT product and project development. In: Seminar on Information Systems Theory and Practice (ISTP), Saint Petersburg, Russian Federation, pp. 103–105 (2023). https://doi.org/10.1109/ISTP60767.2023.10426998
8. Mesquita, B.G.V.: Guia PMBOK e as modificações da 7ª ed. Revista Inovação, Projetos e Tecnologias **10**(1), 123–125 (2022). https://doi.org/10.5585/iptec.v10i1.22195
9. Neves, R.N., Müller, D.L., Araujo, T.C.M., Boeira, M.B.D., Rodrigues, E., Bernardino, M.: Avaliação de um Software para o Ensino de Gerenciamento de Projetos com base no PMBOK – Um Grupo Focal. In: Escola Regional de Engenharia de Software (Eres). Sociedade Brasileira de Computação, pp. 51–60 (2023). https://doi.org/10.5753/eres.2023.237743
10. Nebesnyi, R., Kunanets, N., Vaskiv, R., Veretennikova, N.: Formation of an IT project team in the context of PMBOK requirements. In: IEEE 16th International Conference on Computer Sciences and Information Technologies (CSIT), Lviv, Ukraine, pp. 431–436 (2021). https://doi.org/10.1109/CSIT52700.2021.9648612
11. Oliveira, D.: Sistema de gestão de compras e estoque da Distribuidora 3C. Graduação em Administração de Empresas - Instituto de Desenvolvimento Econômico Rural e Tecnológico Dados da Amazônia (2019)
12. Rosenberger, P., Tick, J.: Suitability of PMBOK 6th edition for agile-developed IT projects. In: IEEE 18th International Symposium on Computational Intelligence and Informatics (CINTI), Budapest, Hungary, pp. 241–246 (2018). https://doi.org/10.1109/CINTI.2018.8928226

13. Santos, F.O.: Gerenciamento das Aquisições em Projetos - Engineering, Procurement and Construction (EPC) como modalidade de contrato Turnkey para empreendimentos de engenharia. Pós-graduação em Gerenciamento de Projetos. Universidade Católica de Minas Gerais (2013)
14. Shiang-Jiun, C., Yu-Chun, P., Yi-Wei, M., Cheng-Mou, C., Chi-Chin, T.: Trustworthy software development - practical view of security processes through MVP methodology. In: 24th International Conference on Advanced Communication Technology (ICACT), PyeongChang, Kwangwoon_Do, Korea, pp. 412–416 (2022). https://doi.org/10.23919/ICACT53585.2022.9728811
15. Souza, A.C.C., Souza, F.C.M., Vilela, R.F., Valle, P.H.D.: PMBOK Game II: Um Jogo Educacional para Apoiar o Ensino de Gestão de Projetos de Software. In: Workshop Sobre Educação em Computação (WEI), SBC, João Pessoa/PB, vol. 31, pp. 454–464 (2023). https://doi.org/10.5753/wei.2023.229976
16. Sadeh, A., Rogachevsky, K., Dvir, D.: The role of the project manager in the agile methodology. In: Portland International Conference on Management of Engineering and Technology (PICMET), Portland, OR, USA, pp. 1–5 (2022). https://doi.org/10.23919/PICMET53225.2022.9882761
17. Project dotProject – Homepage. https://dotproject.net/. Accessed 21 Mar 2024
18. Project dotProject+ Modules add-on. https://sourceforge.net/projects/dotmods/. Accessed 21 Mar 2024
19. Krause, W.: ISO 21500: orientações sobre gerenciamento de projetos - diretrizes para o sucesso. Editora Brasport, Rio de Janeiro (2014)
20. Dinsmore, P.C., Cabanis-Brewin, J.: The AMA Handbook of Project Management, 4th edn. AMACOM, New York (2014)

Plant Disease Classification Through Image Representations with Embeddings

Arturo Álvarez-Sánchez, Diego M. Jiménez-Bravo[✉], Luís Augusto Silva, Álvaro Lozano Murciego, and André Sales Mendes

Expert Systems and Applications Lab, Faculty of Science, University of Salamanca, Plaza de los Caídos s/n, 37008 Salamanca, Spain
{id00802439,dmjimenez,luisaugustos,loza,andremendes}@usal.es

Abstract. In response to the significant challenges faced by the agricultural sector due to climate change, this research focuses on enhancing plant disease detection and management through advanced computer vision techniques. The study introduces an innovative Artificial Intelligence (AI) model leveraging image classifications and embeddings to predict plant diseases, utilizing a public dataset comprising 27 combinations of plants and diseases. This research integrates Natural Language Processing (NLP) techniques to generate embeddings from images, which are then utilized to train a classification model based on classical machine learning algorithms. The proposed system employs a novel approach by transforming a typical computer vision problem into a feature-based classification problem, allowing the use of diverse machine learning models. The efficiency of the model is demonstrated through its ability to accurately classify various plant diseases, with the best-performing model achieving promising results in terms of accuracy and reliability. The research highlights the potential of combining technological innovations with agricultural expertise to address the complexities of modern agriculture. Future directions include refining the model by exploring different embedding techniques and increasing the dataset using data augmentation methods, to further improve the system's performance.

Keywords: artificial intelligence · embedding · image classification · natural language processing · plant disease

1 Introduction

The agricultural sector is currently grappling with significant challenges, exacerbated by the escalating impacts of climate change. Climate volatility, alongside unpredictable environmental conditions, has led to a crisis in crop productivity, leaving farmers susceptible to the detrimental effects of plant diseases. In the region of Castilla y León (Spain), farmers confront numerous adversities, including droughts, floods, and the persistent threat of crop diseases, necessitating urgent solutions to safeguard agricultural sustainability.

Recognizing the urgency of the situation, our research endeavors to address the pressing need for effective disease detection and management strategies. Rather than solely focusing on innovative tools, our approach aims to offer comprehensive solutions that empower farmers to tackle the complexities of crop health more effectively. Through the integration of advanced technologies, including computer vision, we aim to provide farmers with timely insights into crop health status, facilitating informed decision-making and proactive disease management practices.

This research represents a significant stride towards enhancing the resilience of agricultural systems amidst the challenges posed by climate change. By leveraging a combination of technological innovations and agricultural expertise, we aspire to equip farmers with the knowledge and tools necessary to navigate the uncertainties of modern agriculture successfully. Finally, our goal is to foster sustainable practices that ensure the long-term viability of agricultural communities and contribute to global food security efforts.

As a result, this study makes use of computer vision technologies to propose an Artificial Intelligence (AI) model that is capable of predicting crop diseases. For the training and construction of this model, a public dataset is used. This dataset includes a total of 27 combinations of plants and diseases.

The rest of the article is structured as follows: Sect. 2 discusses the state of the art, Sect. 3 presents the proposed system, Sect. 3 presents the case study of this work, Sect. 5 presents the results and finally Sect. 6 discusses the conclusions and future directions of this research.

2 Background

In 2015, Khirade et al. [10] tackled the problem of plant disease detection using digital image processing techniques and a back propagation neural network (BPNN). Authors have elaborated different techniques and approaches for the detection of plant disease using the images of leaves. The following are different datasets, their implementation, and results on the plant disease area. But first, we will talk about the preferred technique used in image detection, which is CNN (Convolutional Neural Network).

2.1 Convolutional Neural Network in Computer Vision

A convolutional neural network [14] (CNN) is a category of machine learning model, namely a type of deep learning algorithm well suited to analyzing visual data. CNNs use principles from linear algebra, particularly convolution operations, to extract features and identify patterns within images. Although CNNs are predominantly used to process images, they can also be adapted to work with audio and other signal data.

CNN architecture is inspired by the connectivity patterns of the human brain, in particular, the visual cortex, which plays an essential role in perceiving and processing visual stimuli. The artificial neurons in a CNN are arranged to efficiently interpret visual information, enabling these models to process entire images. CNNs are frequently used for computer vision tasks such as image recognition and object detection, with common use cases including self-driving cars, facial recognition, and medical image analysis.

2.2 Plant Disease Detection

In 2015 Hughes and Salathé [9] from Penn State University USA, created PlantVillage, an open-access repository of images on plant health. They created a library of open-access information on over 150 crops and over 1,800 diseases, accessible on the same website. This content has been written by plant pathology experts, reflecting information sourced from the scientific literature.

With this in mind, in 2016, Mohanty et al. [12] used this previous dataset to create an image-based plant disease detector. They used convolutional neural networks to achieve this task and they reported on the classification of 26 diseases in 14 crop species using 54,306 images. They measured the performance of their models based on their ability to predict the correct crop-disease pair, given 38 possible classes. The best-performing model achieved an overall accuracy of 99.35%. Hence, they demonstrated the technical feasibility of an assisted plant disease diagnosis system.

Later on, in 2021, the Vishwakarma Institute of Technology used the PlantVillage dataset for a different study [11]. They experimented with their algorithms on the dataset using RGB images, later converted to grayscale ones, and then a Gaussian filter was used to smooth the images. They achieved an average accuracy of 93.

In 2019, Singh and his team [16], developed "PlantDoc: a Dataset for Visual Plant Disease Detection". This dataset was made with the intent to improve previous diseased plant datasets. The main objective was to create a dataset that represents reality conditions. In their words "PlantVillage [previous dataset] limits the effectiveness of detecting diseases because, in reality, plant images may contain multiple leaves with different types of background conditions with varying lighting conditions". Their dataset is made of downloaded images from the internet since collecting large-scale plant disease data through fieldwork requires enormous effort, and they collected about 20,900 images by using scientific and common names of 38 classes mentioned in the PlantVillage dataset. Later on, they filtered them based on the APSnet [1] and removed inappropriate (such as non-leaf plants, lab-controlled and out-of-scope images) and duplicate samples. The final dataset contains 27 classes spanning over 13 species with 2,598 images.

Later in 2020, the Hosei University, Japan, developed LeafGAN [5]. This method consisted of an image generation network that was specially designed to mitigate the serious overfitting problem in image-based plant diagnosis tasks via the effective generation of high-quality and widely varying pseudo-training images. LeafGAN was built on CycleGAN [17] and their own proposed label-free leaf segmentation module to guide the network in transforming the relevant regions while preserving the backgrounds. They used the fine-tuned ResNet-101 [8] model as the backbone of their module, and replaced the last layer of the network with a three-node layer. Using a dataset of 1,000 leaf images, their module achieved an F1-score of 83.9%.

In 2022, Schuler et al. proposed a Color-aware two-branch DCNN for efficient plant disease classification [15]. They proposed feeding a DCNN CIE Lab [3] instead of RGB color coordinates. modifying an Inception V3 architecture to include one branch specific for achromatic data and another branch specific for chromatic data. This modification takes advantage of the decoupling of chromatic and achromatic information. Besides, splitting branches reduces the number of trainable parameters and computation load by up to 50% of the original figures using modified layers. They achieved a state-of-the-art classification accuracy of 99.48% on the PlantVillage [9] dataset and 76.91% on the Cropped-PlantDoc [16] dataset.

3 Proposed System

In this section, a new system for plant disease detection is proposed. We propose a system that makes use of novel and recent techniques to solve a classical image classification problem. The proposed system makes use of Natural Language Processing (NLP) techniques to obtain embeddings from images that afterward are used to train a classification model based on classical machine learning algorithms. A scheme of the proposed system can be seen in Fig. 1 and is explained in the following subsections.

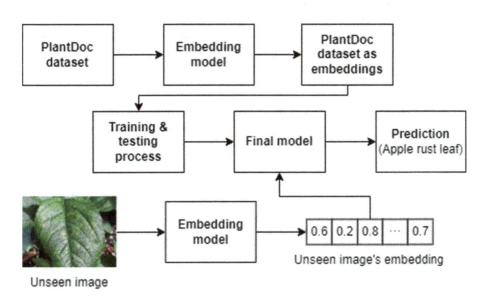

Fig. 1. Schematic view of the proposed system.

3.1 NLP Module

The NLP module is in charge of processing the images of the dataset and generating for every one of them an embedding vector with the most important features considered by the model. This embedding vector is associated with the class of each of the images.

In recent years, NLP models, such as those that obtain embeddings from different sources of information, have gained relevance [4,7]. These techniques have also been used for image processing. Among the models that accept this type of data, the DINOv2 [13] model stands out. DINOv2 is a set of models that are capable of solving 8 different types of tasks related to computer vision. Among these tasks is image classification. DINOv2 has been evaluated on 30 different benchmarks and its efficacy has been confirmed.

Therefore, DINOv2, specifically the version DINOv2_vits14, is the model that will be used to obtain the embedding of the images of the dataset and also to obtain the embeddings of the images that will be predicted with the model. The DINOv2_vits14 model obtains a vector of 384 features for each of the images it processes.

3.2 Image Classification Approach

The proposed image classification process makes use of the embedding model to generate the image features. This generates a new structure for the dataset, where each image is represented as a vector of 384 dimensions, along with its corresponding label. Thus, an image classification problem that is typically solved with computer vision techniques has been transformed into a classical feature-based classification problem.

Once the original dataset (with images) has been processed and the new dataset structure (table model) has been obtained, the latter can be used to train different classical machine learning algorithms. The complete list of machine learning algorithms used in this study is detailed in Sect. 4.2. These algorithms are trained and from each of them, we will obtain a model that could be used to make predictions about unknown data.

3.3 Final Prediction

Following the selection of classifier models, the prediction process is carried out in two simple steps: (I) obtaining the embedding vector of the image to be classified; and (II) this embedding vector of features is sent to the model to perform the classification/prediction of the label that corresponds to the image.

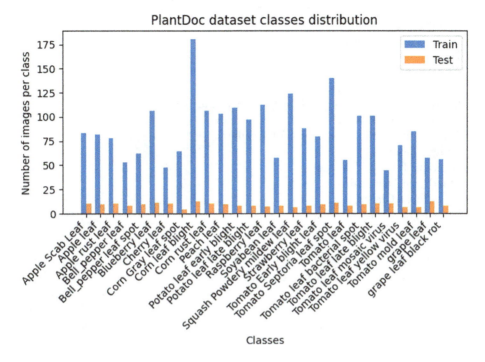

Fig. 2. PlanDoc dataset classes distribution.

4 Case Study

4.1 Data Description

We will be using the PlantDoc [16] dataset, as it presents the best samples for the problem at hand. By using samples taken in outdoor environments, we will achieve a model capable of recognizing diseases more optimally, because most of the images will be taken in the crop field itself. This will help the model to be able to better recognize patterns surrounded by noise and other factors that hinder its training.

This dataset includes 2,598 samples across multiple plant species. Every plant species section is divided into healthy and diseased subsections with multiple images per disease subsection. Every plant has at least two disease sample folders. Figure 2 shows the class distribution.

4.2 Machine Learning Algorithms

To set up the machine learning algorithms, we have chosen to use the Python PyCaret library [2]. This library allows us to train several models based on classical machine learning algorithms. The following algorithms have been used to set up this training:

- Dummy Classifier.
- Logistic Regression.
- K Neighbors Classifier.
- Naive Bayes.
- Decision Tree Classifier.
- SVM (Support Vector Machines) - Linear Kernel.
- SVM - Radial Kernel.
- MLP (Multi-Layer Perceptron) Classifier.
- Ridge Classifier.
- Random Forest Classifier.
- Quadratic Discriminant Analysis.
- Ada Boost Classifier.
- Linear Discriminant Analysis.
- Extra Trees Classifier.

These algorithms are trained with the same settings. For this purpose, a training is configured in which the training set is balanced to avoid overtraining in certain classes. For this, the SMOTE method [6] is used and an evaluation method is defined during the training stage based on the Stratified10Fold. Afterward, all these models are evaluated with the training set of the dataset.

4.3 Evaluation Metrics

To evaluate the different models and therefore define the best one for the problem addressed in this research, it is necessary to define the evaluation metrics that will be used. Therefore, the main metric to be used for the evaluation is the accuracy. Its formulation is available in 1,

$$Accuracy = \frac{TP + TN}{TP + TN + FP + FN} \quad (1)$$

where TP (True Positives) is the number of positive instances classified as positive, TN (True Negatives) is the number of negative instances classified as negative, FP (False Positives) is the total number of instances wrongly classified as positive instances, and FN (False Negatives) is the total number of positive instances classified as negatives.

Nonetheless, we will use other metrics to discern the performance of the different classifiers. These metrics are: (I) recall, (II) precision, (III) F1 score, (IV) Kappa coefficient, and (V) the Matthews Correlation Coefficient (MCC).

5 Results

In the following, we present the results obtained from the system proposed in this article. Firstly, a results table, Table 1, with the evaluation metrics for the different trained machine learning models are shown.

Table 1. Machine learning algorithms' results.

Model	Accuracy	Recall	Prec	F1	Kappa	MCC
Linear Discriminant Analysis	**0.7223**	**0.7223**	**0.7407**	**0.7232**	**0.7103**	**0.7110**
MLP Classifier	0.7030	0.7030	0.7166	0.7001	0.6901	0.6908
Ridge Classifier	0.6911	0.6911	0.6950	0.6846	0.6780	0.6788
Logistic Regression	0.6880	0.6880	0.6990	0.6864	0.6744	0.6750
SVM - Linear Kernel	0.6765	0.6765	0.7010	0.6764	0.6624	0.6636
Extra Trees Classifier	0.6761	0.6761	0.6728	0.6647	0.6617	0.6626
Random Forest Classifier	0.6718	0.6718	0.6717	0.6625	0.6574	0.6582
Naive Bayes	0.6641	0.6641	0.6735	0.6596	0.6496	0.6503
K Neighbors Classifier	0.6277	0.6277	0.6715	0.6233	0.6130	0.6158
SVM - Radial Kernel	0.5961	0.5961	0.7057	0.5965	0.5760	0.5868
Decision Tree Classifier	0.4305	0.4305	0.4404	0.4264	0.4061	0.4068
Quadratic Discriminant Analysis	0.0783	0.0783	0.0264	0.0141	0.0020	0.0175
Ada Boost Classifier	0.0663	0.0663	0.0453	0.0269	0.0341	0.0675
Dummy Classifier	0.0355	0.0355	0.0013	0.0024	0.0000	0.0000

According to the table of results, the algorithm that obtains the best results across all the selected evaluation metrics is the Linear Discriminant Analysis. The algorithm obtained an accuracy of 0.7223, a promising result given the high number of classes and the difficulty of discerning in many cases about the different diseases on the plants' leaves.

Although other classifiers also obtain results close to 70% accuracy, given the considerable difference with the Linear Dicriminat Analysis, it seems that Linear Dicriminat Analysis seems to be the most accurate model. Similar conclusions can also be drawn when comparing other metrics. As a result, Cohen's Kappa index, which allows us to evaluate the reliability of a classification model, should also be taken into account in this problem. Here, the closer to 1, the more reliable the model is. Hence, the decision to select the Linear Discriminant Analysis as the model for the proposed system is justified.

As a way to better understand this model, Fig. 3 shows the confusion matrix of the Linear Discriminant Analysis algorithm, where we can see how this model works in detail. It can be seen that the vast majority of the errors that occur are between the different diseases of the same species. This can be seen in the predictions made for "Tomato leaf", "Tomato leaf mosaic virus", and "Corn leaf blight" classes. Also, to a minor degree, errors occur between different types of plants.

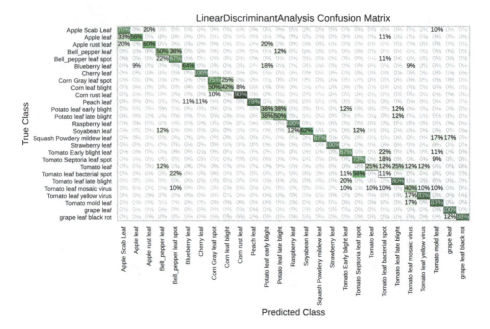

Fig. 3. Liniar Discriminant Analysis confusion matrix.

Table 2. Results comparison in the PlantDoc dataset.

Study	Accuracy
Schuler et al. [15]	0.7691
Ours	0.7223

Finally, the results obtained in this research are also compared with the results obtained in previous research. This comparison can be seen in Table 2. Note that the proposed system does not outperform the recent results of other researchers with the PlantDoc dataset. Nonetheless, the results of this research are preliminary and we intend to continue improving the proposed system to improve the results that constitute the state of the art.

6 Conclusions

In this study, a new approach to solving the problem of classifying images of diseases of different types of plant species has been explored. We have used an innovative solving method based on advances in the area of NLP, such as embedding models. These models allow us to process the images and extract a series of features for each one of them. These features are then used to solve the problem using a more classical machine learning approach. The results obtained are promising, although they fall short of the current state-of-the-art results. Nevertheless, one advantage of the proposed solution is the time needed to perform the training. While the state-of-the-art solutions are based on CNNs and deep learning, which require large amounts of time and resources for training, our proposal requires fewer resources and time to obtain a robust model.

Nevertheless, we believe that there is still room for improvement in this research. Therefore, we propose to work on the following lines of research. Firstly, we propose to test different versions of the DINOv2 model as well as other models for image embeddings. Further, we intend to explore the finetuning of this type of model to specialize them even more in the detection of plant image features and their diseases. Lastly, another proposal is to carry out a study in which the number of images in the PlantDoc dataset will be increased by using data augmentation techniques. We consider these lines of research will improve the current results.

References

1. APSnet. https://www.apsnet.org/Pages/default.aspx
2. Home - PyCaret. https://pycaret.org/
3. Atwood, J., Towsley, D.: Diffusion-Convolutional Neural Networks (2015). http://arxiv.org/abs/1511.02136
4. Bojanowski, P., Grave, E., Joulin, A., Mikolov, T.: Enriching Word Vectors with Subword Information. http://www.isthe.com/chongo/tech/comp/fnv
5. Cap, Q.H., Uga, H., Kagiwada, S., Iyatomi, H.: LeafGAN: An Effective Data Augmentation Method for Practical Plant Disease Diagnosis (2020). http://arxiv.org/abs/2002.10100
6. Chawla, N.V., Bowyer, K.W., Hall, L.O., Kegelmeyer, W.P.: SMOTE: synthetic minority over-sampling technique. J. Artif. Intell. Res. **16**, 321–357 (2011). https://doi.org/10.1613/jair.953
7. Conneau, A., Lample, G., Ranzato, A., Denoyer, L., Jégou, H.: Word Translation Without Parallel Data. https://github.com/facebookresearch/MUSE
8. He, K., Zhang, X., Ren, S., Sun, J.: Deep Residual Learning for Image Recognition (2015). http://arxiv.org/abs/1512.03385
9. Hughes, D.P., Salathé, M.: An open access repository of images on plant health to enable the development of mobile disease diagnostics. Technical report. http://www.fao.org/fileadmin/templates/wsfs/docs/expert_paper/How_to_Feed_the_World_in_2050.pdf!!!
10. Khirade, S.D., Patil, A.B.: Plant disease detection using image processing. In: Proceedings - 1st International Conference on Computing, Communication, Control and Automation, ICCUBEA 2015, pp. 768–771. Institute of Electrical and Electronics Engineers Inc. (2015). https://doi.org/10.1109/ICCUBEA.2015.153
11. Kulkarni, P., Karwande, A., Kolhe, T., Kamble, S., Joshi, A., Wyawahare, M.: Plant disease detection using image processing and machine learning. Technical report
12. Mohanty, S.P., Hughes, D., Salathé, M.: Using deep learning for image-based plant disease detection. Technical report
13. Oquab, M., et al.: DINOv2: Learning Robust Visual Features Without Supervision (2023). https://arxiv.org/abs/2304.07193v2
14. O'Shea, K., Nash, R.: An Introduction to Convolutional Neural Networks (2015). http://arxiv.org/abs/1511.08458

15. Schuler, J.P.S., Romani, S., Abdel-Nasser, M., Rashwan, H., Puig, D.: Color-aware two-branch DCNN for efficient plant disease classification. Mendel **28**(1), 55–62 (2022). https://doi.org/10.13164/mendel.2022.1.055
16. Singh, D., Jain, N., Jain, P., Kayal, P., Kumawat, S., Batra, N.: PlantDoc: A Dataset for Visual Plant Disease Detection (2019). https://doi.org/10.1145/3371158.3371196. http://arxiv.org/abs/1911.10317
17. Zhu, J.Y., Park, T., Isola, P., Efros, A.A.: Unpaired Image-to-Image Translation using Cycle-Consistent Adversarial Networks (2017). http://arxiv.org/abs/1703.10593

Explaining Social Recommendations Using Large Language Models

Md. Ashaduzzaman[✉], Thi Nguyen, and Chun-Hua Tsai

College of Information Science and Technology, University of Nebraska Omaha, Omaha, NE 68182, USA
aashaduzzaman@unomaha.edu

Abstract. This paper introduces an innovative approach to recommender systems through the development of an explainable architecture that leverages large language models (LLMs) and prompt engineering to provide natural language explanations. Traditional recommender systems often fall short in offering personalized, transparent explanations, particularly for users with varying levels of digital literacy. Focusing on the Advisor Recommender System, our proposed system integrates the conversational capabilities of modern AI to deliver clear, context-aware explanations for its recommendations. This research addresses key questions regarding the incorporation of LLMs into social recommender systems, the impact of natural language explanations on user perception, and the specific informational needs users prioritize in such interactions. A pilot study with 11 participants reveals insights into the system's usability and the effectiveness of explanation clarity. Our study contributes to the broader human-AI interaction literature by outlining a novel system architecture, identifying user interaction patterns, and suggesting directions for future enhancements to improve decision-making processes in AI-driven recommendations.

Keywords: Recommender Systems · LLMs · Prompt Engineering · Natural Language Explanations · Human-AI Interaction

1 Introduction

Social recommendation systems facilitate decision-making processes by filtering out irrelevant information, thereby assisting users in online shopping or forming social connections on media platforms [17]. However, the opacity of artificial intelligence (AI) systems enables powerful predictions but compromises direct explainability, which can undermine user trust and experience [2]. In response to this challenge, recent research in recommender systems has focused on developing more sophisticated interfaces, including enhancing the explainability and controllability of recommendations [13, 18, 19]. Despite these advancements, existing designs and solutions rely on predefined rules that lack interactive elements with users and adapt to their desires, needs, and literacy levels. Consequently, there remains a pressing need for a dynamic, user-centered approach to delivering AI-based recommendations that can accommodate users' diverse backgrounds and foster transparency and informed decision-making.

The rapid development of Large Language Models (LLMs), such as Bard, GPT-4, LLaMA, and others, has ushered in a new era characterized by both challenges and opportunities [10,14]. These models have demonstrated exceptional proficiency in comprehending and generating human-like text. Prompt engineering provides reusable solutions to common challenges in output generation and interaction when working with LLMs. It enables the creation of more accurate, context-specific, and nuanced responses from LLMs, particularly supporting advancements in conversational AI [12]. To facilitate the delivery of recommendations accompanied by natural language explanations, we propose an architecture for an explainable recommender system that leverages the human-like conversational capabilities of generative AI. Our proposed architecture enables users to receive personalized suggestions and gain insight into the rationale underlying these recommendations.

Specifically, this paper aims to answer the research question of *how to utilize large language models (LLMs) to enhance explanatory capabilities within social recommendation systems.* To this end, we designed and developed a prototype of an *Advisor Recommender System*, which integrates two content-based recommendation models - text similarity and topical similarity - to generate personalized advisor recommendations for graduate students based on their research interests. Moreover, our system enables users to engage in follow-up inquiries regarding the given recommendations, i.e., allowing users to ask questions about the recommended advisors, and the system could explain why a particular scholar or faculty is recommended to them. These communications are based on LLM-powered responses providing detailed, context-sensitive explanations. We hypothesize that these interactive explanatory capabilities can foster enhanced user comprehension and trust in the recommender system.

To assess the efficacy of our proposed system and the user experience, we undertook pilot user studies involving 11 graduate students majoring in computer sciences or information systems. Through a qualitative thematic analysis of the data collected, three primary themes emerged: firstly, the types of information sought by participants when engaging with the system's recommendations; secondly, the usability and feasibility of integrating a chatbot within the recommendation system; and thirdly, how participants interacted with the chatbot to obtain explanatory rationales for the recommendations. Through these user feedback, we identified potential avenues for system improvement to enhance user satisfaction and facilitate informed decision-making. Based on these experimental findings, we then discuss the design implications and insights of using LLMs along with explainable recommendation systems.

2 Related Works

The concept of collaborative filtering was first introduced by Goldberg et al. [7] in 1992 as a method to filter mail by tracking individuals' responses to the documents they read. Resnik and Varian [15] first introduced the term 'recommender system' in 1997 to describe systems that not only filter out less relevant items but also suggest particularly interesting ones. In October 2006, Netflix's release of a movie rating dataset [3] and subsequent challenge spurred increased research interest in recommender systems. Advances in machine learning deliver AI systems that operate independently but cannot explain their decisions. To bridge this gap, DARPA initiated the explainable artificial intelligence (XAI) program in May 2017 [8]. The black-box nature of AI systems enables powerful predictions but lacks direct explainability. XAI offers significant potential to enhance trust and transparency in AI-based systems. Explanation in recommendations is crucial for justifying outcomes, increasing visibility to swiftly identify and correct errors, and continuously improving the system [2]. Parra and Brusilovsky [13] sought to enhance the controllability and transparency of hybrid recommendations by integrating sliders for fusion control and a Venn diagram to visualize the outcomes. Several research studies have focused on enhancing explainability by augmenting interactive hybrid recommender interfaces with various types of explanations [18,19]. Tsai et al. [19] discussed four types of explanatory interfaces: a two-way bar chart for illustrating publication similarity; a topical radar for showing topic similarity; an enhanced strength network for visualizing co-authorship networks; a venn tag for depicting CN3 interest similarity. Several studies have utilized NLP models to generate textual explanations for specific recommendations [6,9]. P5 [6] demonstrates that learning multiple recommendation-related tasks simultaneously is feasible through a unified sequence-to-sequence framework by treating these tasks as prompt-based natural language tasks, integrating user-item information and features with personalized prompt templates as inputs, highlighting a promising approach for unified, instruction-based recommendation. However, a significant limitation of these NLP-related models is their inability to generate human-like text. Recent LLMs like GPT-4 [1] excel at generating coherent and contextually relevant text, and are adept at holding conversations, answering questions, providing detailed explanations, and engaging in nuanced, context-aware dialogues, surpassing earlier versions in complexity and understanding. Our research is driven by the potential to leverage the advanced capabilities of recent LLMs to develop an explainable recommender system.

3 Explainable Recommender System Architecture

In this section, we describe the proposed architecture of an explainable recommender system, as illustrated in Fig. 1. The architecture is based on an advisor recommender system that utilizes LLM for generating explanations. Initially, users must input research interest keywords into the system. These keywords are processed by both a Text Similarity and a Topic Similarity model. Each model identifies the top three advisors and displays them to the user. The results along with similarity scores, LDA topics, and other relevant model data are saved for the prompt engineering process. When a user requests an explanation of the

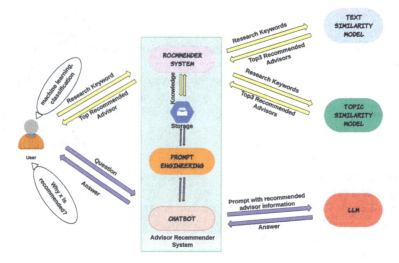

Fig. 1. Proposed architecture of Advisor Recommendation System leveraging LLM for explanation.

recommendations, the chatbot interacts with the feature engineering process to create a suitable prompt. This prompt is then sent to the LLM, which provides an explanation for the user query. The response is delivered to the user, and the conversation history is stored to enhance future interactions through refined prompt engineering. Details of each step are outlined below:

3.1 System Implementation and Data

An advisor recommender system designed for prospective graduate students seeking a research supervisor can significantly streamline the process of finding the right mentor for advanced studies. This system matches students with potential supervisors based on shared research interests, publications, and other relevant criteria. It generates a list of potential supervisors whose research aligns closely with the student's interests. It reduces the time and effort needed for students to find compatible supervisors. By aligning students with supervisors who share their research interests, the system promotes more effective and fulfilling research collaborations. The recommendations produced by our proposed system originate from information obtained from the Citation Network Dataset [16]. This citation information is pulled from several sources including DBLP, ACM, and the Microsoft Academic Graph (MAG), among others. The dataset includes information from the conference proceedings, detailing the conference papers (author(s), title, keywords, and abstract) and providing extensive profiles on the authors and attendees, including names, publications, affiliations, and positions. While the dataset includes a vast quantity of data (80435 authors), only a limited portion of it was selected randomly for our research. We extracted the names, IDs, and research keywords from the publications of 2,500 researchers who we are considering as potential advisors for the user. We selected a limited

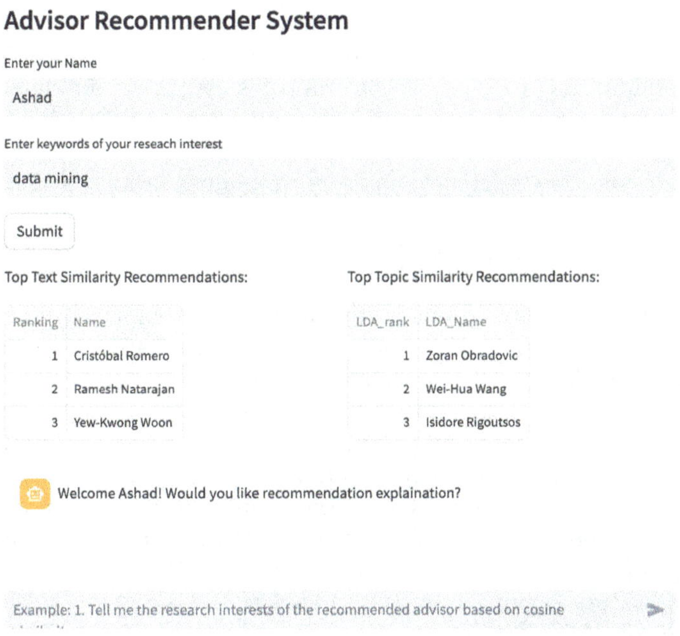

Fig. 2. User interface of the Advisor Recommendation System.

number of researchers to ensure the system responds quickly and to avoid complexity, as our main goal is to develop a system that can clearly explain its recommendations.

We developed the Advisor Recommender system using the Streamlit framework. We hosted this app file, along with the relevant datasets and trained models, in a GitHub repository. After setting up the repository, we connected it to the Streamlit cloud, allowing us to deploy the app seamlessly. GitHub repository can be found at[1]. Figure 2 displays the user interface of the Advisor Recommender system. It prompts users to enter their name and research interest keywords. Users may type any research interests related to computer and information science. If users have multiple research interests, they can enter them in the text field, separated by commas or semicolons. When a user clicks the submit button after entering their name and research interest keywords, the interface interacts with text and topic similarity-based models to recommend the top three advisors for each model. Initially, it prompts the user to wait while the top-recommended advisors are being retrieved. The output is then presented in two columns of information. On the left side of the image, the top three advisors are listed based on text (Cosine) similarity, and on the right side, based on topic (LDA) similarity, displayed in a table format. The first column of each table shows the ranking based on similarity score, and the second column lists the

---

[1] https://github.com/asadayon/Advisor_Recommendation.

name of the advisor. At the bottom of the interface, an AI chatbot is integrated, asking whether an explanation of the recommendations is needed, followed by a user input field for queries. This chatbot captures user input through an interactive interface, generates a relevant prompt using prompt engineering techniques, communicates with LLM via API, and delivers the response back to the user. Users can ask questions related to advisors, recommendations, and models.

3.2 Recommendation Models

Text Similarity Model. Previous research [18,19] employed text similarity models based on scholars' publications to recommend attendees with similar interests. This model operates by initially generating count vectors from the keywords associated with all authors' publications within our comprehensive publication database. These count vectors serve as numerical representations of the keywords, allowing us to quantify and analyze the textual data effectively. Subsequently, we employ cosine similarity to measure the degree of similarity between the keywords input by the user (representing their research interests) and the keywords associated with other authors' publications. Cosine similarity provides a score ranging from 0 to 1, where a score closer to 1 signifies a higher degree of similarity, indicating closely aligned research interests between the user and the author in question. Based on the cosine similarity scores, the system then identifies and recommends the top three authors whose publication keywords most closely match the user's specified interests. These authors are suggested as the top three recommended advisors, underpinning the system's goal to facilitate meaningful academic collaborations and mentorships grounded in shared research interests.

Topic Similarity Model. Given the abstract of the article, a list of authors, and known past collaborators, Topic Models, such as Latent Dirichlet Allocation (LDA), can generate a list of highly relevant authors who would be suitable reviewers for this article [5]. By categorizing keywords from a vast array of publications into 30 distinct topics, the system effectively maps out the research landscape into thematic clusters, each defined by a set of representative keywords. Authors in the database are analyzed to ascertain their thematic engagement across these topics, with each author assigned a probability score reflecting their involvement in each topic. This creates a thematic profile for each author. When a user inputs interest keywords, the system dynamically matches these against the predefined topics to identify the top three topics most aligned with the user's interests, quantified by probability scores. Using the gensim library, the system compares the user's topic preferences with the thematic profiles of authors to recommend the top three authors whose research interests best match the user's. This approach ensures that recommendations are grounded in thematic, rather than merely textual, similarities, fostering more relevant and intellectually aligned collaborations.

3.3 Generating Explanations

LLM models are trained on vast amounts of text data, allowing them to learn language patterns, context, and semantics at a sophisticated level. These models excel in various natural language processing tasks, including text generation, summarization, translation, and question-answering. Our system's chatbot, powered by the LLM model (specifically, GPT-3.5 Turbo), provides explanations for the system's recommendations in a conversational and user-friendly manner. When supplied with context such as recommended outputs and pertinent background knowledge as input prompts, the model crafts explanations that are both informative and easily understandable. This approach enhances user engagement by encouraging deeper exploration of the recommendations and offering clear insights into the underlying rationale.

Prompt engineering involves carefully crafting and refining the input prompts to steer the responses of these models, ensuring that the output is accurate, relevant, and coherent [11]. This technique is essential for fully utilizing the capabilities of LLM models, making them more versatile and effective across various fields. The prompt must be clear and precise to guide the AI in generating relevant and accurate responses. Ambiguities or vague descriptions can lead to unpredictable or irrelevant outputs. Including the right context in the prompt can help the AI understand the scope and depth of the response needed. The prompt should be aligned with the end goal while it's generating text or providing explanations.

In the proposed prompt engineering architecture, we specify that the system is designed to operate as an advisory recommendation system, elucidating the rationale behind recommending specific advisors. We then gather pertinent data such as the top recommended advisors, their similarity scores, and the LDA topics derived from both the Text and Topic similarity models. Additionally, the system retrieves research interest keywords of the top recommended advisors from its storage. To enhance user interaction, we incorporate contextual details such as the user's name, research interests, recommendations from both models, advisors' research interests, similarity scores, and topics into prompts. We also stipulate that the system should refrain from providing responses beyond its designated scope to mitigate hallucination. This augmented prompt framework is employed to facilitate communication between the LLM and the chatbot, ensuring effective handling of user queries.

4 Experiment

We conducted pilot user studies of the Advisor recommendation system involving 11 participants (P1 through P11), all of whom are graduate students with multiple interests in computer science related fields, except for one undergraduate. These participants had prior experience working with research advisors. All participants resided in Omaha, United States, and were invited via email to voluntarily participate in the pilot studies, which involved using an advisor recommendation system followed by a post-study questionnaire. Upon agreeing to participate, each was sent a Zoom invitation. The user studies varied in length from 13 to 20 min. Each session began with a brief presentation by the first author about the advisor recommendation system, followed by providing the participants with a link to access the system. After using the system, participants engaged in a semi-structured interview as part of the post-study questionnaire. All sessions were recorded on the cloud via Zoom, and the interviews were conducted in English. Consent for recording was obtained from each participant.

Interview recordings were transcribed using Zoom. We employed an inductive approach to generate coding categories [4]. Two researchers independently reviewed the interview transcripts and manually identified initial codes. These initial codes were then discussed collectively by the research team, leading to the formulation of a more comprehensive code list. The finalized codes were organized into three main themes. These themes include the types of information users wish to follow up on based on system recommendations, the usability of integrating an AI-powered chatbot, and user responses to the chatbot's reasoning capabilities and its ability to impart new knowledge. Quotations have been lightly edited for clarity.

5 Findings

The Findings section is structured around three central themes: firstly, the types of information students seek when receiving recommendations from the system. Secondly, the usability of integrating a chatbot within the recommendation system. Lastly, the nature of user interactions with the chatbot when seeking explanations for the recommendations.

5.1 Theme 1: Information Needs of Advisor Seeking

There are different types of information that the users want to know after getting the recommended advisor lists from the system. The most in-demand type of information is closely related to the top recommended advisors, including their academic institutions or affiliations, the publications that they had on the requested topics, and any personal website if available. For instance, user P5 posed questions after getting the list of advisors from the chatbot: *"Give a bio about [X]"* and *"Where is this person [the advisor] affiliated?"*. User P3 inquired: *"What are the papers he published?"*. Meanwhile, user P6 requested: *"Can you share his profile with me?"*. However, our current system design only incorporates the advisors' publication research keywords as a means of prompt engineering. Consequently, the chatbot is unable to respond to queries about other aspects of the advisors' profiles. The chatbot responded with: *"I'm not sure about [X]'s current affiliation..."*, *"I apologize, but I do not have access to a specific list of publications for individual advisors..."* and *"I'm unable to provide real-time or specific information about..."*. The majority of users (9 out of 11) believe that information on advisor affiliation is essential for the recommendation system, as it facilitates the subsequent task of contacting them. User P1 commented during the post-study questionnaires, *"Unless and until it shows the affiliation of that particular advisor to some university or similar institution, it is like... how can we use this recommendation?"*. According to this user, the recommendation is not useful unless the affiliation is provided. User P2 commented, *"If I found a source [weblink] about [advisor], it would be much better... I wish it could be implemented in the system"*. According to this user, a personal website or link to the recommended advisor is very useful. When User P3 was asked about their experience while using the system, he commented, *"It is quite good, but one recommendation I can give is to add the description of the particular advisor, and what are the papers and the publications..."* The user suggested that the system would be better if it also provided descriptions of advisors and their publication lists.

5.2 Theme 2: User Perceptions of Usability

The term usability in this subsection refers to the system's ability to provide users with practical information that helps them effectively find advisors aligned with their research interests. Among 11 participants, there are 7 users who think that the chatbot is able to provide them with helpful information when finding advisors for their research. *"It is helpful to narrow down the supervisor names based on my research interest"*, user P7 added when being asked if they perceived any benefit from using the system. P8 reported that after requesting an explanation of the model used, he received a convincing answer. He commented, *"I feel like I'm having a human-like chat with the system"*, indicating that he believed he was conversing with a real person. However, the remaining 4 users suggested that it would be more beneficial if more detailed information about individual advisors were provided. P2 expressed confusion while using the system, stating, *"I was a bit confused about how much it knows; I didn't realize the scope was really limited. I expected the advisor system to provide more detail"*. The user, P2, suggests that the system should provide more detailed information, as the current scope is too limited. When asked about interface improvements to enhance usefulness and interactivity, most users were satisfied with the current interface. However, we received several specific suggestions. These included adding tooltips with quick instructions to guide users on how to begin and the functionalities available. Additionally, it was suggested that user inputs should remain visible at the top of the screen while allowing the conversation section below to be scrollable. This would prevent users from having to scroll all the way back up to adjust research keywords as needed. Again, incorporating thumbs-up and thumbs-down icons would enable users to rate the responses they receive. Another point of praise was the system's quick response time, as noted by three users, indicating that users appreciate prompt replies from the chatbot and dislike waiting for responses.

5.3 Theme 3: Transparency and Explainability

For each user study, the users are encouraged to ask the chatbot any questions that come to their minds related to the recommended list of advisors. There are 7 out of 11 users who explicitly asked the chatbot to give an explanation on its recommendations or previous answers, such as *"Why was [X] recommended as number 1?"*, *"Why [X] is recommended in both similarities"*, *"...explain (to) me in simple terms"*, while the rest of the users asked follow-up questions in relevance to the previous conversation they had with the chatbot. Users found the concept of explaining recommendations both interesting and intuitive. P9 commented, *"Yeah, it was very interesting... when I purchase anything on Amazon, it suggests similar products. However, it never explains the basis on which those products are recommended. When I searched for my topic, like image processing, this system not only gave recommendations but also explained why these advisors were suggested. So, that was really a good idea"*. Thus, the user found it very interesting to receive reasons for the recommendations. User P10 commented, *"In this scenario, where any recommendation given by [the system] can be verified by, you know, the algorithm inside it, I think it's very unique and helpful"*. This implies that they appreciate transparency in how the recommendations are generated, as it allows for greater trust and reliability in the system's suggestions. P11 commented, *"I really like the way it was trying to explain, even to any user, even unaware of the technical aspect, but still can get a glimpse of what the model is trying to recommend. It gives you more confirmation, more clarity on why something is being recommended or prescribed to you, instead of being blindly recommended to you"*, The main point of the comment is that user appreciates the system's ability to explain its recommendations in a way that is accessible to all users, regardless of their technical knowledge. In fact, knowing that there is a logical explanation behind the recommendations gives users more confidence and trust to adopt them into their decision-making process. Some users noted that the answers were clear, short, and concise, making them easy to follow. One user specifically requested explanations in simple terms, which the system provided. This suggests that users generally prefer explanations that are both concise and straightforward.

6 Discussion

In addressing the research question on how to utilize LLMs to enhance explanatory capabilities within social recommendation systems, our study demonstrated the effectiveness of leveraging LLMs in providing personalized, transparent explanations. Through prompt engineering, our Advisor Recommender System provided personalized explanations, citing shared research interests and high cosine similarity scores when users inquired about specific advisor recommendations. While the system successfully addressed basic queries, limitations were noted when detailed affiliation information was lacking. These findings highlight the potential of LLMs in contextualizing recommendations and improving user understanding, while also indicating areas for further refinement in prompt design.

Usability emerged as a critical factor in the adoption and effectiveness of the system. Participants appreciated the system's ability to narrow down potential advisors based on their research interests. However, there were notable concerns about the depth of information provided. The mixed feedback on the system's knowledge scope highlights the importance of setting clear expectations for users about the system's capabilities and limitations. Enhancing the utility and satisfaction of recommendation systems requires comprehensive data integration, which can be accomplished by implementing prompt engineering to provide detailed information about recommended advisors. Enhancements such as tooltips, persistent visibility of user inputs, and interactive elements like rating icons could further improve the user experience and engagement.

A significant number of users have explicitly requested explanations for the recommendations or previous answers provided by the chatbot, highlighting a strong desire for transparency and understanding of the system's suggestions. Comments from participants P9, P10, and P11 underscore the importance of explanation and transparency in recommendation systems, with examples illustrating common frustrations when recommendations lack accompanying explanations and the value users place on understanding algorithms for credibility. However, there's a concern about whether users genuinely comprehend the explanations provided, suggesting a potential gap between expressed interest and actual understanding. Further research is needed to investigate users' comprehension levels and develop strategies to support them in formulating informed inquiries effectively, thus enhancing their learning from AI recommendations.

7 Conclusion

Our research on the Advisor Recommender System demonstrates the value of integrating LLMs to provide natural language explanations within recommender systems. The findings from our pilot study indicate that users value transparency and personalized explanations, which enhance their understanding and trust in the recommendations. While the system shows promising results in user interaction and satisfaction, there are opportunities for further refinement, particularly in enriching the data sources. Future work will focus on these improvements to better meet user needs and further advance the field of human-AI interaction. We also plan to assess the extent to which the explanation of recommendations enhances user knowledge and how individuals with different levels of digital literacy understand these explanations.

Acknowledgements. This material is based upon work supported by the National Science Foundation under Grant No. 2153509.

References

1. Achiam, J., et al.: GPT-4 technical report. arXiv preprint arXiv:2303.08774 (2023)
2. Adadi, A., Berrada, M.: Peeking inside the black-box: a survey on explainable artificial intelligence (XAI). IEEE Access **6**, 52138–52160 (2018)
3. Bennett, J., Lanning, S., et al.: The Netflix prize. In: Proceedings of KDD Cup and Workshop, New York, vol. 2007, p. 35 (2007)
4. Braun, V., Clarke, V.: Using thematic analysis in psychology. Qual. Res. Psychol. **3**(2), 77–101 (2006)
5. Chauhan, U., Shah, A.: Topic modeling using latent Dirichlet allocation: a survey. ACM Comput. Surv. (CSUR) **54**(7), 1–35 (2021)
6. Geng, S., et al.: Recommendation as language processing (RLP): a unified pretrain, personalized prompt & predict paradigm (P5). In: Proceedings of the 16th ACM Conference on Recommender Systems, pp. 299–315 (2022)
7. Goldberg, D., et al.: Using collaborative filtering to weave an information tapestry. Commun. ACM **35**(12), 61–70 (1992)
8. Gunning, D., Aha, D.: DARPA's explainable artificial intelligence (XAI) program. AI Mag. **40**(2), 44–58 (2019)
9. Li, L., Zhang, Y., Chen, L.: Generate neural template explanations for recommendation. In: Proceedings of the 29th ACM International Conference on Information & Knowledge Management, pp. 755–764 (2020)
10. Liu, Y., et al.: Summary of ChatGPT-related research and perspective towards the future of large language models. Meta-Radiology 100017 (2023)
11. Lu, Y., et al.: Fantastically ordered prompts and where to find them: overcoming few-shot prompt order sensitivity. arXiv preprint arXiv:2104.08786 (2021)
12. Marvin, G., Hellen, N., Jjingo, D., Nakatumba-Nabende, J.: Prompt engineering in large language models. In: Jacob, I.J., Piramuthu, S., Falkowski-Gilski, P. (eds.) ICDICI 2023. AIS, pp. 387–402. Springer, Singapore (2024). https://doi.org/10.1007/978-981-99-7962-2_30
13. Parra, D., Brusilovsky, P.: User-controllable personalization: a case study with SetFusion. Int. J. Hum. Comput. Stud. **78**, 43–67 (2015)
14. Radford, A., et al.: Better language models and their implications. OpenAI Blog **1**(2) (2019)
15. Resnick, P., Varian, H.R.: Recommender systems. Commun. ACM **40**(3), 56–58 (1997)
16. Tang, J., et al.: ArnetMiner: extraction and mining of academic social networks. In: KDD 2008, pp. 990–998 (2008)
17. Tang, J., Xia, H., Liu, H.: Social recommendation: a review. Soc. Netw. Anal. Min. **3**, 1113–1133 (2013)
18. Tsai, C.-H., Brusilovsky, P.: Explaining recommendations in an interactive hybrid social recommender. In: Proceedings of the 24th International Conference on Intelligent User Interfaces, pp. 391–396 (2019)
19. Tsai, C.-H., Brusilovsky, P.: The effects of controllability and explainability in a social recommender system. User Model. User-Adap. Inter. **31**, 591–627 (2021)

Utilizing Retrieval-Augmented Large Language Models for Pregnancy Nutrition Advice

Taranum Bano[1](✉), Jagadeesh Vadapalli[1], Bishwa Karki[1], Melissa K. Thoene[2], Matt VanOrmer[2], Ann L. Anderson Berry[2], and Chun-Hua Tsai[1]

[1] College of Information Science and Technology, University of Nebraska Omaha, Omaha, NE 68182, USA
tbano@unomaha.edu
[2] College of Medicine, University of Nebraska Medical Center, Omaha, NE 68188, USA

Abstract. The importance of nutrition during pregnancy cannot be overstated, as it profoundly impacts maternal and fetal health outcomes. Optimal fetal growth and development are contingent upon adequate nutrition throughout gestation, which in turn requires that expectant mothers possess a high level of nutritional literacy. This latter factor may serve as a valuable predictor of pregnancy outcomes. This paper seeks to leverage the capabilities of a retrieval-augmented large language model to provide personalized prenatal nutrition guidance. We employed Meta's LLAMA 2 model and integrated an expert-curated dataset of nutrition information. Our evaluation, conducted using ChatGPT-based metrics, revealed that while the augmented model did not yield significant improvements in overall response quality, it could generate more thoughtful and specific responses easily comprehensible to users. We conclude by discussing the challenges encountered and lessons learned from our investigation.

Keywords: Chatbot · Large Language Models · RAG · Health

1 Introduction

Nutrition is paramount during pregnancy due to its profound impact on maternal health and fetal development. The dietary choices made by expectant mothers significantly influence the growth and well-being of the fetus, as well as the overall health status of the mother [19]. Ensuring sufficient intake of essential nutrients like proteins, carbohydrates, fats, vitamins, and minerals is crucial for promoting fetal organ development, skeletal growth, and normal physiological function. Insufficient consumption of these vital nutrients can lead to developmental delays, restricted perinatal outcomes, and reduced newborn weight [15]. Of note, some women may consume more calories than recommended but experience malnutrition from a nutrient deplete diet of ultraprocessed foods [3].

Furthermore, maternal nutritional deficiencies may continue to affect the child's growth and health even after birth. Studies have linked maternal malnutrition during pregnancy to various chronic illnesses such as obesity, diabetes, cardiovascular disease, and impaired cognitive development in offspring [15]. Adequate nutrition not only mitigates the risk of adverse pregnancy outcomes but also contributes to long-term health trajectories for both mother and child. Thus, optimizing dietary intake during pregnancy is imperative to ensure optimal health outcomes and lay a foundation for lifelong well-being [15]. Despite the critical role of nutrition during pregnancy, expectant mothers face significant challenges in accessing reliable advice. The vast amount of conflicting and non-evidence-based information available, particularly online, often leaves them overwhelmed and uncertain about which sources to trust [14]. Additionally, the structure of prenatal care-with its infrequent appointments that need to cover a range of health concerns-may not allow for in-depth discussions on nutrition, thus impeding the absorption of necessary dietary information. Implementing supplementary support systems specifically designed to address pregnancy nutrition queries is therefore crucial. Such systems would provide a consistent and reliable framework for guidance, helping to alleviate the burden of navigating complex dietary recommendations on one's own. Moreover, enhancing health literacy, which involves understanding and interacting effectively with the healthcare system, is essential for empowering individuals to make informed decisions about their health [5,18]. This highlights the necessity for enhanced accessibility to accurate and comprehensible health information, which is crucial for effective healthcare engagement.

The potential of Large Language Models (LLMs) as a tool for providing personalized nutrition advice to pregnant patients has been explored, with models such as ChatGPT demonstrating promising results [20]. However, concerns have been raised regarding the accuracy and reliability of LLM-generated responses, which can be random, mistaken, or misleading [2]. To address this limitation, we propose combining LLMs with retrieval-augmented generation (RAG) techniques, which has shown promise in optimizing language model performance [10]. Integrating LLMs with RAG makes it possible to deliver users contextually relevant health information. Furthermore, by leveraging these advanced AI capabilities and expert-approved data embedding, we argue that high-quality, personalized health content can be generated, tailored to individual needs and preferences. The success of this design has the potential to bridge the gap in health literacy, empowering individuals to make informed decisions about their health and well-being [7].

This paper investigates whether LLMs enhanced with RAG can improve the quality of responses provided in pregnancy nutrition advice. To address this research question, we leveraged Meta's LLAMA 2 models and integrated an expert-approved, self-collected dataset on pregnancy nutrition. We employed the state-of-the-art ChatGPT 4 model to evaluate the performance of our augmented model, which has been demonstrated to surpass human worker label quality [6]. Our evaluation revealed that the RAG-enhanced LLM may not significantly enhance the general information quality in the responses; however, the responses generated by this model were found to be more concise and easier to comprehend, potentially benefiting individuals with limited health literacy. This finding provides valuable insights into the efficacy of utilizing LLMs in nutrition advising and the impact of incorporating self-collected data into this specific use case in health care. We subsequently discuss the implications, lessons learned, and conclusions drawn from our findings.

2 Related Works

2.1 Nutrition Literacy and Advice in Pregnancy

Research highlights the crucial role of nutrition during pregnancy, yet many women in the US struggle to meet recommended standards due to challenges like obesity and inadequate dietary intake [19]. Adolescent pregnant women, in particular, exhibit less healthy dietary patterns, emphasizing the need for enhanced support to optimize health outcomes for this vulnerable population [12]. Adequate intake of essential nutrients such as proteins, carbohydrates, fats, vitamins, and minerals is crucial for promoting fetal organ development, skeletal growth, and normal physiological functioning [15]. Moreover, maternal nutrition has long-term implications, influencing the risk of chronic diseases in offspring [9]. Additionally, a mother's diet during pregnancy is linked to her mental well-being, with certain nutrients playing a role in promoting positive mental health [21]. Despite its importance, access to nutrition counseling remains limited, with systemic barriers, socioeconomic disparities, and high rates of unplanned pregnancies contributing to variations in access [8].

The importance of nutrition literacy, or the ability to understand available nutrition information, is highlighted by [16]. The study emphasizes that pregnant women are more likely to seek and follow nutrition advice as their nutrition literacy increases. Nutrition literacy among pregnant women is a critical aspect of maternal and fetal health during pregnancy, as it enables informed decisions about dietary choices impacting both mother and fetus. [16] further analyzes that despite evidence-based recommendations, limited health literacy poses challenges, leading to inadequate nutrient intake and delayed prenatal care, potentially risking neural tube defects and compromised maternal-fetal health.

In response to the challenges faced by pregnant women in receiving accurate and easy-to-understand nutrition advice, various tools and interventions have been developed to provide support and information. [14] discuss the role of platforms like Nutripedia in providing evidence-based nutritional information, aiming to combat misinformation and promote positive health behaviors. Similarly, [1] introduce chatbots of *Dr. Joy*, which offer continuous digital support to perinatal women and their partners. [14] suggest that ongoing research is necessary to address limitations and enhance the effectiveness of these interventions, particularly in leveraging advanced technologies such as LLMs in conversation.

2.2 Retrieval-Augmented Large Language Models

Several studies have delved into the exploration of RAG techniques and its implications across different domains. The study [4] explores the incorporation of background explanation to enhance the accessibility of complex medical texts to a broader audience. By employing RAG techniques, the study investigates how external knowledge sources, such as definitions and explanations from resources like Wikipedia, can be utilized to augment the understanding of expert-authored content. The findings reveal the effectiveness of RAG in improving summary quality and simplicity while maintaining factual correctness. One of the challenges results indicated was that while these LLMs can generate simplified content, the quality of the summaries falls short of ideal standards.

Similarly, [11] explores the efficacy of RAG in incorporating external knowledge to enhance response generation within knowledge-intensive Natural Language Processing (NLP) tasks. It highlights the growing importance of integrating external knowledge sources into NLP systems for generating contextually relevant and informative responses. The effectiveness of RAG is evaluated through a comprehensive assessment that compares its design with current techniques and evaluates the retriever's ability to select relevant passages. Through these evaluations, the study demonstrates RAG's capability to utilize external information effectively, leading to state-of-the-art performance in knowledge-intensive NLP tasks. Another similar study [13] explores the application of RAG strategies to enhance the accuracy and relevance of LLMs in healthcare, particularly in nephrology. The research delves into the challenges posed by the imperfect accuracy and tendency to produce inaccuracies or irrelevant outputs inherent in LLMs, which are critical in medical applications requiring precision for decision-making processes. The study suggest the potential of RAG in enhancing the accuracy of specialized ChatGPT models tailored to align with specific medical guidelines. One such example includes the KDIGO 2023 guidelines, which provide guidance for chronic kidney disease.

The Work in Progress research [20] explored the potential of LLMs as a tool for providing personalized nutrition advice to pregnant patients with models such as ChatGPT demonstrating promising results. An illustrative example of this research is creation of NutritionBot, eliciting input and offering personalized nutrition recommendations based on demographic and personal history. Google Dialogflow is utilized to implement and host the chatbot, which integrates domain-specific nutrition questions and employs ChatGPT to generate contextually appropriate responses. Building upon the insights gained from [20], this research aims to further explore the efficacy of RAG in enhancing chatbot response quality, particularly in the context of nutrition advising for pregnant women, while also addressing limitations identified by previous research employing LLMs for response generation.

3 Method

3.1 Retrieval-Augmented Large Language Models

In order to develop a robust RAG system capable of providing personalized nutrition advice in a production-ready environment, meticulous design and implementation are essential. The proposed architecture is illustrated in Fig. 1. The

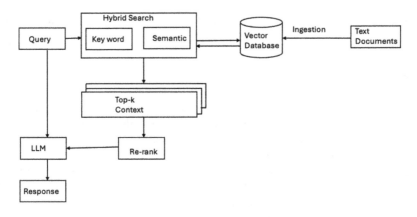

Fig. 1. RAG architecture

initial crucial step involves the indexing process, wherein a vast amount of relevant nutritional data is systematically organized, processed, and stored in a structured format to facilitate efficient querying. This indexing phase lays the foundation for the subsequent retrieval or search process, enabling the RAG system to identify and retrieve pertinent information from the indexed database rapidly.

The retrieval or search process generates accurate and personalized nutrition advice within the RAG system. Upon receiving a user's query or input, the RAG system swiftly searches the indexed database to retrieve relevant information, including nutritional facts, dietary recommendations, and health guidelines. The retrieved data is subsequently utilized by the generation component of the RAG system to produce coherent, informative, and tailored nutrition advice that addresses the user's specific needs and concerns.

The indexing process commences with the segmentation of text into smaller, manageable chunks, which may be as granular as a sentence, paragraph, or specific sequence of 500 characters. Each chunk is then transformed into a vector of predetermined length using an embedding algorithm. The resultant vectors are subsequently stored in a vector database optimized for efficient retrieval.

The development of a RAG system critically depends on the ability to locate relevant contextual information. Retrieving pertinent documents or knowledge significantly aids the generation process. While traditional vector-based search methods can capture semantic relevance, they sometimes struggle with precise term matches and brief queries. To address these challenges, we investigate the implementation of hybrid search and rerank methods to improve retrieval effectiveness.

Hybrid Search: In our RAG framework, hybrid search involves the concurrent use of vector and keyword searches. By establishing both vector and keyword indices in the database beforehand, the system capitalizes on the strengths of each method when processing a user query. This dual approach ensures a more comprehensive and precise retrieval capability, thereby enhancing the quality of information supplied to the generative model.

Rerank Method: The rerank method plays a pivotal role in refining the search process, thus increasing the relevance and accuracy of the generated responses. Empirical evidence suggests that employing a reranker can significantly improve the results of initial retrieval efforts. In this method, a hybrid search is conducted to retrieve a broad set of candidate documents. Each document is then assessed for its relevance to the query using a scoring mechanism that evaluates semantic alignment. Subsequently, documents are reordered based on their relevance scores, with the top K documents selected for final output. This procedure prioritizes the retrieval of semantically pertinent documents. For this process, we employ Cohere's multilingual rerank model.

This study leverages the Dify framework, which incorporates Nomic's advanced indexing techniques to embed textual chunks [17]. This framework offers a multifaceted approach, enabling not only the visualization of embedded vectors but also providing flexibility to experiment with diverse indexing configurations. Users can modify indexing methods, explore metadata details, and adjust retrieval counts, among other settings, thereby rendering it an indispensable tool in optimizing the indexing and retrieval processes.

The development of production-ready applications utilizing LLMs poses several challenges, including computational difficulties, data privacy concerns, integration complexities, latency issues, and others. However, the Dify framework facilitates the orchestration of LLMs in production environments by supporting various RAG features, ETL capabilities, Vector Databases, and other functionalities. In our experiments, we employed different sizes of the Llama 2 model, specifically 7B, 13B, and 70B, all quantized to a 4-bit length. We utilized Dify's model for indexing and Cohere's multilingual model for reranking. Our experimental setup consisted of a server equipped with two NVIDIA RTX4090 GPUs, 128 MB DDR5 RAM, and an Intel 14th Gen Core i9 CPU.

3.2 Dataset and Evaluations

The datasets used to construct the RAG model in this study were rigorously curated from diverse sources, including federal agencies and scientific publications, to establish a comprehensive knowledge base for nutrition-related factors during pregnancy. An initial review of 20 information sources commonly used to inform patients during office visits was conducted independently by three co-authors with expertise in healthcare and nutrition. This search was subsequently expanded to incorporate 20 online nutrition e-books, augmenting our dataset's size. The primary datasets were further supplemented by collecting data from reputable sources, including the Food and Drug Administration (FDA), the Centers for Disease Control and Prevention (CDC), and academic papers indexed in the PubMed database. Ultimately, 12,000 words (comprising 40 documents) were indexed in the vector database.

To assess the effectiveness of the proposed model, we generated a dataset comprising 117 unique sample questions from ChatGPT-4. The prompt used was *"Provide 100 common questions that pregnant women frequently ask about nutrition"*. This prompt was repeated twice, resulting in 117 distinct questions that were utilized as the evaluation dataset. These questions were subsequently submitted to various candidate models to elicit responses. The sample questions included *"How much water should I drink while pregnant?"*, *"Are there specific nutrients that support the immune system during pregnancy?"*, and *"What nutritional considerations are there for women experiencing multiple pregnancies (twins, triplets, etc.)?"* among others.

The evaluation of the responses from the candidate models focused on two key aspects: general quality and specificity to the posed questions. To facilitate this assessment, the following prompt template was used to rate the quality of the responses. This approach aimed to measure both the accuracy and the applicability of the responses provided by the models. We modified the definition of quality to assess scores from these two aspects. The definition of general quality is presented below.

Prompt: On a scale from 1 to 100, where 1 represents the lowest quality and 100 signifies the highest quality, please evaluate the quality of the response provided to a user. The user's description is enclosed within <user></user> tags, the question posed by the user within <question></question> tags, and the corresponding answer within <answer></answer> tags. *Quality is defined by the answer's reliability, conciseness, and clarity, ensuring it conveys the necessary information without superfluous details or misleading the user.* Please provide only the scale, no explanation or other text.
<user>The user is a pregnant woman seeking nutrition advice.</user>
<question>How much water should I drink while pregnant?</question>
<answer>During pregnancy, it's important to stay properly hydrated to support your health and your baby's growth. The recommended daily intake of water for pregnant women is eight to twelve 8-ounce glasses, which can increase depending on factors such as physical activity, hot weather, or fever. Proper hydration can help reduce swelling, constipation, and fatigue, and improve digestion, elimination, and overall well-being. It's best to drink water regularly throughout the day, aiming for at least 50 ounces of fluid intake from water, milk, juice, and other beverages. You can also monitor your urine output to ensure you're drinking enough - if it's light yellow or clear, you're on the right track!</answer>

We revised the definition of specificity to state: *"Quality is defined by the answer's specificity. The shorter and easier-to-read answers should receive higher scores. Answers should be summarized from the context, rather than presented as bullet points"*.

In this study, we submitted a comprehensive set of prompts to the OpenAI API and retrieved the corresponding quality scores. Our experimental design involved evaluating the performance of different LLaMA2 model sizes (7B, 13B, and 7B) with and without embedding context, resulting in a total of 702 prompts being sent for evaluation. To ensure the accuracy and reliability of our results, we employed both ChatGPT 3.5 and 4 models in our assessment[1]. A paired t-test was conducted to compare the mean values of the returned quality scores between models of the same size.

4 Findings

The results of our experiment are presented in Tables 1 and 2. Our analysis indicates that the RAG models consistently produced significantly shorter responses than the baseline models, a trend observed across all three model sizes. Furthermore, incorporating context into the prompt via the extrinsic result tended to yield more concise responses. Specifically, the context embedding approach generated responses that were approximately 50% shorter in length. Below is a sample response to the prompt 'How much water should I drink while pregnant?

LLAMA-2 70b with an embedded RAG context During pregnancy, it's important to stay properly hydrated to support your health and your baby's growth. The recommended daily intake of water for pregnant women is eight to twelve 8-ounce glasses, which can increase depending on factors such as physical activity, hot weather, or fever. Proper hydration can help reduce swelling, constipation, and fatigue, and improve digestion, elimination, and overall well-being. It's best to drink water regularly throughout the day, aiming for at least 50 ounces of fluid intake from water, milk, juice, and other beverages. You can also monitor your urine output to ensure you're drinking enough - if it's light yellow or clear, you're on the right track!

LLAMA-2 70b without an embedded RAG context It's important to stay well-hydrated during pregnancy, as your body will be working hard to support the growth of your baby. The amount of water you should drink while pregnant will depend on several factors, including your age, weight, activity level, and the climate you live in.
The Institute of Medicine (IOM) recommends that pregnant women drink at least 8–10 cups (64–80 ounces) of water per day. However, some healthcare providers may recommend more or less water depending on individual needs.
In addition to plain water, you can also consume water-rich foods such as fruits, vegetables, and broth-based soups to help meet your fluid needs. It's also important to note that you can overhydrate, so it's important to listen to your body and drink when you feel thirsty rather than forcing yourself to drink a certain amount of water each day.
It's also important to note that some women may experience swelling or edema during pregnancy, which can be a sign of excess fluid retention in

---

[1] All experiments were completed in April 2024. We used the latest version of the ChatGPT subscription and the APIs.

the body. In this case, your healthcare provider may recommend limiting your fluid intake to help manage the swelling.

Overall, it's important to work with your healthcare provider to determine the right amount of water for you during pregnancy. They can help you assess your individual needs and make recommendations based on your specific situation.

Table 1. Length of Response (Number of Characters)

Model	Context M(SD)	Baseline M(SD)	
Llama2-7b	1227.27 (518.39)	2548.75 (496.83)	**
Llama2-13b	1373.79 (582.46)	2367 (476.10)	**
Llama2-70b	1202.55 (571.24)	2289.92 (615.57)	**

Mean (M), Standard Deviation (SD), *$p < 0.05$, **$p < 0.01$

Table 2. The results of general quality & specificity test

| | Model | ChatGPT 3.5 | | | ChatGPT 4 | | |
		RAG M(SD)	Baseline M(SD)		RAG M(SD)	Baseline M(SD)	
Quality	7b	78.93(12.55)	82.26(11.01)	*	95.12(8.81)	97.90(2.87)	**
	13b	76.28(13.94)	83.27(9.80)	**	95.25(9.80)	98.29(4.41)	**
	70b	79.35(12.79)	83.37(8.85)	**	96.02()5.23)	98.84(2.21)	**
Specificity	7b	52.86(22.28)	37.54(16.06)	**	81.66(14.18)	64.10(14.39)	**
	13b	51.06(21.44)	41.36(18.14)	**	80.72(13.93)	67.26(13.99)	**
	70b	56.40(23.29)	41.15(19.35)	**	85.51(9.59)	66.79(15.95)	**

In our exhaustive assessment of model performance, we observed a consistent pattern of superior performance by the 70b model compared to smaller models, including the 7b and 13b models. Notably, the 13b model exhibited enhanced performance relative to the 7b model when evaluated using ChatGPT 3.5, although this advantage was not replicated with ChatGPT 4. Across all candidate models, our results demonstrate that baseline models consistently outperformed RAG models in terms of overall quality, a finding supported by statistical testing confirming these differences' significance. Furthermore, our analysis suggests that longer and more informative responses tend to receive higher rating scores for general quality, likely due to their increased content density, as exemplified by the sample response provided above.

In our evaluation of specificity quality, we observed that the 70b model with RAG embedding outperformed the 7b and 13b models. Conversely, the 13b model emerged as the top performer in the baseline model, surpassing the larger 70b model. This finding suggests that smaller models can also generate high-quality responses that rival those of larger models when equipped with contextualized embeddings. Moreover, our results reveal a significant effect of RAG embedding, with the enhanced model outperforming the baseline model in generating shorter responses to questions. These findings imply that incorporating contextual information through embeddings can lead to more concise, accurate, and informative responses.

5 Conclusions

This study investigates whether Large Language Models (LLMs) augmented with Retrieval-Augmented Generation (RAG) can enhance the quality of responses provided in pregnancy nutrition advice. Our experimental results reveal that RAG models consistently generated shorter responses compared to baseline models across all model sizes, with context embedding yielding even more concise responses (approximately 50% shorter). Notably, the 70b model outperformed smaller models in terms of overall response quality, whereas the 13b model demonstrated superior performance when evaluated using ChatGPT 3.5. Furthermore, our analysis shows that the 70b model with RAG embedding excelled in specificity quality, while the 13b baseline model surprisingly surpassed the larger 70b model. These findings suggest that incorporating contextual information through embeddings could lead to more concise, accurate, and informative responses and that smaller models can still produce high-quality responses when equipped with contextualized embeddings.

Reading and comprehending health information is a fundamental skill for patients to make informed decisions about their care [18]. In healthcare settings, healthcare providers, educators, and patients often prefer receiving concise and easily digestible information. Our experimental results suggest that RAG-enhanced LLMs can generate concise information that may be more accessible to users. This finding sheds light on the potential for adopting this technology in healthcare to improve patient health literacy. In our context, a pregnant woman could utilize this tool to access and understand information about nutrition during pregnancy. Based on doctor-approved data as context, the model could provide useful responses to her queries at any time and from any location. This highlights a novel interaction and use case for pregnant women to improve nutritional advice during pregnancy.

Our experimental results suggest that, in constrained contexts, smaller models can provide similar or even superior responses to users when the context is properly embedded. This finding has significant implications, as larger models require substantial computing power (i.e., powerful GPUs), which can be a major accessibility issue for users in rural areas or those with limited access to high-performance machines or the financial means to afford such equipment. In contrast, smaller models like 7B could be executed on regular personal computers or mobile devices, thereby increasing accessibility to more users and leveraging the power of large language models. Further studies could continue to explore the area of edge computing and its deployment to end-users.

We plan to address several limitations of this study in future research endeavors. One key aspect crucial to ensuring the success of RAG embedding is data quality, which was compromised in this experiment due to the limited dataset of only 40 documents collected as proof-of-concept. A more comprehensive dataset would be necessary to test the effectiveness of RAG embedding further. Another essential consideration is indexing performance, a critical factor in RAG that requires proper contextualization to avoid reproducing results from the original model. Unfortunately, we did not monitor the performance of indexing in this study, which constitutes a limitation. Furthermore, our evaluation relied on a GPT-powered assessment to test information quality, which may not guarantee quality for healthcare providers and patients. These issues necessitate further exploration and research to realize the full potential of RAG embedding.

References

1. Chung, K., Cho, H.Y., Park, J.Y.: A chatbot for perinatal women's and partners' obstetric and mental health care: development and usability evaluation study. JMIR Med. Inform. **9**(3), e18607 (2021)
2. Floridi, L., Chiriatti, M.: GPT-3: its nature, scope, limits, and consequences. Mind. Mach. **30**, 681–694 (2020)
3. Graciliano, N.G., da Silveira, J.A.C., de Oliveira, A.C.M.: Consumo de alimentos ultraprocessados reduz a qualidade global da dieta de gestantes. Cadernos de Saúde Pública **37**(2), e00030120 (2021)
4. Guo, Y., et al.: Retrieval augmentation of large language models for lay language generation. J. Biomed. Inform. **149**, 104580 (2024)
5. He, K., et al.: A survey of large language models for healthcare: from data, technology, and applications to accountability and ethics. arXiv preprint arXiv:2310.05694 (2023)
6. He, Z., et al.: If in a crowdsourced data annotation pipeline, a GPT-4. arXiv preprint arXiv:2402.16795 (2024)
7. Jiang, F., et al.: Artificial intelligence in healthcare: past, present and future. Stroke Vasc. Neurol. **2**(4) (2017)
8. Killeen, S.L., et al.: Using FIGO nutrition checklist counselling in pregnancy: a review to support healthcare professionals. Int. J. Gynecol. Obstet. **160**, 10–21 (2023)
9. Koletzko, B., et al.: Nutrition during pregnancy, lactation and early childhood and its implications for maternal and long-term child health: the early nutrition project recommendations. Ann. Nutr. Metab. **74**(2), 93–106 (2019)
10. Kresevic, S., et al.: Optimization of hepatological clinical guidelines interpretation by large language models: a retrieval augmented generation based framework. NPJ Digit. Med. **7**(1), 102 (2024)
11. Lewis, P., et al.: Retrieval-augmented generation for knowledge-intensive NLP tasks. In: Advances in Neural Information Processing Systems, vol. 33, pp. 9459–9474 (2020)

12. Marshall, N.E., et al.: The importance of nutrition in pregnancy and lactation: lifelong consequences. Am. J. Obstet. Gynecol. **226**(5), 607–632 (2022)
13. Miao, J., et al.: Integrating retrieval-augmented generation with large language models in nephrology: advancing practical applications. Medicina **60**(3), 445 (2024)
14. Montenegro, J.L.Z., da Costa, C.A., Janssen, L.P.: Evaluating the use of chatbot during pregnancy: a usability study. Healthcare Anal. **2**, 100072 (2022)
15. Naaz, A., Muneshwar, K.N.: How maternal nutritional and mental health affects child health during pregnancy: a narrative review. Cureus **15**(11) (2023)
16. Nawabi, F., et al.: Health literacy in pregnant women: a systematic review. Int. J. Environ. Res. Public Health **18**(7), 3847 (2021)
17. Nussbaum, Z., et al.: Nomic embed: training a reproducible long context text embedder (2024). arXiv: 2402.01613 [cs.CL]
18. Rudd, R.E., Rosenfeld, L., Simonds, V.W.: Health literacy: a new area of research with links to communication. Atlantic J. Commun. **20**(1), 16–30 (2012)
19. de Seymour, J.V., Beck, K.L., Conlon, C.A.: Nutrition in pregnancy. Obstet. Gynaecol. Reprod. Med. **29**(8), 219–224 (2019)
20. Tsai, C.-H., et al.: Generating personalized pregnancy nutrition recommendations with GPT-powered AI chatbot. In: 20th International Conference on Information Systems for Crisis Response and Management (ISCRAM), vol. 2023, p. 263 (2023)
21. Yelverton, C.A., et al.: Diet and mental health in pregnancy: nutrients of importance based on large observational cohort data. Nutrition **96**, 111582 (2022)

Vineyard Leaf Disease Prediction: Bridging the Gap Between Predictive Accuracy and Interpretability

Noor E. Mobeen[1], Sarang Shaikh[2](✉), Livinus Obiora Nweke[2], Mohamed Abomhara[2], Sule Yildirim Yayilgan[2], and Muhammad Fahad[1]

[1] Department of Computer Science, Norwegian University of Science and Technology (NTNU), Gjøvik, Norway
{noore,muhamfah}@stud.ntnu.no

[2] Department of Information Security and Communication Technology, Norwegian University of Science and Technology (NTNU), Gjøvik, Norway
{sarang.shaikh,livinus.nweke,mohamed.abomhara,sule.yildirim}@ntnu.no

Abstract. Balancing the accuracy and interpretability of predictive models has been a persistent challenge in traditional approaches. In this study, we advance this field by integrating cutting-edge artificial intelligence (AI) techniques with Explainable AI (XAI) methodologies to significantly enhance both the accuracy and interpretability of vineyard leaf disease predictions. We employ state-of-the-art convolutional neural networks (CNNs) and introduce a fine-grained model architecture featuring, adept at discerning subtle disease indicators in vineyard leaves. This innovative approach not only boosts the diagnostic performance of the models but also provides clear visualizations of the decision-making processes. This study utilizes a focused dataset strategy, incorporating one specialized grape disease dataset (Esca) and a subset of the general PlantVillage dataset, specifically selecting categories relevant to Apple and Grape diseases. The obtained results have demonstrated our model's exceptional capability in accurately identifying and classifying various leaf diseases, showcasing its practical applicability in real-world vineyard management. Furthermore, our approach addresses the vital need for transparency and trust in AI applications within agriculture, particularly in viticulture.

Keywords: vineyard disease detection · artificial intelligence · deep learning · explainable AI (XAI) · fine-grained-classification

1 Introduction

Viticulture, the science and practice of grape cultivation, serves as a cornerstone in the global wine industry, contributing significantly to agricultural economies worldwide [1]. However, the health and yield of vineyards are constantly threatened by various leaf diseases that pose a particularly pervasive challenge. For instance, as reported in [2] one of the oldest disease "Esca" has reached upto 80% in various old vineyards in central Italy and its southern parts. This implies that the presence of these diseases, and lack of effective strategies to mitigate them, could cause a severe loss in production. [3]. Traditional pest and disease detection methods in vineyards exhibit inefficiencies, potentially leading to delayed diagnoses and subsequent yield losses [4]. Figure 1, 2, and 3 show the sample images for both healthy and diseased vineyard leaves. The figures show the different diseases in grape leaves like Esca (Esca dataset), BlackRot (PlantVillage dataset), and for apple leaves like AppleScab, and CedarAppleRust (PlantVillage dataset).

Fig. 1. Sample images from Esca dataset

Fig. 2. Sample images from PlantVillage dataset (Grapes)

Fig. 3. Sample images from PlantVillage dataset (Apple)

Recognizing these limitations, recent advancements in digital image processing, particularly using AI-based techniques, promise to revolutionize vineyard management practices [5]. These techniques have the potential to expedite anomaly detection within grapevine yields, enabling early intervention strategies to mitigate disease spread and associated financial losses for wine producers [4]. As previously mentioned, the advancements about the use of AI-based techniques in vineyard disease prediction; most of the state-of-the-art (SOTA) studies have recently used deep learning based image analysis techniques such as Convolutional Neural Network (CNN), and its variants like Residual Neural Network (ResNet), and Densely Connected Neural Network (DenseNet) [4]. These techniques will be further discussed later in the paper in Sect. 2. Additionally, transformer-based technique is becoming more popular these days specially for image classification tasks. The most common model in this category is the VisionTransformer (ViT) model [6].

To the best of our knowledge, there exists only single study which used ViT model for leaf disease classification [7]. Hence, in this study we proposed fine-grained model that incorporates a swin-transformer architecture as its backbone to capture detailed image features critical for the accurate classification/predictions of vineyards disease leaf images. Furthermore, this research is distinct in its application, utilizing two public datasets, Esca and PlantVillage as none of the studies have used them together before. Finally, our approach is further enhanced by the integration of XAI techniques, including both Grad-CAM and LIME. Because, we recognize the importance of not only achieving high predictive accuracy but also showing insights against the decision-making procedure of our proposed fine-grained model.

The rest of the paper is organized as follow: Sect. 2 discusses the relevant SOTA studies and background information; Sect. 3 shows and explains our proposed methodology; Sect. 4 discusses the experiments performed with our proposed approach; Sect. 5 presents our results & findings, and Sect. 6 shows the conclusion & future work.

2 Related Work

There has been an increasing interest in recent years toward the application of machine learning and deep learning techniques for the early detection and classification of grapevine diseases. In general, there are some new achievements in the early detection and classification of diseases in vineyards. However, most of these studies have focused on broad classifications or have been limited to specific types of diseases without a deeper, fine-grained analysis or robust interpretability mechanisms that are crucial for practical applications. For example, few advancements particularly in the application of CNNs, have facilitated significant progress in the analysis of grape leaf diseases. Alessandrini et al. in [8] proposed a new grapevine image dataset to classify between two classes: healthy and unhealthy grape images affected by Esca disease. The dataset is suitable for various machine-learning tasks, including image segmentation and synthesis. Furthermore, Carraro et al. address the significant challenges of detecting the Esca disease complex in asymptomatic grapevine leaves using CNNs [9]. In their exploration of grapevine diseases, they employ hyperspectral imaging combined with CNNs to differentiate between symptomatic and asymptomatic leaves. While this approach marks a significant step forward, it lacks the deep granularity provided by our proposed model in this study.

Additionally, Zia et al. in [10] further contribute to the enhancement of prediction accuracy and performance in disease diagnosis. They performed using the AlexNet model on the publicly available PlantVillage dataset. As they demonstrated high accuracy using CNNs on the PlantVillage dataset, their approach did not incorporate the critical element of explainability, which is a core component of our proposed approach in our study. The integration of XAI into agricultural AI systems has been in focus of several studies [7,11,12]. Bandi et al. utilized the YOLOv5 model to train two different datasets, PlantDoc and PlantVillage, for disease detection and employed ViT for disease stage classification. Expanding upon previous research on disease detection in grapevine leaves, another study introduced by Mamba et al. [5] discusses the effectiveness of federated learning in crop disease detection using CNN models and those based on attention mechanisms. In general, the experiments have shown that the performance of federated learning is highly affected by factors such as the number of learners involved, communication rounds, total iterations, and data quality. Among the models tested, ResNet 50 demonstrated the highest performance, while ViTB16 and ViTB32 were found to be less suitable for federated learning due to their computational time and cost implications. Hence, while extensive work has been done on explainable AI, research on interpretable methods in the agricultural field remains limited. The authors in [11] focuses on enhancing the interpretability of deep learning models used in classifying leaf diseases across various fruit leaf datasets. By utilizing models such as ResNet, VGG, and GoogLeNet augmented with attention mechanisms, they demonstrate an improvement in the models' ability to focus on relevant features of leaf images.

In contrast to these studies, our research adopts a unique dual-dataset approach, utilizing both the Esca dataset and specifically the Apple and Grape classes from the PlantVillage dataset. This method is innovative and fills several gaps in the current research by offering a technique that not only improves the accuracy of disease diagnostics but also enhances the interpretability of results across different types of data. By combining the proposed fine-grained model with advanced XAI techniques, our model meets the high-accuracy demands of modern agriculture while also providing deeper insights into the decision making processes. This facilitates greater trust and adaptability in real-world vineyard management. Table 1 shows the SOTA summary of several AI techniques applied for vineyard leaf disease detection together with the gap in the techniques which is covered in our research study.

Table 1. SOTA summary of AI models for vineyard disease prediction

Ref.	CNN	Dense Net	Res Net	Mobile Net	Efficient Net	VIT	Alex Net	YOLO v5	VGG	Fine Grained
Paper [8]	✓									
Paper [9]	✓					✓				
Paper [10]	✓			✓		✓				
Paper [7]		✓				✓	✓			
Paper [11]		✓			✓				✓	
Paper [12]		✓	✓	✓			✓	✓		
Our study	✓	✓	✓	✓						✓

3 Proposed Methodology

Figure 4 shows the proposed methodology involving a sequence of steps for processing leaf images for healthy vs disease classification and analysis. Initially, the leaf image undergoes preprocessing steps including resizing, normalization, and augmentation to enhance the dataset and improve the model's robustness. Following this, the fine-grained model combined with Swin-Transformer (Swin T) is employed for the detailed and accurate classification of the leaf images. Finally, the methodology uses Grad-CAM (Gradient-weighted Class Activation Mapping) and LIME (Local Interpretable Model-agnostic Explanations) to interpret and visualize the model's predictions, highlighting important regions of the leaf that contribute to the models' decisions. This approach aims to enhance both the performance and interpretability of the model.

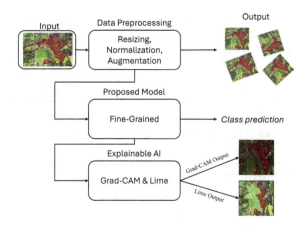

Fig. 4. Proposed Methodology

3.1 Dataset

In this study, we used two SOTA datasets; 1) Esca dataset, and 2) PlantVillage (PV) dataset. From the PV dataset, we selected only two categories grapes and apples. The datasets comprises of healthy and various disease images related to vineyards leaves. The overall distribution as well as train, validation, and test set split is shown in the Table 2.

Table 2. Overall, train, valid, and test set distribution of the datasets

Esca Dataset					PlantVillage (Grapes) Dataset					PlantVillage (Apple) Dataset				
	Train set	Valid set	Test set	Total		Train set	Valid set	Test set	Total		Train set	Valid set	Test set	Total
Healthy	529	132	221	882	Healthy	254	423	106	783	Healthy	987	1645	411	3043
Esca	533	133	222	888	Black Rot	708	1180	295	2183	Apple Scab	378	630	158	1166
					Black Measles	830	1383	346	2559	Black Rot	373	621	155	1149
					Leaf Blight	646	1076	269	1991	Cedar Apple Rust	165	275	69	509
Total	1062	265	443	1770	Total	2438	4062	1016	7516	Total	1903	3171	793	5867

3.2 Preprocessing

This section outlines the preprocessing steps essential for preparing the data for subsequent model training and analysis. 1) **Image Resizing**: In the preprocessing stage, all images from both datasets were initially resized to 1280 × 720 pixels. This standardization was crucial for ensuring consistency across all instances, particularly for the SOTA models such as CNN, DenseNet, and ResNet. For the fine-grained implementation, however, the images were resized to 384 × 384 pixels. The reason for choosing this size was to facilitate feature extraction without compromising consistent information across images.

2) **Data Normalization**: Before feeding the data to the model, we have normalized the data with the mean values 0.4762, 0.3054, 0.2368, and standard deviation values 0.3345, 0.2407, 0.2164. This normalization process helped center the data around zero and scale it to a comparable range, facilitating stable and efficient model training. 3) **Data Augmentation**: We implemented data augmentation techniques to increase the diversity and improve the robustness of the training dataset. The techniques opted for augmenting images include horizontal and vertical flips, rotation of up to 90°, and scaling.

3.3 Fine-Grained Model

Figure 5 illustrates the basic architecture of fine-grained model used as a part of the proposed methodology. The proposed model utilizes the Swin-Transformer as its backbone, due to its effectiveness in capturing fine details through its hierarchical architecture. The Swin-Transformer is composed of four integral components, Patch Partitioning, Swin Transformer Blocks, Shifted Window, and Feature Hierarchy. It first segments the image into non-overlapping patches. Further, these patches are processed through a series of Swin-Transformer Blocks with a shifted window framework, which enhances the model to capture local features with the global context in a flexible and efficient fashion. [13]. This approach will be very good in complex tasks of image classification for tasks like vineyard leaf disease prediction.

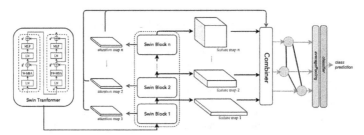

Fig. 5. Model Architecture for fine-grained model with Swin Transformer as a backend.

3.4 Explainable AI (XAI)

In fact, the increased use of AI in the treatment of healthcare and agricultural management is leading to increased demand for transparency and understandability in such systems [14]. The subsequent section discusses ways in which we have implemented XAI techniques, such as Grad-CAM and LIME, in order to increase the transparency and trustworthiness of our predictive models. 1) **Grad-CAM**: In our experiments, we have used the Grad-CAM model to visualize the explanation of the decisions and predictions from our models. The ability to show us the points of focus in an image makes it a great tool for providing such information [15]. Thus, this serves to highlight parts of the input that most affect the model's decision—valuable interpretability [16].

2) **LIME**: LIME helps in understanding complex model decisions by changing the input data and looking at how those changes affect the output. This approach adds another perspective by showing what features contribute to bringing about a prediction. In disease detection from vineyards, for example, LIME can show how features of the image of leaves, such as spots or color gradations, lead to the identification of a specific disease. Thus, with LIME, we can embellish the interpretability of our model, ensuring that the decisions are precise and comprehensible at a granular level.

4 Experiments

This section discusses the experimental detail which we performed to evaluate the performance of SOTA models vs our proposed fine-grained model on the task of leaf disease prediction. Our approach employed a series of pre-trained models that includes DenseNet121, DenseNet169, ResNet50, MobileNetV2, and the proposed FineGrained model across two distinct datasets: Esca, which is specific to grapevine leaves, and selected classes from the PlantVillage dataset, namely Apple and Grapes.

We selected several SOTA models such as DenseNet121, DenseNet169, ResNet50, MobileNetV2 to compare their performance with our proposed fine-grained model using swin-transformer. We trained, validated, and tested all the SOTA as well as our proposed model on the respective selected datasets' using the distribution shown in the Table 2 with the following hyperparameters shown in the Table 3.

Table 3. Hyperparameters for models' training

Batch size	8 to 32
Learning rate	0.0001
Epochs	30

To evaluate the performance of our predictive models, we used a comprehensive set of key metrics that include accuracy, precision, recall, and f1-score. The below four equations show how all of these metrics are calculated.

$$\text{Accuracy} = \frac{TP + TN}{TP + TN + FP + FN} \quad (1)$$

$$\text{Precision} = \frac{TP}{TP + FP} \quad (2)$$

$$\text{Recall} = \frac{TP}{TP + FN} \quad (3)$$

$$\text{F1-Score} = 2 \times \frac{P \times R}{P + R} \quad (4)$$

*TP = True Positive; TN = True Negative; FP = False Positive;
*FN = False Negative; P = Precision; R = Recall

5 Results and Discussion

This section discusses the results obtained using the experiments performed based on experimental settings explained in the Sect. 4.

5.1 Fine-Grained Model Results

The results shown in Tables 4, and 5 clearly show the proposed fine-grained models' improved performance compared to other SOTA models. The proposed model in this study achieved 100% scores for all the evaluation metrics discussed in the Sect. 4. The use of the fine-grained model with swin-transformer backbone significantly enhanced its capability to discriminate between closely related disease states, providing high precision and recall.

Table 4. Performance Metrics for Esca Dataset

Model	Accuracy	Precision	Recall	F1 Score	Support
DenseNet121	1.00	1.00	1.00	1.00	444
ResNet50	0.99	0.99	0.99	0.99	443
DenseNet169	1.00	1.00	1.00	1.00	444
MobileNetV2	0.99	0.99	0.99	0.99	443
FineGrained	1.00	1.00	1.00	1.00	444

Table 5. Performance Metrics for PlantVillage Dataset

Model	Apples				Grapes			
	Acc.	Prec.	Recall	F1	Acc.	Prec.	Recall	F1
DenseNet121	1.00	1.00	1.00	1.00	1.00	1.00	1.00	1.00
ResNet50	1.00	1.00	1.00	1.00	1.00	1.00	1.00	1.00
DenseNet169	1.00	1.00	1.00	1.00	1.00	1.00	1.00	1.00
MobileNetV2	0.99	0.99	0.99	0.99	0.99	0.99	0.99	0.99
FineGrained	1.00	1.00	1.00	1.00	1.00	1.00	1.00	1.00

Furthermore, Fig. 6, 7, and 8 shows the confusion matrices of the proposed fine-grained models' performance on the test set of the datasets shown in the Table 2. Figure 6 shows the performance of fine-grained model for predicting the test dataset images either as "Healthy" or affected by "Esca". The confusion matrix shows perfect classification performance, with no false positives or false negatives. The fine-grained model correctly identified all 221 healthy samples and all 222 Esca-affected samples. This indicates a highly accurate model for this dataset, as every prediction made was correct.

Figure 7 shows the performance of fine-grained model, specifically for predicting various diseases of grapes leaves from PlantVillage dataset. The diseases being predicted are "Black rot", "Esca black measles", "Leaf blight isariopsis leaf spot", and "Healthy". The confusion matrix indicates that the model has perfectly classified all samples across all the target classes. The model demonstrates perfect accuracy for the given dataset, correctly classifying all samples into their respective classes without any errors. This suggests highly effective performance of the model for predicting these specific grapes leaves diseases from the PlantVillage dataset. Figure 8 shows the fine-grained models' performance specifically for predicting various diseases of apple leaves from PlantVillage dataset. The diseases being predicted are "Apple scab", "Black rot", and "Cedar apple rust". The confusion matrix indicates that the model has perfectly classified all samples across all the target disease classes together with "Healthy" class. This demonstrates the model's high effectiveness and accuracy in distinguishing between Apple scab, Black rot, Cedar apple rust, and Healthy classes.

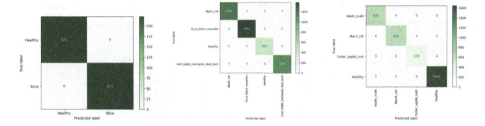

Fig. 6. Confusion matrix for Esca dataset using fine-grained model

Fig. 7. Confusion matrix for grape leaves from PlantVillage dataset using fine-grained model

Fig. 8. Confusion matrix for apple leaves from PlantVillage dataset using fine-grained model

5.2 XAI Results

Figure 9 illustrates the application of XAI techniques, specifically LIME and Grad-CAM, to various datasets of leaf images affected by different diseases. The image is structured in a tabular format with three main columns: "Original Image", "LIME", and "Grad-CAM". In this visual representation, we observe how two prominent XAI techniques, LIME and Grad-CAM, explain the decision-making process of the proposed fine-grained model tasked with classifying vineyard leaf diseases. The Original Image column shows the raw images of leaves, each exhibiting distinct disease symptoms. In the LIME column, the model's predictions are explained by highlighting regions of the leaves that significantly influence its decision. LIME, on the other hand, uses a colorised overlay, with light green areas highlighting the sections that agree with the result and dark red areas highlighting those which disagree with it, making it easier to trace out which parts of a leaf contributed most towards predicting whether or not it was disease.

The Grad-CAM column, on the other hand, is a more comprehensive one which involves overlaying heatmaps on the images. These heatmaps typically show the areas where the model pays more attention, with the warmer colors implying that something is far much important than the others. For the ESCA dataset, both LIME and Grad-CAM point out discolored parts that are affected by disease on the leaf, thereby affirming these areas as essential for predicting by the model. Like LIME, Grad-CAM emphasizes the spotted areas in the images in line with the exhibited symptoms. Both methods clearly show the damage patterns in the images which the model is interested in when diagnosing a disease. This side-by-side visualization is centered on the potential uses of LIME and Grad-CAM in the process of understanding and verifying the outcomes of models as well as shedding light on what the model takes into account and why it makes decisions. In this regard, implementing these XAI techniques helps make AI models that aid in agricultural diagnostics more understandable and trustworthy because their main focus can be broken down into simple words while showing how some regions in the pictures contribute towards identifying diseases.

Fig. 9. XAI results for fine-grained model performance

6 Conclusion and Future Work

The study has managed to bridge the gap between high prediction accuracy attainment and model interpretation within vineyard leaf disease prediction. The detection of leaf disease has been greatly improved through the application of fine-grained models as well as interpretability techniques like LIME and Grad-CAM. We did not only improve the accuracy of detection but also conveyed actionable knowledge to vineyard owners through our findings, who can use it for disease management strategies that form basis for decision making. Our research progress is of great significance for AI application in farming especially in connection with growing grapes. Users grasp and rely on system predictions better hence increasing their use in vineyard operations. This research helps in developing AI models that are beneficial in cultivating vineyards sustainably through provision of in-depth interpretations for model predictions, important in ascertaining onset symptoms of diseases for prompt remediation.

In future work, we are planning to further improve the proposed approach by applying it to real-world dataset by collecting leaf images directly from vineyards. Further, we would also like to use federated learning to retain data privacy and security. This method can be used to train federated models across a good number of decentralized devices or servers, holding local data samples, without the need for data sharing—thereby keeping sensitive information at source. This approach will enhance not only the strength and generalization of our model but also meet regulations and industry standards for data privacy, making it more suitable for field practice within agricultural settings.

Acknowledgement. This study is a part of "VIPA-DELF" sub-project and has indirectly received funding from the European Union's Horizon Europe research and innovation action programme, via the CHAMELEON Open Call #1 issued and executed under the CHAMELEON project (Grant Agreement no. 101060529). The technical work done in this study has benefited from the Experimental Infrastructure for Exploration of Exascale Computing (eX3), which is financially supported by the Research Council of Norway under contract 270053

References

1. Ethan, C.: The art and science of grape growing: a comprehensive guide to viticulture (2023)
2. Romanazzi, G., Murolo, S., Pizzichini, L., Nardi, S.: Esca in young and mature vineyards, and molecular diagnosis of the associated fungi. Eur. J. Plant Pathol. **125**, 277–290 (2009). https://api.semanticscholar.org/CorpusID:39052945
3. Gallo, R., et al.: New solutions for the automatic early detection of diseases in vineyards through ground sensing approaches integrating lidar and optical sensors. Chem. Eng. Trans. **58**, 673–678 (2017). https://api.semanticscholar.org/CorpusID:56052818
4. Attri, I., Awasthi, L.K., Sharma, T.P., Rathee, P.: A review of deep learning techniques used in agriculture. Ecol. Inform. 102217 (2023)
5. Mamba Kabala, D., Hafiane, A., Bobelin, L., Canals, R.: Image-based crop disease detection with federated learning. Sci. Rep. **13**(1), 19220 (2023)
6. Bazi, Y., Bashmal, L., Rahhal, M.M.A., Dayil, R.A., Ajlan, N.A.: Vision transformers for remote sensing image classification. Remote Sens. **13**(3), 516 (2021)
7. Bandi, R., Swamy, S., Arvind, C.: Leaf disease severity classification with explainable artificial intelligence using transformer networks. Int. J. Adv. Technol. Eng. Explor. **10**(100), 278 (2023)
8. Alessandrini, M., Calero Fuentes Rivera, R., Falaschetti, L., Pau, D., Tomaselli, V., Turchetti, C.: A grapevine leaves dataset for early detection and classification of esca disease in vineyards through machine learning. Data Brief **35**, 106809 (2021). https://www.sciencedirect.com/science/article/pii/S2352340921000937

9. Carraro, A., Saurio, G., López-Maestresalas, A., Scardapane, S., Marinello, F.: Convolutional neural networks for the detection of esca disease complex in asymptomatic grapevine leaves. In: Foresti, G.L., Fusiello, A., Hancock, E. (eds.) ICIAP 2023. LNCS, vol. 14365, pp. 418–429. Springer, Cham (2024). https://doi.org/10.1007/978-3-031-51023-6_35
10. Zia, M.A., Akram, A., Mumtaz, I., Saleem, M.A., Asif, M.: Analysis of grape leaf disease by using deep convolutional neural network. Agric. Sci. J. **5**(1), 25–36 (2023)
11. Wei, K., et al.: Explainable deep learning study for leaf disease classification. Agronomy **12**(5), 1035 (2022)
12. Arvind, C., et al.: Deep learning based plant disease classification with explainable AI and mitigation recommendation. In: 2021 IEEE Symposium Series on Computational Intelligence (SSCI), pp. 01–08. IEEE (2021)
13. Liu, Z., et al.: Swin transformer: hierarchical vision transformer using shifted windows. In: Proceedings of the IEEE/CVF International Conference on Computer Vision, pp. 10012–10022 (2021)
14. Nahiduzzaman, M., et al.: Explainable deep learning model for automatic mulberry leaf disease classification. Front. Plant Sci. **14**, 1175515 (2023)
15. Ashoka, S., et al.: Explainable AI based framework for banana disease detection (2024)
16. Quach, L.D., Quoc, K.N., Quynh, A.N., Thai-Nghe, N., Nguyen, T.G.: Explainable deep learning models with gradient-weighted class activation mapping for smart agriculture. IEEE Access **11**(August), 83752–83762 (2023)

Application of Unmanned Aerial Vehicles for Autonomous Fire Detection

José Silva[1(✉)], David Sousa[2], Paulo Vaz[1], Pedro Martins[1], and Alfonso López-Rivero[3]

[1] CISeD – Research Centre in Digital Services, Instituto Politécnico de Viseu, Viseu, Portugal
{jsilva,paulovaz,pedromom}@estgv.ipv.pt
[2] Department of Informatics Engineering, Instituto Politécnico de Viseu, Viseu, Portugal
estgv18748@alunos.estgv.ipv.pt
[3] Computer Science Faculty, Pontifical University of Salamanca, Salamanca, Spain
ajlopezri@upsa.es

Abstract. This paper describes the development and implementation of an autonomous system for fire detection using Unmanned Aerial Vehicles (UAVs). This project aims to explore new technologies for quick and effective responses, thereby reducing the damage caused by natural disasters. The approach utilizes computer vision techniques, with the UAV equipped with a camera to identify signs of fire and smoke.

Using UAVs in fire detection presents a promising alternative to conventional methods, offering the potential to cover vast areas quickly and transmit real-time data to command centers. The methodology includes designing and implementing a UAV equipped with computer vision technology. The YOLO object detection model was used, configured to identify signs of fire and smoke in images captured during flight. The UAV was programmed to navigate autonomously following predefined routes, with the capability to adjust its trajectory based on real-time data collected.

Results indicate that the UAV could correctly identify signs of fire and smoke with significant accuracy. The system's capabilities were validated under controlled conditions, where the UAV demonstrated effectiveness in detecting and sending alerts to a base station.

This study concludes that the use of UAVs for fire detection is feasible and effective, providing a valuable tool for firefighting efforts.

Keywords: Unmanned Aerial Vehicles · Fire Detection · Computer Vision · YOLO

1 Introduction

The application of Unmanned Aerial Vehicles (UAVs) in various sectors of society has demonstrated a transformative impact, especially in areas requiring rapid, scalable, and efficient solutions [1, 2]. Among these applications, deploying UAVs for environmental monitoring and emergency response, particularly for detecting and managing fires, presents a significant opportunity to enhance public safety and resource management [3, 4]. This research explores the integration of UAV technology in fire detection, a critical issue in places where wildfires are a frequent and devastating occurrence, causing extensive environmental damage and loss of life.

Mediterranean climate, characterized by hot, dry summers, makes it highly susceptible to wildfires. Traditional fire detection methods, which often rely on human observation and satellite monitoring, are hindered by delays in detection time and can be affected by cloud cover or the availability of resources. As a result, the response to these fires is often reactive rather than proactive. With the introduction of UAVs equipped with advanced sensing and imaging technologies, it is possible to achieve quicker detection times, providing a more immediate response to emerging threats.

The primary aim of this project is to develop an autonomous UAV system capable of early fire detection, utilizing computer vision algorithms. This approach allows UAVs to independently patrol high-risk areas and identify potential fire outbreaks through real time image processing. By integrating the YOLO (You Only Look Once) object detection model, the UAVs are trained to recognize specific patterns of smoke and fire from visual data captured during flight. This method not only enhances the speed of detection but also improves the accuracy of the surveillance system.

Unlike traditional object detection systems that process images at a slower rate, YOLO applies a single neural network to the entire image, enabling it to predict multiple bounding boxes and class probabilities for those boxes simultaneously [5]. This method drastically reduces processing time, making it ideal for applications where speed is critical, such as detecting fast-spreading wildfires [6].

To implement this solution, the project was carried out using the DJI drone, a model selected for its balance of affordability, ease of use, and sufficient technological capability to support the required computational processes. The drone was programmed to navigate predetermined routes over areas typically vulnerable to fire outbreaks, with real-time data processing facilitated through an onboard system that analyzes the captured images for signs of fire and smoke.

The system's architecture was designed to ensure that data collected during flights could be immediately transmitted to a central server, where further analysis is conducted to confirm potential fires. Alerts can then be generated for faster identification. This proactive approach helps combat the spread of fires and significantly reduces the potential damage and cost associated with these disasters.

Moreover, the study addresses the technical challenges encountered in developing the autonomous UAV system, including data accuracy and the operational range of the drones.

This project contributes to emergency management through technological innovation and sets a foundation for future research in the application of UAVs in disaster response and environmental monitoring. As UAV technology continues to evolve, its potential to support and enhance emergency response mechanisms becomes increasingly significant, promising a future where technology and strategic management converge to effectively mitigate the impacts of natural disasters [7].

For future work, additional sensors, such as air quality sensors, should be integrated and advanced communication technologies 5G, should be explored to improve real time data transmission.

The rest of this article is organized as follows. Section 2 presents the theoretical concepts associated with the topic addressed are presented. Section 3 briefly describes the methodology used. In Sect. 4, the real application of the proposed methodology is shown. Finally, the conclusion about this case study is presented in Sect. 5.

2 Background

This chapter provides an in-depth look at the foundational technologies and frameworks that underpin the development of an autonomous UAV system for fire detection. Understanding these components is crucial for appreciating their integration and functionality within the project.

2.1 Computer Vision

Computer vision is a facet of artificial intelligence that enables computers and systems to derive meaningful information from digital images, videos, and other visual inputs and act on that information [8]. It involves developing algorithms to detect, classify, and track objects or attributes in images and video. This technology is crucial in various applications ranging from autonomous driving to content tagging in social media platforms and, critically, in scenarios like the UAV-based detection systems used in fire monitoring [9–11].

The operational backbone of this UAV fire detection project utilizes computer vision to analyze aerial imagery captured in real-time. The primary objective is identifying early fire signs such as smoke or unusual heat patterns. This technology allows UAVs to perform autonomous monitoring over large and inaccessible areas, providing faster response times than traditional methods.

YOLO (You Only Look Once)

YOLO stands out within object detection models due to its speed and efficiency, making it particularly suited for real-time applications [12, 13]. YOLO frames object detection as a single regression problem, from image pixels to bounding box coordinates and class probabilities. It divides the image into a grid and predicts bounding boxes and probabilities for each grid cell. The model applies a single neural network to the entire image, enabling it to simultaneously predict multiple bounding boxes and their respective class probabilities. This approach allows YOLO to detect objects significantly faster than region-proposal-based methods like R-CNN, making it ideal for applications where speed is crucial [14].

Other traditional object detection methods, such as R-CNN (Region-based Convolutional Neural Networks) and its more advanced variants, like Fast R-CNN and Faster R-CNN, involve multiple stages to detect objects [15]. These models generate region proposals (candidate bounding boxes), then extract features using a convolutional network, and then classify those regions using the extracted features. While these methods are highly accurate, they are computationally intensive and slow in processing, which limits their utility in real-time scenarios.

SSD (Single Shot MultiBox Detector), another popular model, attempts to balance speed and accuracy better than R-CNN [16]. SSD skips the proposal generation step and directly predicts bounding box coordinates and class scores via a single network. It is faster than R-CNN but generally less accurate than the more refined Faster R-CNN.

In selecting YOLO for the UAV project, the critical deciding factor was its ability to provide the best trade-off between speed and accuracy. The ability to quickly process and analyze imagery is paramount for fire detection, as it directly influences the response time in critical situations. YOLO's capability to perform detections at high frame rates ensures that UAVs can monitor vast areas swiftly and detect emergent fires in their early stages, potentially saving lives and property by allowing quicker responses.

2.2 Roboflow

Roboflow is an essential tool in computer vision that streamlines the process of developing and deploying machine learning models by enhancing how data is prepared and managed [17]. It offers a robust platform that simplifies the transformation of raw images into formats suitable for training advanced machine learning algorithms. This capability is particularly valuable in projects involving complex image recognition tasks, such as the UAV-based fire detection system described in this research [18].

In the context of UAV fire detection, Roboflow's role extends beyond mere image processing; it facilitates a comprehensive workflow that includes annotating, organizing, and augmenting the visual data captured during UAV flights. This preparation is crucial as the quality and arrangement of training data significantly impact the performance of the detection models. Roboflow enables precise annotation of images, allowing researchers to accurately label features of interest such as smoke or flames. These labels are essential for training the YOLO model to recognize and react to these specific fire indicators in diverse environmental conditions.

Furthermore, Roboflow supports data augmentation, a technique used to artificially expand the size of a dataset by creating modified versions of images in the dataset [19]. This process helps improve the robustness of the model by exposing it to a broader range of variations in image data, such as different lighting conditions, angles, and occlusions, which are common in dynamic aerial scenarios.

By leveraging Roboflow, the project enhances the training phase of the YOLO model, ensuring that the UAV system is accurate in detecting signs of fire and efficient in differentiating between false alarms and genuine fire outbreaks. The tool's ability to manage and preprocess data effectively reduces the time and effort required to prepare large datasets, accelerating the development cycle and enabling quicker deployment of the UAV system in real-world fire surveillance operations.

2.3 MAVLink

MAVLink (Micro Air Vehicle Link) is a communication protocol for drones and other unmanned vehicles [20]. It facilitates the seamless exchange of critical data between the UAV and ground control systems, essential for real-time operations and autonomous decision-making processes. MAVLink supports a wide array of messages that handle everything from vehicle status updates to specific commands and mission-related data.

MAVLink is crucial in enabling robust and efficient communication pathways in the UAV-based fire detection project. The protocol allows the UAV to transmit real-time telemetry data back to the control station, including its location, speed, and camera feeds. This real-time data transmission is vital for monitoring the UAV's operational status and adjusting flight paths or detection parameters on the fly, enhancing the UAV's effectiveness in dynamic and possibly hazardous environments.

Furthermore, MAVLink aids in the precise execution of autonomous flight plans. It sends commands that control the UAV's movements, ensuring that the vehicle adheres to predetermined flight paths over areas at high risk for fires. This controlled navigation is crucial for covering designated zones thoroughly and systematically [21].

The lightweight yet powerful nature of MAVLink makes it ideal for use in applications requiring minimal latency and high reliability, such as in emergency response scenarios where UAVs are tasked with rapid fire detection and reporting.

3 Application Architecture

In this section, the global vision regarding the design of the algorithm is demonstrated, including the import of the route defined by the user, its communication to the UAV and the analysis of each image from the stream received by the drone.

Figure 1 provides a visual representation of the various steps involved in the operational process, starting from route planning to the final step of notifying the operator about potential fire detections.

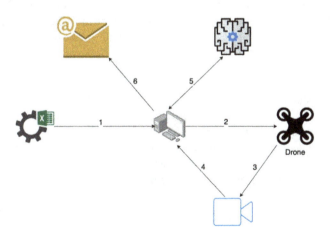

Fig. 1. System Structure.

These six steps represent a closed-loop system designed to maximize the efficiency and responsiveness of UAV-based fire detection. This system architecture ensures rapid detection and notification of wildfires and enhances the capability to monitor large areas with precision, ultimately contributing to effective disaster management and response strategies.

1. Route Importation:

    This initial step involves importing the pre-planned flight route into the UAV system. The route, usually stored in a CSV file format, includes detailed instructions about the flight path the UAV should follow. This file specifies the coordinates, altitude, speed, and other relevant flight parameters. The data from this file is loaded into the UAV's navigation system, setting up the autonomous flight path that the UAV will follow during its mission.

2. Sending Route to UAV:

    Once the route is loaded and confirmed, it is sent to the UAV through a communication link, typically using the MAVLink protocol. This step ensures that the UAV is fully aware of the mission specifics, including waypoints and behaviors at each point, such as hover or circle for better surveillance of specific areas.

3. Video Capture by Drone:

    With the flight plan in action, the UAV begins its autonomous navigation along the designated path. Simultaneously, it starts capturing real-time video footage of the surveyed area using its onboard cameras. This video is crucial for the detection of any smoke or fire signs from above the terrain.

4. Video Stream to Control Station:

    The video captured by the UAV is streamed back to the control station in real-time. This step is vital as it allows the operators and the fire detection algorithms to analyze the footage immediately, looking for any abnormalities that could indicate a fire.

5. Analysis by YOLO Model:

    As the video stream is received at the control station, it is processed on the fly by the YOLO object detection model. This model has been trained to identify specific features associated with fire and smoke. Each frame of the video is analyzed, and potential threats are identified based on the learned characteristics of fire and smoke in various conditions and environments.

6. Notification to User:

    The final step in the workflow occurs when the YOLO model detects a potential fire. An alert is generated within the system, and a notification is immediately sent to the system operator. This notification includes details about the detection, such as the location, time, and a snapshot or video clip of the detected event. The operator can then take necessary actions, such as alerting local fire services, deploying firefighting teams to the location, or activating other emergency response measures.

4 Results

This case study aims to validate the use of an unmanned aerial vehicle (UAV) with aerial surveillance capabilities, as well as fire detection and supervision. This UAV will carry a camera for real-time image transmission. The UAV must follow a route on autopilot, thus being able to execute a mission autonomously by providing GPS coordinates of the area over which it must fly.

4.1 Dataset

Training the computer vision model requires a dataset, so finding suitable images was the first step in training this model. To search for public images, the RoboFlow application was used, and the dataset was found (Roboflow-fire, 2023). This dataset consists of 5949 images, which had to be processed.

The first step in preparing the dataset included the elimination of duplicate classes classified as "Fire", "fire" and "nf", which represented the fire class. Maintaining the first "Fire" class. In the end, the dataset included only two classes: "Fire" and "Smoke". Subsequently, this dataset's size was reduced due to the lack of memory to train the model.

The second step was to select the areas of the images where these classes could be found, as shown in Fig. 2.

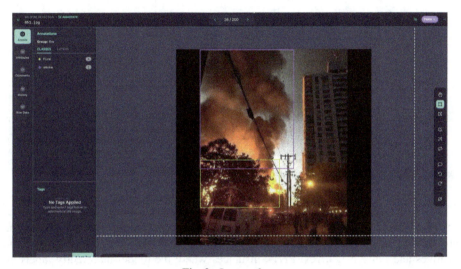

Fig. 2. Image class.

In the third step, a version of the dataset was generated. The dataset comprises over 2100 annotated images, categorized into two distinct classes: fire and smoke. The pre-processing techniques were auto-orientation and resizing to 640x640 pixels.

4.2 Experimentation and Results

To train the model for fire detection, the YOLOv8 architecture was used, with the dataset partitioned into 70% for training, 20% for testing and 10% for validation.

In Fig. 3, we can examine the confidence curve, which plots the F1 score on the y-axis against the confidence level on the x-axis. The F1 score ranges from 0 to 1, where 0 indicates no detection of the targeted classes across all images, and 1 indicates perfect detection of all classes across all images. The confidence level, a crucial parameter during inference, ranges from 0 (0% confidence) to 1 (100% confidence).

Values between 0.3 and 0.7 on the confidence axis generally suggest that the model performs adequately in recognizing what is intended, yet there is room for improvement. Specifically, the confidence curve reveals that the optimal threshold value for this model lies between 0.3 and 0.5. Setting the confidence threshold within this range balances sensitivity and precision, enabling effective detection while minimizing false positives.

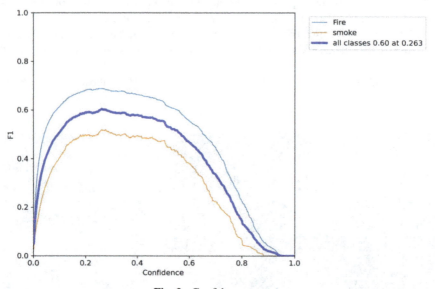

Fig. 3. Confidence curve.

The second key metric of the model involves analyzing the Recall curve, which correlates recall values to different confidence thresholds. Recall is calculated as the ratio of True Positives (TP), which represent correctly detected items, to the sum of True Positives and False Negatives (FN)—instances where the model incorrectly labels an actual positive as negative. The ideal recall value is 1, indicating that all positive cases are identified correctly.

Analysis reveals that for confidence thresholds between 0.3 and 0.5 (or 30% to 50%), the model achieves a recall value above 0.6, as shown in Fig. 4. This range suggests the optimal confidence levels to maximize the model's sensitivity without compromising its ability to accurately discern true positives from false negatives.

From this analysis, it can be concluded that the model meets the required accuracy standard. This level of precision ensures that the model is effectively balancing between identifying true fire incidents and minimizing the rate of false alerts.

4.3 Real-Time Inference

Once the model is trained, it can be used to analyze images received from the drone's video stream. To enable this, the NumPy and YOLO libraries were integrated into the streaming file (see Fig. 5). Additionally, in the *"notifier.py"* file, a function was added to facilitate the sending of notifications via email.

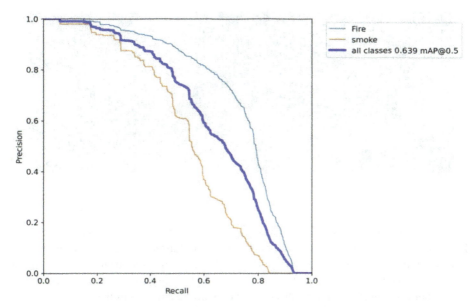

Fig. 4. Recall curve.

```
import socket
import cv2
import numpy as np
from ultralytics import YOLO
import notifier
model = YOLO('./runs/detect/train/weights/best.pt')

cap = cv2.VideoCapture('./fogo-fogueira.mov')
tello_video = cv2.VideoCapture('udp://@0.0.0.0:11111')

presence = 0
sentEmail = False
while True:
    try:
        ret, frame = tello_video.read()
        results = model(frame, device='mps')

        result = results[0]
        bboxes = np.array(result.boxes.xyxy.cpu(), dtype="int")
        classes = np.array(result.boxes.cls.cpu(), dtype="int")

        if len(bboxes) > 0:
            presence = presence + 1
        else:
            if(presence > 0):
                presence = presence - 1
        if presence > 5 and not sentEmail:
            sentEmail = True
            notifier.sendEmail()
        for cls, bbox in zip(classes, bboxes):
            (x, y, x2, y2) = bbox
            cv2.rectangle(frame, (x, y), (x2, y2), (0, 0, 225), 2)
            cv2.putText(frame, str(cls), (x, y - 5), cv2.FONT_HERSHEY_PLAIN, 2, (0, 0, 255), 2)
        cv2.imshow("Img", frame)
```

Fig. 5. Video stream inference code.

The YOLO function imports the trained model, and during each cycle of the while loop, it calls the "*model()*" function (shown in Fig. 6). This function requires the current video frame as its input image and the graphics processing unit (GPU) as an additional parameter to handle the computation. The average inference speed with the GPU active is 22 ms, allowing the system to process over 25 frames per second effectively.

```
Tello > ● notifier.py > ...
  1   import smtplib, ssl
  2
  3   def sendEmail(sender_email, receiver_email, port, app_password, message):
  4
  5
  6       context = ssl.create_default_context()
  7
  8       with smtplib.SMTP_SSL("smtp.gmail.com", port, context=context) as server:
  9           server.login(sender_email, app_password)
 10           server.sendmail(sender_email, receiver_email, message)
```

Fig. 6. Notification code.

The output includes the coordinates (i.e., the corners of the bounding boxes) and the identified class of each detected object. This information is then overlaid on the live video feed for enhanced visualization, aiding in the real-time monitoring and assessment of the situation by the operators. This setup not only improves the responsiveness of the system but also ensures that crucial visual data is immediately actionable.

The figure shows how the fire alert notification is carried out by sending the email. An email is sent notifying the existence of a fire.

5 Conclusions and Future Work

This project explored the feasibility of using drones for early detection of fires. To achieve this objective, a Yolo v8 model was developed, capable of accurately identifying signs of fire or smoke in real time. Additionally, a drone was selected based on its ability to fulfill the project's stringent requirements. Advanced algorithms were implemented to import and execute flight routes autonomously, minimizing the need for human intervention and enhancing the system's efficiency.

The fire detection capabilities of the system were rigorously tested using a video stream transmitted by the drone. This setup allowed the model to analyze the footage and determine the presence of fire or smoke in real time. In the event of a fire detection, the system is configured to alert users immediately via email, ensuring swift communication that could potentially lead to quicker emergency response times.

Several enhancements are proposed to increase the effectiveness and applicability of the drone system. Future developments could include integrating additional sensors, such as air quality sensors, to detect elevated levels of CO and $CO2$, which are indicative of fires. Furthermore, incorporating a 5G module would improve communication capabilities with the UAV, ensuring faster data transmission and more reliable control in varied environments.

Another prospective improvement is establishing a direct communication channel with local emergency services. This would streamline the process of relaying critical information during fire outbreaks, potentially speeding up the deployment of firefighting resources and improving overall emergency response strategies.

Acknowledgements. This work is funded by National Funds through the FCT – Foundation for Science and Technology, I.P., within the scope of the project Ref. UIDB/05583/2020. Furthermore, we would like to thank the Research Centre in Digital Services (CISeD) and the Instituto Politécnico de Viseu for their support.

References

1. Mohsan, S.A.H., Othman, N.Q.H., Li, Y., Alsharif, M.H., Khan, M.A.: Unmanned aerial vehicles (UAVs): practical aspects, applications, open challenges, security issues, and future trends. Intel. Serv. Robot. **16**(1), 109–137 (2023)
2. Shakhatreh, H., et al.: Unmanned aerial vehicles (UAVs): a survey on civil applications and key research challenges. IEEE Access **7**, 48572–48634 (2019)
3. Sudhakar, S., Vijayakumar, V., Kumar, C.S., Priya, V., Ravi, L., Subramaniyaswamy, V.: Unmanned Aerial Vehicle (UAV) based forest fire detection and monitoring for reducing false alarms in forest-fires. Comput. Commun. **149**, 1–16 (2020)
4. Akhloufi, M.A., Couturier, A., Castro, N.A.: Unmanned aerial vehicles for wildland fires: sensing, perception, cooperation and assistance. Drones **5**(1), 15 (2021)
5. Diwan, T., Anirudh, G., Tembhurne, J.V.: Object detection using YOLO: challenges, architectural successors, datasets applications. Multimed. Tools Appl. **82**, 9243–9275 (2023)
6. Talaat, F.M., ZainEldin, H.: An improved fire detection approach based on YOLO-v8 for smart cities. Neural Comput. Appl. **35**(28), 20939–20954 (2023)
7. Wang, Y., et al.: The application of UAV remote sensing in natural disasters emergency monitoring and assessment. In: International Conference on Digital Image Processing, vol. 11179, pp. 340–346 (2019)
8. Khan, A.A., Laghari, A.A., Awan, S.A.: Machine learning in computer vision: a review. EAI Endorsed Trans. Scalable Inf. Syst. **8**(32), e4–e4 (2021)
9. Al-Kaff, A., Martin, D., Garcia, F., de la Escalera, A., Armingol, J.M.: Survey of computer vision algorithms and applications for unmanned aerial vehicles. Expert Syst. Appl. **92**, 447–463 (2018)
10. Ettalibi, A., Elouadi, A., Mansour, A.: AI and computer vision-based real-time quality control: a review of industrial applications. Procedia Comput. Sci. **231**, 212–220 (2024)
11. Chai, J., Zeng, H., Li, A., Ngai, E.W.: Deep learning in computer vision: a critical review of emerging techniques and application scenarios. Mach. Learn. Appl. **6**, 100134 (2021)
12. Li, Y., Fan, Q., Huang, H., Han, Z., Gu, Q.: A modified YOLOv8 detection network for UAV aerial image recognition. Drones **7**, 304 (2023)
13. Silva, J., Coelho, P., Saraiva, L., Vaz, P., Martins, P., López-Rivero, A.: Validating the use of smart glasses in industrial quality control: a case study. Appl. Sci. **14**(5), 1850 (2024)
14. Ren, Y., Zhu, C., Xiao, S.: Small object detection in optical remote sensing images via modified Faster R-CNN. Appl. Sci. **8**, 813 (2018)
15. Ren, S., He, K., Girshick, R., Sun, J.: Faster R-CNN: towards real-time object detection with region proposal networks. IEEE Trans. Pattern Anal. Mach. Intell. **39**, 1137–1149 (2017)
16. Kumar, A., Zhang, Z.J., Lyu, H.: Object detection in real time based on improved single shot multi-box detector algorithm. EURASIP J. Wirel. Commun. Netw. **2020**(1), 204 (2020)
17. Protik, A.A., Rafi, A.H., Siddique, S.: Real-time personal protective equipment (PPE) detection using YOLOv4 and TensorFlow. In: Proceedings IEEE Region 10 Symposium (TENSYMP), Jeju, Republic of Korea, 23–25 August, pp. 1–6 (2021)
18. Kholiya, D., Mishra, A.K., Pandey, N.K., Tripathi, N.: Plant detection and counting using YOLO based technique. In: IEEE Conference on Innovation in Technology (ASIANCON), Ravet, India, 25–27 August, pp. 1–5 (2023)

19. Abdulghani, A.M., Abdulghani, M.M., Walters, W.L., Abed, K.H.: Multiple data augmentation strategy for enhancing the performance of YOLOv7 object detection algorithm. J. Artif. Intell. 2579-0021 **5** (2023)
20. Mogili, U.R., Deepak, B.: An intelligent drone for agriculture applications with the aid of the MAVlink protocol. In: Deepak, B., Parhi, D., Jena, P. (eds.) Innovative Product Design and Intelligent Manufacturing Systems. LNME, pp. 195–205. Springer, Singapore (2020). https://doi.org/10.1007/978-981-15-2696-1_19
21. Kwon, Y.M., Yu, J., Cho, B.M., Eun, Y., Park, K.J.: Empirical analysis of MAVlink protocol vulnerability for attacking unmanned aerial vehicles. IEEE Access **6**, 43203–43212 (2018)

Track on Disruptive Technologies

Smart City Air Quality Monitoring: A Mobile Application for Intelligent Cities

Pedro Martins[1(✉)], Diogo Silva[2], João Pinto[2], José Varanda[2], Paulo Váz[1], José Silva[1], and Maryam Abbasi[3]

[1] CISeD - Research Centre in Digital Services, Polytechnic of Viseu, Viseu, Portugal
{pedromom,paulovaz,jsilva}@estgv.ipv.pt
[2] Polytechnic Institute of Viseu, Viseu, Portugal
{estgv2907,estgv1776,estgv1774}@alunos.estgv.ipv.pt
[3] Applied Research Institute, Coimbra Polytechnic, Coimbra, Portugal
maryam.abbasi@ipc.pt

Abstract. Cities are growing fast, and we need to keep an eye on how it affects the environment, especially the air we breathe. This study introduces a fancy app that works on phones and uses sensors to track pollution in cities. It's part of making cities smarter. The app collects data on different pollutants and shows them in an easy-to-understand way on your phone. It also sends you alerts in real-time if there is a problem with the air quality. This app helps people take action to make their cities cleaner and healthier. The study explains how the app works, how it is built, and what it finds out. In the future, they want to test it more, make the app better, use fancy math to predict air quality issues, and keep up with new technology. This research is a great step in managing air quality in cities, giving a way to make cities cleaner and safer for everyone.

Keywords: Air Pollution Monitoring · Mobile Environmental Dashboard · Internet of Things (IoT) Technologies · Real-time Environmental Alerts · Public Health Intervention · Data-driven Decision-making

1 Introduction

Air pollution, a global problem emphasized by the World Health Organization (WHO), is the cause of a staggering seven million premature deaths annually, manifested through the increased incidence of cardiovascular and respiratory diseases, as well as cancers [10]. The predominant culprit in this health crisis is particulate matter, an amalgamation of non-gaseous elements suspended in the air, including sulfate, nitrates, ammonia, sodium chloride, black carbon, mineral dust, and water [10].

Contemporary solutions to combat air pollution are mainly based on the use of static sensors that process data within predefined areas, limiting accessibility to the wider population [9]. In light of the expanding world of Internet of

Things (IoT) technologies, there is a compelling need for an advanced monitoring solution [4].

In response to this imperative, our objective was to devise a novel solution that goes beyond conventional approaches, offering users a dynamic and comprehensive view of pollution data through an intuitive mobile dashboard. By empowering users to actively monitor various pollution parameters, our solution ensures user agency in environmental awareness. Notably, the system employs real-time alerts triggered by significant deviations in environmental indicators, providing users with timely information for informed decision-making in their daily activities.

This document unfolds across six sections. Section 2 delves into existing research related to our proposed solution, providing a contextual foundation. Section 3 elucidates the methodology employed in crafting and developing our solution. The architecture of our implementation is expounded in Sect. 4, complemented by a detailed account of the experimental setup. Section 5 meticulously presents and analyzes the results of our efforts. The conclusions are drawn in Sect. 6, which paves the way for future work as outlined in the conclusion section.

2 Related Work

In the landscape of pollution monitoring, a pervasive challenge pertains to sensor calibration, a critical concern addressed by Maag *et al.* [6]. Their work underscores the ubiquity of portable air pollution sensors, enabling widespread deployments with high spatiotemporal resolution. However, the inherent limitation lies in the susceptibility of low-cost sensors to errors, particularly in demanding environmental conditions. Maag *et al.* delves into various calibration models that successfully mitigate these limitations, improving the data quality of low-cost air pollution sensors. Notably, their research advocates for active recalibration strategies, acknowledging the dynamic nature of air pollution monitoring.

As elucidated in Sect. 1, prevailing solutions often rely on fixed sensors for data analysis, restricting accessibility to the broader public. Addressing the cost and complexity of current systems, Chowdhury *et al.* [3] propose an affordable method for the detection of air pollution. Their approach integrates diverse gas sensors, a GSM module, a cloud server, and a mobile application, offering a streamlined yet effective solution for real-time monitoring.

Alver *et al.* introduce EcoSensor [1], another cost-effective solution leveraging integrated sensors for air pollution data collection. This device transmits real-time pollution levels to Android-based devices while concurrently recording data on a Cloud-based server for comprehensive pollution distribution analysis.

Montanaro *et al.* [7] contribute a solution focused on public engagement and validation through a combination of user feedback, reports, and real-time data from strategically placed mobile IoT sensors. This approach facilitates citizen participation, offering a platform for reporting and visualizing data, thereby creating an informed perspective on the city's environmental conditions. This inclusive system fosters environmental awareness and supports sustainable practices.

Taking a distinctive approach, Okokpujie et al. [8] present a system implemented with an Arduino microcontroller for real-time air pollution monitoring. This project utilizes a Wi-Fi module, an MQ135 Gas Sensor, and an LCD screen to record air quality metrics in Parts per Million (PPM) and disseminate the information through continuous updates on the Internet. Their work contributes to fostering awareness of daily air quality, offering a practical and real-time solution for air quality measurement.

3 Methodology

In the pursuit of developing the mobile application and IoT components for this project, the Agile methodology emerges as the chosen framework, aligning seamlessly with the dynamic nature of software development projects. This methodology, renowned for its adaptability, collaboration, and customer-centric focus, proves particularly well-suited to the ever-evolving landscape of emerging technologies such as the Internet of Things (IoT) [4].

As a comprehensive approach to project management and software development, Agile prioritizes flexibility, collaboration, and customer satisfaction. Central to its operational philosophy is an iterative development process, wherein cross-functional teams dynamically evolve requirements and solutions collaboratively.

Key strengths of the Agile methodology include its flexibility and adaptability, enabling teams to respond to shifting priorities and emerging insights. The emphasis on customer-centricity ensures regular feedback and active participation, aligning the delivered product closely with user expectations. The methodology promotes collaboration and communication between team members, stakeholders and customers, promoting a high degree of interaction throughout the development process. Incremental delivery, a hallmark of Agile, emphasizes the provision of incremental functional product increments in short iterations, known as sprints.

However, Agile does present certain challenges. Resistance to change is a significant hurdle, requiring a fundamental shift in thinking and organizational structure. Concerns about documentation and rigidity may arise, as Agile prioritizes working software over extensive documentation. Resource allocation can pose challenges, especially in larger organizations with pre-existing structures. Additionally, the iterative and adaptive nature of Agile may be unsettling for stakeholders accustomed to more predictable timelines and deliverables, necessitating a delicate balance between agility and predictability. Managing dependencies in larger projects, especially between different Agile teams or between Agile and non-Agile teams, requires careful planning and communication.

The fundamental principles guiding Agile include prioritizing individuals and interactions over processes and tools, valuing working software over comprehensive documentation, seeking active collaboration with customers over contract negotiation, embracing change over rigidly following a plan, and consistently delivering value to the customer through incremental delivery.

In conclusion, the Agile methodology stands as a dynamic and customer-centric approach to project management and software development, offering a framework that aligns with the ever-evolving landscape of emerging technologies.

4 Architecture

This section presents the comprehensive architecture employed in this work. The subsequent subsections provide a detailed exploration of individual components, each designed to cater to specific functionalities and diverse requirements.

Figure 1 visually encapsulates the essence of our proposed architecture. Here, IoT sensors play a pivotal role in establishing robust connections with the underlying database infrastructure. Subsequently, an API acts as a bridge, facilitating the transformation of raw sensor data into comprehensible information accessible to the mobile application.

A noteworthy aspect of our architecture involves the strategic substitution of physical sensors with simulated datasets. This simulation methodology streamlines the data aggregation process, as these datasets are consolidated into a single CSV file. This file serves as a repository that encapsulates a comprehensive record of all information similar to what physical sensors would transmit.

In essence, the architecture optimally integrates the functionalities of IoT sensors, database connections, API-based data transformation, and simulated datasets to create a cohesive and efficient system. The following subsections will provide an in-depth exploration of each architectural facet, elucidating their roles, interactions, and contributions to the overall functionality of the proposed solution.

Sensors play a key role in capturing and emitting data related to various air pollutants, as detailed in Table 1. These electrochemical sensors, equipped with internal heaters, induce a change in resistance when exposed to specific gases, thus facilitating the detection process [4].

- MQ-7: Designed for detecting carbon monoxide in the air due to its high sensitivity.
- MQ-135: Versatile in its function, capable of detecting LPG, propane, methane, alcohol, hydrogen, and smoke.
- MQ-131: Specialized in detecting ozone gas concentrations in the air.
- ME3-SO2: Tailored for detecting and measuring sulfur dioxide (SO2) gas concentrations.
- PMS5003: Designed to detect and measure the concentration of suspended particles in the air, particularly those with diameters of 2.5 and 10.0 μm.
- DHT22: A combined humidity and temperature sensor.

In the contemporary era, with ubiquitous mobile device ownership, the research endeavors to develop a mobile application. This application, leveraging Firebase services, will show processed information on pollutant levels worldwide.

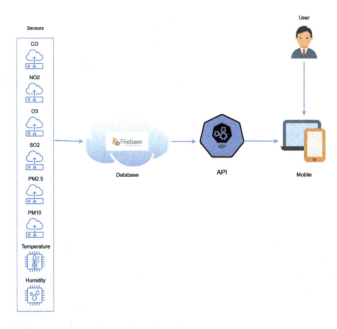

Fig. 1. Proposed Architecture

Table 1. Electronic Components

Electronic Components	
MQ-7	Carbon Monoxide
MQ-135	Nitrogen Dioxide
MQ-131	Ozone
ME3-SO2	Sulphur Dioxide
PMS5003	PM2.5 and PM10
DHT22	Temperature and Humidity

Mobile Application: The proposed app interfaces with the Firebase database, retrieving stored data via Wi-Fi or mobile networks. For app development, Flutter, Google's Open Source UI Software Development Kit, will be utilized within the Android Studio environment.

Firebase: A comprehensive platform that offers various services, including hosting, a real-time database, and authentication. The application uses Firebase's Real-Time Database capabilities [5].

Real-time Database: A cloud-hosted NoSQL database within Firebase, recognized for seamlessly synchronizing data between on-line and off-line states. Its features cater to applications that require real-time updates and offline capabilities [5].

API: A vital component that facilitates communication between sensors and the database to support the mobile application.

Flutter: An open-source UI software development toolkit from Google, enabling the creation of natively compiled mobile, web, and desktop applications from a unified codebase.

Flask: A lightweight Python web framework adhering to the WSGI standard, offering essential tools for web development without imposing excessive conventions. Developers often choose Flask for its flexibility in various web server environments.

5 Results and Analysis

This section presents the outcomes derived from the comprehensive evaluation of the performance of the proposed architecture.

The Firebase Real-Time Database emerges as a robust solution adept at managing vast amounts of concurrently transmitted data. Its exceptional handling of substantial data loads positions it as a powerhouse, particularly suited for applications tied to IoT devices. The architecture excels in seamlessly managing concurrent connections and real-time data updates without compromising performance.

The scalability inherent in the system is a pivotal attribute, facilitating the effortless accommodation of increased loads as user engagement and data interactions expand. This scalability proves indispensable in scenarios where multiple users engage in simultaneous data access or updates, mirroring the demands of real-time collaboration tools and dynamic content-sharing platforms.

The distributed infrastructure of the database optimizes data flow, efficiently handling massive volumes of data and numerous interactions. This configuration ensures a responsive and reliable system across diverse applications.

The API, crafted with Flask for its lightweight and flexible nature, stands as a key component in our architecture. Its rapid development capabilities, requiring minimal code, allow for a focused approach to essential tasks. The flask, designed for RESTful APIs and equipped with built-in support for handling JSON data, aligns precisely with our database structure. This strategic alignment ensures smooth communication throughout all interactions within our application.

The culmination of these architectural components guarantees seamless handling of large data sets, minimizing potential issues that could arise during data-intensive operations.

In practical terms, our mobile application serves as the interface, presenting users with informative graphs detailing specific pollution indicators. This synthesis of technology and user experience underscores the effectiveness of the proposed architecture in delivering relevant real-time data visualizations to the end-user.

5.1 Stress Testing Our Architecture

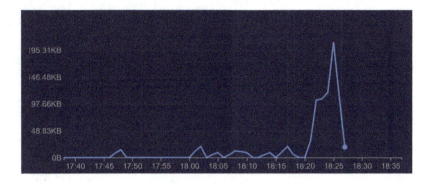

Fig. 2. Data transfer

To rigorously evaluate the robustness of our infrastructure, we conducted extensive tests to measure data transfer efficiency to the cloud-hosted database. Figure 2 visually depicts the data uploaded during these tests, showcasing the system's ability to handle significant data volumes promptly.

In a remarkably brief timeframe, our infrastructure seamlessly processed the upload of 2000 objects to the database, amounting to a mere 400 KB of data. This result underscores the exceptional efficiency of our application, demonstrating its ability to accommodate substantial data inputs from IoT sensors and manage these volumes with optimal resource utilization.

Our focus on the analysis of PM2.5 and PM10 is particularly crucial, as these metrics encapsulate airborne particles that have substantial implications for human health, particularly in the context of air quality. The finer nature of PM2.5 particles enables them to infiltrate deep into the respiratory system, posing potential health risks. Laden with pollutants like heavy metals and organic compounds, these particles pose a daily inhalation risk, potentially compromising overall life quality. Given their origin in combustion processes and industrial emissions, understanding and monitoring these particles becomes imperative in our current environment, where such emissions are prevalent.

The trajectory of PM2.5 levels, depicted in Fig. 3, illustrates fluctuations over a span of 50 days. Notably, there is a discernible spike in the particle levels on a specific day. This anomaly serves as a crucial indicator, leading to an investigation of the events on that particular day to discern the cause behind the surge. Such detailed insights aid in devising targeted strategies to mitigate these levels, ultimately contributing to an enhanced quality of life.

Similarly, PM10, when present at elevated levels, serves as an additional marker of compromised air quality. Inhalation of PM10 particles can lead to adverse respiratory and cardiovascular effects, underscoring the importance of monitoring and addressing these particles for the well-being of individuals and the broader community.

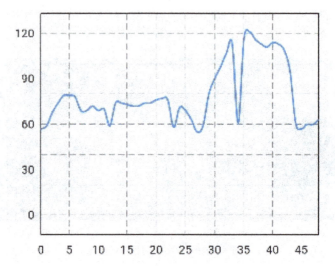

Fig. 3. PM2.5 levels

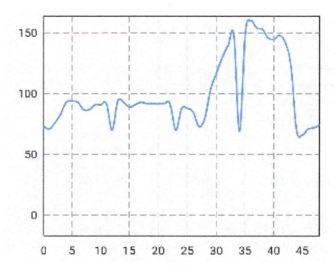

Fig. 4. PM10 levels

Examining Fig. 4, it becomes evident that the levels of PM10 particles experienced a parallel increase during the same time frame as PM2.5. This synchronous escalation indicates a correlation between these two particles, signifying the potential dual impact they pose on our well-being. By leveraging our solution, we gain the ability to identify and address this disturbing trend, offering a valuable tool to mitigate pollution. The application serves as a practical resource, aiding in pinpointing the root causes behind the concurrent rise in both

PM2.5 and PM10 levels, facilitating informed strategies to reduce environmental pollutants and enhance overall air quality.

6 Conclusions

In summary, our developed solution provides a novel approach to monitor and regulate air quality in specific regions. Using sensor-collected data, the system visualizes pollutant levels and various parameters through comprehensive graphs, empowering users to gauge ambient air quality. The broader objective is to promote increased awareness of potential environmental consequences and provide a tool to mitigate these impacts.

Given the reliance on preexisting datasets for data collection, our solution serves as a proof of concept, addressing constraints such as time limitations and the unavailability of physical sensors. It sets the stage for a scalable implementation in a natural environment.

The analysis of the results demonstrates the feasibility of visualizing pollution fluctuations within large data sets. Cloud-based data storage enhances scalability, enabling concurrent monitoring across multiple regions.

6.1 Future Work

Subsequent efforts could involve a thorough real-world evaluation of the proposed architecture, incorporating the actual deployment of sensors alongside datasets for more robust insights.

Improvements to the mobile application are another avenue for future work. Incorporating features like alerts and expanding data representation in graphs would provide users with a more intuitive and comprehensive understanding, empowering them to address environmental concerns in their city or locality.

The use of the proposed sensors for real-time data collection could facilitate the integration of machine learning algorithms. Predicting alerts and abrupt changes in pollution levels through algorithms like Artificial Neural Networks (ANN), as explored by Chitra *et al.* [2], could offer valuable information. This predictive capability, informed by both current and historical data, could trigger timely alerts within the mobile application, enhancing user awareness of potentially hazardous conditions.

Continuously monitoring emerging technologies and trends remains imperative, ensuring that the proposed architecture remains at the forefront of advancements in the dynamic field of air quality monitoring.

Acknowledgments. "This work is funded by National Funds through the FCT - Foundation for Science and Technology, I.P., within the scope of the project Ref. UIDB/05583/2020. Furthermore, we would like to thank the Research Center in Digital Services (CISeD) and the Instituto Politécnico de Viseu for their support".

Maryam Abbasi thanks the national funding by FCT - Foundation for Science and Technology, Pi.I., through the institutional scientific employment program-contract (CEECINST/00077/2021).

References

1. Alvear, O., Zamora, W., Calafate, C.T., Cano, J.-C., Manzoni, P.: EcoSensor: monitoring environmental pollution using mobile sensors. In: 2016 IEEE 17th International Symposium on A World of Wireless, Mobile and Multimedia Networks (WoWMoM), pp. 1–6. IEEE (2016)
2. Chitra, P., Abirami, S.: Smart pollution alert system using machine learning. In: Research Anthology on Machine Learning Techniques, Methods, and Applications, pp. 1072–1085. IGI Global (2022)
3. Chowdhury, S., Islam, M.S., Raihan, M.K., Arefin, M.S.: Design and implementation of an IoT based air pollution detection and monitoring system. In: 2019 5th International Conference on Advances in Electrical Engineering (ICAEE), pp. 296–300. IEEE (2019)
4. Dhingra, S., Madda, R.B., Gandomi, A.H., Patan, R., Daneshmand, M.: Internet of things mobile-air pollution monitoring system (IoT-Mobair). IEEE Internet Things J. **6**(3), 5577–5584 (2019)
5. Li, W.-J., Yen, C., Lin, Y.-S., Tung, S.-C., Huang, S.: JustIoT internet of things based on the firebase real-time database. In: 2018 IEEE International Conference on Smart Manufacturing, Industrial & Logistics Engineering (SMILE), pp. 43–47. IEEE (2018)
6. Maag, B., Zhou, Z., Thiele, L.: A survey on sensor calibration in air pollution monitoring deployments. IEEE Internet Things J. **5**(6), 4857–4870 (2018)
7. Montanaro, T., Sergi, I., Basile, M., Mainetti, L., Patrono, L.: An IoT-aware solution to support governments in air pollution monitoring based on the combination of real-time data and citizen feedback. Sensors **22**(3) (2022)
8. Okokpujie, K., Noma-Osaghae, E., Modupe, O., John, S., Oluwatosin, O.: A smart air pollution monitoring system. Int. J. Civ. Eng. Technol. (IJCIET) **9**(9), 799–809 (2018)
9. Ramos-Romero, A., Garcia-Yataco, B., Andrade-Arenas, L.: Mobile application design with IoT for environmental pollution awareness. Int. J. Adv. Comput. Sci. Appl. **12**(1) (2021)
10. Roser, M.: Data review: how many people die from air pollution? Our World in Data (2021). https://ourworldindata.org/data-review-air-pollution-deaths

On the Use of Message Brokers for Real-Time Monitoring Systems

Manuel Lopes[1], Luciano Correia[1], João Henriques[1,2(✉)], and Filipe Caldeira[1,2]

[1] Polytechnic of Viseu, Viseu, Portugal
{estgv17025,pv22382}@alunos.estgv.ipv.pt,
{joaohenriques,caldeira}@estgv.ipv.pt
[2] CISeD - Research Centre in Digital Services, Viseu, Portugal

Abstract. Safety and efficiency of construction sites are increasingly reliant on using Real-Time Monitoring Systems (RTMSS) to collect and process data from sensors and devices. Selecting the Message Broker (MB) responsible for efficiently routing this data is crucial for the RTMSS's performance and reliability.

This work evaluates different MBs technologies that can be used in the context of an RTMSS applied to the domain of civil construction. Such evaluation was focused on proposed key performance indicators (KPIs) relevant to real-time data processing, including latency, throughput, and concurrency, under various test scenarios. The results denote each MB's strengths and weaknesses, aiding in selecting the most suitable option for the RTMSS.

Keywords: Message broker · Performance evaluation · Real-time monitoring system · Construction safety · Latency · Throughput · Concurrency

1 Introduction

Safety risks in construction workspaces are significant due to intrinsic dynamics and evolving conditions, eventually resulting in hazards ranging from the operation of heavy machinery to working at heights. Lieck et al. [1] highlighted an evolving landscape of workplace safety, emphasizing the heightened risks in construction sectors. Acknowledging its status as one of the more accident-prone industries, mainly due to the transient nature of construction sites, the range of trades and tasks involved, and weather reliance, highlights the necessity of robust safety measures. Integrating advanced real-time monitoring systems is critical in mitigating these risks and reducing the frequency of accidents [2]. These systems monitor worker safety, environmental conditions, and equipment status, enabling informed decision-making and proactive risk mitigation. MB are crucial in RTMSS as they efficiently route and distribute collected data to other components, such as supporting analysis and visualization.

Selecting the most suitable MB ensures the RTMSS's performance and reliability. MBs offer different capabilities and performance under workloads. This paper comprehensively evaluates three widely used MB - Kafka [3,4], RabbitMQ [5,6], and ActiveMQ [7,8] - in the context of a construction site RTMSS. KPIs were crucial for evaluating RTMSS under various test scenarios, including latency, throughput, and concurrency. The findings provide valuable insights into the strengths and weaknesses of each broker, aiding in selecting the most suitable option for specific RTMSS requirements.

This research includes the following contributions aiming to improve RTMSS:

- Comprehensive performance evaluation of different MB in a controlled environment replicating real-world construction site scenarios.
- Analysis of the strengths and weaknesses of each broker regarding latency, throughput, and concurrency, crucial factors for real-time data processing. Such an approach drives the selection of the most suitable message broker based on the specific requirements of the RTMSS.
- Evaluation of each message broker performance, this paper aims to provide valuable guidance on selecting the optimal solution for efficient and reliable real-time data processing in construction site RTMSS deployments.

The construction industry faces significant challenges in ensuring worker safety and maintaining efficient operations [9]. RTMSS have emerged as valuable tools to address these challenges by providing real-time insights into various aspects of the construction site, such as worker location, environmental conditions, and equipment status. These systems heavily rely on MB to efficiently route and distribute data collected from various sensors and devices to designated applications for analysis and visualization. Thus, selecting the most suitable message broker is critical for the RTMSS's performance and reliability. Different MB offer varying capabilities and exhibit distinct performance characteristics under various workloads.

Beyond this section, this paper is organized as follows. Section 2 presents the related work. Section 3 presents the RTMSS architecture. Section 4 describes the adopted methodology. Section 5 describes the validation and evaluation work. Section 6 offers a discussion of the achieved results. Section 7 concludes the paper.

2 Related Work

This section examines existing RTMSSs to understand current approaches and identify potential gaps. Research in Real-time Location Systems (RTLS) has explored various technologies [10–12] for tracking worker location and alerting them to nearby hazards. Additionally, fall detection systems employing Machine Learning (ML), Deep Learning (DL), and sensor data have been developed to detect falls and trigger prompt interventions [13–15].

Several existing RTMSS solutions offer valuable insights. The ViPER+ system [16] utilizes ultra-wideband radio for proximity alerts between workers and machinery in construction environments. BIM-based mobile applications [17] integrate mobile hazard detection with web interface monitoring, particularly suited for indoor construction safety tracking. Leica Geosystems [18] focuses on deformation and site monitoring to enhance safety in construction through real-time data analysis. VANTIQ [19] offers a comprehensive safety system utilizing real-time data analytics, event-driven architecture, and a combination of IoT devices for hazard detection and response. CoreTech [20] is a developing solution using wearable neurotechnology and AI to predict potential safety incidents from existing wearable data.

While these solutions offer valuable insights, they often target specific industries or lack a dedicated messaging hub for optimized data flow. This research project addresses this gap by proposing an RTMSS that is versatile, cost-effective, scalable, and reliable. The proposed system offers applicability across various sectors and prioritizes stringent safety standards and robust performance, all at an affordable cost. Combining these features, this RTMSS strives to contribute significantly to real-time safety monitoring.

[21] proposed an MB-based architecture for an RTMSS in smart factories. Their work highlights the potential benefits of MB in enabling modularity, scalability, and real-time data communication within RTMSS.

While existing research provides valuable insights, this work addresses some critical gaps, comparing the performance of three widely used MB (Kafka, RabbitMQ, and ActiveMQ) across various proposed KPIs under controlled and realistic scenarios relevant to construction site RTMSS. The evaluation and discussion are tailored to the unique requirements of construction site RTMSS, considering factors like sensor data volume, real-time processing needs, and potential scalability challenges. By addressing these gaps, our research offers a more comprehensive and practical understanding of MB performance in the context of construction site RTMSS, aiding informed decision-making for selecting the most suitable option for specific project requirements.

3 RTMSS Architecture

An RTMSS relies on a specific architecture to efficiently collect, process, and visualize data from various sources within the construction site. Figure 1 presents the architecture with the RTMSS components and flows between them. The current RTMSS was initially proposed in [22] having undergone improvements, as this section notes. It illustrates the integrated workflow since data was captured to the final notification during fall detection, encompassing the system's three primary layers: the producer, middleware, and consumer layers to be discussed.

Producer Layer: Users are distributed over different locations, denoted as Site A and Site B, and are equipped with IoT sensors, including accelerometers and gyroscopes for motion detection and additional sensors for localization. These sensors acquire real-time data to be delivered through an Internet gateway and ingested into an RTMSS.

Middleware Layer: The MB takes a central location within the architecture, managing each user's communication channels - topics or queues -. It acts as the pivotal relay point, ensuring that the delivered data from producers are efficiently queued for consumption.

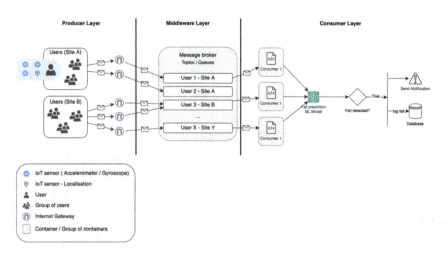

Fig. 1. RTMSS Architecture.

Consumer Layer: Multiple consumer processes, corresponding to the number of users, are continuously connected to the MB to retrieve messages. Each consumer runs the fall prediction ML model by processing the incoming data to classify a fall accident.

Notification and Logging: After detecting a fall, an alert is triggered by a notification to the intended recipients. Concurrently, the fall classified event is logged into a database to keep a record of the incident for further analysis. Such an approach will contribute to the system's learning and refinement.

4 Validation and Evaluation

This section is focused on validating and evaluating the RTMSS MB, assessing its performance across various dimensions critical for real-time safety monitoring. A series of tests employing Docker containers, Apache JMeter, and a Python-based testing framework were conducted to evaluate the system's scalability, latency, throughput, and concurrency capabilities.

Deploying MB in Docker containers provided a controlled and isolated testing environment. This way, consistency is ensured and reduces discrepancies between deployment stages [23,24]. Each MB container was limited to 2.0 CPUs and 2048 MB of memory to ensure fair comparison. Regarding the hardware used in the tests, a MacBook Pro 2021 14-inch model with 16 GB RAM and 512 GB storage was used, and it was equipped with an Apple M1 processor.

4.1 Methodology

To ensure a controlled evaluation, the work employed a Docker containerized environment. This approach minimized external factors impacting performance and contributed to a consistent assessment across MB.

The evaluated MB were Apache Kafka (version 2.8.1), RabbitMQ (version 3.9.19), and ActiveMQ (version 5.17.3). The evaluation focused on five Key Performance Indicators (KPIs), crucial for real-time data processing in the context of RTMSS:

Concurrency: The system's capability to accommodate an increasing number of users (producers and consumers) and evaluate the impact on the system's performance. This impact was assessed under different loads in scenarios that included a single user to hundreds of users. This KPI is crucial in evaluating the scalability of the system. A system that can effectively handle a growing number of users indicates its success in real-world deployments. It highlights the system's ability to adapt to increased demands, ensuring consistent performance and reliability as the user base expands.

Message Latency: This KPI measures the average time taken for a message to be delivered from the producer to the MB, recorded in milliseconds. In real-time systems like RTMSS, lower latency is critical as it directly impacts the effectiveness of the system's response mechanism and the time to action when an accident occurs.

Latency Consistency: Beyond the average latency, the consistency of this latency is vital. Latency consistency, given by the average latency deviation, evaluates the variability in message processing times, offering insights into the predictability and reliability of the system.

Throughput: Representing the number of messages delivered per second, throughput is a crucial indicator to assess the system's ability to handle high volumes of data. Higher throughput rates denote the system's capability to efficiently manage traffic without impacting performance.

Availability: This KPI was evaluated by monitoring the system's status, particularly under high-load conditions. Instances where the MB stopped or failed under heavy loads were noted as critical to the systems' resilience. In the context of RTMSS, keeping the system's availability ensures continuous operation and prompt responses. The ability of the system to remain operational, especially under challenging conditions like high message volumes and numerous producers, is a significant indicator of its robustness and suitability for critical applications.

Table 1 summarizes the KPIs, their descriptions, and the evaluation methods employed:

The evaluation employed different test configurations to simulate various real-world construction site scenarios. Message sizes ranged between 44 Bytes (low message payload), 440 bytes (typical sensor data), and 4.400 bytes (high-volume data). The number of concurrent producers in the different tests was predefined as 1, 10, 25, 50, 100, 150, and 200. A Python-based test platform was developed to evaluate the MB in an environment more similar to the current producer and consumer components in the proposed RTMSS. On the other hand, JMeter produced loads, simulating message producers delivering data to the MB under various test configurations. Performance data was collected and analyzed using JMeter's built-in monitoring capabilities and custom scripting.

Table 1. Proposed KPIs

KPI	Description	Evaluation
Concurrency	Assessment of the system's capability to handle an increasing number of active users (producers and consumers) and their impact on performance	Systematically increase the number of users and observe performance impact from a single user to hundreds
Message Latency	Measures the average time taken for a message to be delivered from the producer to the MB	Recording the average time (in milliseconds) taken for message delivery in various tests
Latency Consistency	Evaluates the variability and predictability of message latency beyond the average latency	Measuring the average latency deviation to assess consistency in message delivering times
Throughput	Indicates the system's capacity to manage high data volumes, represented by the time needed to deliver different message loads	Monitoring the elapsed time delivering different message loads
Availability	Measures the system's operational status and resilience, especially under high-load conditions	Monitoring operational status during tests, noting any failures or stops in MB under heavy loads

This adopted methodology contributes to a systematic and controlled evaluation, enabling a fair comparison of the performance characteristics of each MB under conditions relevant to construction site RTMSS applications.

4.2 Evaluation Methodology

The adopted evaluation methodology in this work combined manual tests and Apache JMeter simulations to assess the RTMSS MB performance. Manual tests were made to test mainly the performance of the complete flow, from message production to consumption, reflecting the combined efficiency of production and consumption processes. A set of synthetic messages was simulated and used in this test for each MB. On the other hand, JMeter tests focus on evaluating the producer latency, deviation, and throughput in different concurrency scenarios. Such scenarios employed different numbers of messages and message size combinations.

A Python-based framework is used to interact with the different MB. The framework utilizes the following libraries: confluent_kafka (version 2.2.0) for Apache Kafka; pika (version 1.3.2) for RabbitMQ; paho.mqtt (version 1.6.1) for ActiveMQ. This framework focuses on evaluating three KPIs: latency, throughput, and concurrency. To assess these KPIs a set of different tests was run, which are described in the Table 2:

Table 2. Python-based tests

Test Metric	Description
Average Latency	Evaluated by sending 1,000 messages and measuring the average time per message
Throughput	Measured by sending varying message sequences (1,000, 5,000, 10,000, 50,000, 100,000) and recording the elapsed time
Concurrency	Assessed by sending message sequences with increasing user counts (1, 25, 50, 100, 150, 200) and measuring the elapsed time per user

Apache JMeter [25] was used to assess MB performance under various conditions. So, the first factor to consider is the message size varying between tests. The baseline is a dummy data capture from the sensors utilized to detect a fall: "41,-251,-110, -68,-1449,-270, 71,-953,-270;". The size of this message is about 44 Bytes. Since the system will potentially use more sensors and, consequently, will have more data to transmit, it is important to evaluate the system under different message sizes scenarios. In that purpose, more tests were made with baseline messages with a factor of 10 and 100, resulting in message sizes of approximately 440 Bytes and 4.400 Bytes. The test's message size variation was baseline, baseline x10, and baseline x100. Secondly, the number of concurrent users varied from 1 to 200 simultaneous users. The third factor was the number of messages produced, which ranged between 10.000, 100.000, and 1.000.000. The metrics evaluated in these tests were throughput, average latency, consistency analyzed by latency deviation, and availability analyzing error rates under each scenario.

In addition to manual tests, Apache JMeter was employed to simulate diverse workload scenarios and comprehensively evaluate the MB's performance under various conditions.

Message Size is a factor that plays a crucial role in message transmission efficiency. The evaluation began with a baseline message size of approximately 44 Bytes, representing a dummy data capture from fall detection sensors. The message content: "41,-251,-110, -68,-1449,-270, 71,-953,-270;" reflects the sensor data format. To simulate potential system expansion with additional sensors and increased data transmission, tests were conducted with baseline message sizes multiplied by factors of 10 and 100, resulting in approximate message sizes of 440 Bytes and 4.400 Bytes, respectively. Therefore, the message size variations in these simulations included baseline, baseline x10, and baseline x100. To assess the system's scalability and ability to handle simultaneous user interactions, tests varied the **number of concurrent users** from 1 to 200, simulating a realistic range of potential user activity. Different numbers of Messages were produced in each test, to evaluate the system's capacity to process large data volumes. The **number of messages produced** per test was varied between 10.000, 100.000, and 1.000.000.

The JMeter tests focused on analyzing KPIs under each scenario: Throughput, represented by the number of messages delivered per second, is a crucial indicator of the system's ability to handle high data volumes. Higher throughput rates denote the system's capability to manage dense traffic efficiently without performance degradation. Average Latency corresponds to the average time (in milliseconds) taken to deliver a message from the producer to the MB. In real-time systems like RTMSS, lower latency is critical as it directly impacts the effectiveness of the system's response mechanism and the time to action when an accident occurs. Latency Consistency: This aspect was analyzed through latency deviation. Lower latency deviation suggests a more consistent and predictable message delivery time. Availability: This metric was assessed by analyzing error rates under each scenario. Lower error rates indicate higher system availability and reliability in message transmission. By systematically varying these workload factors and analyzing the corresponding impact on performance metrics, this JMeter evaluation provides valuable insights into the MB suitability for the RTMSS.

4.3 Results

This section presents the evaluation's findings in a clear and organized manner, using tables, graphs, and other visual aids to communicate the results effectively.

Table 3 presents the results of manual tests conducted on three MB: Kafka, RabbitMQ, and ActiveMQ. The tests evaluated three KPIs: latency, throughput, and concurrency. The test load (number of messages produced) and the number of concurrent producers were varied to assess the brokers' performance under different conditions.

Regarding the latency test, all three brokers exhibited low average latency, measured in milliseconds (ms), with Kafka demonstrating the lowest average latency of 0,00156 ms. This observation suggests that all brokers can handle individual message processing with minimal delays.

Throughput tests were measured by elapsed time to produce the number of messages pre-defined for each test scenario. Considering this, the highest throughput should be approximated to 0 once it demonstrates that all the messages were delivered in less time. Kafka consistently showed the highest throughput across all test loads, followed by RabbitMQ and ActiveMQ. As the test load increased, the throughput for all brokers increased, indicating their ability to handle larger message volumes. However, the rate of increase varied: Kafka's throughput exhibited a linear relationship with the test load, suggesting scalability under increasing message traffic. RabbitMQ's throughput growth accelerated at lower test loads but plateaued at higher loads, indicating potential limitations in handling extremely high message volumes. ActiveMQ's throughput growth was slower than the other two brokers, but it displayed a more consistent upward trend across all test loads.

Concurrency tests measured the elapsed time to deliver a sequence of messages per user, with multiple users running in parallel. Results generally grew for all brokers as the number of concurrent producers increased, as expected,

Table 3. Manual Test Results (Python-based framework)

KPI	Test Load	Kafka	RabbitMQ	ActiveMQ
Latency (ms)	1.000	0,00156	0,07794	0,021758
Throughput (s)	1.000	0,000491381	0,056292868	0,008664227
	5.000	0,003901100	0,181825256	0,041674757
	10.000	0,006659508	0,369703150	0,103834152
	50.000	0,027277136	1,832922554	0,493202448
	100.000	0,051109505	3,619038439	1,086379147
Concurrency (s)	1	40,86	42,37	44,06
	25	36,57	68,53	40,81
	50	37,73	126,37	63,28
	100	45,34	240,14	105,38
	150	56,75	361,42	143,93
	200	61,31	Failed	Failed

due to increased competition for system resources. Kafka maintained the lowest elapsed time across all scenarios, followed by ActiveMQ and RabbitMQ. Both RabbitMQ and ActiveMQ failed at 200 concurrent producers, indicating limitations in handling extreme concurrency loads.

Figure 2 shows the throughput results of JMeter tests conducted to measure the performance of the three MB. It was estimated by the number of produced messages per second for different message payloads and sizes and the number of concurrent users.

The conducted tests are divided into three main sections considering the message payload: "Low", which corresponds to sending the basic message presented before with the size of 44 Bytes; "Moderate", which alters the size of the sent messages in a factor of ten, establishing each message size in about 440 Bytes; and "High", which multiplies the initial message size by 100, resulting in a message size of 4.440 Bytes. The second column presents the different numbers of concurrent producers on each test. Finally, in the third column, the number of messages produced on each test varies from 10.000 to 1.000.000 delivered messages.

The JMeter tests revealed the impact of message payload size and message rate on the throughput of the three MB: Kafka, RabbitMQ, and ActiveMQ. As expected, increasing message size generally led to decreased throughput for all brokers due to the higher processing overhead associated with more significant messages. This observation aligns with previous research findings.

Furthermore, throughput generally increased under higher message rates, pinpointing the systems' ability to handle increased loads. However, this trend plateaued or decreased at higher message rates, particularly for larger payloads. This suggests potential resource limitations or queuing behavior within

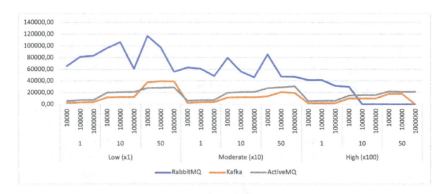

Fig. 2. JMeter Throughput results.

the tested systems, highlighting the importance of considering scaling strategies for high-traffic scenarios.

While RabbitMQ and ActiveMQ generally exhibited higher throughput than Kafka in most test scenarios, the specific performance characteristics of each broker varied depending on message size and rate.

- **Kafka:** While offering lower overall throughput compared to the other two brokers, it demonstrated more consistent performance across different message sizes and rates. This suggests that Kafka may be suitable for applications requiring predictable performance even under diverse workloads, even if some raw throughput is sacrificed. This MB only presented an error rate of 1,22% in the test with a high message payload of 50 concurrent users and 1.000.000 produced messages.
- **RabbitMQ:** Highlighted significantly higher throughput under low and moderate message payloads. However, its performance dropped significantly under high message payload scenarios, highlighting eventual limitations in handling heavy messages. In an increased number of concurrent users (100 and 200) with moderate message payload, RabbitMQ demonstrated an error rate of 95,63%.
- **ActiveMQ:** Demonstrated higher throughput than Kafka in most scenarios, but lower than RabbitMQ. In the scenario with 200 producers and moderate message payload, 7% of the produced messages were not delivered due to errors during production. Like Kafka, ActiveMQ showed relatively consistent performance across different message sizes and rates. This observation suggests that ActiveMQ may be preferred for applications prioritizing predictable performance even under diverse workloads.

It should be noticed that the achieved results are intertwined with the setup configurations. Factors such as hardware specifications, software versions, and specific use case details can significantly impact MB' performance. Therefore, further testing under different conditions and message patterns is recommended to better understand the brokers' behavior in real-world scenarios and facilitate informed selection based on specific application requirements.

Figure 3 presents the average latency results of the JMeter-based tests.

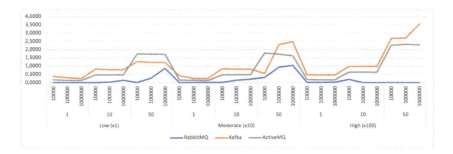

Fig. 3. JMeter Latency results.

Regarding average latency, RabbitMQ generally depicted the lowest latency across most scenarios, particularly for smaller messages. However, latency increased significantly under high concurrency and large message sizes, resulting in error rates from 86,31% Kafka demonstrated moderate latency, with an increase observed under high concurrency, particularly in the last scenario. On the other hand, ActiveMQ showed comparable latency to RabbitMQ for smaller messages but performed better under high concurrency and large message sizes.

Consistency is also a valuable metric. In real-time systems, consistency refers to the property of ensuring that system operations are performed in a coherent, reliable manner, maintaining the correctness and predictability of data and processes (Wang et al., 2012). In Fig. 4, consistency is given by the latency deviation of the executed tests.

Notably, ActiveMQ was the MB that presented less deviation. Both RabbitMQ and Kafka had similar deviations, but RabbitMQ showed more limitations in the scenarios with more messages and concurrent producers. Kafka presented the highest deviation in the last scenario, behind ActiveMQ, and forward RabbitMQ, which did not handle the heaviest tests, so it was impossible to calculate the latency deviation.

5 Discussion

The conducted manual and JMeter tests provided valuable insights to assess the MB performance characteristics, Kafka, RabbitMQ, and ActiveMQ under varying message sizes, rates, and concurrency levels.

All three MB achieved low average latency, indicating their suitability for fast message processing applications. However, RabbitMQ generally denoted the lowest latency, particularly for smaller messages. This advantage comes with caution, as its latency increased significantly under high concurrency and large

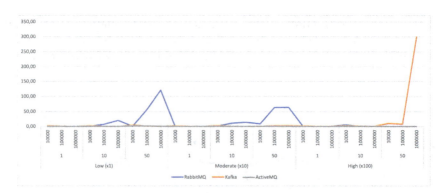

Fig. 4. JMeter Deviation results.

message sizes, resulting in high error rates. Kafka denoted a moderate latency, with an increase observed under high concurrency, suggesting potential resource limitations under heavy workloads. ActiveMQ offered similar latency to RabbitMQ for smaller messages but performed better under high concurrency and large message sizes. This highlights ActiveMQ's ability to keep lower latency even under increased pressure.

Regarding throughput, RabbitMQ consistently demonstrated the highest throughput across diverse test scenarios. However, its performance dropped significantly with increasing message size, suggesting limitations in handling large messages. This makes RabbitMQ a suitable choice for applications requiring high message volume processing, with a smaller message size. Kafka and ActiveMQ displayed similar moderate throughput under all the message payload conditions, exhibiting more consistent performance. MB are a potential choice for applications requiring stable throughput under varying workloads.

Kafka kept the lowest elapsed time (faster processing) across all concurrency levels, suggesting its efficiency in handling concurrent message producers. Both RabbitMQ and ActiveMQ failed at 200 concurrent producers, denoting their constraints on handling extreme concurrency loads. However, ActiveMQ demonstrated a higher consistent performance than RabbitMQ under high concurrency, with lower error rates, suggesting its potential for handling moderate to high concurrency scenarios.

ActiveMQ exhibited the lowest latency deviation and higher consistent performance across different test conditions. This makes it a potential choice for applications requiring predictable behavior even under fluctuating workloads. RabbitMQ and Kafka showed similar deviation, but RabbitMQ experienced limitations in high-load scenarios, leading to incomplete data for deviation calculations.

6 Conclusions

This work evaluated the performance of Kafka, RabbitMQ, and ActiveMQ as message brokers for a RTMSS. It were conducted manual python-based and JMeter tests assessing factors like message size, concurrency, and message count. The results showed trade-offs between latency, throughput, and consistency for each MB, Kafka, RabbitMQ, and ActiveMQ.

According to the evaluation, Apache Kafka denoted a higher throughput and performance under high concurrency scenarios. These results reveal Apache Kafka as a good match under real-time data processing use cases demanding low latency. RabbitMQ, while exhibiting limitations under high-load scenarios, also proved to be an efficient and much faster option when managing smaller message payload lengths and revealed a consistent performance. That means a good choice under extreme scenarios that involve volumes of data. ActiveMQ showcased an impressive throughput and moderate latency, keeping its performance relatively stable under moderate workloads but indicating scalability challenges at higher concurrency levels.

Using containerized components helped keep a consistent, controlled evaluation environment. That revealed a suitable approach for deploying and testing RTMSS.

This work also proposed a set of KPIs to be considered when selecting the MB to address the RTMSS requirements, such as data volume, concurrency, message payload, and workload. From the results, it was possible to have a clear landscape of the MB strengths and weaknesses to support well-informed decisions regarding RTMSS design and deployment.

Theoretically, this work aims to contribute to the existing state-of-the-art of RTMSS by evaluating the suitability of different MB technologies deployed under a container virtualized environment. It contributes to understanding how these systems can be optimized for reliability and performance under different workloads and concurrent users, particularly in safety-critical applications like construction site monitoring.

The development of the RTMSS offers several promising avenues for future exploration. A crucial next step is to execute a PoC where all components of the RTMSS, especially the producer layer, are fully operational with real-world data. This PoC would involve deploying the system in a work environment, such as a construction site, to capture and process real-time data from IoT devices and sensors. The focus would be validating the system's efficacy in a real scenario, ensuring it can reliably and accurately detect and respond to safety incidents. In the future, dynamic load-balancing strategies will also be developed and implemented to enhance the scalability and efficiency of the RTMSS. This would involve using cloud computing resources to manage varying loads effectively. The applicability of edge computing technologies will also be tested to process data closer to the source, thereby reducing latency and improving the real-time response capabilities of the system. The next step will be the continuing research and integration with more sophisticated machine learning algorithms for enhanced data analysis and prediction capabilities. This can include deep learning techniques for more accurate fall detection and anomaly identification. Exploring additional MB, particularly those emerging in the market would provide a comprehensive understanding of the evolving landscape of MB and their applicability to RTMSS.

Applying the RTMSS to different high-risk sectors such as mining or healthcare is also marked as future work. This would validate the system's versatility and effectiveness in different environments with varied safety challenges. The aim of these future endeavors is not only to refine and enhance the RTMSS but also to contribute significantly to the field of real-time safety monitoring. The goal is to extend the system's applicability and effectiveness across various industries, thereby improving safety standards and response mechanisms in high-risk environments.

Acknowledgements. This work is funded by National Funds through the FCT - Foundation for Science and Technology, I.P., within the scope of the project Ref. UIDB/05583/2020. Furthermore, we would like to thank the Research centre in Digital Services (CISeD) and the Instituto Politécnico de Viseu for their support of the internal project Ref.PIDI/CISeD/2023/010.

References

1. Lieck, L., et al.: Occupational safety and health in Europe: state and trends 2023. European Agency for Safety and Health at Work (2023). https://doi.org/10.2802/574458
2. Chan, A.P.C., Guan, J., Choi, T.N.Y., Yang, Y., Wu, G., Lam, E.: Improving safety performance of construction workers through learning from incidents. Int. J. Environ. Res. Public Health **20**(5), 4570 (2023). https://doi.org/10.3390/ijerph20054570
3. Kreps, J., Narkhede, N., Rao, J.: Kafka: a distributed messaging system for log processing. In: Proceedings of the NetDB, pp. 1–7 (2011)
4. Apache: Apache Kafka. https://kafka.apache.org/. Accessed 12 Mar 2024
5. Videla, A., Williams, J.J.W.: RabbitMQ in Action: Distributed Messaging for Everyone. Simon and Schuster (2012)
6. RabbitMQ: RabbitMQ. https://www.rabbitmq.com. Accessed 12 Mar 2024
7. Snyder, B., Bosnanac, D., Davies, R.: ActiveMQ in Action. Manning Greenwich Conn. (2011)
8. ActiveMQ: ActiveMQ. https://activemq.apache.org. Accessed 12 Mar 2024
9. European Commission: Eurostat - Accidents at work statistics. https://ec.europa.eu/eurostat/statistics-explained/index.php?title=Accidents_at_work_statistics#Analysis_by_activity. Accessed 16 Mar 2024
10. Lim, D.Z., et al.: Development of a machine learning-based real-time location system to streamline acute endovascular intervention in acute stroke: a proof-of-concept study. J. Neurointervet. Surg. **14**(8), 799–803 (2023). https://doi.org/10.1136/neurintsurg-2021-017858
11. Capraro, F., Segura, M., Sisterna, C.: Human real time localization system in underground mines using UWB. IEEE Lat. Am. Trans. **18**(02), 392–399 (2020). https://doi.org/10.1109/TLA.2020.9085295

12. Cheng, C.H., Kuo, Y.H., Lam, H., Petering, M.: Real-time location-positioning technologies for managing cart operations at a distribution facility. Appl. Sci. **11**(9), 4049 (2021). https://doi.org/10.3390/app11094049
13. Sarkar, S., Chang, C.K., Wang, X., Ellul, J., Azzopardi, G.: Elderly fall detection systems: a literature survey. Front. Robot. AI **7**, 71 (2020). https://doi.org/10.3389/frobt.2020.00071
14. Xu, T., Zhao, Y., Zhu, J.: New advances and challenges of fall detection systems: a survey. Appl. Sci. **8**(3), 418 (2018). https://doi.org/10.3390/app8030418
15. Mauldin, T.R., Canby, M.E., Metsis, V., Ngu, A.H.H., Rivera, C.C.: SmartFall: a smartwatch-based fall detection system using deep learning. Sensors **18**(10), 3363 (2018). https://doi.org/10.3390/s18103363
16. Ansaripour, A., Heydariaan, M., Kim, K., Gnawali, O., Oyediran, H.: ViPER+: vehicle pose estimation using ultra-wideband radios for automated construction safety monitoring. Appl. Sci. **13**(3), 1581 (2023). https://doi.org/10.3390/app13031581
17. Hossain, M., et al.: BIM-based smart safety monitoring system using a mobile app: a case study in an ongoing construction site. Constr. Innov. (2023). https://doi.org/10.1108/CI-11-2022-0296
18. Leica: Leica Geosystems. https://leica-geosystems.com/products. Accessed 12 Mar 2024
19. VANTIQ: VANTIQ - Real-Time Safety Monitoring & Response. https://vantiq.com/connect/solution/real-time-safety-monitoring-response/. Accessed 12 Mar 2024
20. CoreTech: CoreTech, by FCLABS. https://www.brainfitfirst.com. Accessed 12 Mar 2024
21. Nguyen, C.N., Lee, J., Hwang, S., Kim, J.S.: On the role of message broker middleware for many-task computing on a big-data platform. Cluster Comput. **22**, 2527–2540 (2019). https://doi.org/10.1007/s10586-018-2634-9
22. Ferreira, C., et al.: An intelligent and scalable IoT monitoring framework for safety in civil construction workspaces. In: de la Iglesia, D.H., de Paz Santana, J.F., López Rivero, A.J. (eds.) DiTTEt 2022. AISC, vol. 1430, pp. 69–78. Springer, Cham (2023). https://doi.org/10.1007/978-3-031-14859-0_6
23. Bernstein, D.: Containers and cloud: from LXC to docker to kubernetes. IEEE Cloud Comput. **1**(3), 81–84 (2014). https://doi.org/10.1109/MCC.2014.51
24. Rad, B.B., Bhatti, H., Ahmadi, M.: An introduction to docker and analysis of its performance. IJCSNS Int. J. Comput. Sci. Netw. Secur. **173**, 8 (2017)
25. Apache: Apache JMeter. https://jmeter.apache.org. Accessed 14 Mar 2024

Incorporating Electric Vehicles in Strategic Management or Value Creation Initiatives with a Focus on Sustainability?

Sónia Gouveia[1,2(✉)], Daniel H. de la Iglesia[3], José Luís Abrantes[1,2], Alfonso J. López Rivero[4], Elisabete Silva[1,2], Eduardo Gouveia[1,2], and Vasco Santos[1]

[1] Superior School of Technology and Management, Polytechnic Institute of Viseu, Viseu, Portugal
`sgouveia@estgv.ipv.pt`
[2] CISeD – Research Centre in Digital Services, Instituto Politécnico de Viseu, Viseu, Portugal
[3] Faculty of Science, University of Salamanca, Salamanca, Spain
[4] Facultad de Informática, Universidad Pontificia de Salamanca, Salamanca, Spain

Abstract. Electric vehicles represent an emerging topic in organizations and society in general. At a time when extreme climate changes are occurring and being discussed, it is imperative to look for sustainable technologies from an economic and environmental point of view. The proposed research aims to elucidate evolving trends, emerging paradigms and critical in-sights related to electric vehicles and their relationship with value creation, strategic management, and sustainability by examining academic literature. This study utilizes a bibliometric analysis method to explore the dynamic interaction between these essential elements of literature. The source is the Scopus database. The analysis was conducted using VOS viewer software 1.6.20, focusing on the bibliographic coupling of documents, which was presented as a network visualization map. It emphasizes search terms demonstrating the emerging need for innovation strategies in automotive industry organizations, consumers, and public entities to create value within a global sustainability logic. This study highlights the relationship between the electric vehicle industry and the value creation strategy and sustainability in a digital world.

Keywords: Electric Vehicle · Strategic Management · Value Creation · Sustainability · Digital

1 Introduction

The dynamics imprinted on the electric mobility market, coupled with environmental and technological issues, lead to a paradigm involving various stakeholders in the energy chain, including producers, consumers, marketers, and network operators. In this context, the European electric vehicle (EV) market has experienced remarkable growth in recent years. This occurs due to government support, private investments, and political initiatives [1, 2]. On the other hand, some challenges compromise the widespread adoption of

EVs, namely consumer behavior, the availability of electrical infrastructure, and accessibility [3, 4]. Several variables are associated with technical and performance parameters in this process, including battery capacity, range, and charging speed [5]. Moreover, the impact of EVs on energy systems and the environment is of utmost importance, emphasizing the need for efficient charging infrastructure and regulatory measures to optimize energy consumption.

Regarding strategic management, value creation, and sustainability, EVs can be essential in public and private organizations. Based on concerns such as global warming or urban pollution, EVs align with a shift towards a more sustainable economy [6].

By considering existing and established models such as Porter's Five Forces, organisations can leverage strategic management frameworks to assess the attractiveness of the EVs industry and enhance value creation through innovation. Additionally, EVs contribute to sustainability in organisations through strategic management and value creation via energy efficiency [7].

Strategic sustainability is a way to integrate environmental science into organisations to obtain a competitive advantage, resulting in value creation and sustainability, and EVs align with this strategic management approach in organisations [8]. Tiwari [9] proposes a mathematical model to study the proliferation of EVs and analyses value creation opportunities for organisations, specifically in strategic management and sustainability achieved through the adoption of EVs.

The organization's adoption of EVs can also influence business processes. Nanjundaswamy et al. [10] argue that in this context, adopting electric vehicles adds value by reducing operating costs and simultaneously promotes ecological practices, contributing to achieving sustainable development objectives. Authors analyse the impact of EVs on the overall cost and sustainable development.

Value creation is present in various areas, including public transportation. Several factors influence the adoption of electric vehicles, including social and demographic, political, economic, technological, and environmental factors, and examine the feasibility and sustainability of integrating EVs into public transportation through park-and-ride stations [11]. In particular, value creation in the automotive production industry is crucial as it can occur through sharing resources and market knowledge in the EVs industry to create more sustainable value chains [12].

Digital transformation is present in all areas, and EVs are no exception. The digital transformation in the automotive ecosystem is reshaping value creation in organisations, focusing on electric vehicles and emerging trends for young companies with intensive digital use [13].

EVs represent a technological change that impacts organisations' value creation, strategic management, and environmental sustainability. Organisations that recognise and adapt to these changes can position themselves more competitively in the market. The development and promotion of EVs require a multifaceted approach involving technological advancements, supportive policies, and investment in charging infrastructure to drive sustainable mobility.

This paper is organised as follows: Sect. 2 outlines the research methodology. Section 3 addresses the bibliometric analysis; Sect. 4 proposes a systematic literature review (SLR); Sect. 5 concludes the paper by presenting findings and future work.

2 Research Methodology

Systematic literature reviews (SLR) are essential in consolidating scientific knowledge, providing a consolidated and critical view of existing research in a given field. Furthermore, the ability to replicate plays a crucial role in validating the results, allowing other researchers to reproduce the process and obtain similar conclusions [14].

The present study combines qualitative and quantitative analysis through bibliometric analysis followed by a SLR. It was used the PRISMA protocol (Preferred Reporting Items for Systematic Reviews and Meta-Analyses) [15].

It developed a literature search in the Scopus database in April 2024 using the terms "electric vehicles" and "strategic management" or "value creation" in "Title, Abstract, and Keywords" and "sustainable" in "All fields". It resulted in the selection of 74 papers (identification step, Fig. 1). These terms are emerging and have been systematically referenced in the literature [16] and are present in various areas of the Scopus database [17].

Next (screening step, Fig. 1), it applied the following criteria: document type (paper), language (English), and research areas. Some research areas were excluded from the study, namely "Physics and Astronomy", "Psychology", "Multidisciplinary", "Biochemistry, Genetics, and Molecular Biology" and "Chemistry". These areas were not considered because it was understood that they are unrelated to the research that aims to

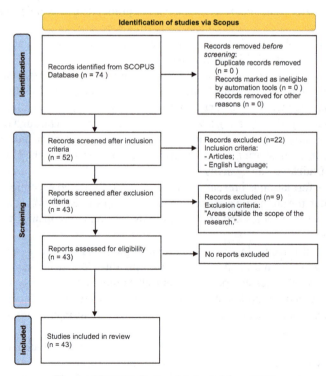

Fig. 1. PRISMA Protocol (adapted from [14])

analyse EVs economic, business, and environmental aspects and their integration into the context of value creation and strategic management. Therefore, they were considered outside the scope of the SLR. Combining all the criteria resulted in 43 papers (included step, Fig. 1).

The authors sought to answer the following research question (RQ):

What is the current state of the literature on incorporating electric vehicles in strategic management or value-creation initiatives with a focus on sustainability?

3 Bibliometric Analysis

Descriptive statistics from peer-reviewed research articles were used to address the RQ, starting with the distribution of papers per year. The distribution of papers published yearly in a given research field provides an overview of research trends and can offer insights into potential future trends. Globally, the distribution of papers per year is shown in Fig. 2 before applying the criteria. As can be seen, there was a significant increase starting in 2017, with over 16 papers in recent years. This evolution explains the current high interest in using the terms "electric vehicles" and "value creation" or "strategic management" and "sustainability". The low number of papers in 2024 is justified by the short time, only 4 months, for the accounting.

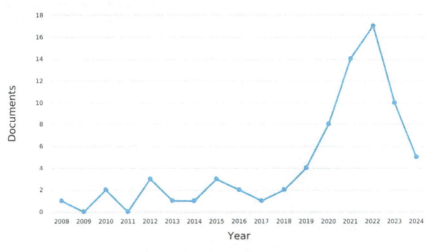

Fig. 2. Number of papers by year [17].

Considering the geographical analysis, 35 countries are the origin of the published documents. Figure 3 shows the top 10 papers by country or territory. It is observed that Germany has the highest number of publications (14), followed by China (7), India, Sweden, and the United Kingdom (6), Canada, France, Italy, and the Russian Federation (4), and Brazil (3).

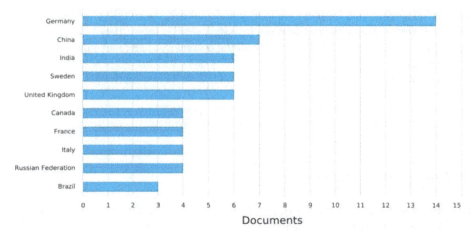

Fig. 3. Documents by country/territory [17].

The data shows (Fig. 4) that the most representative research areas are Engineering and Energy, totalling about 40%. The Environmental Sciences area, which reflects sustainability, presents a gap, representing only about 16% of the research areas. Considering the recent evolution in RQ (especially since 2019 - Fig. 2), it is evident that the areas of "Business, Management, and Accounting" or "Economics, Econometrics, and Finance" are deficient, totalling approximately 15.5% together. These areas have a solid connection to the search terms "value creation" and "strategic management" and allow the authors to identify them as areas deserving attention in the future when EVs are included in the organisations' vehicle fleets.

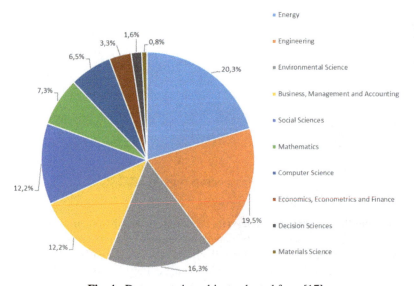

Fig. 4. Documents by subject, adapted from [17].

A content analysis was conducted to provide a more comprehensive understanding of the leading research areas resulting from applying criteria to the 43 papers related to the RQ. The authors identified a research focus for each cluster, defined by bibliographic coupling analysis using VOSviewer [18] to map research areas and related articles. It was defined as a minimum of one citation per document. Consequently, the coupling will include 21 articles, as shown in Fig. 5. After a brainstorming session and a content analysis process, a final configuration of the groupings for these research areas was defined, resulting in 4 clusters:

Cluster 1 (red): Business model strategies;
Cluster 2 (light green): Innovation and value creation through the adoption of EVs;
Cluster 3 (blue): EV and technology;
Cluster 4 (green): Sustainability and Value Chain in the Battery Industry.

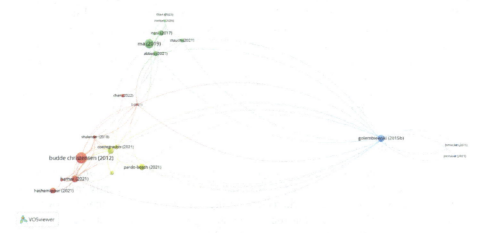

Fig. 5. Bibliographic coupling [18].

4 Systematic Literature Review

In this section, the SLR was developed by considering the research terms and finding the most relevant contributions. The selected search terms allowed for the identification of 74 articles in Scopus database, and after using VOSviewer, four clusters were obtained, as mentioned. The subsequent analysis aims to indicate the contribution of each search term (value creation, strategic management, sustainable, electric vehicle) to the production of these clusters, to analyze and understand the interconnections between different aspects of EVs and their impact on business and society.

4.1 Value Creation

Afentoulis et al. [19] anticipate issues with the reliability of electrical grids due to unregulated mass EV charging, suggesting that EV aggregator companies develop well-defined business models to harness the potential of smart charging. Coenegrachts et al.

[20] identify challenges associated with mobility, including low integration with public transportation, lack of charging infrastructure, and regulatory barriers, and further discuss business models for shared mobility hubs. They explore ways to create value through shared mobility services' physical and digital integration. Golembiewsk et al. [21] analyse trends in battery technologies for electric mobility, identifying opportunities for value creation along the battery value chain. EVs batteries can represent a business opportunity through their circularity [22]. However, identifying such opportunities requires collaboration among various stakeholders. The authors propose a framework to assist stakeholders in identifying circular opportunities that lead to multi-directional value creation. Reinhardt et al. [23] explore archetypes of sustainable business models considering a second use of EVs batteries as a form of sustainable value creation. Real-time energy trading services for privately owned EVs are depicted by an e-vehicle provider, by a provider of energy trading skills and technology [24]. The authors propose a fully reproducible simulation model of the value creation process of real-time trading services. They show that energy trading services, including solar energy generation, have the most significant economic potential but sometimes may only maximize the use of renewable energy in households. Christensen et al. [25] analyze how innovative business models can overcome resistance to electric vehicles. In the context of value co-creation, a process in which both suppliers and consumers actively participate in value creation, several publications are aggregated [26–28].

4.2 Sustainability

Abbasi et al. [29] conducted a study using the unified theory of acceptance and technology (UTAUT) model. They concluded that motivational factors significantly influence EV purchase intentions, particularly when considering perceived environmental knowledge and technophilia. They argue that marketing strategies should promote carbon emissions reduction and the achievement of sustainable development goals. Pirmana et al. [30] examine the economic and environmental implications of EV production. They conclude that EV production increases productivity, gross value-added, and job creation with a relatively small environmental impact. The environmental impact is related with emissions of SO_x, CO_2, and NO_x. Chirumalla et al. [22] address the creation of a circular business model for EV batteries, supporting the environmental benefits of resource reuse.

Similarly, the business model for the second use of electric vehicle batteries, as proposed by Reinhardt et al. [23], focuses on the sustainability of product life cycles. Stauch [31] analyses the added value of using solar energy to EVs and how this combination can reduce carbon emissions. Thies et al. [32] propose a model for the automotive production industry to plan project portfolios efficiently and economically, ensuring compliance with CO_2 emissions regulations subject to regulatory pressure. Li et al. [33] analyze the impact of a luxury EV sales business model on brand competitiveness.

4.3 Strategic Management

Pardo-Bosch et al. [34] address the effectiveness and feasibility of city strategic plans for establishing public charging infrastructure networks to encourage EV adoption and use.

A holistic analysis based on the Value Creation Ecosystem (VCE) and the City Model Canvas (CMC) is employed to visualize how such plans can propose public value with a long-term and sustainable approach. Meisel and Merfeld [24] discuss strategies for integrating EVs into urban infrastructures and emphasize the role of strategic management in urban mobility and sustainability challenges. Hashemipour et al. [35] propose a strategic approach to dynamic allocation for EV aggregators in electricity markets. They consider Peer-to-peer (P2P) clusters, i.e., groupings or clusters of electric vehicles that can exchange energy among themselves or the electrical grid. In this context, P2P" EVs can "communicate" and "negotiate" with each other, sharing energy according to electricity needs and prices. Hamwi et al. [36] examine strategies adopted in managing EV charging infrastructure, highlighting the importance of strategic management to ensure effective and sustainable deployment of this infrastructure. The study focuses on demand response business models (DRBMs), which have great potential for efficiently and sustainably promoting energy flexibility. Stauch [31] analyses buyers' acceptance of acquired products combining EVs with solar energy technology, indicating a strategic approach to adding value to EVs. Shalender [37] studies how entrepreneurship-oriented companies drive sustainable mobility through EV adoption. The analysis highlights the importance of strategy in value creation by driving EV adoption and acceptance. Still, from a strategic perspective in the automotive industry, it concludes that management inefficiency in an industrial context contributes more to industrial innovation inefficiency than technological gap inefficiency [38].

4.4 Electric Vehicle

As per bibliometric analysis, EVs are the focal point connecting all the references that led to the various clusters. All clusters are directly related to EVs, as they discuss various aspects of the automotive industry for EVs, including business model strategies, innovation and value creation, the technology of primary batteries, and sustainability. In the context of EVs and digital society, it is essential to mention that digital transformation takes on special prominence, whether in promoting innovations in business models [25] or technological infrastructures [29].

5 Conclusions and Future Work

The bibliometric analysis of research trends in the areas of "Electric Vehicles", "Value Creation", "Strategic Management" and "Sustainability" demonstrates significant growth since 2019. In the period from 2008 to 2014, the predominant areas were "Engineering". The "Environmental Sciences" field related to sustainability was not as developed. In a digital society, consumers are increasingly informed and concerned about environmental issues, often making choices based on options with lower environmental impact. Therefore, sustainability is a crucial aspect to consider and could be decisive for the implementation of EVs.

Other research areas more closely related to management, strategy, and value creation, such as "Business, Management, and Accounting" or "Economics, Econometrics, and Finance", are deficient. This evidence highlights the need for more attention to business models, implementation strategies, and value-creation methods for organizations.

In the literature review, various initiatives were identified across different research areas, including the co-creation of value between consumers and mobility-sharing companies, business strategies to promote sustainability (including technology marketing, among others), studies on the economic and environmental impacts of EV production and circulation, and the need for effective public policies for EV adoption. The reviewed articles provide a comprehensive view of the strategies, challenges, and opportunities related to EVs and emphasize the importance of strategic approaches to value creation and sustainability.

This study revealed that the areas of "Electric Vehicles" contribute to "Value Creation" and "Strategic Management" are emerging fields with interconnectedness and a global geographic distribution. In future research, the goal is to expand the study, optimize the analysis of identified clusters, and consider other criteria such as co-occurrence, co-authorship, or co-citations to solidify further this theme that is revolutionizing society as a whole. It is essential to highlight that digital transformation is fundamental to support the transition to EVs in a digitally transformed context.

Acknowledgements. Regarding authors identified by number 2, this work is funded by National Funds through the FCT - Foundation for Science and Technology, I.P., within the scope of the project Ref. UIDB/05583/2020. Furthermore, we would like to thank the Research Centre in Digital Services (CISeD) and the Instituto Politécnico de Viseu for their support.

References

1. Mariasiu, F., Chereches, I. A., Raboca, H.: Statistical analysis of the interdependence between the technical and functional parameters of electric vehicles in the European market. Energies (2023). https://doi.org/10.3390/en16072974
2. Razmjoo, A., et al.: A comprehensive study on the expansion of electric vehicles in Europe. Appl. Sci. (2022). https://doi.org/10.3390/app122211656
3. Bairrão, D.R., Soares, J., Canizes, B., Lezama, F., Vale, Z.: Retail electricity tariffs for electric vehicles in Europe: a multivariate analysis in 4 European countries. Adv. Control Optim. Dyn. Syst. (2022). https://doi.org/10.1016/j.ifacol.2022.07.053
4. Pollák, F., et al.: Promotion of electric mobility in the European union—overview of project PROMETEUS from the perspective of cohesion through synergistic cooperation on the example of the catching-up Region. Sustainability (2021). https://doi.org/10.3390/SU13031545
5. Ivanenko, N.I.: Overview of trends and prospects of electric transport development in the EU and assessment of economic/«climate» efficiency of electromob operation (2022). https://doi.org/10.15407/srenergy2022.02.013
6. Adams, C.A.: Sustainability reporting and value creation. Soc. Environ. Account. J. (2020). https://doi.org/10.1080/0969160X.2020.1837643
7. Fedotov, P.K.: Critical analysis of the electric vehicle industry. Exchanges: Warwick Res. J. (2022). https://doi.org/10.31273/eirj.v10i1.3628

8. Fogel, D.S.: Strategic Sustainability: A Natural Environmental Lens on Organizations and Management, pp. 3–319. Routledge Taylor & Francis Group, New York (2016)
9. Tiwari, A., Farag, H.: Analysis and modeling of value creation opportunities and governing factors for electric vehicle proliferation. Energies (2022). https://doi.org/10.3390/en16010438
10. Nanjundaswamy, A., Kulal, A., Dinesh, S., Divyashree, M.S.: Electric vehicles in business processes and sustainable development. Manag. Matters **20**(1), 95–113 (2023). https://doi.org/10.1108/MANM-11-2022-0111
11. Fawakherji, M., Benhaddou, D., Race, B.: Determinants of electric vehicle adoption and integration with public transportation through park-and-ride (2023). https://doi.org/10.1109/IWCMC58020.2023.10182837
12. Richard, C., Hülsmann, M., Brintrup, A.: The structure of the value creation network for the production of electric vehicles. In: Windt, K. (ed.) Robust Manufacturing Control. Lecture Notes in Production Engineering, pp. 47–61. Springer, Heidelberg (2013). https://doi.org/10.1007/978-3-642-30749-2_4
13. Dechezleprêtre, A., Diaz, L., Fadic, M., Lalanne, G.: How the green and digital transitions are reshaping the automotive ecosystem. OECD Sci. technol. Ind. Policy Pap. (2023). https://doi.org/10.1787/f1874cab-en
14. Cooper, C., Booth, A., Varley-Campbell, J., et al.: Defining the process to literature searching in systematic reviews: a literature review of guidance and supporting studies. BMC Med. Res. Methodol. **18**, 85 (2018). https://doi.org/10.1186/s12874-018-0545-3
15. Mishra, V., Mishra, M.P.: PRISMA for review of management literature – method, merits, and limitations – an academic review. In: Rana, S., Singh, J., Kathuria, S. (eds.) Advancing Methodologies of Conducting Literature Review in Management Domain (Review of Management Literature, vol. 2, pp. 125–136. Emerald Publishing Limited, Leeds (2023). https://doi.org/10.1108/S2754-586520230000002007
16. Anderson, E.G., Bhargava, H.K., Boehm, J., Parker, G.: Electric vehicles are a platform business: what firms need to know. Calif. Manage. Rev. **64**(4), 135–154 (2022). https://doi.org/10.1177/00081256221107420
17. Scopus. (2024). https://www.scopus.com/home.uri
18. Van Eck, N.J., Waltman, L.: Software survey: VOSviewer, a computer program for bibliometric mapping. Scientometrics **84**, 523–538 (2010). https://doi.org/10.1007/s11192-009-0146-3
19. Afentoulis, K.D., Bampos, Z.N., Vagropoulos, S.I., Keranidis, S.D., Biskas, P.N.: Smart charging business model framework for electric vehicle aggregators. Appl. Energy **328**, 120179 (2022). https://doi.org/10.1016/j.apenergy.2022.120179
20. Coenegrachts, E., Beckers, J., Vanelslander, T., Verhetsel, A.: Business model blueprints for the shared mobility hub network. Sustainability **13**, 6939 (2021). https://doi.org/10.3390/su13126939
21. Golembiewski, B., vom Stein, N., Sick, N., Wiemhöfer, H.-D.: Identifying trends in battery technologies with regard to electric mobility: evidence from patenting activities along and across the battery value chain. J. Clean. Prod. **87**, 800–810 (2015). https://doi.org/10.1016/j.jclepro.2014.10.034
22. Chirumalla, K., Reyes, L.G., Toorajipour, R.: Mapping a circular business opportunity in electric vehicle battery value chain: a multi-stakeholder framework to create a win–win–win situation. J. Bus. Res. **145**, 569–582 (2022). https://doi.org/10.1016/j.jbusres.2022.02.070
23. Reinhardt, R., Christodoulou, I., García, B.A., Gassó-Domingo, S.: Sustainable business model archetypes for the electric vehicle battery second use industry: towards a conceptual framework. J. Clean. Prod. **254**, 119994 (2020). https://doi.org/10.1016/j.jclepro.2020.119994

24. Meisel, S., Merfeld, T.: Assessing the financial value of real-time energy trading services for privately owned non-commercial electric vehicles. Transp. Res. Part D: Transp. Environ. **80**, 102229 (2020). https://doi.org/10.1016/j.trd.2020.102229
25. Christensen, T.B., Wells, P., Cipcigan, L.: Can innovative business models overcome resistance to electric vehicles? Better place and battery electric cars in Denmark. Energy Policy **48**, 498–505 (2012). https://doi.org/10.1016/j.enpol.2012.05.054
26. Ma, Y., Rong, K., Luo, Y., Wang, Y., Mangalagiu, D., Thornton, T.F.: Value co-creation for sustainable consumption and production in the sharing economy in China. J. Clean. Prod. **208**, 1148–1158 (2019). https://doi.org/10.1016/j.jclepro.2018.10.135
27. Nosi, C., Pucci, T., Silvestri, C., Aquilani, B.: Does value co-creation really matter? An investigation of italian millennials intention to buy electric cars. Sustainability **9**, 2159 (2017). https://doi.org/10.3390/su9122159
28. Chen, C.-F., Lee, C.-H.: Investigating shared e-scooter users' customer value co-creation behaviors and their antecedents: perceived service quality and perceived value. Transp. Policy **136**, 147–154 (2023). https://doi.org/10.1016/j.tranpol.2023.03.015
29. Abbasi, H.A., et al.: Consumer motivation by using unified theory of acceptance and use of technology towards electric vehicles. Sustainability **13**, 12177 (2021). https://doi.org/10.3390/su132112177
30. Pirmana, V., Alisjahbana, A.S., Yusuf, A.A., et al.: Economic and environmental impact of electric vehicles production in Indonesia. Clean Technol. Environ. Policy **25**, 1871–1885 (2023). https://doi.org/10.1007/s10098-023-02475-6
31. Stauch, A.: Does solar power add value to electric vehicles? An investigation of car-buyers' willingness to buy product-bundles in Germany. Energy Res. Soc. Sci. **75**, 102006 (2021). https://doi.org/10.1016/j.erss.2021.102006
32. Thies, C., Hüls, C., Kieckhäfer, K., Wansart, J., Spengler, T.S.: Project portfolio planning under CO2 fleet emission restrictions in the automotive industry. J. Ind. Ecol. **26**, 937–951 (2022). https://doi.org/10.1111/jiec.13228
33. Li, Z., Liang, F., Cheng, M.: Research on the impact of high-end Ev sales business model on brand competitiveness. Sustainability **13**, 14045 (2021). https://doi.org/10.3390/su132414045
34. Pardo-Bosch, F., Pujadas, P., Morton, C., Cervera, C.: Sustainable deployment of an electric vehicle public charging infrastructure network from a city business model perspective. Sustain. Cities Soc. **71**, 102957 (2021). https://doi.org/10.1016/j.scs.2021.102957
35. Hashemipour, N., del Granado, P.C., Aghaei, J.: Dynamic allocation of peer-to-peer clusters in virtual local electricity markets: a marketplace for EV flexibility. Energy **236**, 121428 (2021). https://doi.org/10.1016/j.energy.2021.121428
36. Hamwi, M., Lizarralde, I., Legardeur, J.: Demand response business model canvas: a tool for flexibility creation in the electricity markets. J. Clean. Prod. **282**, 124539 (2021). https://doi.org/10.1016/j.jclepro.2020.124539
37. Shalender, K.: Entrepreneurial orientation for sustainable mobility through electric vehicles: insights from international case studies. J. Enterp. Commun.: People Places Glob. Econ. **12**(1), 67–82 (2018). https://doi.org/10.1108/JEC-05-2017-0032
38. Chen, Y., Ni, L., Liu, K.: Innovation efficiency and technology heterogeneity within China's new energy vehicle industry: a two-stage NSBM approach embedded in a three-hierarchy meta-frontier framework. Energy Policy **161**, 112708 (2022). https://doi.org/10.1016/j.enpol.2021.112708

A Framework for Monitoring Pollution Levels in Smart Cities

Diogo Silva[1], João Pinto[1], José Varanda[1], João Henriques[1,2(✉)], Filipe Caldeira[1,2], and Cristina Wanzeller[1,2]

[1] Polytechnic of Viseu, Viseu, Portugal
{pv29075,estgv17744,estgv17764}@alunos.estgv.ipv.pt,
{joaohenriques,caldeira,cwanzeller}@estgv.ipv.pt
[2] CISeD - Research Centre in Digital Services, Polytechnic of Viseu, Viseu, Portugal

Abstract. *Smart cities* emerged as an approach addressing the rising challenges of urbanization, including pollution (air, water, or sound), impacting the environment, public health, and overall quality of life.

This work aims to improve pollution quality within urban environments, ultimately promoting sustainability and enhancing the well-being of urban residents by exploring smart city approaches and employing innovative strategies and technologies to monitor and manage air quality effectively. Our approach uses real-time air quality sensors, IoT (Internet of Things) networks, and data analytics in container virtualized environments to generate accurate and timely information about pollution levels. It enables city administrators to make informed decisions to mitigate its impact at scale.

Keywords: Smart cities · pollution · IoT · urbanization · sustainability

1 Introduction

Pollution represents a relevant concern in this modern age where almost everything is either polluted or contaminated in some way, impacting human health and reducing lifespan. This fact, demands for awareness on defining new strategies to reduce the pollution levels.

As the global population continues its trend toward urbanization, cities face the challenge of maintaining livability, so the need to develop new solutions for cities to stay sustainable is getting more prominent over time [1].

This work aims to contribute to streamlining the analysis of pollution levels, looking into particles like PM2.5 to provide a more effortless and cleaner report of the situation at hand and give a better perspective of where and what could be the cause and even prevent unavoidable accidents from happening.

Although various solutions exist for the problem above, some limitations are notable. Wong et al. [2] highlight the need for high rigor in the calibration of specific data, and the project cost could be an issue if the intention is to build a scalable solution, as it would involve a high-volume production.

The Internet of Things solution will facilitate monitoring pollution levels in urban areas, generate various reports, and issue alerts when surpassing predefined limits. It will trigger public action or alert authorities to intervene before the situation escalates, addressing incidents such as fires or other emergencies.

This solution aims to extract insights into the pollution areas and use AI to analyze and make predictions. Therefore, it can identify the areas with higher pollution levels and alert authorities in case a specific threshold exceeds its limit.

Considering the enormous amount of data provided by the sensors throughout the site, we designed this solution to scale and adapt to the large volumes of data gathered. We also implemented technologies like Apache Kafka as an application where users can view alerts and reports.

By using sensors in the context of smart cities, it will be possible to enable people to find solutions to the issue of pollution and also help mitigate accidents or even avoid them. With all the data collected, it is possible to do incredible things and help our planet improve again, leaving it habitable for our children and grandchildren.

2 Related Work

With the growth of urbanization in many cities, the need for innovative sustainability solutions is increasing. The smart cities paradigm meets such demand since it considers factors such as minimizing the environmental footprint of urban activities, efficiently managing energy resources, and creating inventive services and solutions for citizens [3]. By these means, Belli *et al.* analyses the principal features of IoT infrastructures for smart cities, going into detail about the innovative and sustainable management example in the town of Parma. In addition, the city has obtained a top 10 in the iCityRank ranking, which aims to measure the progress of different municipalities to become a smart city. The city of Parma has implemented several intelligent solutions, such as a Smart City Government Plan, which intended to formulate a vision, chart a roadmap, devise an investment strategy, and craft an action plan; the project Parma Futuro Smart is a political strategy tool that aids the city in making decisions to outline an intelligent city Action and Investment Plan by the year 2030. All this to innovation and the union and grouping of different authorities and responsible political entities to promote better city functioning and greater sustainability [3].

Besides the continuous increase in the urbanization of cities, there is also an increase in the levels of different pollutants, where Alías *et al.* review article appears [4], which highlights the fact that the constant urbanization of cities results in the development and expansion of new transport systems, which consequently leads to enormous noise pollution. The article also highlights the emergence of recent projects based on WASN (Wireless Acoustic Sensor Networks). These devices facilitate the automated creation of dynamic noise maps throughout urban areas by relying extensively on controlled measurements. The subsequent visualization of this data allows responsible authorities to analyze and act on noise pollution levels in certain regions.

Closer to this topic, the work from [5] was identified. This project consists of an IoT system composed of many monitoring devices, such as pollution sensors and a GPS, which is connected to an ESP8266 air pollution module, enabling the identification of the whereabouts and level of pollution of a particular set of data. Developing an interface interconnected to the IoT system allowed us to visualize the data collected by the sensors. With the integration of the Google Maps tool, the project was able to present good results, where it is possible to verify the exact location of areas with high pollution levels. Finally, the fact that the solution alerts environmental authorities allows us to adopt a stance of more excellent prevention and caution towards the state of the environment.

Also similar to our topic, the approach of Miles et al. [6] consisted of an IoT decision support system to help the authorities identify a severe level of atmospheric pollution and initiate a response through the implementation of complete road closures. This system constructs a foundational traffic model, serving as input for an atmospheric dispersion model. The DSS subsequently aims to alleviate pollution by executing road closures and assessing the anticipated outcomes. The conclusion is that the design of the decision system establishes a robust foundation for future efforts focused on integrating vehicle interactions into the foundational traffic model and improving the atmospheric dispersion model to simulate more complex areas.

Since one of the most common aspects in projects involving this theme is the act of supporting authorities and promoting awareness among all citizens about pollution levels, it is vital to refer to the solution proposed by Garzon et al. [7]. They built a Proof-of-Concept (PoC) of a service for monitoring air pollution. It is context-aware and proactively notifies citizens through mobile devices when they enter an area with air pollution surpassing specific thresholds defined by the user. The work also highlighted the need to attain a crucial density and widespread distribution of sensors across the city to form dependable estimates regarding particulate matter concentration in areas that do not have that many sensors to collect data. Regarding implementation, it is worth noticing that they used the Spring Cloud framework for workload distribution, a mobile application and a web dashboard for data visualization, and the Kafka message broker, which receives JSON files with data from the sensors, which, therefore, are stored and persisted in a MongoDB database. During the evaluation, they took measurements and collected data in Rangiora on the South Island of New Zealand. Considering the continuous tracking of the mobile device's location, they also measured mobile battery consumption within different parameters to assess the potential for battery drain.

3 Methods

This section outlines the methods and approaches employed to investigate integrating intelligent technologies and data-driven solutions to address pollution quality, explicitly focusing on air pollution and temperature and humidity in the context of smart cities. It also highlights the methodology used to explore the role of smart cities in promoting sustainability and mitigating risks associated with pollution and disasters.

In implementing the proposed Architecture, it adopted the Agile methodology, as its working principles align perfectly with software development projects. In addition to being an organized process, it consists of good practices, allows teamwork, and is open to change, giving this added value since emerging technologies are constantly evolving (IoT).

Agile methodology is a project management and software development approach that prioritizes flexibility, collaboration, and customer satisfaction. The process relies on a system of iterative development, where cross-functional teams organize themselves to evolve requirements and solutions collaboratively.

The strengths of the Agile methodology are:

- Flexibility and Adaptability: allows teams to adapt to shifting priorities, evolving requirements, and emerging insights;
- Customer-Centricity: regular customer feedback and involvement ensure that the delivered product aligns closely with user expectations;
- Collaboration and Communication: Agile promotes a high level of collaboration among team members, stakeholders, and customers;
- Incremental Delivery: Agile emphasizes delivering a functional product incrementally in short iterations, known as sprints;

The challenges are:

- Resistance to Change: Agile methodologies require major thinking and organizational structure transformation;
- Documentation and Rigidity Concerns: traditional project management methods often emphasize extensive documentation, while Agile favors working software over comprehensive documentation;
- Resource Allocation: Agile requires dedicated and cross-functional teams, which may pose challenges in terms of resource allocation, especially in larger organizations with pre-existing structures and responsibilities;
- Lack of Predictability for some Stakeholders: Agile's iterative and adaptive nature can be unsettling for stakeholders who are used to more predictable timelines and deliverables. Balancing agility with the need for predictability can be challenging;
- Dependency Management: managing dependencies between different Agile teams or between Agile and non-Agile teams can be complex in larger projects. Coordinating dependencies requires careful planning and communication.

The fundamental principles are:

- Individuals and Interactions over Processes and Tools: Agile prioritizes the human element, emphasizing the importance of communication and collaboration among team members;

- Working Software over Comprehensive Documentation: Agile values delivering a functional product and encourages minimal but sufficient documentation to support development;
- Customer Collaboration over Contract Negotiation: Agile seeks active involvement and collaboration with customers throughout the development process to ensure the final product meets their needs;
- Responding to change over Following a Plan: Agile acknowledges that change is inevitable and welcomes it, allowing teams to adapt and respond to evolving requirements;
- Continuous Delivery of Value: Agile aims to continuously deliver value to the customer through incremental delivery, ensuring the first implementation of features with the highest priority.

In conclusion, Agile methodology offers a dynamic and customer-centric approach to project management and software development.

4 Architecture

This section presents the proposed architecture. The following subsections will describe their components and their roles. Section 4 will present the architecture of the proposed framework, while Subsect. 4.1 presents the containers supporting the implementation of their components. Finally, the Subsect. 4.2 shows the purpose of developing a mobile application in this context.

Figure 1 illustrates an architecture where IoT sensors establish database connections. After that, an API was used to convert the sensor data into information that the mobile and web applications could access. The proposed architecture system designs scalability and maintainability, leveraging Docker containers to separate various functionalities. In the proposed architecture, those sensors were replaced as datasets to simulate the data collected from sensors. Those datasets were defined as a single CSV file gathering all the received information from sensors.

4.1 Docker Containers

The PoC of the present framework included three primary containers:

- **Python API Container:** Python is a flexible, high-level programming language known for its readability, simplicity, and extensive community support. It was created by Guido van Rossum and first released in 1991. Python has become one of the most popular programming languages in the world because it is versatile and user-friendly and has a wide range of applications. Creating a Docker container for a Python-based API simplifies deployment, enhances scalability, and ensures consistency across different environments. By encapsulating the application within a container, it was created a portable and reproducible package that can be easily shared and deployed, making it an integral part of modern software development workflows;

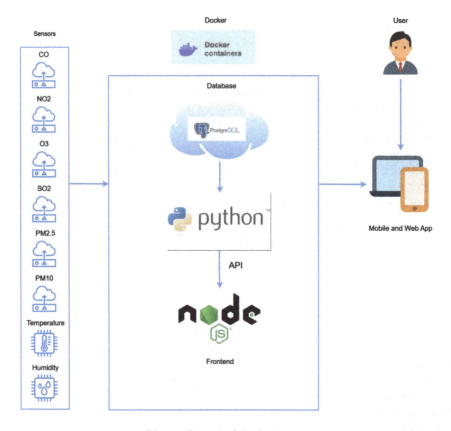

Fig. 1. Proposed Architecture

– **PostgreSQL Container:** PostgreSQL is a powerful open-source relational database management system (RDBMS). It is known for its advanced features, extensibility, and adherence to SQL standards. Designers have crafted PostgreSQL to handle various workloads, from small-scale applications to large and complex enterprise systems. Dockerizing PostgreSQL provides numerous benefits, including portability, scalability, and ease of management. Encapsulating the database within a container creates a consistent environment that can be easily shared, replicated, and deployed across different environments. This approach enhances flexibility and efficiency in database management, making it a valuable practice in modern software development workflows.
– **Node.Js Frontend Container:** Node.Js framework was used for front-end development. Node.Js is a runtime environment that allows the execution of JavaScript code on the server side-built on the V8 JavaScript runtime, the engine that powers Google Chrome. Node.Js enables developers to use JavaScript for both client-side and server-side development, providing a unified language across the entire web application stack. The main characteristics

of Node.Js are Asynchronous and non-blocking events, single-threaded and event-loop models, cross-platform, and many others. Dockerizing a Node.Js front end offers the advantages of consistency, reproducibility, and ease of deployment. Encapsulating the front end within a container creates a self-contained environment that simplifies the development workflow and ensures the application runs consistently across different environments. This approach contributes to the efficiency and reliability of the front-end development process in modern software development practices.

4.2 Mobile Application

Due to the significant adoption of mobile devices, it will be considered a mobile application to report the information processed by the Firebase services, including the level of pollutants in the air anywhere in the world. The proposed mobile app communicates with the Firebase database to retrieve the saved data via Wi-Fi or mobile. The development team will design the mobile application to enable users to check information about the pollution level of a designated smart city and access all current values triggered by the sensors, such as temperature, humidity, ozone, carbon monoxide, nitrogen dioxide, and sulfur dioxide. Additionally, the application will include alerts that notify users if any values fall outside the predefined range. This feature will help users to respond to any issues that may arise quickly. Overall, the mobile application will provide users with an easy and convenient way to ensure the health and well-being of the inhabitants of a particular city.

5 Experimental Work

This section will describe the experimental work in implementing the PoC of the proposed framework.

5.1 Hardware and Sensors

For this project, we utilized a Windows 11 computer with an Intel I10 processor, 16 GB of RAM and a 1 TB SSD as the host machine for the Docker containers. We evaluated sensors' use as shown in Table 1, where we list these components that meet the established requirements. These sensors are electrochemical and vary their resistance when exposed to certain gases. Internally, it has a heater in charge of increasing the internal temperature, and with this, the sensor can react with the gases, causing a change in the resistance value. However, as we could not access the hardware, all the sensor values were simulated. We use the datasets and some air Quality Sensors as shown in Table 1 for the effect. The sensors provide accurate and timely information on pollution levels. Continuous data transmission to a central repository involves temperature and humidity sensors; IoT communication networks, which send data from the air quality sensors; and data analytics, utilizing advanced tools and algorithms to process and analyze the collected air quality data).

- MQ-7: its function is to detect carbon monoxide in the air since it has a high sensitivity;
- MQ-135: its function is to detect LPG, propane, methane, alcohol, hydrogen, and smoke;
- MQ-131: its function is to detect ozone gas concentrations in the air;
- ME3-SO2: its function is to detect and measure the concentration of sulfur dioxide (SO2) gas in the air;
- PMS5003: its function is to detect and measure the concentration of suspended particles in the air. It can detect particles with diameters of 2.5 and 10.0 µm;
- DHT22: its function is a combined humidity and temperature sensor.

Table 1. Eletronic Components

Eletronic Components	
MQ-7	Carbon Monoxide
MQ-135	Nitrogen Dioxide
MQ-131	Ozone
ME3-SO2	Sulphur Dioxide
PMS5003	PM2.5 and PM10
DHT22	Temperature and Humidity

5.2 Procedure

A Python script was developed to acquire sensor data and simulate sensor values similar to those of specific intelligent cities. This approach aims to improve the system's precision, as the script generates values for different sensors (humidity, temperature, and ozone) within a realistic range. During active testing in a simulated smart-city environment, the system gathered and stored sensor data in the PostgreSQL database container every five minutes. Real-time notifications were sent to users through the notifications container for events such as high-temperature alerts or significant risks to public health if one of the gases appeared in high quantities.

5.3 Setup Guide

The pollution monitoring portal is a solution deployed using Docker containers, ensuring each component's streamlined and isolated environment. The portal's backend is developed in Flask, a lightweight and efficient Python web framework that provides the necessary API endpoints and logic for data processing. This backend seamlessly connects to a containerized PostgreSQL database, ensuring secure and efficient data management. For the front end, it used React, a robust JavaScript library for building user interfaces, enabling an interactive and user-friendly experience for monitoring pollution levels. Each component - the Flask backend, PostgreSQL database, and React front-end - is encapsulated in its own Docker container, ensuring easy scalability, quick deployment, and consistent performance across different environments. This containerized approach simplifies the development and deployment process, allowing for a high degree of flexibility and reliability in the operation of our pollution monitoring system. Docker files were used to build and run all the containers necessary for this solution.

6 Results and Discussion

The smart city approach revolutionized how urban environments manage and monitor environmental health. Results denote the feasibility of integrating advanced sensors and data analytics into urban infrastructure, significantly enhancing the capability to track, analyze, and respond to pollution levels in different city areas.

This work found that deploying IoT sensors across various city locations has enabled real-time air quality monitoring. This achievement allows for immediate detection of pollution spikes, making it easier for prompt responses, allowing authorities to take action before it is too late, and, of course, making sure that the city in question is a safe place to live.

Considering that it used data from a dataset, it must address some points. Once we implement this solution to its fullest potential, the collected data will unveil significant spatial variation in pollution levels. For instance, industrial areas will most likely exhibit higher levels of airborne particulates, PM2.5 and PM10, whereas residential areas are more affected by vehicular emissions. Moreover, we could look into the times of the day when the levels are higher or spiked. Let us think about rush hour and the number of cars that will be all at the same spot emitting pollutant gases into the atmosphere, or perhaps specific activities on a particular day that can also cause that spike, and that indicates a strong correlation between human activities and pollution levels.

It is also possible to check the green areas and stricter environmental regulations that will lower pollution levels, underscoring the effectiveness of urban planning and policy interventions.

Overlaying pollution data with public health records will reveal a clear correlation between high pollution periods and increased respiratory and cardiovascular conditions among the city's population.

With all of these findings, some points should be highlighted. The dynamic nature of pollution emphasized by real-time monitoring underscores the need for equally dynamic policy responses. This policy should be adaptable, based on real-time data, to manage environmental health risks effectively, a severe problem in today's society.

Given that this analysis can help us determine the time and place of higher pollution levels, it can also help us with strategies and interventions to fight back, such as implementing traffic control measures or enriching the available public transportation data to the population.

With all this data and knowledge, using the right tools can improve the future of urban development and environmental health.

Making pollution data available through online portals is also possible, promoting transparency and enhancing public engagement. Educating citizens about pollution levels and health implications can lead to more environmentally responsible behavior. Moreover, it will be possible to involve the collaboration of multiple sectors, such as the government, private industry, and community groups, to implement and develop sustainable solutions.

Fig. 2. Temperature chart

As shown in Fig. 2, the temperature stayed more or less the same throughout that period. However, if the temperature drops for some reason, it is possible to pinpoint where it happened and the time, making it easier to detect the eventual incident.

It also analyzed PM2.5 and PM10, which refer to airborne particles that have significant implications for human health, primarily air quality. PM2.5 particles are finer and can enter deep into the respiratory system, posing a potential health risk. They may carry pollutants such as heavy metals and organic compounds that could be inhaled daily without realizing them, leading to poor quality of life. These particles come from combustion processes and industrial emissions.

As denoted in Fig. 3 the curve of particle PM2.5, and throughout five days, the level of this particle in the air was altering. On a specific day, this level spiked. With this, it is possible to find out what happened that day and determine the cause for this rise, helping us reduce this level for a better quality of life.

Fig. 3. PM2.5 and PM10 levels

The PM10 particle can also indicate poor air quality, and when inhaled, it can cause respiratory and cardiovascular issues.

As presented in Fig. 3, the level of the particle PM10 also increased in the same time frame that the PM2.5 increased, which means that these two particles are related, and both will cause harm to our lives. With our solution, we can notice this issue and find a way to reduce pollution. This portal gives us a tool and helps us to pinpoint the cause of this level increase.

7 Conclusion

This work proposed a framework aimed at increasing the efficiency of reducing pollution levels through IoT technologies, such as real-time air quality sensors, IoT networks, and data analytics components able to be deployed in container virtualized environments to generate accurate and timely information about pollution levels. This way, it will be possible to support administrators on informed decisions to mitigate its impact at scale.

The proposed approach aims to increase awareness and reduce pollution in living spaces. This way, it is possible to support defining appropriate actions on monitoring environments and contribute to improving human health in cities.

In future work, it is intended to scale this solution by using Kubernetes. Although Docker allows efficient management of every component by deploying them into different containers, in Kubernetes, it would be possible to manage an even more significant number of containers in the container runtime. Moreover, the frontend can be extended by increasing the diversity of graphs to improve the insights from generated and collected data.

Moreover, in the future, we aim to demonstrate the applicability of the proposed framework under more realistic scenarios by using physical sensors, collecting data from the air quality, and subsequently carrying out a more truthful analysis.

Acknowledgements. This work is funded by National Funds through the FCT - Foundation for Science and Technology, I.P., within the scope of the project Ref UIDB/05583/2020. Furthermore, we thank the Research Centre in Digital Services (CISeD), the Polytechnic of Viseu, for their support.

References

1. Gupta, H., et al.: An IoT based air pollution monitoring system for smart cities. In: 2019 IEEE International Conference on Sustainable Energy Technologies and Systems (ICSETS), pp. 173–177. IEEE (2019)
2. Wong, M.S., et al.: Towards a smart city: development and application of an improved integrated environmental monitoring system. Sustainability **10**(3), 623 (2018)
3. Belli, L., et al.: IoT-enabled smart sustainable cities: challenges and approaches. Smart Cities **3**(3), 1039–1071 (2020)

4. Alías, F., Alsina-Pagès, R.M., et al.: Review of wireless acoustic sensor networks for environmental noise monitoring in smart cities. J. Sens. **2019** (2019)
5. Spandana, G., Shanmughasundram, R.: Design and development of air pollution monitoring system for smart cities. In: 2018 Second International Conference on Intelligent Computing and Control Systems (ICICCS), pp. 1640–1643. IEEE (2018)
6. Miles, A., Zaslavsky, A., Browne, C.: IoT-based decision support system for monitoring and mitigating atmospheric pollution in smart cities. J. Decis. Syst. **27**(sup1), 56–67 (2018)
7. Garzon, S.R., et al.: Urban air pollution alert service for smart cities. In: Proceedings of the 8th International Conference on the Internet of Things, pp. 1–8 (2018)

Value Creation and Strategic Management in the Era of Digital Transformation: A Bibliometric Analysis and Systematic Literature Review

Sónia Gouveia[1,2(✉)], Daniel H. de la Iglesia[3], José Luís Abrantes[1,2], and Alfonso J. López Rivero[4]

[1] Superior School of Technology and Management, Polytechnic Institute of Viseu, Viseu, Portugal
`sgouveia@estgv.ipv.pt`
[2] CISeD – Research Centre in Digital Services, Instituto Politécnico de Viseu, Viseu, Portugal
[3] Faculty of Science, University of Salamanca, Salamanca, Spain
[4] Facultad de Informática, Universidad Pontificia de Salamanca, Salamanca, Spain

Abstract. The discourse on the essential factors determining the survival of modern organizations has firmly placed digital transformation in a prominent and pivotal role. Some of these elements naturally include the strategic management of organizations for value creation. This study uses a bibliometric analysis method to explore the dynamic interaction between these essential elements in the digital transformation era. This research seeks to elucidate evolving trends, emerging paradigms, and key insights related to value creation and strategic management in digital transformation by examining a wide range of academic literature. Web of Science (WoS) database is used to obtain sample documents. The analysis was conducted using VOSviewer software (version 1.6.20), focusing on the bibliographic coupling of documents and countries, which were presented as network visualization maps. Five clusters were identified, and the authors defined emerging research areas based on the information gathered.

Keywords: Digital Transformation · Strategic · Management · Value Creation

1 Introduction

Digitalization is creating disruptive changes in organizations, their businesses, and consequently in their strategies for value creation. Digital transformation has been described as the challenge to address the threat of digital disruption [1]. This challenge has gained such importance that some argue that the real challenge will not be to develop new strategies, business models, or new organizational projects but to manage the transition between the current stage and the intended future stage through artificial intelligence [2].

The literature addresses the digital transition from various perspectives and associates this theme with organizational value creation and strategic management. Vial [3]

states that digital technologies trigger strategic responses from organizations seeking to alter their paths of value creation while managing structural changes and organizational barriers that affect outcomes. Additionally, proponents argue that digital transformation facilitates significant improvements in organizational efficiency and customer engagement, making it an appropriate tool for value creation [4]. At the level of business models, researchers also mention that digital transformation leads to creating new distribution channels, enabling value creation for the customer [5].

Yar Hamidi [6] addresses the analysis of the value chain and its underlying belief that all activities of a company must be analyzed and quantified for the contribution each brings to build and ensure a competitive advantage. Naturally, digital transformation will be in this path, as if we consider Porter's value creation model [7], encompassing a company's infra-structure, procurement, inbound logistics, operations, outbound logistics, marketing and sales, services, human resource management, and technology development, we presume that digital transformation can be a pivotal point in the value creation process.

Goedhart and Koller [8] even consider digital transformation the ultimate value-creation process in the long term. They mention that the desired benefits of more efficient customer engagement, optimized operational costs, employee empowerment, and even transformation of products/services through digital technologies facilitate value creation.

Organizations will try to adjust to a "new normality [9]", influencing and changing the environment in which they operate, identifying, creating, and exploring possibilities. With this goal in mind, they will develop strategic management approaches that consider digital transformation aimed at value creation.

Digital transformation represents a specific type of change to the organization's strategy [10] as it can alter the value creation, potentially even contributing to changing the organization's scope. Digital transformation necessitates significant organizational changes guided by digital technologies, leading to profound organizational strategies and routine alterations [11]. In this framework, digital transformation is not merely a technological process but a broader investment in skills, projects, and infrastructure [12]. It is worth noting that even in the presence of technology, expected results are often not achieved due to deficiencies in organizational practices and individuals' resistance to change [13]. Therefore, organizations must manage technology knowledge to facilitate digital transformation initially and subsequently generate value. This fact is significant and varies from one organization to another. The acceptance of SMEs for technologies related to digital transformation is lower when compared to larger organizations [14].

Considering the literature underpinning the importance of digitization in value creation and its impact on strategic management in organizations, the main objective of this study is to enhance our understanding of digital transformation and its effect on organizations' strategies for value creation.

2 Research Methodology

This study comprises quantitative and qualitative analyses in the form of a bibliometric analysis and a Systematic Literature Review (SLR). The bibliometric analysis is a quantitative technique involving the visual and logical analysis of articles and evaluating, mapping, and identifying structural patterns within a research domain using mathematical models, visualization clusters, and algorithms [15]. Bibliometric analysis involves the assessment of scientific publications, patents, and conference proceedings using various indicators (such as co-authorship, co-occurrence, citations, and bibliographic coupling, among others). Applying these methods, researchers can identify influential research works, prominent authors, and connections within a specific research field. A systematic review adheres to standardized methodologies/guidelines in formulating and exploring a research question, systematic searching, filtering, reviewing, evaluating, interpreting, synthesizing, and reporting findings from multiple publications on a topic/domain of interest [16]. This research will apply bibliographic coupling by documents and bibliographic coupling by countries.

To better understand trends in organizations' strategies for value creation through digital transformation, we utilized a consolidated approach, the PRISMA protocol (Preferred Reporting Items for Systematic Reviews and Meta-Analyses) [17].

The research was conducted on the Web of Science database in February 2024, considering the following search terms: "Digital Transformation", "Value Creation", and "Strategy" in the Topic field, resulting in a total of 190 papers. Subsequently, we applied successive criteria, including document type (paper), language (English), and research areas "Business Economics, Computer Science, and Engineering and Operations". The combination of all criteria resulted in a total of 72 papers.

The selected research areas have a direct relevance to the search terms "Digital Transformation", "Value Creation", and "Strategy", as identified in the literature [18]. These areas are fundamental to understanding how organizations implement technological changes, compete in markets, and revise their strategies for value creation. Although the database offers other areas that may have some indirect relevance, we chose the mentioned areas because we consider them to have a direct and comprehensive connection with the concepts in question.

The authors attempted to answer the following Research Question (RQ): *What is the current development of the literature on digital transformation concerning value creation and management strategy?*

3 Bibliometric Analysis

In order to address the research question, descriptive statistics were analyzed from research articles, beginning with the distribution of papers per year. The distribution of papers published yearly in a given research field provides an overview of research trends and can offer insights into potential future trends.

Refining according to the established criteria and after applying the PRISMA protocol, we obtained 72 papers distributed, as shown in Fig. 1. As can be seen, there was a significant increase from 2017 onwards, surpassing 40 papers in more recent years. In the research areas of "Business Economics, Computer Science, and Engineering and Operations", there is some expression starting from 2017, with a boom occurring from 2020 onwards. This evolution explains the current high interest in using the terms "digital transformation", "value creation", and "strategy/strategies". With the growth of publications, the number of citations naturally increases, albeit with a delay.

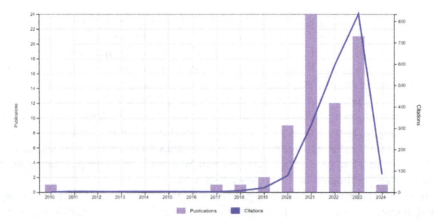

Fig. 1. Number of papers by year. (papers – with criteria application). Source: WoS Database

As can be observed, the choice of research areas "Business Economics, Computer Science, and Engineering and Operations" proved to be appropriate as they represent the most significant percentages of available areas before the application of criteria (Fig. 2 a). Notably, the analysis considers that more than one research area may be present in the same paper. Regarding the chosen research areas, the Business Economics area is present in about 93% of the papers after applying criteria's (Fig. 2b).

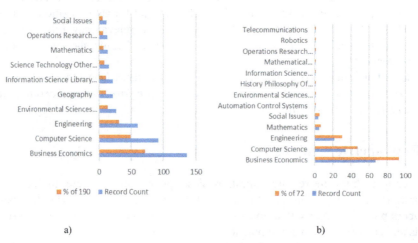

Fig. 2. Documents by subject (papers – a) without criteria application b) with criteria application). Source: author's own creation

The authors conducted a content analysis to provide a more comprehensive understanding of the leading research areas resulting from applying criteria to the 72 papers related to the RQ. Thus, the authors identified a research focus for each cluster, defined through bibliographic coupling analysis per document using VOSviewer [19] to map research areas and related articles. We considered a minimum of two citations as minimal for solid relationships between branches, and we did not limit the cluster sizes. The bibliographic coupling strength of two articles is the number of elements in the intersection of their reference lists. As a result, we obtained a total of 46 strong connections resulting in 5 clusters. In Fig. 3, clusters are formed based on articles determined by author keywords. This analysis acknowledges articles that have made important intellectual contributions in the areas of the RQ. We will delve further into these clusters in the SLR, which we will address in Sect. 4.

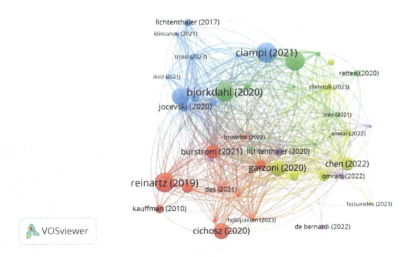

Fig. 3. Bibliographic coupling by documents. Source: Authors' own creation

Research found a total of 118 countries as the origin of the published documents. The majority of publications were produced in Germany (18), Italy (12), the United Kingdom (7), Finland (6), the USA (6), Brazil (5), PRC (5), and Sweden (5).

4 Systematic Literature Review (SLR)

Following the results of the SLR content analysis and after a brainstorming session and a content analysis process, a final configuration of groupings for these research areas was defined, resulting in 5 clusters (Fig. 3).

Cluster 1 (red): Strategic Management in Digital Transformation
Cluster 2 (green): Emerging Trends in Digital Entrepreneurship and Sustainability
Cluster 3 (blue): Digital Capabilities and Business Model Innovation
Cluster 4 (yellow): Digitalization of SMEs
Cluster 5 (purple): Value Creation through Innovation

Cluster 1: Strategic Management in Digital Transformation
In this cluster, is outlined a scenario where organizations face significant challenges in the era of digital transformation. They explore management strategies aimed at guiding

organizations through this complex and ever-evolving process. By discussing aspects such as business model innovation, adoption of digital technologies, and development of digital capabilities, the articles highlight the importance of a strategic and results-oriented approach in managing digital transformation.

Cluster 2: Emerging Trends in Digital Entrepreneurship, and Sustainability
This cluster focuses on the intersection between virtual entrepreneurship and sustainability, highlighting the emergence of new ventures that utilize virtual technologies to address demanding social and environmental situations. The papers within this cluster analyze the evolving landscape of social entrepreneurship, the role of digital systems in promoting sustainable practices, and the ability of solutions to promote financial inclusion and environmental responsibility. By exploring how digital innovation intersects with sustainability needs, this cluster offers insight into the future of entrepreneurship with an emphasis on profitability and social impact.

Cluster 3: Digital Capabilities and Business Model Innovation
At the heart of this cluster lies the exploration of digital capabilities and their role in driving business model innovation. The articles within this cluster investigate how organizations harness big data analytics, dynamic managerial capabilities, and digital platforms to transform their business models and create value in an increasingly digital landscape. By emphasizing the importance of agility, customer-centricity, and organizational learning, this cluster provides actionable insights for companies seeking to thrive amidst digital disruption and capitalize on new opportunities for growth and competitiveness.

Cluster 4: Digitalization of SMEs
Focused on small and medium-sized enterprises (SMEs), this cluster delves into the unique challenges and opportunities associated with digitalization and transformation. The articles inside this cluster observe elements influencing SMEs' adoption of digital technologies, the role of entrepreneurial persistence in using digital transformation, and the potential profits of digital platforms for increasing market competitiveness. By addressing issues such as resource constraints, regulatory compliance, and skill development, this cluster offers practical guidance for SMEs navigating the complexities of digitalization and leveraging technology for sustainable growth.

Cluster 5: Value Creation through Innovation
In this cluster, the spotlight is on value creation through innovation and digital transformation across various sectors and industries. The articles within this cluster explore omnichannel practices for enhancing customer value, the evolution of social entrepreneurship in the digital age, and the drivers of successful digital transformation initiatives. By highlighting the interconnectedness of technology, strategy, and societal impact, this cluster underscores the importance of embracing innovation and digitalization as drivers of value creation and competitive advantage in today's rapidly evolving business landscape.

An interesting note is that, among the analyzed documents, the ones with the strongest links are associated with the digitalization and transformation of SMEs (Table 1). This situation highlights a considerable trend closer to the digitization and transformation of SMEs despite fewer articles on this subject matter than different clusters diagnosed within the evaluation. This observation also may indicate a growing recognition of the

Table 1. Main contributions for each cluster.

Cluster	Main References	Average total link strength per cluster
Cluster 1: Strategic Management in Digital Transformation	[20–22]	52,8
Cluster 2: Emerging Trends in Digital Entrepreneurship, and Sustainability	[23–25]	39,1
Cluster 3: Digital Capabilities and Business Model Innovation	[26–28]	68,5
Cluster 4: Digitalization of SMEs	[29–31]	71,5
Cluster 5: Value Creation through Innovation	[32, 33]	45,2

importance of digitalization for SMEs, both among researchers and in the context of organizations.

5 Conclusions and Future Work

The bibliometric analysis of research trends in the areas of "Digital Transformation", "Value Creation", and "Strategy" demonstrates that this topic has grown dramatically in the last 3 years. During this period, "Business Economics, Computer Science, and Engineering and Operations" are the predominant areas relative to the total areas available in WoS.

The bibliometric analysis and the SLR findings enable systematizing the current literature to answer the RQ. Conducting bibliographic coupling using the VOSviewer software allowed the determination of bibliographic coupling strength among the 72 papers, resulting in 5 clusters in emerging areas: Strategic Management in Digital Transformation of Organizations; Emerging Trends in Digital Entrepreneurship, and Sustainability; Digital Capabilities and Business Model Innovation; Digitalization and Transformation of SMEs and Value Creation through Innovation and Digital Transformation.

In future work, the objective is to extend the research, optimize the analysis of the identified clusters, and consider other criteria such as co-occurrence, co-authorship, or co-citations to solidify further this theme that is revolutionizing society as a whole. As a final note, it is essential to highlight that while the digitalization and transformation of SMEs are receiving increasing attention, it is no less important to consider that other areas, such as Emerging Trends in Digital Entrepreneurship, Sustainability, and Value Creation through Innovation and Digital Transformation may be overlooked. Therefore, these will be analyzed and considered in future work.

Acknowledgements. Regarding authors identified by numbers 1 and 2, this work is funded by National Funds through the FCT - Foundation for Science and Technology, I.P., within the scope of the project Ref. UIDB/05583/2020. Furthermore, we would like to thank the Research Centre in Digital Services (CISeD) and the Instituto Politécnico de Viseu for their support.

References

1. Li, F.: Leading digital transformation: three emerging approaches for managing the transition. Int. J. Oper. Prod. Manag. **40**(6), 809–817 (2020)
2. Ross, J.W., Beath, C.M., Mocker, M.: Designed for Digital—How to Architect Your Business for Sustained Success. MIT Press, Cambridge (2019)
3. Vial, G.: Understanding digital transformation: a review and a research agenda. J. Strateg. Inf. Syst. **28**(2), 118–144 (2019). https://doi.org/10.1016/j.jsis.2019.01.003
4. Tabrizi, B., Lam, E., Girard, K., Irvin, V.: Digital transformation is not about technology (2019). https://bluecirclemarketing.com/wp-content/uploads/2019/07/Digital-Transformation-Is-Not-About-Technology.pdf
5. Matarazzo, M., Penco, L., Profumo, G., Quaglia, R.: Digital transformation and customer value creation in Made in Italy SMEs: a dynamic capabilities perspective. J. Bus. Res. **123**, 642–656 (2021)
6. Yar Hamidi, D.: On value and value creation: perspectives from board research and practice in SMEs. In: Gabrielsson, J., Khlif, W., Yamak, S. (eds.) Research Handbook on Boards of Directors, pp. 420–443. Edward Elgar, Cheltenham (2019). https://wlv.openrepository.com/handle/2436/622917
7. Porter, M.E.: Competitive Advantage: Creating and Sustaining Superior Performance. The Free Press, New York (1985)
8. Goedhart, M., Koller, T.: The value of value creation. McKinsey Q. (2020). https://www.mckinsey.com/business-functions/strategy-and-corporate-finance/ourinsights/the-value-of-value-creation
9. Bratianu, C.: Toward understanding the complexity of the COVID-19 crisis: a grounded theory approach. Manage. Mark. Challenges Knowl. Soc. **15**(S1), 410–423 (2020). https://doi.org/10.2478/mmcks-2020-0024
10. Hess, T., Matt, C., Benlian, A., Wiesboeck, F.: Options for formulating a digital transformation strategy. MIS Q. Exec. **15**(2), 123–139 (2016)
11. Alzamora-Ruiz, J., del Mar Fuentes-Fuentes, M., Martinez-Fiestas, M.: Together or separately? Direct and synergistic effects of effectuation and causation on innovation in technology-based SMEs. Int. Entrep. Manag. J. **17**, 1917–1943 (2021). https://doi.org/10.1007/s11365-021-00743-9
12. Davenport, T.H., Westerman, G.: Why so many high-profile digital transformations fail. Harv. Bus. Rev. **9**, 15 (2018)
13. Venkitachalam, K., Bosua, R.: Perspectives on effective digital content management in organizations. Knowl. Process. Manag. **26**(3), 202–209 (2019)
14. Min, S., Kim, B.Y.: SMEs' digital transformation competencies on platform empowerment: a case study in South Korea. J. Asian Financ. Econ. Bus. **8**, 897–907 (2021)
15. Olawumi, T.O., Chan, D.W.M.: A scientometric review of global research on sustainability and sustainable development. J. Clean. Prod. **183**, 231–250 (2018). https://doi.org/10.1016/j.jclepro.2018.02.162
16. Cooper, C., Booth, A., Varley-Campbell, J., et al.: Defining the process to literature searching in systematic reviews: a literature review of guidance and supporting studies. BMC Med. Res. Methodol. **18**, 85 (2018). https://doi.org/10.1186/s12874-018-0545-3

17. Mishra, V., Mishra, M.P.: PRISMA for review of management literature – method, merits, and limitations – an academic review. In: Rana, S., Singh, J., Kathuria, S. (eds.) Advancing Methodologies of Conducting Literature Review in Management Domain (Review of Management Literature), vol. 2, pp. 125–136. Emerald Publishing Limited, Leeds (2023). https://doi.org/10.1108/S2754-586520230000002007
18. Qiao, W., Ju, Y., Dong, P., Tiong, R.: How to realize value creation of digital transformation? A system dynamics model. Expert Syst. Appl. **244**, 122667 (2023). https://doi.org/10.1016/j.eswa.2023.122667
19. Van Eck, N.J., Waltman, L.: Software survey: VOSviewer, a computer program for bibliometric mapping. Scientometrics **84**, 523–538 (2010). https://doi.org/10.1007/s11192-009-0146-3
20. Holopainen, M., Saunila, M., Ukko, J.: Value creation paths of organizations undergoing digital transformation. Knowl. Process. Manag. **30**(2), 125–136 (2023). https://doi.org/10.1002/kpm.1745
21. Klos, C., Spieth, P., Clauss, T., Klusmann, C.: Digital transformation of incumbent firms: a business model innovation perspective. IEEE Trans. Eng. Manage. **70**(6), 2017–2033 (2023). https://doi.org/10.1109/TEM.2021.3075502
22. Olsson, H.H., Bosch, J.: Going digital: disruption and transformation in software-intensive embedded systems ecosystems. J. Softw.: Evol. Process **32**, e2249 (2020). https://doi.org/10.1002/smr.2249
23. Kasperovica, L., Lace, N.: Factors influencing companies' positive financial performance in digital age: A meta-analysis. Entrep. Sustain. Issues **9**(2), 312–332 (2021)
24. Ratten, V., Jones, P.: New challenges in sport entrepreneurship for value creation. Int. Entrep. Manag. J. **16**, 961–980 (2020). https://doi.org/10.1007/s11365-020-00664-z
25. Rusly, F., Talib, Y., Hussin, M., et al.: Modelling the internal forces of SMEs digital adaptation strategy towards industry revolution 4.0. Pol. J. Manage. Stud. **24**(1), 306–321. https://doi.org/10.17512/pjms.2021.24.1.18Ratten
26. Ciampi, F., Demi, S., Magrini, A., Marzi, G., Papa, A.: Exploring the impact of big data analytics capabilities on business model innovation: the mediating role of entrepreneurial orientation. J. Bus. Res. **123**, 1–13 (2021). https://doi.org/10.1016/j.jbusres.2020.09.023
27. Cozzolino, A., Verona, G., Rothaermel, F.T.: Unpacking the disruption process: new technology, business models, and incumbent adaptation. J. Manage. Stud. **55**, 1166–1202 (2018). https://doi.org/10.1111/joms.12352
28. Tavoletti, E., Kazemargi, N., Cerruti, C., Grieco, C., Appolloni, A.: Business model innovation and digital transformation in global management consulting firms. Eur. J. Innov. Manag. **25**(6), 612–636 (2022). https://doi.org/10.1108/EJIM-11-2020-0443
29. Anwar, M., Scheffler, M.A., Clauss, T.: Digital capabilities, their role in business model innovativeness, and the internationalization of SMEs. IEEE Trans. Eng. Manage. **71**, 4131–4143 (2024). https://doi.org/10.1109/TEM.2022.3229049
30. Christofi, M., Khan, H., Zahoor, N., Hadjielias, E., Tarba, S.: Digital transformation of SMEs: the role of entrepreneurial persistence and market sensing dynamic capability. IEEE Trans. Eng. Manage. (2022). https://doi.org/10.1109/TEM.2022.3230248
31. Garzoni, A., De Turi, I., Secundo, G., Del Vecchio, P.: Fostering digital transformation of SMEs: a four levels approach. Manag. Decis. **58**(8), 1543–1562 (2020). https://doi.org/10.1108/MD-07-2019-0939
32. Costa Climent, R., Haftor, D.M., Chowdhury, S.: Value creation through omnichannel practices for multi-actor customers: an evolutionary view. J. Enterp. Commun.: People Places Glob. Econ. **16**(1), 93–118 (2022). https://doi.org/10.1108/JEC-07-2021-0100
33. Taneja, S., Siraj, A., Ali, L., Kumar, A., Luthra, S., Zhu, Y.: Is FinTech implementation a strategic step for sustainability in today's changing landscape? An empirical investigation. IEEE Trans. Eng. Manage. (2023). https://doi.org/10.1109/TEM.2023.3262742

A Framework for Wood Moisture Control in Industrial Environment

Ricardo Cláudio[1], Francisco Soares[1], Jorge Leitão[1], João Henriques[1,2(✉)], Filipe Caldeira[1,2], and Cristina Wanzeller[1,2]

[1] Polytechnic of Viseu, Viseu, Portugal
{pv29086,estgv14018,estgv15934}@alunos.estgv.ipv.pt,
{joaohenriques,caldeira,cwanzeller}@estgv.ipv.pt
[2] CISeD - Research Centre in Digital Services, Viseu, Portugal

Abstract. Wood can quickly degrade without adequate preservation care. Several different chemicals and processes can extend the life of the wood. These treatments bust the wood's durability and shield it from insects or fungi. Additionally, physical processes like drying reduce moisture content, discouraging microbial growth.

In the scope of wood moisture control, this work surveys the literature with the efforts undertaken as part of a Master's degree in Computer Engineering. The focus is on creating an intelligent control system to improve control mechanisms for the natural wood drying process. This method is acknowledged for its lower labor and energy costs but is associated with the drawback of slower and unstable drying cycles.

This work aims to improve control over the drying process by monitoring and recording humidity levels in each wooden stack.

Results denote the suitability of the proposed framework in acquiring accurate insights into the humidity levels, which is possible by triggering alerts about any anomaly during the process. This way, the need for constant on-site supervision is decreased, work efficiency is optimized, costs are reduced, and repetitive work is removed.

Keywords: Intelligent control system · natural wood drying · real-time humidity monitoring · IOT · predictive algorithms

1 Introduction

Wood requires an adequate preservation care in order to avoid its degradation. To extend its lifetime, chemicals and processes can be applied in that purpose. Treatments can contribute to improve wood's durability and shield it from insects or fungi. Moreover, physical processes like drying also reduce moisture content, discouraging microbial growth.

In the scope of wood moisture control, this work designs, implements, and evaluates a framework supported by a control system for the air-drying process of wood. This framework is devised to record and forecast moisture content throughout the natural drying process, ensuring optimal humidity levels during chemical treatment.

This section provides a comprehensive context for each crucial aspect, from the treatment method, chemicals, and types of wood to our proposed solution, experimental setup, and conclusions.

Copper naphthenate (CuNap) has been industrially used as a wood preservative in North America for at least 100 years and is considered one of the best alternatives due to European legislation. It's typically used to preserve dimensional lumber, utility poles, railroad crossties, posts, fences, and guardrails. Its properties allow more efficacy against fungi and wood-destroying insects. Its low toxicity is why it keeps gaining market acceptance and is classified as a chemical for "General Use" by the US EPA. Copper naphthenate sold for pressure treatment of wood is typically supplied as an 8% copper (as metal) concentrate for dilution. The properties of the treating solutions are dependent on the type of oil used as the carrier to ensure lower costs. The wood industry opts to use water as a solvent, increasing copper content to ensure the target [1].

Pinus pinaster (pine), *Eucalyptus globulus* (eucalyptus) are the predominant species used in the wood industry in Portugal, excluding *Quercus suber* also known as cork oak used specifically to produce cork, Eucalyptus being a fast-growing tree is hard to work with because of its density and hardness making it accepting less impregnation during the treatment process, so commonly *Pinus pinaste* is widely used for this type of applications by having a lower density allowing it to dry faster and having better absorption over the other two species [2].

Autoclave wood treatment is when chemical preservatives are forcefully applied to wood within a pressurized vessel. This method is widely regarded as the best and most effective approach, ensuring deeper penetration and retention of preservatives. It guarantees the retention values necessary to meet the appropriate wood standards.

The entity responsible for developing European Standards in the field of wood preservation is the European Wood Preservative Manufacturers Group (WEPM). This group standardizes the Timber Use Class system, defined in UNE EN 335, widely adopted across the timber industry. It helps determine the appropriate timber treatment, ensuring it meets the requirements for its intended use. The Timber Use Class system categorizes typical service situations into up to five use classes, guiding the treatment needed for optimal performance [3].

In this work, as a reference, a User Class (UC) of 4 is considered a Penetration Class (NP) level of 5, which means that 100% of the sapwood is impregnated by the wood preservative. The preservative retention is a minimum of 16,7 kg of CuNap per cubic meter of sapwood treated. Generally, wood is often dried to a moisture content of around 20%.

In the industry, kiln and air drying are two primary wood drying techniques. Each one presents advantages and limitations. Kiln drying is a controlled and faster alternative that utilizes heated chambers to regulate temperature, humidity, and airflow. This framework puts a strong emphasis on keeping the drying

Table 1. Use class and typical service situations

Use Class	Typical Service Situations
UC 1	Above ground, covered. Permanently dry
UC 2	Above ground, covered. Occasional risk of wetting
UC 3	Above ground, not protected. Exposed to frequent wetting
UC 4	In contact with ground or fresh water. Permanently exposed to wetting
UC 5	Permanently exposed to wetting by salt water

Table 2. Treatment recommendations for wood poles

Component	Poles
User Class	UC 4
Penetration Class	NP 5
Reference Retention	16,7 kg
Reference Moisture Content	20%

conditions. On the other hand, the air-drying method is a traditional approach that involves stacking and exposing lumber to natural environmental conditions. Although cost-effective and energy-efficient, this process is relatively slow, and the final moisture content can be influenced by fluctuations in weather conditions (Tables 1 and 2).

IoT devices capable of monitoring the moisture levels in each stack will be implemented to create a more efficient way to control the process. These devices will move this data via WiFi to a mobile Android device using the MQTT protocol. The data is transmitted to the back end, which is processed, analyzed, and stored.

The front end, included in the mobile app, will analyze the data from the back end and present a dashboard with the needed information by the operator to understand if a stack is eligible to be sent to the treatment facility, with the use of a prediction AI model, the operator will also be able to predict according to the weather conditions when each stack reaches the optimum level of moisture.

The back-end component will analyze data retrieved from the device, presenting a dashboard that consolidates information collected for each sensor.

This document is organized into five sections. Section 2 surveys the existing literature and background on intelligent control systems, wood drying processes, and integrating sensors in industrial settings. Section 3 showcases the proposed architecture used to implement a solution, the technical challenges, design decisions, and implementation details. Section 4 presents the practical setup implementation of the proposed three-layer framework. Section 5 presents and interprets the results obtained from the experimental work, including quantitative data, visualizations, and comparisons with traditional methods. Finally, Sect. 6 summarizes the essential findings and highlights potential paths for future research and improvements to the proposed framework.

2 Related Work

With the evolution of the Internet of Things (IoT) in wood management, there has been a significant advancement over the last several years concerning integrating IoT devices for timber moisture monitoring. This section will provide a detailed overview of the most recent developments and associated work, showing synergies with IoT devices and weather data to effectively monitor water content in wood stacks and predict optimum production times.

Saban, M. et al. [4] presented an IoT system for monitoring the moisture content of wood using Bluetooth Low Energy (BLE) technology. The proposed system comprises three main components: a compact and low-power BLE moisture sensor, a wireless BLE gateway, and a cloud-based monitoring platform. The BLE moisture sensor accurately measures the moisture content of wood and transmits the data wirelessly to the BLE gateway. The BLE gateway is an intermediary between sensor nodes and the cloud platform. It collects and aggregates the moisture data from multiple sensors and relays it to the cloud platform for real-time visualization and analysis. The cloud-based monitoring platform provides a user-friendly interface for managing moisture data, enabling users to monitor wood moisture levels remotely.

The performance evaluation made by J. Niklewski et al. [5] of a simple numerical moisture transport model, based on Fick's second law of diffusion, [6] as a tool to provide input to decay prediction models when assessing the durability of rain-exposed wood. The model is calibrated against high-resolution data from a previous study on rain-exposed wood joints and tested against field-test data presented by Isaksson and Thelandersson [7]. The results show that the model can reproduce the influence of rain on the moisture content in wooden specimens with sufficient accuracy for decay prediction. The error between the numerical result and the measurements tends to increase at high moisture contents and with decreasing temperature. However, the total error is reduced when the moisture content history is post-processed in a decay-prediction model, as the decay rate tends to decrease with decreasing temperature. The estimated service life varied with depth and the different decay models.

3 Architecture

This section describes the proposed architecture aimed at optimizing the wood drying process, outlining the core layers of the control system. Similarly to the interior approach in the case of the work of Saban, M. et al. [4], our framework adopts an exterior approach.

The proposed framework is built upon a three-layer approach to enhance the efficiency of the wood drying process. These layers include the Data Collection Layer, Data Relay Layer, and Data Analysis Layer.

3.1 Data Collection Layer

The primary objective of the data collection layer revolves around getting precise and real-time information concerning humidity levels in individual wood stacks. We intend to leverage low-computational-power devices strategically placed across the storage areas, ensuring accurate data collection without compromising efficiency. These devices play an essential role in monitoring and transmitting humidity data.

3.2 Data Relay Layer

This layer's primary goal is transmitting collected data for later analysis. To achieve this, our proposal involves developing a dedicated mobile application. This application is an efficient channel, facilitating data collection and transmission between the low-computing devices and the analysis layer. It ensures a continuous flow of essential data throughout the wood drying process.

3.3 Data Analysis Layer

The data analysis layer is at the framework's core, as it is essential for processing and thoroughly examining the collected humidity data. Python was planned to be used for solid data processing and detailed analysis. This layer is supported by algorithms extracting practical insights, enabling proactive decision-making and precise control of the wood drying process.

The three-layer, represented by Fig. 1 architecture, aims to provide a structured approach to capture, transmit, and analyze crucial data during wood drying. This modular framework is designed to optimize operational efficiency, reduce costs associated with manual labor, and ensure greater control over the drying process.

The mentioned architectural structure represents a modern and integrated approach to controlling the wood drying process. By implementing layers for data collection, transmission, and analysis, we aim to achieve an intelligent and efficient system for this process in the wood industry.

4 Experimental Work

Wood drying is a crucial yet slow and unstable process, incurring lower labor and energy costs. This section explores the experimental work designed to update this process, focusing on enhancing control mechanisms through an intelligent system.

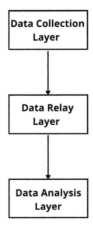

Fig. 1. Three-layer architecture diagram

4.1 Contextualization

The traditional method of natural wood drying is composed of prolonged cycles and inconsistent results. This Experimental Setup aims to establish an efficient control system that monitors humidity levels within individual wood stacks while using weather forecasting data for predictive analysis.

4.2 Objectives

The primary goal is to set a three-layer system gathering data collection, relay, and analysis processes. This system is focused on implementing low-computing devices, like Raspberry Pi, to measure humidity levels in wood stacks and relay data effectively. Besides that, the system is focused on developing an Android application to collect and transmit data and creating a back-end system with Django that manages stacks, sensors, and real-time data. This experimental setup works towards providing insights into humidity levels, optimizing work efficiency, and reducing costs associated with manual labor.

The three-layer Experimental Setup design to optimize wood drying processes, shown in Fig. 2, is, as mentioned before, composed of a data collection layer, a data relay layer, and a data storage and analysis layer.

4.3 Data Collection Layer

The Data Collection Layer covers the deployment of a low-computing device. The Raspberry Pi is strategically placed to monitor and measure humidity levels within individual wood stacks. This device collects real-time data from sensors embedded in the wood storage areas. Bluetooth technology transmits the collected humidity data from the Raspberry Pi to the subsequent Data Relay Layer. This transmission ensures the transfer of data.

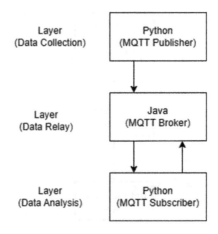

Fig. 2. Development Experimental Setup

Technical Implementation Details. Raspberry Pi Deployment for Data Collection:Raspberry Pi devices were strategically located to measure humidity within individual wood stacks and served as the foundational data collection mechanism. These devices collected real-time data from embedded sensors in the wood storage areas.

4.4 Data Relay Layer

The Data Relay Layer is an intermediary hub that receives normalized data from various sensors across the wood stacks. An Android-based application is responsible for processing, aggregating, and normalizing this data before transmission.

Technical Implementation Details. MQTT Library Implementation in Android: The system's communication and data transmission in the Android application were facilitated by leveraging the MQTT library. This choice enabled a streamlined and efficient data relay between sensors and the relay layer. The usage of MQTT can be depicted in Fig. 3.

UI Construction with Jetpack Compose: The Android application's user interface (UI) was constructed using Jetpack Compose, providing a declarative way to build UIs and enabling dynamic and interactive user experiences.

MVVM Model-View-ViewModel: The system's Experimental Setup follows the MVVM design pattern, facilitating a clear separation of concerns and enhancing the scalability and maintainability of the application.

Kotlin as Android Programming Language: The development was executed using Kotlin, capitalizing on its concise syntax, strong type safety, and interoperability with existing Java code, thereby enhancing development productivity.

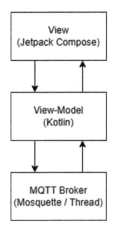

Fig. 3. Android Development Experimental Setup

4.5 Data Analysis Layer

At the core of the Experimental Setup, the back-end system is developed using Django. This layer serves as the repository for all collected and processed data. It offers a robust infrastructure capable of managing a large amount of data, sensor parameters, and historical records of humidity levels in wood stacks.

Technical Implementation Details. Django Back-End for Data Storage and Analysis: The core back-end system, developed using Django, was the repository for all collected and processed data. It offered a robust infrastructure capable of managing vast data, sensor parameters, and historical humidity records from wood stacks.

Django Back-End for Data Storage and Analysis using MVC: The data storage and analysis layer leverages Django, implementing the Model-View-Controller (MVC) design pattern. This architectural pattern within Django ensures efficient management and organization of collected data, enhancing scalability and maintainability.

In conclusion, the development of the Experimental Setup for the wood drying control system involved the implementation of a three-layer model, each playing an essential role in the process.

The data collection layer relied on strategic Raspberry Pi deployment, ensuring real-time data collection regarding humidity levels in individual wood stacks.

Utilizing the MQTT library in Android and implementing the MVVM pattern in the data relay layer ensured efficient communication, smooth data handling, and organized transmission. Additionally, implementing Jetpack Compose provided a dynamic user interface.

The data storage and analysis layer centered around Django, incorporating the Model View Controller model to efficiently manage large volumes of data, guaranteeing secure storage and structured analysis of humidity records. This architecture integrates advanced technologies and robust design patterns and aims to optimize the wood drying process.

5 Results and Analysis

This section describes the defined requirements for the system to run, and in each layer, its implementation and evaluation details are performed to understand the feasibility of the proposal.

5.1 Defining the Requirements

The requirements are defined as a particular value or minimum level confirming that the generated results are valid and imposed by the limitations. Various requirements are set according to each layer, and each aims to confirm the expected result.

Take, for example, a one-hectare drying area. Wood stacks can range from 2 m used for fences to 12 m in length (used for telephone poles). It is, therefore, considered that the average occupancy value of a stack at any given time is 14 m^2. For natural drying to take place using the air-drying method, one m^2 is required for air circulation.

Considering the WiFi sensors available on the market at affordable implementation costs, the WiFi 5/802.11ac standard was used. Given the maximum performance of 92 linear meters of range, with the natural obstacles created by each stack, a value of 90 m^2 of the relay system range is deemed appropriate [8].

We conclude that each stack's average occupancy is 16 m^2 and that the average Wi-Fi sensor covers an area of 90 m^2. With this, the physical limitations outline the requirements can be estimated as in Table 3.

Table 3. System requirements

Requirement (Rq)	Value
Number of sensors (NoS)	1250
Sensors near relay (SnR)	12
Reads per day (RpD)	15000

These three values will serve as the foundation for the subsequent section. To substantiate the presented values, each calculation has been conducted using the following formula:

- NoS = 1 hectare/16 m^2 per stack.
- SnR = 90 m^2 Wifi range/16 m^2 per stack.
- RpD = NoS * 1 read per 1 h * 0.5 coverage.

The final formula used the average value of one reading per hour, considering the gradual process of wood drying, which spans up to six months. This timeframe is deemed appropriate and can be shortened if necessary.

5.2 Data Collection Results and Evaluation

In this phase, we get into the outcomes of our data collection efforts and evaluations aimed at implementing the outlined system requirements. This stage played a crucial role in validating the feasibility of the proposed system, relying on observed data and calculations to support the initial specifications. The objective was to assess the practical approach of the defined system requirements through a meticulous data acquisition and analysis process.

This section addresses three key aspects: determining the number of sensors required, establishing their proximity to the Wi-Fi relays, and defining the frequency of data readings. Careful computations supported by real-life situations have been performed to ensure the effective operation of our system within its designated area.

We focus on testing each sensor's ability to collect, store, and relay data to validate the system's capacity to meet the specified requirements. The upcoming test case specifically aims to confirm the system's capability to manage a substantial volume of stored data. It is worth noticing that data quality is taken for granted by the sensor vendor, and the relay aspect will be addressed in the subsequent test case.

This test is conducted from the perspective of a sensor without any connection. All recorded humidity values are stored and will be published when a new connection to a broker becomes available.

As depicted in Fig. 4, a sample of 1000 readings taken by the sensor was used to verify the system's capability. The disk space and time for each read and write were recorded. As a result, an average usage of

- Disk space per message: 143 kb
- Average read and write time to disk: 6.25 ms

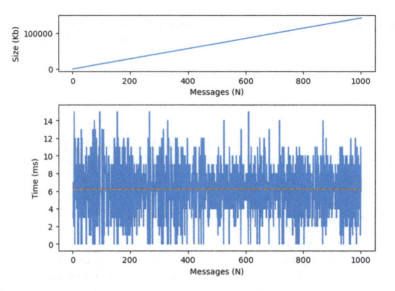

Fig. 4. Messages relay stress

5.3 Data Relay Results and Evaluation

The data relay layer, implemented in Kotlin and assisted by Moquette's lightweight JWM implementation, serves as an MQTT broker within an Android application. This application runs on a Helio G95 platform supported by 2×2.05 GHz & 6×2.0 GHz processors and 6 GB of RAM. The leading hardware and software configurations used to run the proposed system are shown in Table 4.

Table 4. Data relay hardware and software configurations

Item	Version
OS	Android 13
Chipset	Mediatek MT6785V/Helio G95
RAM	6 GB
WLAN	Wi-Fi 802.11 a/b/g/n/ac

Scalability. Since we focus on processing many connections and data transfer, we evaluated our framework concerning scalability and throughput, considering the increasingly connected sensors.

Figure 5 shows the efficiency result comparing the processing time and the number of connected clients. It's expected that with the increasing number of

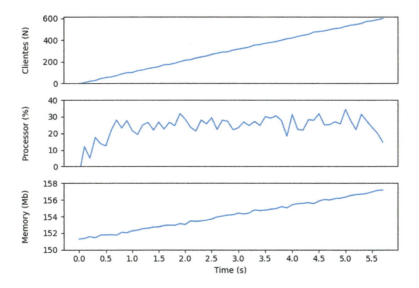

Fig. 5. Clients connection stress

clients and data transference, the amount of processing time and resources used will also rise. However, the proposal can accommodate a huge number of connections compared to what is needed by the requirement threshold. To prove this, the requirements are multiplied by 100 in the test case presented, thus guaranteeing the application's behavior regarding many connections.

The trial was completed in 5.8 s, with the ability to connect more than 100 clients per second. The average processing capacity used was around 27%, with each client connected to the broker using approximately 10kb of memory.

Throughput. With the last test, it was possible to prove the framework's capability to handle many client connections. The next step is to assess the system's ability to transport the data published by each client. The Throughput test is run with the help of a client that connects and subscribes to the messages published by each sensor.

To validate the ability to meet the minimum requirements, each sensor publishes 100 messages at each 100-millisecond interval. The test is performed through threads that execute simultaneously to simulate simultaneous message publication between sensors. Additionally, in the middle of the test, the broker is turned off for 2 s to imitate the adverse conditions to which the application will be subjected.

The results of this test are presented in Fig. 6. They denote the transaction capacity of around 550 messages per second, with an average processing capacity utilization of 45%. A total of sixty thousand messages were transmitted during the test, which concluded in 13.1 s, including a two-second interruption in the transmission service.

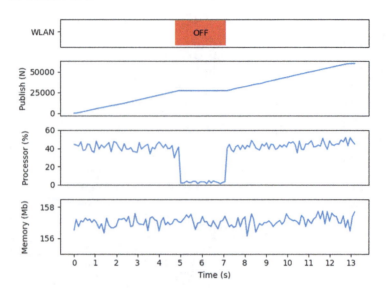

Fig. 6. Messages relay stress

6 Conclusions and Future Work

The proposed framework has been proven to cover the implementation requirements, demonstrating robust quality in all evaluated aspects.

The architecture revealed remarkable scalability, allowing it to efficiently fit the demand for different volumes of data or answer clients. In addition, the resilience of the system to data loss during collection and transactions was excellent, guaranteeing the integrity of the information throughout the process.

The results obtained from the analysis of the predictive models reinforced the system's effectiveness, consolidating its ability to provide accurate and valuable forecasts for the operators. In this way, the proposed solution not only met expectations but exceeded them, standing out as an efficient framework.

The provided architecture implements a solution to the problem presented, and with the increasingly low cost of sensors available on the market, the possibility of implementation arises. The system itself also serves as an example for the implementation of other similar systems involving data acquisition, processing, and presentation.

6.1 Future Work

Implementing this Proof of Concept of the proposed framework only gathers the necessary steps. However, several improvements must be considered before being deployed in an industrial environment. They range from improvements regarding energy savings, data transition performance, or even enhancements in the analysis of the data collected.

A low-power Bluetooth data transition protocol could be implemented between the first two layers of the system, thus achieving considerable energy savings in each sensor. Another option could be to use the ZigBee Protocol to transition between the sensors, allowing for a constant data flow.

The data analysis layer cannot only consider the weather forecast. Other factors include the speed at which the pile is drying up to the time the forecast is computed. It will subsequently consider the season in which the drying is taking place. The implementation of live data via the MQTT protocol can also be considered in this layer.

The mobile application can be expanded by creating a native implementation, allowing the sensors to be configured by proximity, or by subscribing to the MQTT server to launch alerts for sensor failures or stacks with proper humidity for treatment.

Acknowledgements. This work is funded by National Funds through the FCT - Foundation for Science and Technology, I.P., within the scope of the project Ref UIDB/05583/2020. Furthermore, we thank the Research Centre in Digital Services (CISeD), the Polytechnic of Viseu, for their support.

References

1. Brient, J.A., Manning, M.J., Freeman, M.H.: Copper naphthenate: protecting America's infrastructure for over 100 years and its potential for expanded use in Canada and Europe. Wood Mater. Sci. Eng. **15**(6), 368–376 (2020)
2. Uva, J., et al.: National forestry inventory - IFN6°. ICNF (2015)
3. Regulation (EU) no 305/2011 of the European parliament and of the council (2011). https://eur-lex.europa.eu/legal-content/EN/TXT/?uri=CELEX%3A32011 R0305. En 335:2013 - Durability of Wood and Wood-based Products - Use Classes: Definitions, Application to Solid Wood and Wood-based Products
4. Saban, M., Medus, L.D., Casans, S., Aghzout, O., Rosado, A.: Sensor node network for remote moisture measurement in timber based on bluetooth low energy and web-based monitoring system. Sensors **21**(2), 491 (2021)
5. Niklewski, J., Fredriksson, M., Isaksson, T.: Moisture content prediction of rain-exposed wood: test and evaluation of a simple numerical model for durability applications. Build. Environ. **97**, 126–136 (2016)
6. Nicolin, D.J., Rossoni, D.F., Jorge, L.M.M.: Study of uncertainty in the fitting of diffusivity of Fick's second law of diffusion with the use of bootstrap method. J. Food Eng. **184**, 63–68 (2016)
7. Isaksson, T., Thelandersson, S.: Experimental investigation on the effect of detail design on wood moisture content in outdoor above-ground applications. Build. Environ. **59**, 239–249 (2013)
8. Park, M.: IEEE 802.11ac: dynamic bandwidth channel access. In: Proceedings of the IEEE International Conference, pp. 1–5 (2011)

EgiCool - IoT Solution for Datacenter Environmental Control and Energy Consumption Monitoring

Vicente Gonçalves[1,2], José Daniel[1,2], Celestino Gonçalves[1(✉)], Filipe Caetano[1], and Clara Silveira[1]

[1] Instituto Politécnico da Guarda, Guarda, Portugal
{celestin,caetano,mclara}@ipg.pt

[2] Agência de Desenvolvimento para a Sociedade da Informação e do Conhecimento, Guarda, Portugal
{vicente.goncalves,jose.daniel}@adsi.pt

Abstract. The increasing demand for digital services and cloud computing has brought the problem of energy consumption of datacenters and the quest for innovative solutions. Among the emerging approaches, free cooling stands out as a promising technique that allows a significant advancement in the operational efficiency of datacenters, contributing to environmental sustainability. The objective of this work is to detail the development of an IoT solution for monitoring the physical space of a datacenter, with the main purpose of evaluating solutions that led to a reduction in its energy consumption, through the implementation of free cooling. We intend to monitor the energy consumption, temperature and humidity inside and outside the datacenter. The data will be collected by sensors that are connected to ESP-32 microcontrollers, sent via Wi-Fi, using the MQTT protocol, to Amazon Web Service (AWS) IoT, stored in AWS DynamoDB, processed and presented in a web application. We concluded the first phase of development of the proposed IoT solution with the construction of a prototype and a feasibility study analysis.

Keywords: datacenter energy efficiency · free cooling · IoT solution · AWS

1 Introduction

In recent years, Web 2.0 and cloud computing have changed the paradigm of the internet. The web has evolved from being just a hypertext system to a complex ecosystem, composed of billions of nodes from countless distinct entities distributed across the planet. The demand for cloud services has substantially increased, primarily due to the low cost when compared to the classical system. It is predicted that the global datacenter market will reach $143.4 billion USD by 2027, with a 13.4% global growth rate compared to $59.3 billion USD in 2020 [1]. This significant growth directly impacts global energy consumption, and efforts have been made to improve the energy efficiency of equipment and infrastructures. Major technology firms are increasingly acknowledging the inherent benefits of high-latitude Arctic regions for datacenter placement. These areas offer abundant renewable energy, along with naturally cold air and optimal humidity levels. Rather than relying on artificial cooling methods, modern datacenters are strategically designed or being constructed to leverage local weather patterns, aiming to minimize energy consumption associated with cooling systems.

The paper is structured as follows. In Sect. 2, we address the state of the art, where we discuss the problem, the proposed solution, an exploratory analysis of the solution's feasibility, and related work. In Sect. 3, we delve into the system design, referencing the development methodology and requirements analysis. Section 4 covers system development, presenting the architecture and the different components comprising it. In Sect. 5 we outline the tests and results obtained. Finally, in Sect. 6, we present the conclusions.

2 State of the Art

The increasing demand for digital services has brought energy consumption of datacenters and the quest for innovative solutions to mitigate their environmental impact and associated operational costs to the forefront. Among the emerging approaches, free cooling stands out as a technique that utilizes external air to cool equipment, reducing reliance on conventional cooling systems and consequently lowering energy consumption. This strategy not only represents a significant advancement in the operational efficiency of datacenters but also contributes to the environmental sustainability of the technology industry.

2.1 The Problem: Energy Consumption of a Datacenter

A datacenter is a physical facility designed to centralize the technological resources of one or more organizations. It provides physical security, electrical power, cooling, and network infrastructure to the IT equipment housed within it. The equipment housed in the datacenter, in turn, provides processing capacity, storage, and network communication. Therefore, a datacenter can be defined as a facility with all the necessary resources for the storage and processing of digital data and its supporting areas [2].

Energy in datacenters is consumed by two main categories of equipment: IT equipment and the infrastructure supporting IT equipment, which provides the thermal environment necessary for the operation of IT equipment [3]. Between 2010 and 2018, energy consumption of datacenters in the EU28, calculated using the Borderstep Institute model, increased from 53.9 TWh/a to 76.8 TWh/a. In 2018, the energy consumption of datacenters accounted for 2.7% of total energy consumption in the EU28 [4].

Within a datacenter, we can distinguish between the energy consumption of IT components (servers, network, storage, etc.) and the energy consumption of the infrastructure itself (cooling, redundancy, etc.). During the period under analysis, the energy consumption of IT equipment increased by approximately 65%, from 26.5 TWh/a in 2010 to 43.8 TWh/a in 2018. During the same period, the energy consumption of the infrastructures increased by only about 20%, which translates into an effective improvement in energy efficiency. The average value of PUE (Power Usage Effectiveness) in the EU28 decreased from 2.03 in 2010 to 1.75 in 2018 [4]. Among the various components of a datacenter, air conditioning and other cooling systems are among the largest energy consumers.

The datacenter is installed at the School of Technology of the Polytechnic Institute of Guarda and serves the Agency for the Development of the Information Society and Knowledge (ADSI) and the Polytechnic Institute of Guarda. It comprises 12 racks housing IT equipment, each powered by 2 independent circuits and backed by 2 UPS units. Power is sourced from the grid or a generator during outages. Presently, there's no available estimate for the annual energy consumption.

2.2 The Solution: Free Cooling

Cooling equipment is critical in a datacenter. Heat dissipation is a determining factor for the availability and reliability of hardware. Traditionally, datacenter cooling relies on air conditioning systems based on mechanical refrigeration. The energy consumed by the cooling system represents between 30 to 50% of total consumption [5]. Free cooling involves using natural climate to cool the datacenter, as opposed to the more traditional method using conventional systems like air conditioning [6]. Direct free cooling is the simplest. It is an effective method that improves the energy efficiency of datacenter cooling systems, taking advantage of the cooler external air. Air ducts, fans, and filters enable the effective use of external air to fully or partially replace air conditioning cooling [7]. The conceptual principle of free cooling is to use external air as a cooling source, taking advantage of naturally lower temperatures available in certain regions and times of the year. Utilizing favorable weather conditions to reduce the need for traditional mechanical refrigeration systems enables more efficient energy use, reduced operational costs, and a lower carbon footprint for datacenters [8, 9].

Free cooling techniques in datacenter cooling have been widely used around the world. Cloud data centers are increasingly being built in cold, dry areas, or near bodies of water, utilizing free cooling systems like dry coolers for access to cold air or water. For instance, Iceland has emerged as one of the world's most cost-effective locations for data centers due to its favorable climate [10]. In 2011, Google established a datacenter in Hamina, Finland. It is one of Google's most advanced and efficient datacenters. It uses seawater from the Bay of Finland in the cooling system and will be powered by 100% renewable energy. In 2013, Facebook installed a datacenter in Sweden, utilizing outside air to cool servers and repurposing excess heat to warm office spaces [11]. In 2016, they announced plans to install a new datacenter in Ireland [12]. In 2018, the Microsoft Project Natick team installed a datacenter off the coast of the Orkney Islands (Scotland), 117 feet deep on the seafloor [13].

2.3 Exploratory Viability Analysis of the Solution

The Advanced Computing Laboratory at the University of Coimbra installed a free cooling system in its datacenters. When outside temperatures are lower than inside temperatures, the datacenter utilizes cold ambient air for equipment cooling. This free cooling system resulted in an energy saving of approximately 90 MWh/year, corresponding to a reduced PUE by 0.2 points, while providing similar environmental conditions to the old system (same temperatures). The installation costs of the system were around €6,040.

Taking into account the financial incentive, the payback period was estimated to be about one year [4].

Our datacenter is located in the city of Guarda. Based on the data collected from the IPMA Portal [14], we conducted a comparative study between the normal air temperatures in the city of Guarda and the city of Coimbra, for the period between 1981 and 2010. The objective was to establish a comparison regarding normal temperatures, per month of the year, in each city. Throughout the study period, temperatures remained consistently lower in Guarda. In the graphs of Fig. 1, we can observe that the average number of days with a maximum temperature exceeding 25 °C is significantly higher in Coimbra. For example, in July in Coimbra, 25.4 days had a maximum temperature above 25 °C and 9 days above 30 °C, while in Guarda, in the same month, 20.1 days had a maximum temperature above 25 °C and 8.4 days above 30 °C. We can also observe that Guarda had 34 days per year with a minimum temperature equal to or below 0 °C, while Coimbra had only 5 days. Furthermore, the number of days with a maximum temperature exceeding 25 °C is lower in the city of Guarda, with 41 days compared to 117 days in the city of Coimbra.

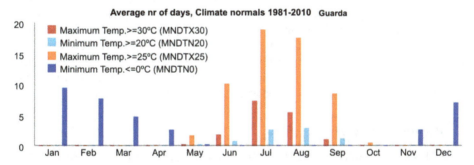

Fig. 1. Average Number of Days, Climatological Normals 1981–2010 - Guarda

Taking into account the observed data and comparing it with the case of the University of Coimbra, and assuming that the equipment should be maintained at a temperature below 25 °C, we estimate that our datacenter could operate on free cooling for approximately 240 days per year. With a similar investment, it would be possible to achieve a reduction of around 66% in energy consumption spent on cooling, representing 26.4% of the total consumption.

3 System Design

At this section we address the design of the system and how the choice of a structured development methodology provided us with effective development and an efficient response to the proposed challenges. We also reflect on requirements analysis and system characterization.

3.1 Development Methodology

Implementing a free cooling system in a datacenter can be a technical challenge with some complexity where it would be very easy to lose objectivity without an efficient development methodology. The Agile Scrum methodology has stood out for its efficiency in structuring and managing Internet of Things (IoT) projects [15]. It offers flexibility and adaptability, through iterative processes, which enable a quick response to changes in project requirements. In the specific context of this project, by adopting this methodology, we were able to divide the project into short sprints, which allowed us to continuously assess progress, integrate new features and make recurring adjustments. We were able to optimize the entire implementation process, allowing quick adaptation to changes and efficient collaboration. We divided the project into four weekly sprints, with the aim of producing tangible improvements in each of them. The defined routine includes daily meetings to evaluate progress, adjust strategies and carry out micro tests. The process was necessarily iterative and based on constant adaptations.

In the first Sprint the main objective was the integration of humidity, temperature, electric current and voltage sensors with the microprocessor. The sprint 1 backlog included tasks such as defining the sensors to be used, defining what data would be collected and developing the corresponding code for implementation. During the Sprint the focus was placed on the efficient implementation of the sensors and the development of the code to collect and process the data obtained by them. We carried out tests to ensure the accuracy of the collected data. Daily meetings were used to evaluate progress, adjust strategies as necessary and carry out micro tests. At the end of Sprint 1 we already had the data collection system implemented and ready to send data to the cloud.

3.2 Requirements Analysis

The analysis of functional and non-functional requirements was carried out to ensure that all specificities of the system were defined. With functional requirements we define the critical functionalities that the system must offer, with non-functional requirements we define the characteristics that could facilitate the use and accuracy of the system. We consider that the ubiquity of the system gives it efficiency in different scenarios.

The developed system presents characteristics of a ubiquitous system. It is context-aware because it is aware of its surroundings as it is capable of measuring temperature, measuring humidity and detecting whether the datacenter door is open or closed. It has an implicit interaction (iHCI) because it is capable of carrying out interactions implicitly and naturally, sending alerts to users automatically and without the need for direct intervention. It is distributed because it is made up of several modules connected together in a network. It is intelligent because it makes use of weather forecasts to estimate free cooling duration for the upcoming days. It is autonomous because it has the ability to autonomously decide when to activate the refrigeration systems and humidity control systems.

Functional Requirements. The system must monitor and control the temperature inside the datacenter and activate the cooling system when the inside temperature exceeds 25 °C. When the outside temperature is at least 5 °C lower than inside, free cooling starts. Otherwise, the air conditioning turns on. The two cooling systems can never operate simultaneously and should only be deactivated when the inside temperature is below 22 °C. The system must send abnormal temperature notifications when the interior temperature exceeds 27 °C and send an alert when it detects that the door is open. The system must monitor and control the relative humidity inside the datacenter, activating the dehumidifier if it exceeds 60% or the humidifier if it drops below 40%. The dashboard must be automatically adjusted based on user interactions, organizing reports according to descending order of access. The system must calculate and present an estimate of free cooling usage based on weather forecasts for the next 5 days.

Non-functional Requirements. The system must carry out activation, deactivation and alert sending actions within a maximum time of 1 min, ensuring a quick response to environmental conditions. Temperature, humidity, voltage and current measurements must be highly accurate in order to guarantee the reliability of the system's actions. The dashboard must be responsive, in order to adapt to the resolution of the different devices used by system users.

4 System Development

At this section we address the development of the system and constituent modules.

4.1 System Architecture

The system consists of two main components: The data collection component and the data processing component. In Fig. 2 we can see a representative diagram of the proposed architecture. Data processing is carried out using two ESP32 microcontrollers and sent to the cloud using the MQTT protocol via Wi-Fi.

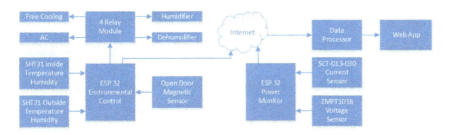

Fig. 2. Architecture of the system

The data collection component is divided into two modules. The environmental data collection module (Environmental Control), which collects data through the SHT31 sensors and the magnetic door sensor, and the energy consumption monitoring module (Power Monitor), where data is collected using SCT-013-030 and ZMPT101B sensors. In the data processing component, data is submitted via the MQTT protocol to the AWS IoT platform, stored in a NoSql DynamoDb database, processed using Lambda functions and made available through the API Gateway to be presented in a WebApp.

4.2 Environmental Control

The implemented environmental control system consists of SHT31 temperature and humidity sensors, outside and inside the datacenter, and the door sensor. The data collected is processed by an ESP32 microcontroller connected to 4 relays that control the actuators: the free cooling system, the humidifier, the dehumidifier and the air conditioning. The SHT31 sensors provide accurate data on temperature and humidity, allowing the ESP32 to make decisions autonomously.

The door sensor allows the system to detect and send an alert when the door is open for more than one minute. The decision to make the environmental control module an autonomous unit capable of acting independently of the internet connection was taken to ensure that the system continues to function reliably even when the internet connection is not available. The four relays act as electronically controlled switches to activate the corresponding actuators. The free cooling system is activated when external conditions are favorable, allowing outside air to refrigerate the equipment.

The need to activate the cooling systems is assessed only once every minute. As we can see in Fig. 3, the air conditioning is activated when the internal temperature reaches a value above 25 °C and the outside temperature is not at least 5 °C lower than the inside temperature. If the outside temperature is at least 5 °C lower than the inside temperature, free cooling will be activated instead of air conditioning. To prevent cooling systems from always turning on and off, a hysteresis interval of 3 °C was defined that ensures that the system only turns off when the internal temperature is below 22 °C. When the internal temperature exceeds 27 °C, the temperature alert is activated. Whether or not humidity control systems are activated is assessed once every 5 min. The humidifier is activated when the internal humidity is below 40% and the dehumidifier is activated when the internal humidity is above 60%.

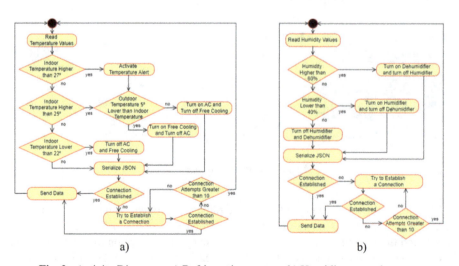

Fig. 3. Activity Diagrams: a) Refrigeration system, b) Humidity control system

The collected data is sent, via WiFi using MQTT, to the AWS Cloud IoT service every minute. We also use a 16 × 2 LCD that allows the values to be displayed in real time on the device. In Fig. 4 we can see the implemented system.

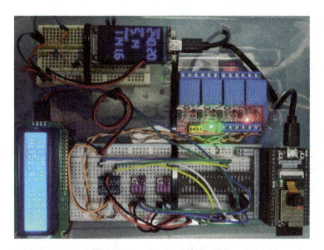

Fig. 4. Implemented System

4.3 Power Monitor

Data relating to energy consumption is collected using SCT-013-030 clamp meters and ZMPT101B voltage sensors. The voltage sensors are capable of measuring electrical line voltage and the clamp meters are capable of measuring electrical currents in a conductor. They are accurate in measurement and non-invasive. With the combination of these two devices, it is possible to obtain real-time data on electrical current and voltage and calculate the energy consumption of the datacenter. On the other hand, the voltage sensor connects directly to an analog input pin of the ADC of the ESP32 microcontroller because its output signal is a sine wave always with positive values. The sensor's built-in potentiometer was used to adjust the output signal, making it range from 0 V to 3.3 V. In order to connect the current sensor to the analog input pin of the ADC of the microcontroller, it was necessary to create an offset circuit (Fig. 5) in order to shift the sinusoidal signal so that it always has positive values.

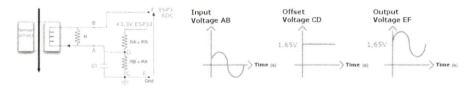

Fig. 5. Connection of the SCT sensor with the 1.65V offset circuit

Similar to the environmental control module, the data is transmitted via WiFi using MQTT to the AWS Cloud IoT service every minute. The ESP32, equipped with an LCD, dynamically displays real-time values.

4.4 Data Processor

AWS [16] is a cloud computing platform that provides services in more than 190 countries [17]. For this project, we opted for AWS, its services enabled us to develop an integrated and centralized solution for data collection, storage, and accessing information. In Fig. 6 we can observe the flow followed through the AWS services used.

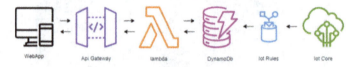

Fig. 6. AWS Service Flow

AWS IoT Core is a cloud service that enables IoT devices to connect to other AWS services through a device gateway and a message broker. Registration of IoT devices within the AWS console was required for both the "ESP32 Environment Control" and the "ESP32 Power Monitor". The tables to store the data were created in AWS DynamoDB. Data is forwarded from IoTCore to DynamoDB through the creation of Rules. To access the information stored in DynamoDB we used AWS Lambda. We created two Lambda functions for the API backend. These functions read data from DynamoDB and use API Gateway events to determine how to interact with that data. To create an API and respective endpoints, we used the AWS API Gateway.

4.5 Web App

The developed Web App consumes the created API in AWS API Gateway and provides interfaces for monitoring and generating reports. The initial interface offers a comprehensive view of the system with dynamic graphs of temperature, humidity, energy consumption and alerts. Notifications are sent via email and push notifications.

Fig. 7. Developed monitoring platform

Hosting is on AWS Lightsail's Virtual Private Server (VPS) instances. Figure 7 shows the initial interface displaying real-time data graphs for inside/outside temperatures, relative humidity levels, power metrics, and alerts for critical temperature and datacenter door alerts. The developed platform also provides access to the reporting interface, such as the free cooling usage estimate report based on the weather forecast (Fig. 8).

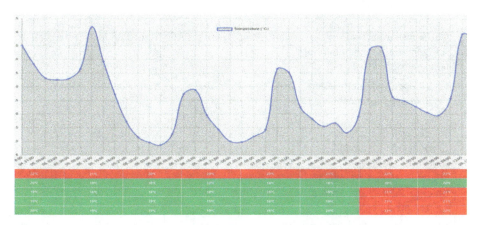

Fig. 8. Free cooling usage estimate report based on weather forecast

5 Tests and Results

We defined 9 scenarios to test and evaluate the effectiveness and reliability of the developed system. Table 1 details the defined scenarios, presenting the specific conditions, expected and obtained results. As can be seen, all tests passed successfully.

Table 1. Test cases to validate system operation.

	Test scenario	Expected outcome	Obtained result
1	Check activation of free cooling - Inside temperature above 25 °C and outside temperature at least 5 °C lower than the inside temperature	Turn on FC & turn off AC	Turned on FC & turned off AC
2	Check activation of the air conditioning - Inside temperature above 25 °C and the outside temperature is not lower than the inside temperature by at least 5 °C	Turn on AC & turn off FC	Turned on AC & turned off FC
3	Check sending of abnormal temperature notification - Inside temperature above 27 °C	Send abnormal temperature alert	Sent abnormal temperature alert
4	Check deactivation of refrigeration systems - Interior temperature below 22 °C	Turn off FC & turn off AC	Turned off FC & turned off AC
5	Check open door alert - Keep the door open for more than 1 min	Send open door alert	Sent open door alert
6	Check automatic adjustment of the dashboard - With the same user access the "Forecast" report 5 times, 4 times to "Alerts" and 3 times to "Temperature"	Order the reports: 1st Forecast, 2nd Alerts, 3rd Temperature	The reports were presented: 1st Forecast, 2nd Alerts, 3rd Temperature
7	Check estimated free cooling usage based on the weather forecast - Enter weather forecast data for 5 days and compare the system estimate with previously calculated values	The system must estimate 78 h of HR use	The system estimated 78 h of HR use

(continued)

Table 1. (*continued*)

	Test scenario	Expected outcome	Obtained result
8	Check voltage values - Compare the voltage values displayed on the system with the values measured using a multimeter	Values must have a variation of less than 5%	235.24 V system, Measured 239.16 V, Variation < 5%
9	Check current values - Compare current values displayed on the system with values measured using a multimeter	Values must have a variation of less than 5%	System 0.26 A, Measured 0.25 A, Variation < 5%

6 Conclusions and Future Work

We started this project with the aim of developing an IoT solution to monitor the datacenter environmental conditions and energy consumption. With the prototype, it was not our objective to create an exact reproduction of our datacenter. The main objective was to evaluate the implementation of a real solution, using the selected sensors and equipment, or equivalents, and verify if it allows us to obtain an estimate of the reduction in consumption. As we mentioned, the collected data is processed by two ESP-32 microcontrollers and sent to AWS IoT through the MQTT protocol, using a Wi-Fi connection. The data is stored in a NoSQL DynamoDB database and published through APIs created in API Gateway, also from AWS. Finally, this data is processed and presented in a web application developed by us and hosted on AWS Lightsail. We find this approach effective for several reasons. The equipment used is precise; the chosen cloud services are robust, simple to use and easily scalable; the implementation cost is residual and recoverable in a short period of time. It is important to note that, in this project, no real data has been collected so far due to the impossibility of assembling the equipment in the datacenter.

The choice of the case study of the University of Coimbra, was related to the similarities, in scale, between the two datacenters, as they are geographically relatively close.

We completed the first phase of development of the IoT solution for monitoring the physical space of the datacenter with the construction of the prototype and feasibility analysis. The collection of real data, for analysis and evaluation of the effectiveness of the solution, will be carried out in the next phase. This collection will allow for a more precise and in-depth assessment of the proposed solution, enabling data analysis and the identification of patterns and behaviors, as well as the implementation of energy consumption optimization strategies based on the results obtained.

References

1. Katal, A., Dahiya, S., Choudhury, T.: Energy efficiency in cloud computing data centers: a survey on software technologies. Clust. Comput. **26**(3), 1845–1875 (2023)
2. Levy, M., Hallstrom, J.: A new approach to data center infrastructure monitoring and management (DCIMM). In: IEEE 7th Annual Computing and Communication Workshop and Conference (CCWC), pp. 1–6 (2017)
3. Lajevardi, B., Haapala, K., Junker, J.: Real-time monitoring and evaluation of energy efficiency and thermal management of data centers. J. Manuf. Syst. **37**, 511–516 (2015)
4. Environment Agency Austria & Borderstep Institute: Energy-efficient Cloud Computing Technologies and Policies for an Eco-friendly Cloud Market, Publications Office of the European Union (2020)
5. Zhang, H., Shao, S., Xu, H., Zou, H., Tian, C.: Free cooling of data centers: a review. Renew. Sustain. Energy Rev. **35**, 171–182 (2014)
6. Pawlish, M., Aparna, S.: Free cooling: a paradigm shift in data centers. In: 2010 Fifth International Conference on Information and Automation for Sustainability, Colombo, Sri Lanka (2010)
7. Daraghmeh, H., Wang, C.: A review of current status of free cooling in datacenters. Appl. Therm. Eng. **114**, 1224–1239 (2017)
8. Cao, Z., Zhou, X., Hu, H., Wang, Z.: Toward a systematic survey for carbon neutral data centers. IEEE Commun. Surv. Tutor. **24**, 895–936 (2022)
9. Zhu, H., et al.: Future data center energy-conservation and emission-reduction technologies in the context of smart and low-carbon city construction. Sustain. Cities Soc. **89**, 104322 (2023)
10. Pei, Q., et al.: CoolEdge: hotspot-relievable warm water cooling. In: Proceedings of the 27th ACM International Conference on Architectural, vol. 3 (2022)
11. Ewim, D., Ninduwezuor-Ehiobu, N., Orikpete, O., Egbokhaebho, B., Fawole, A., Onunka, C.: Impact of data centers on climate change: a review of energy efficient. J. Eng. Exact Sci. **7**, 16397–01e (2023)
12. Liu, Y., Wei, X., Xiao, J., Liu, Z., Xu, Y., Tian, Y.: Energy consumption and emission mitigation prediction based on data center traffic and PUE for global data centers. Glob. Energy Interconnect. **3**, 272–282 (2020)
13. Sooryajith, M., Dinh, A., Chu, T., Nguyen, B., Gangisheety, S.: Enhancing data centers to reduce high energy consumption, vol. 2 (2022)
14. IPMA. https://www.ipma.pt/pt/oclima/normais.clima/1981-2010/. Accessed Nov 2023
15. Fireteanu, V.: Agile Methodology Advantages when delivering Internet of Things projects (2020)
16. Amazon Web Services. https://aws.amazon.com. Accessed Nov 2023
17. Sajee, J.: Overview of amazon web services. Amazon Whitepapers **105**, 1–22 (2014)

Development of an Autonomous Device for People Detection

José Silva[1(✉)], Gabriel Raperger[2], Paulo Vaz[1], Pedro Martins[1], and Alfonso López-Rivero[3]

[1] CISeD – Research Centre in Digital Services, Instituto Politécnico de Viseu, Viseu, Portugal
`{jsilva,paulovaz,pedromom}@estgv.ipv.pt`
[2] Department of Informatics Engineering, Instituto Politécnico de Viseu, Viseu, Portugal
`estgv18509@alunos.estgv.ipv.pt`
[3] Computer Science Faculty, Pontifical University of Salamanca, Salamanca, Spain
`ajlopezri@upsa.es`

Abstract. The growing interest in autonomous security technologies and advanced monitoring systems has driven the development of innovative robotic solutions. This study addresses the development of an autonomous rover, named ROID, specifically designed for the detection of people in diverse environments. The device combines state-of-the-art sensors, such as 3D cameras and LIDAR systems, with advanced algorithms for computer vision and artificial intelligence, operating on a robust platform that integrates the Robotic Operating System (ROS). ROID was developed to perform mapping tasks, autonomous navigation, and human pattern recognition, facilitating continuous and reliable operations in complex contexts such as industrial facilities and public areas. Test results demonstrated the rover's efficacy in automatically identifying and tracking individuals, providing an essential tool for applications ranging from public safety to managing people flow at events or commercial spaces. The integration of modular components and scalable technology suggests broad possibilities for future adaptations of ROID, including expansion to other applications in security and environmental monitoring.

Keywords: Autonomous Robotics · People Detection · Computer Vision · Security Monitoring

1 Introduction

The evolution of autonomous security technologies and advanced monitoring systems marks a significant milestone in our approach to managing security and public safety in urban and commercial environments [1, 2]. These systems harness the potential of cutting-edge technologies, including artificial intelligence, machine learning, and robotics, to enhance our capability to monitor, analyze, and respond to various situations autonomously and efficiently [3, 4]. Such innovations are proving indispensable in areas ranging from public safety to the management of large public gatherings and commercial operations.

Autonomous security technologies are fundamentally reshaping the landscape of surveillance and monitoring. Traditional surveillance systems, which often rely on human monitoring of video feeds, are limited by workforce, cost, and human endurance and focus [5]. In contrast, autonomous systems offer continuous, consistent monitoring without susceptibility to human error. These systems can analyze vast amounts of data from various sources in real time, making them highly effective at identifying patterns or anomalies that may indicate a security issue or operational inefficiency [6].

Moreover, the integration of machine learning algorithms allows these systems to improve over time, learning from past incidents and data to enhance their predictive capabilities. This adaptability makes autonomous systems particularly valuable in dynamic environments where conditions change rapidly, such as during large public events or in busy commercial spaces [7].

The application of these technologies extends beyond mere surveillance. In public safety, autonomous devices are used to patrol areas, recognize faces or behaviors linked to criminal activities, and provide real-time alerts to law enforcement [8–10]. In commercial settings, these systems manage the flow of people, track customer behavior, and optimize the layout and operations of the space to improve user experience and operational efficiency [11, 12].

Autonomous monitoring systems also contribute significantly to crowd management during events. They can analyze movement patterns, predict points of congestion, and suggest adjustments to crowd control measures on the fly. This capability is crucial in enhancing safety and ensuring the smooth operation of events attended by large numbers of people.

Building on this foundation of innovative autonomous technologies, the ROID project introduces a specialized autonomous rover designed for the detection and monitoring of individuals. This rover integrates advanced sensing technology and artificial intelligence to autonomously navigate diverse environments and perform complex monitoring tasks.

The rest of this article is organized as follows. Section 2 presents the theoretical concepts associated with the topic addressed. Section 3 briefly describes the methodology used. In Sect. 4, the real application of the proposed methodology is shown. Finally, the conclusion about this case study is presented in Sect. 5.

2 Background

The successful development of autonomous devices hinges on the careful selection and integration of both software and hardware components. This section delves into the tools and technologies that form the backbone of the ROID project, detailing how each component contributes to the device's capabilities.

2.1 Software Tools

The development of autonomous devices for people detection relies heavily on sophisticated software tools that enable intelligent data processing and decision-making capabilities. Central to this project is the use of the Robotic Operating System (ROS), which provides a flexible framework for writing robot software. ROS is a collection of tools and libraries designed to simplify the task of creating complex and robust robot behavior across a wide variety of robotic platforms [13–15]. For vision processing and object detection tasks, the project utilizes OpenCV (Open-Source Computer Vision Library), a library of programming functions aimed at real-time computer vision [16, 17]. This library facilitates the processing and analysis of visual data, enabling the rover to detect and track human figures effectively.

Additionally, the system employs RVIZ, a 3D visualization tool within ROS that provides a powerful interface for visualizing the rover's sensory information and state [18, 19]. RVIZ aids in debugging and developing the machine's perception and understanding of its surroundings by visualizing its environment and the objects within it in real time. For remote control and access, NoMachine software plays a critical role [20]. It provides a fast and secure way to manage the rover remotely, offering capabilities for remote desktop access, which is essential for testing and operating the rover in different environments without needing physical access.

Moreover, machine learning algorithms implemented via Python libraries such as TensorFlow offer the capabilities needed for pattern recognition and predictive modeling, essential for adapting the rover's behavior based on observed data [21].

2.2 Hardware Tools

On the hardware side, the ROID project integrates several advanced components to ensure high performance and reliability in detecting and tracking individuals.

The primary sensor technology employed is LIDAR (Light Detection and Ranging), which provides accurate distance and mapping capabilities. LIDAR sensors generate precise 3D maps of the rover's surroundings, allowing for detailed environmental awareness and obstacle avoidance. Complementing LIDAR, 3D cameras are used to capture depth information, vital for recognizing human shapes and movements within the device's vicinity.

The hardware setup is powered by an embedded system platform, typically an NVIDIA Jetson board, known for its high computational power and efficiency in handling multiple complex tasks simultaneously. This platform supports the intensive computational needs of real-time image processing and machine learning algorithms, ensuring smooth and responsive operation of the autonomous rover.

2.3 Integration and Implementation

The integration of these software and hardware tools is crucial for the success of the ROID project.

By combining ROS's versatile software environment with robust hardware components, the rover is equipped to perform complex tasks required for effective people detection. The software framework allows for the seamless integration of sensor inputs, data processing, and actuator outputs, facilitating a cohesive operation of the rover. This integration ensures that the rover can not only detect and track people in various environmental conditions but also respond appropriately in real-time to dynamic changes in its surroundings.

3 Process Development

The autonomous rover's algorithm driving is crucial to its operation, dictating how it interacts with and navigates through its environment. This section elaborates on the "Mapping Algorithms" such as GMapping and Explore, which are utilized to generate detailed maps and facilitate exploration of unknown territories.

Following this, the "Navigation Algorithm" is detailed, showcasing how the rover autonomously maneuvers and optimizes its trajectories to ensure efficient movement. Furthermore, the "Human Body Recognition" functionality is introduced, highlighting the rover's capability to detect and respond to human presence, a critical feature for security and monitoring applications.

3.1 Mapping Algorithm

This section is dedicated to environmental mapping algorithms, especially Gmapping and Explore, and their role in the development of the autonomous rover for detecting people. These algorithms are responsible for providing an accurate and dynamic representation of the environment, allowing the rover to navigate autonomously and efficiently while mapping the surroundings.

GMapping, or Grid-based Mapping, is a probabilistic mapping algorithm that uses a grid to manually model the environment. It plays a central role in constructing detailed maps and precisely locating the rover within this three-dimensional space. The explore algorithm is for the rover's autonomous exploration strategy, determining efficient routes to cover the inspection area and optimize people detection.

GMapping

GMapping is a component that enables the autonomous rover to create probabilistic maps of the environment around it, which are essential for effective navigation and obstacle detection.

The "gmapping.launch" file is central in configuring and initializing the GMapping algorithm. This file defines crucial parameters, such as reference frames, laser scanner reading topics, and specific algorithm settings.

The code in the "gmapping.launch" file (Fig. 1) presents the configuration and initiation of the GMapping mapping algorithm in the project context.

This algorithm will support the construction of probabilistic maps of the environment, allowing the rover to understand its location and orientation during autonomous operations.

The code uses arguments to promote the adaptability and customization of the algorithm. Arguments, such as "base_frame", "odom_frame", and "map_frame", define fundamental reference frames for coordination between different rover systems, ensuring a cohesive representation of the three-dimensional environment.

Figure 2 shows an excerpt of XML ("extensible Markup Language") code from the "slam.launch" file. This is responsible for integrating the GMapping algorithm into the autonomous rover's simultaneous mapping system (SLAM). This code allows the dynamic configuration of SLAM, enabling the selection of the specific method to be used, being "gmapping" in this case.

```xml
<?xml version="1.0"?>
<launch>
    <!-- Arguments -->
    <arg name="scan"        default="scan"/>
    <arg name="base_frame"  default="base_footprint"/>
    <arg name="odom_frame"  default="odom"/>
    <arg name="map_frame"   default="map"/>

    <!-- Gmapping -->
    <node pkg="gmapping" type="slam_gmapping" name="jetauto_slam_gmapping" output="screen">
        <param name="base_frame" value="$(arg base_frame)"/>
        <param name="odom_frame" value="$(arg odom_frame)"/>
        <param name="map_frame"  value="$(arg map_frame)"/>
        <remap from="/scan" to="$(arg scan)"/>
        <rosparam command="load" file="$(find jetauto_slam)/config/gmapping_params.yaml" />
    </node>
</launch>
```

Fig. 1. Gmapping.launch code.

```xml
<!--(mapping method choice)-->
<group if="$(eval slam_methods == 'gmapping')">
    <include file="$(find jetauto_slam)/launch/$(arg slam_methods).launch">
        <arg name="scan"       value="$(arg scan_topic)"/>
        <arg name="base_frame" value="$(arg base_frame)"/>
        <arg name="odom_frame" value="$(arg odom_frame)"/>
        <arg name="map_frame"  value="$(arg map_frame)"/>
    </include>
</group>
```

Fig. 2. Gmapping slam.launch.

The conditional group <group if = "$(eval slam_methods == 'gmapping')"> acts as a conditional inclusion block, checking whether the selected SLAM method is GMapping. If so, the specific GMapping file is included in the system using the <include> command. The arguments passed in this block, such as scan_topic, base_frame, odom_frame, and map_frame, customize GMapping settings according to the rover's needs.

Explore

This algorithm enables the autonomous rover to automatically map the environment, providing a basis for efficient navigation and detection of unknown areas.

The "explore.launch" file configures and initializes the exploration algorithm. Here, essential parameters are defined, including reference frames, map-related topics, and save paths. These parameters are fundamental to optimizing the rover's performance during its autonomous exploration operations.

The "explore.launch" file, represented in Fig. 3, supports the implementation of the autonomous rover exploration method. In this file, several parameters are configured to optimize the performance of the "explore_lite," algorithm. This is responsible for guiding the rover in identifying unknown areas of the environment.

At the beginning of the file, arguments are defined as "base_frame," "costmap_topic," "costmap_updates_topic," "map_topic," and "map_save_path." These arguments are fundamental to establishing the necessary connections and providing vital information

```
<launch>
    <arg name="base_frame"               default="base_footprint"/>
    <arg name="costmap_topic"            default="map"/>
    <arg name="costmap_updates_topic"    default="map_updates"/>
    <arg name="map_topic"                default="map"/>
    <arg name="map_save_path"            default="$(find jetauto_slam)/maps/explore"/>
    <node pkg="explore_lite" type="explore" respawn="false" name="explore" output="screen">
        <param name="map_topic"             value="$(arg map_topic)"/>
        <param name="map_save_path"         value="$(arg map_save_path)"/>
        <param name="robot_base_frame"      value="$(arg base_frame)"/>
        <param name="costmap_topic"         value="$(arg costmap_topic)"/>
        <param name="costmap_updates_topic" value="$(arg costmap_updates_topic)"/>
        <param name="visualize"             value="true"/>
        <param name="planner_frequency"     value="0.33"/>
        <param name="progress_timeout"      value="30.0"/>
        <param name="potential_scale"       value="3.0"/>
        <param name="orientation_scale"     value="0.0"/>
        <param name="gain_scale"            value="1.0"/>
        <param name="transform_tolerance"   value="0.3"/>
        <param name="min_frontier_size"     value="0.5"/>
    </node>
</launch>
```

Fig. 3. Explore.launch code.

to the algorithm. The "base_frame" specifies the robot's reference frame, while the "costmap_topic" and "costmap_updates_topic" topics are related to the cost maps used in autonomous navigation.

3.2 Navigation Algorithm

This algorithm plays a vital role in autonomous decision-making, allowing the rover to move intelligently through the environment, avoiding obstacles and following predefined routes.

The launch file, as shown in Fig. 4, defines the control and navigation system of the autonomous rover.

Initially, parameters associated with nomenclature, topics and reference frames are configured. Variables such as "cmd_vel_topic" for controlling angular velocity, "scan_topic" for reading the laser scanner and "map_topic" for the map topic stand out. The variable "use_teb" is determined using the local planning method "teb_local_planner," to make autonomous decisions based on the surrounding environment. Furthermore, specifying the "clicked_point" and "move_base_result" topics contributes to the rover's interaction with points of interest and to monitoring the results of the navigation system's actions.

As shown in Fig. 5, the rover uses maps built by GMapping or Explore. If the maps are not updated, RPLIDAR provides new data for object detection. When identifying a new object, the rover instantly adjusts its path to avoid collisions, ensuring safe and responsive navigation in dynamic environments.

When identifying a new object, the rover instantly adjusts its path to avoid collisions, ensuring safe and responsive navigation in dynamic environments.

Development of an Autonomous Device for People Detection 213

```
<launch>
    <arg name="master_name" default="$(env MASTER)"/>
    <arg name="robot_name"  default="$(env HOST)" />
    <arg name="drive_type"  default="diff"/>

    <!--topic frame-->
    <arg      if="$(eval robot_name == '/')"   name="topic_prefix" default=""/>
    <arg  unless="$(eval robot_name == '/')"   name="topic_prefix" default="/$(arg robot_name)"/>
    <arg      if="$(eval robot_name == '/')"   name="frame_prefix" default=""/>
    <arg  unless="$(eval robot_name == '/')"   name="frame_prefix" default="$(arg robot_name)/"/>
    <arg      if="$(eval robot_name == '/')"   name="tf_prefix"    default=""/>
    <arg  unless="$(eval robot_name == '/')"   name="tf_prefix"    default="$(arg robot_name)"/>

    <arg name="cmd_vel_topic"     default="$(arg topic_prefix)/jetauto_controller/cmd_vel"/>
    <arg name="scan_topic"        default="$(arg topic_prefix)/scan"/>
    <arg      if="$(eval robot_name == master_name)" name="map_topic" default="$(arg topic_prefix)/map"/>
    <arg  unless="$(eval robot_name == master_name)" name="map_topic" default="/$(arg master_name)/map"/>
    <arg name="odom_topic"        default="$(arg topic_prefix)/odom"/>

    <arg      if="$(eval robot_name == master_name)" name="map_frame" default="$(arg frame_prefix)map"/>
    <arg  unless="$(eval robot_name == master_name)" name="map_frame" default="$(arg master_name)/map"/>
    <arg name="odom_frame"        default="$(arg frame_prefix)odom"/>
    <arg name="base_frame"        default="$(arg frame_prefix)base_footprint"/>
    <arg name="use_teb"           default="true"/>
    <arg name="clicked_point"     default="$(arg topic_prefix)/clicked_point"/>
    <arg name="move_base_result"  default="$(arg topic_prefix)/move_base/result"/>
```

Fig. 4. Navigation.launch code.

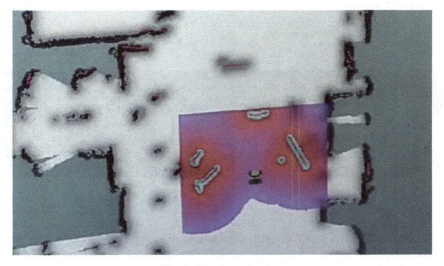

Fig. 5. Result navigation.

3.3 Human Body Recognition

If humans are detected, the publishMarker function is triggered, creating and publishing a visual marker in the topic /jetauto_1/body_marker. This visualization can be observed in tools such as RVIZ, providing a graphical representation of the rover's detection of humans in the ROS environment.

Figure 6 illustrates the inclusion of the libraries necessary to work with ROS and OPENCV and other resources.

```
#include <geometry_msgs/PoseStamped.h>
#include <tf/transform_listener.h>
#include <ros/ros.h>
#include <sensor_msgs/Image.h>
#include <opencv2/opencv.hpp>
#include <cv_bridge/cv_bridge.h>
#include <image_transport/image_transport.h>
#include <visualization_msgs/Marker.h>
#include <string>

int i = 0;
```

Fig. 6. Body_detection_node.

In Fig. 7, the marker_pub_ variable is initialized to facilitate the publication of markers in the "body_marker" topic. This allows the visualization of markers in tools such as RVIZ, providing a graphical representation of the bodies detected by the system. Subsequently, the successful loading of the cascade classifier intended for body detection is verified.

```
// Publica o marcador de corpos
marker_pub_ = nh_.advertise<visualization_msgs::Marker>("body_marker", 10);
// Inicializa o classificador de corpos
if (!body_cascade_.load("cascades/haarcascade_fullbody.xml")) {
    ROS_ERROR("Erro ao carregar o classificador em cascata para corpos.");
} else {
    ROS_INFO("Classificador em cascata para corpos carregado com sucesso.");
}
```

Fig. 7. Body_detection.

A cascade classifier is a trained model capable of recognizing specific patterns in images, in this case, people. It operates in a series of stages, where each stage is responsible for deciding whether or not a specific region of the image contains the object of interest.

The efficiency of this process is critical to detection performance. An error message is generated if the classifier load fails, alerting the problem. On the other hand, if the loading is successful, a success message is displayed, indicating that the classifier is ready to be applied to detect people in the image.

A previously trained cascade classifier (body_cascade_) is used to classify the image. If a body is detected, a message "Body(ies) detected!" in the console, the i counter is incremented, and the publishMarker function is called. The publishMarker function is responsible for creating and publishing a visual marker in ROS.

4 Case Study

This section presents the case study to demonstrate the capabilities of the rover developed in this project in three operating methods: mapping (GMapping), exploration (Explore), navigation (Navigation) including human body recognition.

Figure 8 illustrates the execution of the GMapping method. In this process, the construction of the map is started manually.

The green map represents the construction in progress, identifying objects and walls, and includes a representation of the rover itself. The rover's real-time view is highlighted, allowing the user to view their current route.

Fig. 8. GMapping method.

Figure 9 shows how the Explore method works in execution. Here, it is possible to see how the rover performs autonomous recognition of the environment. There are also several points related to its execution: the green circles represent unexplored areas, indicating potential places to be covered to complete the mapping of the environment.

The purple-red color gradient indicates that RPLIDAR is in a state where it detects objects that the rover could collide with. With this information, the rover adjusts its route to avoid detected objects, thus avoiding collisions.

Figure 10 shows the execution of the Navigation method, where the user only needs to define the desired destination point for the rover. Autonomously, the rover plans its route and then travels it. During navigation, the rover uses RPLIDAR sensors. Even though the map is built, if there is a change in the position of an object, the rover recalculates its route to avoid that object. In the Figure, several markers are marked in green, each created at the rover's position whenever it detects a person.

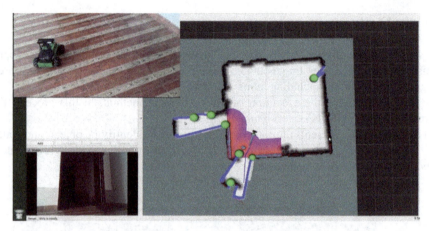

Fig. 9. Explore method.

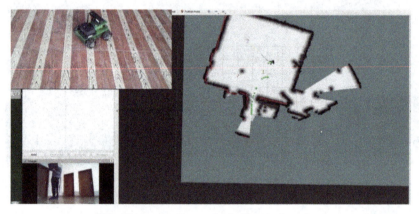

Fig. 10. Navigation method - People detection.

5 Conclusions and Future Work

The results of the autonomous rover developed in this project demonstrate its proficiency in navigation, object identification, and response to the presence of humans, emphasizing its potential in people detection applications. Advanced mapping (GMapping) and exploration (Explore) algorithms, along with the Navigation method, have proven highly effective.

These technologies enabled the rover to create detailed maps of its environment and autonomously explore unknown areas, optimizing paths to circumvent obstacles efficiently. Implementing the navigation method has allowed the rover to perform routes autonomously, navigating around obstacles while operators set destination points, providing enhanced flexibility and control during operations.

The rover's capabilities extend beyond mere navigation and mapping. Its real-time people detection and tracking system presents substantial advantages for security and surveillance applications. The rover can significantly reduce workforce costs and minimize human error by automating routine patrols and monitoring tasks. Its ability to operate continuously over extended periods makes it an invaluable asset for maintaining security in large public spaces, events, or even commercial settings where high foot traffic is expected.

For future work, there are several avenues for further development that could enhance the rover's functionality and applicability.

- Integration with facial recognition technology: implementing facial recognition could allow the rover to identify individuals of interest in scenarios requiring enhanced security measures, such as at airports or sensitive facilities.
- Improved decision-making with AI: enhancing the rover's artificial intelligence systems could provide more advanced decision-making capabilities, allowing it to handle complex interactions and adapt to more dynamic environments without human intervention.
- Multi-rover coordination: developing protocols for multiple rovers to work in tandem could amplify the coverage area and efficiency, creating a network of autonomous units that communicate and coordinate with each other.

Acknowledgements. This work is funded by National Funds through the FCT – Foundation for Science and Technology, I.P., within the scope of the project Ref. UIDB/05583/2020. Furthermore, we would like to thank the Research Centre in Digital Services (CISeD) and the Instituto Politécnico de Viseu for their support.

References

1. Ahmad, K., Maabreh, M., Ghaly, M., Khan, K., Qadir, J., Al-Fuqaha, A.: Developing future human-centered smart cities: critical analysis of smart city security, Data management, and Ethical challenges. Comput. Sci. Rev. **43**, 100452 (2022)
2. Rani, S., et al.: Amalgamation of advanced technologies for sustainable development of smart city environment: a review. IEEE Access **9**, 150060–150087 (2021)
3. Ullah, Z., Al-Turjman, F., Mostarda, L., Gagliardi, R.: Applications of artificial intelligence and machine learning in smart cities. Comput. Commun. **154**, 313–323 (2020)
4. Zhang, M., Wang, X., Sathishkumar, V.E., Sivakumar, V.: Machine learning techniques based on security management in smart cities using robots. Work **68**(3), 891–902 (2021)
5. Fraga-Lamas, P., Fernández-Caramés, T.M., Suárez-Albela, M., Castedo, L., González-López, M.: A review on internet of things for defense and public safety. Sensors **16**(10), 1644 (2016)
6. Erhan, L., et al.: Smart anomaly detection in sensor systems: a multi-perspective review. Inf. Fusion **67**, 64–79 (2021)
7. Van Brummelen, J., O'brien, M., Gruyer, D., Najjaran, H.: Autonomous vehicle perception: the technology of today and tomorrow. Transpor. Res. Part C Emerg. Technol. **89**, 384–406 (2018)
8. Thakur, N., Nagrath, P., Jain, R., Saini, D., Sharma, N., Hemanth, D.J.: Artificial intelligence techniques in smart cities surveillance using UAVs: a survey. In: Ghosh, U., Maleh, Y., Alazab, M., Pathan, A.K. (eds.) Machine Intelligence and Data Analytics for Sustainable Future Smart Cities, pp. 329–353. Springer, Cham (2021). https://doi.org/10.1007/978-3-030-72065-0_18

9. Chen, Z., et al.: Autonomous social distancing in urban environments using a quadruped robot. IEEE Access **9**, 392–403 (2021)
10. Legovich, Y.S., Diane, S.A.K., Rusakov, K. D.: Integration of modern technologies for solving territory patroling problems with the use of heterogeneous autonomous robotic systems. In: Eleventh International Conference Management of large-scale system development (MLSD), pp. 1–5. IEEE (2018)
11. De Bellis, E., Johar, G.V.: Autonomous shopping systems: identifying and overcoming barriers to consumer adoption. J. Retail. **96**(1), 74–87 (2020)
12. Park, D., et al.: Active robot-assisted feeding with a general-purpose mobile manipulator: design, evaluation, and lessons learned. Robot. Auton. Syst. **124**, 103344 (2020)
13. Cañas, J.M., Perdices, E., García-Pérez, L., Fernández-Conde, J.: A ROS-based open tool for intelligent robotics education. Appl. Sci. **10**(21), 7419 (2020)
14. Portugal, D., Iocchi, L., Farinelli, A.: A ROS-based framework for simulation and benchmarking of multi-robot patrolling algorithms. Robot Oper. Syst. (ROS) **3**, 3–28 (2019)
15. Roldán, J.J., et al.: Multi-robot systems, virtual reality and ROS: developing a new generation of operator interfaces. Robot Oper. Syst. (ROS) **3**, 29–64 (2019)
16. Silva, J., Coelho, P., Saraiva, L., Vaz, P., Martins, P., López-Rivero, A.: Validating the use of smart glasses in industrial quality control: a case study. Appl. Sci. **14**(5), 1850 (2024)
17. Sharma, A., Pathak, J., Prakash, M., Singh, J. N.: Object detection using OpenCV and python. In: 3rd International Conference on Advances in Computing, Communication Control and Networking (ICAC3N), pp. 501–505. IEEE (2021)
18. Pütz, S., Wiemann, T., Hertzberg, J.: Tools for visualizing, annotating and storing triangle meshes in ROS and RViz. In: European Conference on Mobile Robots (ECMR), pp. 1–6. IEEE (2019)
19. Pütz, S., Wiemann, T., Hertzberg, J.: The mesh tools package–introducing annotated 3D triangle maps in ros. Robot. Auton. Syst. **138**, 103688 (2021)
20. Dwyer, I., Gerke, K., Do, D. T.: Machine learning-based automated irrigation for indoor: review and a case study. In: International Conference on Communication, Devices and Networking, pp. 405–430, Springer, Singapore (2022). https://doi.org/10.1007/978-981-99-1983-3_38
21. Raschka, S., Patterson, J., Nolet, C.: Machine learning in python: main developments and technology trends in data science, machine learning, and artificial intelligence. Information **11**(4), 193 (2020)

PetWatcher – Ubiquitous Device Proximity Location System

Laura Fernandim, Celestino Gonçalves, Filipe Caetano, and Clara Silveira(✉)

Instituto Politécnico da Guarda, Guarda, Portugal
1000354@sal.ipg.pt, {celestin,caetano,mclara}@ipg.pt

Abstract. This paper presents a solution, PetWatcher, that uses BLE technology in conjunction with the Blynk IoT platform to monitor the proximity of specific devices to a landmark. The main objective is to monitor the presence of devices, placed on animals, through MAC address, in relation to a fixed point, controlling the movement of the devices in a specific area. The system was developed using the ESP32 microcontroller and performs checks whenever it detects movement from the inside of the fixed point, or detects the iTAG signal on the outside of the fixed point, displaying information in real time through the Blynk application. Tests were carried out throughout the development of the system and in the final tests it was observed that the system meets the requirements for which it was developed. The presentation of the solution shows potential for several practical applications, including animal location monitoring, access monitoring and home security.

Keywords: Animal Detection · IoT · BLE · ESP32 · iTAG · Blynk · PetWatcher

1 Introduction

Locating the proximity of devices is an important task in various applications, such as home security, monitoring the elderly, and searching for objects. With the advancement of Bluetooth Low Energy (BLE) technology, it has become possible to carry out this type of monitoring efficiently and with low power consumption.

The Blynk platform [1] enables the integration of Internet of Things (IoT) devices, the visualization and remote control of the data detected by these devices. With the Blynk platform, it is possible to create custom applications that display information in real time. In this article, we describe the process of developing a system, called PetWatcher, composed of the ESP32 that uses BLE and which, together with the Blynk platform, will monitor the proximity of animals in relation to a reference point, in this case, a gate.

This article is structured in seven sections. The first is the introduction section and in Sect. 2 the state of the art is analyzed. Section 3 presents the methodology used for the development of the project. Section 4 presents the design of the system where it talks about the requirements analysis and the characteristics of the ubiquitous system. Section 5 describes the development of the system, where it talks about the architecture, the management platform, the user application and hardware components. Section 6 presents the results and analysis of them. Finally, Sect. 7 presents the conclusions of the project and proposal for future work.

2 Related Works

Monitoring the proximity of devices in pets is a fundamental process for their safety. Several factors must be taken into account, such as the size of the devices, low power consumption, low cost, and synchronization time. BLE technology has been widely adopted in various applications that require low power consumption and wireless communication. It has been used in industries such as healthcare, home automation, asset sourcing, and wearable devices. The use of BLE for proximity monitoring is especially relevant in security, access control, and presence detection applications [2].

With the Blynk platform it is possible to create custom applications for proximity monitoring, control and visualization of data captured by IoT devices. In the specific context of this project, which involves detecting the proximity of BLE devices to a reference point, there are some commercial solutions available, such as:

- Bluvision Beacons – Real-time location [3]. Bluetooth communication, simple installation, without wiring, where there are only 2 physical elements: BEEKs (Beacons) and BluFis (Gateways). Everything is managed from Bluezone Cloud SaaS (Software as a Service). The user's smartphone can become an additional component with the development of an application with or without software (incognito mode).

  - Beacons (BEEKs): Encrypted communication and Comunicam offline; Battery life: 4 years; Precise localization; IP67. Protection against water; Different types of sensors (temperature, light, pulse, vibration, movement); Support for location, tracking and status monitoring; Mobile communications for Beacon; Integrable HID access control; Incorporates iBeacon (Apple) and Eddystone (Google) standards.
  - BluFi Gateways: The BluFi Gateways device connects to a multitude of beacons via Bluetooth (BLE) and via WiFi with BluZone Cloud SaaS.

- Digitanimal: Devices adaptable to different animals [4]. Devices with IoT technology, reduced size and weight from 22gr, prepared for: vultures, eagles, hedgehogs, hares, deer, crows, rebeccos, lynxes, wild boars, bison, wildebeest or elephants. It incorporates multiple sensors with high-density integration.

  - Long battery life and autonomy; Animal activity alerts; Animal temperature alert; Alerts for entering/exiting the premises; Alerts for theft or loss

The commercial solutions mentioned above may involve a financial effort for those who have animals. As an alternative, there are some interesting projects that can address the financial issue. Thus, as related works of monitoring and locating animals, we can consider some examples that use low cost technologies, such as: Mobile animal tracking systems using light sensor for efficient power and cost saving motion detection [5],

Farm Animal Location Tracking System Using Arduino and Global Positioning System (GPS) Module [6] whose objective is the location of animals, through GPS and IoT. Management of Animals on Farms [7], that uses low-cost and energy-efficient technologies, as it is based on BLE technology and the use of iTAG in animals for monitoring. Implementation of an IoT based Pet Care System [8], PetCare: A Smart Pet Care IoT Mobile Application [9], to make up for the lack of time of the owners or the absence of the pets, these projects invested on the implementation of an intelligent system to monitor the basic needs of the animals. Implementation and Analysis of a Wireless Sensor Network-Based Pet Location Monitoring System for Domestic Scenarios [2], monitoring system, operates over the implemented ZigBee network and provides real-time information to Android devices. This integration exemplifies Ubiquitous Computing, enabling continuous interactions between humans and technology, including in the care of pets.

Ubiquitous Computing, also known as pervasive or omnipresent computing, refers to the integration of computer technology into everyday objects and everyday activities, enabling seamless interactions between humans and computers. Characterized by being a distributed system, with implicit interaction, context-aware, autonomous and intelligent, it aims to improve the user experience by simplifying interactions and adapting to the physical environment. By integrating with a wide variety of devices and objects, systems are able to offer more relevant and personalized interactions, making the use of technology as natural as possible. The concept was proposed in 1991 by Mark Weiser [10], where the central idea behind it was to create an environment where technology is deeply integrated into everyday life, without requiring conscious effort by users. This is achieved through the constant availability of computational resources across various devices and surrounding objects, creating an environment of seamless and intuitive interactions.

Table 1 shows the comparison between the above systems and the proposed PetWatcher.

Table 1. Comparison between systems

Project	GPS	Autonomous	Context-aware	Implicit Interaction	Mobile App	Notifications
PetWatcher	–	X	X	X	X	X
[5]	X	X	X	–	X	X
[6]	X	X	X	–	–	–
[7]	–	–	X	–	X	X
[8]	–	–	X	X	X	X
[9]	–	X	X	–	–	X

A variety of technologies can be used to monitor and locate animals. The choice will depend on the needs of each one. In our system we can say that it uses low-cost and low-energy technologies and fulfills the objective of location in relation to a reference point.

3 Methodology

Taking into account the objective of proximity monitoring of devices, the development of the PetWatcher system was carried out using the Agile SCRUM development methodology. Agile methodologies promote collaboration within the team and continuous planning, as well as constant evolution and learning [11], SCRUM methodology is a flexible and efficient Agile framework for the development of complex projects, promoting collaboration, transparency, and continuous adaptation [11]. The project, conducted as a team, begins with the definition of a list of requirements and functionalities requested by the client, known as the Product Backlog. As software is developed and new functionality is requested, it is incorporated into the Product Backlog, maintaining an up-to-date list of all requirements to be implemented. Each item in the list is evaluated with a value that indicates its importance in the context of the project. The work method is defined based on small cycles of project activities, called Sprints, which have a certain period of time that is not too long. In this project, five Sprints of one to two weeks each are used, and were defined in the Product Backlog, as we can see below:

1. Setup and preparation of the Development Environment.
2. Identification of BLE Specific Devices.
3. Development and Testing of the Basic Detection Prototype.
4. Integration with the Blynk Platform.
5. Integrated System Evaluation and Testing.
6. System Tweaks and Improvements.
7. Final Testing and System Validation.
8. Documentation and Dissemination of Results.

For example, a two-week Sprint is described for the initial configuration and preparation of the development environment and identification of BLE devices. The following activities are: prepare and configure the Arduino Integrated Development Environment (IDE) with the necessary libraries, including BLEDevice, BLEUtils, WiFi; perform the initial configuration of the ESP32, ensuring that it is correctly connected to the WiFi network and ready to perform BLE device detection; set the Media Access Control (MAC) addresses of the BLE devices to be monitored, such as "Lola's iTAG" and "owner's iTAG"; perform tests to ensure that BLE device detections are working correctly, allowing you to identify and fix potential teething issues.

Figure 1 shows the SCRUM framework with the integration of the ESSENCE elements and includes the SEMAT [12]. We have the standards that are the roles played by the Product Owner, SCRUM Master and SCRUM Team, who are responsible for the solutions. When we talk about Activities, we refer to everything that is developed and executed by professionals, which are the Daily SCRUM, Sprint Review, Sprint Planning and Sprint Retrospective. Tangible product is everything that professionals produce, they are Product Backlog, Sprint Backlog and Increment. When we refer to the Alphas, we are talking about the essential and relevant elements for evaluating the progress of the project, they are Sprint and Product Backlog Item.

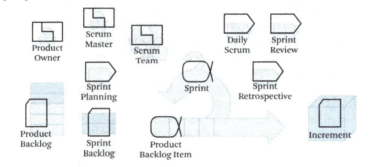

Fig. 1. Integration of ESSENCE elements into the SCRUM methodology

4 Requirements Specification and Characterization of the Ubiquitous System

To represent the opportunities and needs of stakeholders and transform them into a set of requirements, User Cases were used. The User Case describes the functionalities of the system to be built. They have the potential to make the team analyze what they are developing from the point of view of those who will use it. In the testing phase of the project, the acceptance criteria of the User Case are verified, validating and verifying that it complies with the requirements of the users. In the installation phase of the project, the operation of the project is verified in a real context, and it is adjusted if there is any failure, before moving on to the phase of operationalizing the application.

For the development of the PetWatcher system, essential information was collected to define the functionalities of the parties involved (animal protection associations and individuals who need to monitor their pets). The use case diagram, shown in Fig. 2, shows the actors (owner of the house and device on the animal) and documents the actions of the system from the user's point of view.

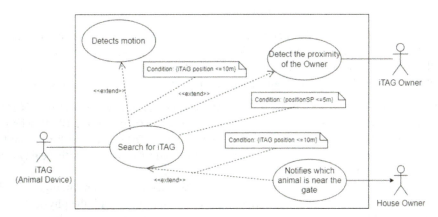

Fig. 2. PetWatcher Use Case Diagram

The system performs a periodic check of the BLE devices around it, previously configured. When motion is detected, it activates the search for the iTAG: if a device is detected near the gate, it searches if the iTAG that is with the user is nearby, and if it is nearby, it sends a message to Blynk. On the other hand, to make up for any failure that may exist in sending the message in a timely manner, there is a search routine between 6:00 pm and 6:30 pm (estimated time for the user to get home) that always sends the message to Blynk. The user, represented by the owner of the house, has the ability to receive alerts about the position of the iTAG (placed on the animal's collar).

The PetWatcher system must check if there is a WiFi failure, case in which the owner of the house must be warned, through a Light-Emitting Diode (LED), for which he has to be attentive, in order to check if the animal is close to the gate.

We can characterize our system as a ubiquitous system [13] as it obeys the following properties:

1. Distributed System - The ESP32 sensor sends messages if it encounters the dog's iTAG. This indicates a level of distributed processing, where the system responds to specific stimuli, such as detecting the owner's iTAG, allowing the system to make checks and send messages to Blynk based on the iTAG detections;
2. Implicit Human Device Interaction (iHCI) - The interaction in the system is mostly explicit, with the system responding to specific stimuli, such as detected movement and the presence of iTAG. It is considered implicit when sending the messages without the need for intervention from the owner of the house;
3. Context-aware - awareness exists with motion detection, iTAG presence, and owner iTAG detection. The system not only monitors physical proximity, but also understands the context of movement, which improves decision-making based on this information;
4. Autonomy - The autonomy is due to the ESP32's ability to activate based on detected motion and iTAG detection. This indicates an enhanced ability of the independent decision-making system in response to specific stimuli;
5. Intelligence - The system demonstrates a certain level of intelligence, with the ability to activate and make proactive decisions based on motion detection, presence of iTAG, and the time set as "owner's arrival time.

Regarding the Smart DEI model [13], We can say that the system is more characterized by the Smart Environment model, since it demonstrates the ability to interact intelligently with the physical environment, with the presence of detection devices and the ability to adapt based on the context. However, it also has features in the other models:

1. Smart Device - The ESP32 with the iTAG acts as the Smart Device, as it is able to activate when it detects motion, scans and sends messages to the Blynk when it finds the iTAG, and also integrates the homeowner's iTAG detection functionality. Performs detection and communication tasks, facilitating interaction with the physical environment and iTAG devices;
2. Smart Environment: The system demonstrates characteristics of an Intelligent Environment, it is able to interact intelligently and autonomously with the physical environment. With motion detection, the system is able to adapt actions based on the context of the physical environment, such as the presence of the iTAG near the gate.

It is able to gather relevant information from the physical environment and make decisions based on that information, making the environment more sensitive and responsive;
3. Smart Interaction: With motion detection, the presence of the animal's iTAG and the owner's iTAG, the system exhibits intelligent and adaptive interaction characteristics. It responds dynamically to different stimuli, making an interaction contextualized, or one that aligns with the intelligent interaction principles represented by the Smart Interaction model.

The characterization of the system in relation to the context [14, 15], systems that are aware of the situation (or context) in their physical, virtual and user environment, and can adapt the system to that reality, benefiting from knowledge of that situation. Our system can be characterized as follows: As primary facets we have the real-time location of the animal and the owner. As secondary facets, we have the device that receives information with time and message if the animal is near the gate. It is a system with context, the animal and the owner have their own location that is monitored in a personalized way. It is active because it sends messages autonomously. Our system acquires context through sensors, network devices and smartphones.

5 System Development

Figure 3 shows the architecture of the developed system: high-level architecture, showing the operation of the system that is based on the detection of the presence of animals near the gate using BLE technology. The TTGO board, consisting of the ESP32 microcontroller, is the central element of the system. It connects to the motion sensor that allows it to detect movement near the gate. Once motion is detected, using the ESP32 BLE, it is checked if known iTAG are detected (up to 10 m). Such recognition is done through the MAC address of the iTAG placed on the animals' collar, which have been previously introduced into the system. Whenever an iTAG of a known animal is detected, the owner's iTAG (MAC address) is detected, when it is also detected, a message is sent to the Blynk platform, informing that an animal is found near the gate. If the WiFi connection fails, a LED lights up.

Using a Passive Infrared (PIR) motion sensor with a microcontroller like the ESP32 offers several advantages:

- Low power consumption: PIR sensors are designed to detect changes in ambient heat. They remain in a low-power state until they detect motion, making them ideal for projects that aim to save energy.

For the implementation of the system, the TTGO ESP32 Lora v1.6.1 board was used, due to its native BLE support and WiFi connection capability. The board consists of the ESP32-PICO-D4 module [16] with the Xtensa® LX6 2-core 32-bit processor at 40 MHz and 448KB RAM from the manufacturer Espressif. The presence detector, which is the HC SR501 PIR motion sensor, was also used [17]. It is based on infrared technology, and can be powered between 5V to 20V DC, with an operation in the temperature range of -15 to + 70 °C (important feature since it is to be placed outdoors). This module detects movement inside the gate in order to trigger the ESP32 to search for the iTAG.

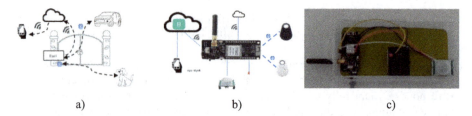

Fig. 3. a) System Architecture; b) High-Level Architecture; c) System PetWatcher.

A LED and a 330 Ω resistance were also used. Two BLE devices, iTAG, were used for proximity monitoring. They are powered by a CR2032 battery that lasts approximately 5 years.

Regarding the software, the system was developed using Arduino IDE using the libraries BLEDevice, BLEUtils, BLEScan, BLEAdvertisedDevice, WiFi, WiFiClient and BlynkSimpleESP32 [6]. Blynk platform provides a customized application where proximity information is displayed in real-time [1]. The application allows the visualization of the status of the BLE device and provides a "user friendly" interface for the owner of the house. Proximity information is updated according to messages received from the Gateway. If the ESP32 is turned off, the app shows on the home screen that the device is offline. If we enter the application, it shows the data of the last communication.

The main modules are motion detection, iTAG search of the animal and the owner, search for the animal's iTAG at the pre-defined time (between 18:00 and 18:30) and illuminate a LED when there is a WiFi failure (see Fig. 4).

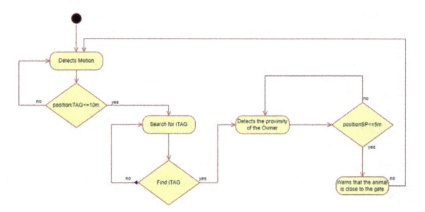

Fig. 4. Activity Diagram for animal detection

6 Results and Discussion

To verify that the system was working and meets the objectives for which it was designed, real-world tests were carried out. Figure 5 illustrates the different situations: a) it did not detect the dog's iTAG, b) it was able to detect the dog's iTAG and c) it detected the dog's and owner's iTAG (who was in the car outside the gate).

Fig. 5. Real-Context Testing: a) Didn't detect the animal's iTAG; b) Animal iTAG Detection; c) Detection of the two iTAG.

The current configuration of the system is capable of accurately detecting the animal's iTAG when it is close to the gate and detecting both iTAG's when the owner's iTAG is also close to the gate, but it still has some limitations in case of WiFi failure or delay when the animal suddenly approaches.

Figure 6 shows the message sent through Blynk to the smartphone:

Fig. 6. Detection of the two iTAG

The results obtained are presented in Table 2. From the analysis of the table, we can affirm that it achieved the objectives in most of what was proposed. As for the inconsistency in the results when we are on the schedule between 6:00 pm and 6:30 pm, it is necessary to carry out more tests to understand what is the reason for the non-functioning in some of the situations. In order to comprehensively validate the performance and efficiency of all individual components of the system, such as delay times for proximity detection and notifications, as well as battery life assessments, additional and specific tests are also needed.

Table 2. Test cases to validate system operation.

Description	Requirements	Expected Outcome	Result Obtained	Test Result
iTAG detection and message sending	The animal is within range of the BLE and so is the owner	Send message "animal name + next to the gate"	Received	Success
	The animal is within range of the BLE and the owner is not	Doesn't send a message	No messages received	Success
	The animal is out of range of BLE	Doesn't send a message	No messages received	Success
	It is between 6:00 p.m. and 6:30 p.m. and the animal is within range of the BLE, research every 5 min	message "animal name + next to the gate"	Some messages received	In 6 tests failed 3
	It is between 6:00 p.m. and 6:30 p.m. and the animal is not in the range of the BLE, it searches every 5 min	Doesn't send a message	No messages received	Success
Time it takes to send the message	Detects the animal's iTAG and detects the owner's iTAG	Sends message "animal name + next to the gate" in less than 1min	Mensagem recebida em menos de 1min,	Success
	Between 6:00 p.m. and 6:30 p.m., do the search every 5 min, find the animal's iTAG	Sends message "animal name + next to the gate" in less than 1min	Some messages received in less than 1min	In 6 tests failed 3

(*continued*)

Table 2. (*continued*)

Description	Requirements	Expected Outcome	Result Obtained	Test Result
In the event of a WiFi failure, ensure that the gate is only opened if the animal is not with it	If the WiFi fails, it is not possible to send the message, signaling will be given on the spot through a light	Lights up an alert LED	LED lit up	Success
Sending Notification and E-mail	The animal is within range of the BLE and so is the owner	Send notification and email "animal name + at the gate"	Notification and e-mail received	Success
	It is between 6:00 p.m. and 6:30 p.m. and the animal is within range of the BLE, research every 5 min	Send notification and email "animal name + at the gate"	Notifications and emails received	Whenever it failed to send the messages, it also failed to notify

7 Conclusion and Future Work

This article presents the development of a system, called PetWatcher, that uses BLE technology and the Blynk IoT platform to monitor the location and proximity of specific devices to a landmark, in this case, a gate. The system is designed to monitor the presence of pets, such as dogs, and alert the owner to the animal's proximity to the gate. The system was successfully implemented using the ESP32 microcontroller, which acts as a BLE gateway. The HC SR501 PIR motion sensor was used to detect movement within the range of the gate and trigger the iTAG search. The iTAG were used as proximity tracking devices, being placed on the animal's collars and another in the owner's possession.

However, there are opportunities for improvement and expansion of the system. As a future work, we can improve the system by integrating other devices, such as devices capable of geolocation through GPS, implement techniques to improve the system's ability to learn the behavior of animals over time and habits of the owner's schedule. When you can't get a WiFi connection, instead of just lighting up the LED, send a message, through BLE, with information if you find the animal. Regarding the search for your dog's iTAG between 6:00 pm and 6:30 pm, as future work, Machine Learning algorithms could be used which, based on the identified patterns, the system would use them to learn and to recognize the signs that the owner is coming home. This could involve techniques such as pattern classification or neural networks to improve accuracy over time. Creation of its own platform, where all movements are recorded and with the capacity to work with the data, with a database to store all the MAC addresses it finds and a smartphone application aimed at monitoring and locating animals. Additionally, the development of a 3D protective case for the ESP32 and the other components, ensuring durability and resistance to adverse weather conditions.

References

1. Blynk: Everything you need to build an IoT project. https://blynk.io/developers. Accessed Nov 2023
2. Aguirre, E., et al.: Implementation and analysis of a wireless sensor network-based pet location monitoring system for domestic scenarios. Sensors **16**(9), 1384 (2016)
3. Kimaldi: Bluvision Beacons – Real-time location. https://www.kimaldi.com/pt-pt/produtos/leitores_e_tags_rfid/tags_rfid-pt-pt/bluvision-beacons-localizacao-em-tempo-real/. Accessed Nov 2023
4. Digitanimal: Locate and monitor. https://digitanimal.pt/investigacao-e-fauna-selvagem/. Accessed Nov 2023
5. So-In, C., et al: Mobile animal tracking systems using light sensor for efficient power and cost saving motion detection. In: 8th International Symposium on Communication Systems, Networks & Digital Signal Processing (CSNDSP) (2012)
6. Ramesh, G., Sivaraman, K., Subramani, V., Vignesh, P.Y., Bhogachari, S.V.V.: Farm animal location tracking system using arduino and GPS module. In 2021 International Conference on Computer Communication and Informatics (ICCCI) (2021)
7. Caetano, F., Brioso, P., Silveira, C.: Management of animals on farms. 17th Iberian Conference on Information Systems and Technologies (CISTI), Madrid, Spain, (2022)
8. Chen, Y., Elshakankiri, M.: Implementation of an IoT based pet care system. In 2020 Fifth International Conference on Fog and Mobile Edge Computing (FMEC), pp. 256–262. IEEE, (2020)
9. Luayon, A.A.A., Tolentino, G.F.Z., Almazan, V.K.B., Pascual, P.E.S., Samonte, M.J.C.: Pet-Care: a smart pet care IoT mobile application. In Proceedings of the 10th International Conference on E-Education, E-Business, E-Management and E-Learning, pp. 427–43, (2019)
10. Weiser, M.: The computer for the 21st century. ACM SIGMOBILE Mob. Comput. Commun. Rev. **265**, 3–11 (1999)
11. Srivastava, A., Bhardwaj, Saraswat, S.: SCRUM model for agile methodology. In: International Conference on Computing, Communication and Automation (ICCCA) (2017)
12. Jacobson, I., Ng, P. W., McMahon, P. E., Goedicke, M.: The essentials of modern software engineering: free the practices from the method prisons!. ACM (2019)
13. Poslad, S.: Ubiquitous Computing: Smart, Devices Environments and Interactions. John Wiley & Sons, Hoboken (2009)
14. Abowd, G.D., Dey, A.K., Brown, P.J., Davies, N., Smith, M., Steggles, P.: Towards a better understanding of context and context-awareness. In Handheld and Ubiquitous Computing: First International Symposium, HUC'99 Karlsruhe, vol. Proceedings, vol. 1, pp. 304–307. Springer, Heidelberg (1999). https://doi.org/10.1007/3-540-48157-5_29
15. Perera, C., Zaslavsky, A., Christen, P., Georgakopoulos, D.: Context aware computing for the internet of things: a survey. IEEE Commun. Surv. Tutor. **16**(1), 414–454 (2013)
16. Espressif Systems: ESP32 - A feature-rich MCU with integrated Wi-Fi and Bluetooth connectivity for a wide-range of apllications. https://www.espressif.com/en/products/socs/esp32. Accessed Nov 2023
17. Findchips: HC-SR501 PIR MOTION DETECTOR. https://datasheetspdf.com/pdf-file/775434/ETC/HC-SR501/1. Accessed Nov 2023

Integration of a Mobile Robot in the ROS Environment. Analysis of the Implementation in Different Versions

Sergio García González[1]([✉]), Vidal Moreno Rodilla[2],
Francisco Javier Blanco Rodríguez[2], Belen Curto Diego[2],
Héctor Sánchez San Blas[1], André Filipe Sales Mendes[1],
and Gabriel Villarrubia González[1]

[1] ESALAB - Faculty of Science, University of Salamanca,
Plaza de los Caídos s/n, 37008 Salamanca, Spain
sergio.gg@usal.es

[2] GROUSAL - Faculty of Science, University of Salamanca,
Plaza de los Caídos s/n, 37008 Salamanca, Spain
https://esalab.es/, http://gro.usal.es/

Abstract. In recent years, the field of robotics has undergone an exponential evolution, and the use of mobile robots has become increasingly common in various sectors of society, such as industry, medicine, agriculture and logistics. As a result, more and more developers are opting for the use of tools or resources that enable agile software development. One of the most prominent and widely used is the Robot Operating System (ROS) framework.

In this article, a detailed analysis will be made of the different versions of ROS, from the first versions to the most recent ones. Furthermore, the process of integrating a mobile robot into ROS 1 and ROS 2 versions will be addressed. The process followed will be indicated, starting from the selection of main components, such as perception, control, software resources and motion planning. In this process, different packages, repositories and tools available in ROS that facilitate this integration will be explored.

Keywords: ROS · Robotics · Turtlebot · Mapping · Navigation

1 Introduction

It is a reality that robotics is currently developing at an exponential rate, and increasingly, it is present in our reality. Within robotics, there are multitude of resources to generate software, some of them created by companies and exclusively private, and others open source. Specifically for open-source software, one of the main options used by developers is the use of the ROS framework.

ROS or **Robot Operative System** is not considered an operating system as such, but it operates in many aspects similarly to how a conventional operating system might. It facilitates the most essential tasks of robotics such as communication and coordination between different components of a robotic system, providing a set of tools, libraries, and conventions that facilitate the programming and development of software for robots and allows for the modularization and reuse of code. In particular, this framework has had two main versions, ROS 1 with which has been working for more than a decade both at an educational and business level, and ROS 2 which is a new version that attempts to improve aspects such as security, performance, and scalability of robotics.

With this work, the key aspects of the integration of ROS in robotics are to be addressed, such as the use of different versions, prototyping, the use of open-source and commercial tools or resources, and current research around it. In addition to focusing on the integration of ROS 1 and ROS 2, this work will emphasize the use of navigation techniques such as **2D SLAM** (Simultaneous Localization and Mapping) **and 3D**, as well as autonomous navigation and precise localization in robotic environments. Just as techniques are very important, *hardware plays a crucial role* in the development of a robotic prototype, so it is important to highlight the tasks of choosing hardware resources in the same way.

In conclusion, all these aspects are essential to achieve autonomous navigation of robots in real environments. These advancements contribute to the puzzle of implementing successful integration of ROS in robotics.

2 State of Art

2.1 Literature Review of Mobile Robotics in the Service Sector

Mobile robotics, especially in the service sector, is experiencing significant growth. The integration of mobile robots in sectors such as hospitality, healthcare, and logistics has been the subject of numerous recent researches [1]. In the hospitality sector, robots are being implemented to streamline processes such as guest registration and service delivery, demonstrating potential to enhance customer experience [2]. Similarly, in the healthcare domain, robots are being used for tasks such as medication delivery and vital sign monitoring, contributing to ensuring proper healthcare, especially in nursing homes [3].

ROS platform has facilitated the development of versatile and customizable robotic applications, as demonstrated in several studies using robots like Turtlebot [4]. These studies have explored everything from human-robot perception and interaction to autonomous navigation. Despite the progress made, technical, ethical, and regulatory challenges persist in the implementation of robots in these sectors. Adapting robots to complex environments, ensuring data security and privacy, and user acceptance are critical aspects that require continuous attention [5]. Therefore, it is necessary to continue researching and addressing the remaining challenges to achieve successful and sustainable integration of robots in these environments.

2.2 Review of the Literature on the Integration of Mobile Robots in ROS

The integration of robots in ROS is essential for the development of advanced robotic applications, covering techniques such as scanning, autonomous navigation, path planning, and SLAM.

Scanning techniques in ROS, such as SLAM, laser scanning [6], RGB-D scanning [7], and point cloud fusion, allow robots to build accurate maps of the environment [8], facilitating their autonomous navigation and safe interaction with the surroundings.

Autonomous navigation in ROS is achieved through the combination of sensors, perception algorithms, route planning, and motion control. Reactive and deliberative approaches are distinguished, as well as global and local navigation [9,10]. Additionally, path planning in ROS involves generating movements to reach destinations while avoiding obstacles, employing algorithms such as A* and Dijkstra [11] among others.

Lastly, SLAM is crucial for real-time localization and mapping, with tools such as Gmapping, Hector SLAM, SLAM Toolbox, and Google Cartographer [12–14].

In summary, the integration of mobile robotics in ROS encompasses various techniques and tools that enable robots to know their environment, plan routes, and move autonomously, being fundamental for advanced robotic applications.

3 Case Studies

In order to analyse a detailed analysis of ROS, different case studies will be presented. Specifically, both physical robots that have been fitted with sensors to scan the environment and robots prepared for simulations have been used to carry out the research:

- Turtlebot 2 mobile robot, consisting of a Kobuki platform and an Orbbec Astra depth camera.
- Prototype robot created from a hoverboard.
- Turtlebot 3 simulated robot (Burger).
- Roomba iCreate 3 with A1 lidar.

3.1 Description of the Hardware Components

This section focuses on providing a concise description of the hardware components used in the work, highlighting their technical specifications and relevant features.

Kobuki robotic platform is a hardware system designed to facilitate the development of mobile robots quickly and easily. This robotic platform is equipped with omnidirectional wheels so that it is considered as a holonomic robot. Also, like other similar commercial platforms, it includes multiple sensors such as infrared and fall sensors.

For this work, the use of a Hoverboard has also been explored as a low-cost option to create an autonomous robot integrated in ROS. In the Fig. 1 it can see the platform built with the board and the two motors.

Fig. 1. Prototype mobile robot with Hoverboard board (LIDAR A1)

The use of the Hoverboard's controller board and motors has been made possible through a reverse engineering process that has unlocked and accessed its functionality. The Hoverboard controller board can provide the necessary capability to perform tasks such as odometry, motor control and communication with other devices, allowing seamless integration into the ROS environment.

Roomba iCreate 3 robotic platform has become a popular choice due to its versatile features and capabilities. In addition, it comes equipped with a number of integrated sensors, such as infrared and shock sensors, which enable obstacle detection and intelligent navigation in different environments.

4 Results

4.1 Robot Integration in ROS 1

For this work, two different mobile robots were integrated into ROS 1. The first was the Kobuki platform that was integrated into the Kinectic version of ROS 1 and the second protype was a Hoveboard controller board that served as a motor controller in the Melodic version of ROS 1.

The objective of this integration in both cases was mapping and navigation with these two mobile robots in 2D and 3D environments.

Kobuki Integration. For the first case study, a 2D and 3D scan of a room was performed with the Orbbec Astra depth camera and the scanning tool Gmapping.

This ROS package implements a SLAM algorithm to build a two-dimensional map of the environment and locate itself on that map. As input data to Gmapping, the internal wheel odometry data and the external environment information collected by the Orbbec Astra camera are used. These data are used to feed a particle filter called *Rao-Blackwellized* (RBPF) [15] to generate the RVIZ grid 2D map. The Orbbec Astra camera performs a 3D scan providing a point cloud. The Gmapping package only uses the distance data in the plane where the 3D camera is located because this package was intended to work with a LIDAR. Figure 2 shows the 2D scan at a given instant, which serves as one of the inputs to build the 2D map of the environment.

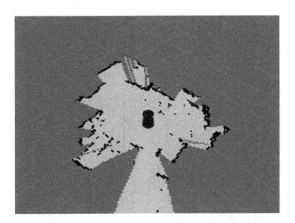

Fig. 2. 2D scanning with Orbbec Astra camera

To build the 2D map of the environment, as the robot moves autonomously, 6 ROS nodes are used: /camara, /scan, /odom, /move_base, /tf and Gmapping. The Fig. 3 shows a graph displaying the interconnection between the different nodes in ROS. In this graph, it can be seen how the nodes of the platform, the camera and the mapping and navigation tools communicate with each other through the corresponding topics and services. This graph provides a clear visual representation of how the connections are established.

The interconnection between nodes of the graph is fundamental for the operation of various functionalities in ROS. In this case, we can observe three important connections:

- Between the nodes **/camera** and **/scan,/camera** node provides visual information, while the **/scan** node represents laser sensor information. These nodes communicate to integrate visual and laser scan data, enabling better environment perception and more accurate map generation in the **Gmapping** node.

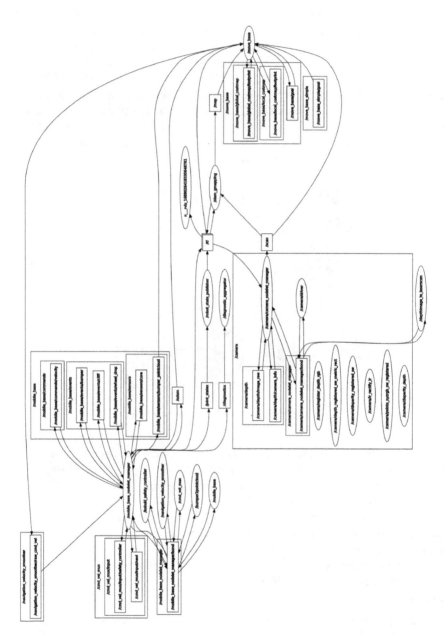

Fig. 3. Graph obtained with the RQT_GRAPH tool of the nodes and topics executed.

- Between the **/odom** node and the **/move_base** node, **/odom** node collects robot odometry information, such as position and orientation. This information is crucial for robot planning and navigation. The **/move_base** node is responsible for route planning and robot motion control. The connection between these nodes allows odometry to be used in the navigation process to estimate the robot's position, orientation and movement.
- Between the **/tf** node and the **Gmapping** node, **/tf** node is responsible for managing transformations between different systems of reference (such as the robot's base frame, camera frame, etc.). In this case, it connects with the Gmapping node to provide the necessary transformations that allow the Gmapping mapping algorithm to locate and correctly adjust the relative positions of objects on the 2D map of the environment.

RTAB-Map has been used for the generation of the 3D map. Therefore, after scanning, we can obtain a three-dimensional color map, as color and depth information captured by the camera is fused.

In 2D mapping, the information is saved in two different files: one in .pgm format, which contains the two-dimensional occupancy map, and another in .yaml format, which stores metadata related to the map, such as its resolution and size. On the other hand, RTAB-Map uses a database to store the data generated during the mapping process, allowing for more efficient storage and the possibility of performing queries and more complex analysis of the data captured by the robot's sensors during exploration. Additionally, RTAB-Map allows for the subsequent visualization at any time of the captured data, as can be seen in the following Fig. 4.

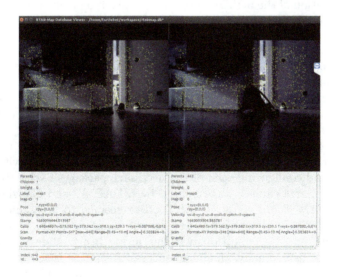

Fig. 4. RTAB-Map database

Hoverborad-Based Prototype. In the integration of the motor-controlled mobile robot with a Hoverboard the most remarkable fact is the change from the Gmapping tool to Google Cartographer. The main difference between Cartographer and Gmapping lies in the algorithms and approaches used. While Gmapping is based on a particle filter to generate 2D maps, Cartographer uses a more advanced approach called SLAM based on graph optimisation. This technique allows a better estimation of the robot's location on map and a higher accuracy in mapping the environment. In addition, Cartographer is able to generate 3D maps, which provides a more detailed representation of the three-dimensional environment, although in this case we have not used 3D scanning.

4.2 Integration in ROS 2

In this section, the transition from ROS 1 to ROS 2 is addressed and the improvements that this new version offers compared to its predecessor are explored. Concerning ROS 1, these have presented a greater degree of difficulty due to the lack of experience with the system and the scarce documentation and community behind these versions of ROS 2. In this ROS 2 ecosystem, the following have been integrated two different mobile platforms, the first one the Kobuki, which had already been successfully integrated in ROS 1, a simulation and the roomba iCreate 3, furthermore a simulated mobile robot.

Before going into the more manual integration process of this work, we managed to achieve the configuration of the camera by means of some practically forgotten packages and a small demonstration of the typical behaviour of a robot that focuses on its target and follows it as long as it does not leave its main focus.

The integration of the mobile robot using the Kobuki platform can be defined in three interactions:

- In the first iteration, the Turtlebot 2 packages that use the Kobuki platform are used to speed up the process, since by making the appropriate configurations and taking care that the nodes communicate properly. However, the first problem appeared and it was that the tree of transformations was not completed correctly and therefore the scanning could not be done if each frame did not communicate its coordinate system to the map.
- In the second iteration, instead of using these packages, a more radical solution is initiated and that is to build the packages manually, which implies installing and configuring each of the necessary packages. Not only that, but the Orbbec Astra camera was replaced by LIDAR A1 and Google Cartographer by SLAM ToolBox, as SLAM ToolBox seemed to be a simpler tool to work with LIDAR. However, the problem of transformations still persists.
- In the third and last iteration, without losing heart, after performing the transformations manually, an almost successful solution is achieved where the map is scanned but only partially because it remains frozen in a discontinuous way and the nodes stop publishing information. This *allowed to establish a suitable TF tree*.

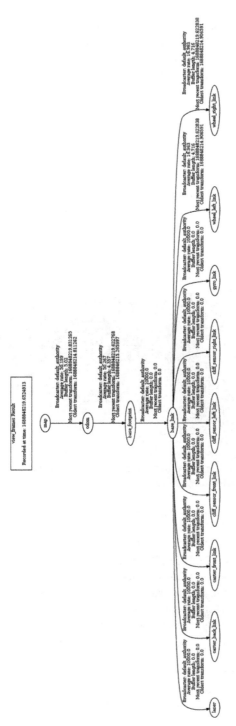

Fig. 5. ROS 2 rqt graph of the transformations tree

In Fig. 5 it can see how the manual transformations, for the passing of information, that was carried out in order to be able to interpret the graph in RVIZ, are seen in a correct way. What we really have to observe is that the transformations are correctly ordered, being the father always a frame like map or world, followed by the odometry data and in turn followed by the frame(s) in charge of the external sensors. The circles or ellipses that can be seen in the graph are the names of the frames, which are the corresponding reference system of the robot components, and the transformations are each of the arrows that relate each frame.

Simulated Robot. To carry out the simulation of the Turtlebot 3 Burger, the environment provided by Gazebo was used. This simulation environment, represented as a virtual hexagon, allowed the Turtlebot 3 Burger to be remotely operated and controlled. Thanks to the use of the A1 LIDAR sensor, a complete mapping of this simulated environment was possible. Figure 6 shows the simulation environment of Turtlebot 3.

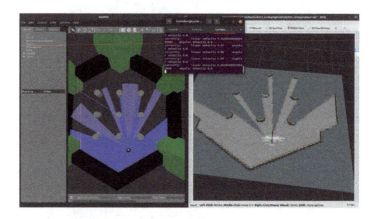

Fig. 6. Mapping of the simulation environment using Google Cartographer

During the simulation, the LIDAR A1 scanned the environment and collected accurate distance and location data. Once the map was generated, it could be used to perform autonomous navigation of the robot. With the map as a reference, the Turtlebot 3 Burger was able to intelligently plan routes and avoid obstacles as it moved through the simulated environment. It can be seen in the Fig. 7, taken from the simulations, how the robot traces a route to the selected target and tries to navigate it.

The comparison between simulation and the real world in ROS 2 is crucial for evaluating system performance. The simulation provides a controlled environment for testing and debugging, but does not fully capture the complexities of the real world.

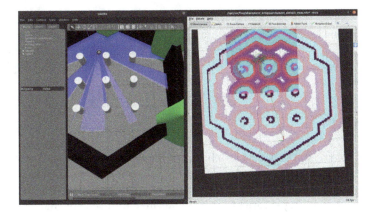

Fig. 7. Autonomous navigation of the Turtlebot 3 Burger

Roomba iCreate 3. Finally, the integration with Roomba iCreate 3 has been successful in terms of teleoperation and the integration of an A1 Lidar to perform a two-dimensional scan of the environment. These functionalities enabled accurate distance and location data to be obtained, which was a major achievement in the development of the work. Unfortunately, however, difficulties were encountered when trying to implement SLAM and navigation techniques on this platform. One of the main challenges was the lack of clear documentation on the current state of the packages used, which were quite recent. In addition, some bugs were found in the packages, which hindered their successful implementation. This kind of information can be found in the Turtlebot 4 github repository, where there are a lot of issues related to mapping and navigation when using these packages.

5 Conclusions

Throughout this work, the integration of a mobile robot into the ROS environment has been investigated, facilitating the development of rapid solutions for various robotic platforms, such as Turtlebot. Scanning, mapping and navigation techniques have been explored, expanding knowledge in these fields.

Integration of both ROS 1 and ROS 2 on mobile platforms such as Kobuki and a hoverboard was achieved, although integration of the Roomba iCreate 3 in ROS 2 was more challenging due to lack of documentation. ROS 1 was found to have more extensive documentation and more stable releases, while ROS 2 offers improvements in navigation, albeit with a significant gap in documentation and community support.

This work has deepened the understanding of ROS and SLAM tools such as Gmapping, in service robotics, and Google Cartographer, accelerating the development of future robotic works. It has reduced the time and resources needed to implement advanced solutions, opening up new possibilities in automation and human-machine interaction.

In conclusion, this work has been enriching, allowing to acquire valuable knowledge about ROS, evaluate different robotic platforms and explore the capabilities of ROS versions. The potential of ROS 2 is highlighted, but the importance of a strong community and thorough documentation for its full exploitation is underlined.

6 Future Lines

In terms of future lines of development, various areas of focus are proposed to advance the improvement of robotic systems' capabilities with ROS.

One of them is the comparative study of algorithms to determine which navigation and mapping algorithm is most suitable for each specific environment, evaluating and comparing different options.

It is suggested to explore the integration of hoverboard-based prototypes into ROS 2, enhancing the performance and stability of the systems.

Furthermore, conducting benchmarks and performance tests on DWA Planners and others is proposed to evaluate their effectiveness in different contexts.

A key line is to achieve the successful integration of Roomba iCreate 3 robot into ROS 2, overcoming previous challenges such as the lack of documentation.

Lastly, considering the use of OMPL (Open Motion Planning Library) to improve real-time path planning is contemplated.

These approaches will continue to enhance ROS-integrated robotic systems and explore new capabilities in robotics.

Acknowledgments. The Spanish Ministry of Universities has supported Héctor Sánchez San Blas' research through a FPU pre-doctoral contract, under grant FPU20/03014

References

1. Li, K., Sun, S., Zhao, X., Jinting, W., Tan, M.: Inferring user intent to interact with a public service robot using bimodal information analysis. Adv. Robot. **33**(7–8), 369–387 (2019)
2. Zhang, Y., Wang, X., Wu, X., Zhang, W., Jiang, M., Al-Khassaweneh, M.: Intelligent hotel ROS-based service robot. In: 2019 IEEE International Conference on Electro Information Technology (EIT), pp. 399–403 (2019)
3. Marín, C.P.: Study of the applicability of automated transport for material transport in hospitals. Master's thesis, Universitat Politècnica de València (2021)
4. Mishra, R., Javed, A.: ROS based service robot platform. In: 2018 4th International Conference on Control, Automation and Robotics (ICCAR), pp. 55–59 (2018)
5. Xu, C., Li, W., Tan, J.T.C., Chen, Z., Zhang, H., Duan, F.: Developing an identity recognition low-cost home service robot based on turtlebot and ROS. In: 2017 29th Chinese Control and Decision Conference (CCDC), pp. 4043–4048 (2017)
6. Wang, R., Li, X., Wang, S.: A laser scanning data acquisition and display system based on ROS. In: Proceedings of the 33rd Chinese Control Conference, pp. 8433–8437 (2014)

7. Pauwels, K., Ivan, V., Ros, E., Vijayakumar, S.: Real-time object pose recognition and tracking with an imprecisely calibrated moving RGB-D camera. In: 2014 IEEE/RSJ International Conference on Intelligent Robots and Systems, pp. 2733–2740 (2014)
8. Labbe, M., Michaud, F.: Appearance-based loop closure detection for online large-scale and long-term operation. IEEE Trans. Rob. **29**(3), 734–745 (2013)
9. Acosta, G., Gallardo, J., Pérez, R.: Reactive control architecture for autonomous navigation of mobile robots. Ingeniare. Revista chilena de ingeniería **24**(1), 173–181 (2016)
10. Rodríguez, A.J.P., et al.: Development of navigation system for terrestrial autonomous vehicles using ROS. Master's thesis, Universidad Politécnica de Cartagena (2017)
11. Goñi, O.E., Fernández León, J.A., Acosta, N.: Architecture of a hybrid controller for robust navigation in partially known environments. In: VIII Workshop of Researchers in Computer Sciences (2006)
12. Zhang, X., Lu, G., Fu, G., Xu, D., Liang, S.: SLAM algorithm analysis of mobile robot based on lidar. In: 2019 Chinese Control Conference (CCC), pp. 4739–4745 (2019)
13. Macenski, S., Jambrecic, I.: SLAM toolbox: SLAM for the dynamic world. J. Open Source Softw. **6**(61), 2783 (2021)
14. Krinkin, K., Filatov, A., yom Filatov, A., Huletski, A., Kartashov, D.: Evaluation of modern laser based indoor slam algorithms. In: 2018 22nd Conference of Open Innovations Association (FRUCT), pp. 101–106 (2018)
15. Li, Q., Kang, J., Cao, X.: Research on SLAM based on RBPF algorithm in indoor environment. J. Phys: Conf. Ser. **012066**(07), 2021 (1971)

The Biomimicry Database: An Integrated Platform to Enhancing Knowledge Sharing in Biomimetics

Vagner Bom Jesus[1], Clara Silveira[1(✉)], and Carlos Carreto[1,2]

[1] Instituto Politécnico da Guarda, Guarda, Portugal
1701172@sal.ipg.pt, {mclara,ccarreto}@ipg.pt

[2] CISE-Electromechatronic Systems Research Centre, Universidade da Beira Interior, Covilhã, Portugal

Abstract. This study describes the implementation of The Biomimicry Database (TBDB), an innovative initiative developed to fill the gap in the accessibility and management of information about biomimetics driven by the need to integrate the principles of sustainability in new technologies. TBDB appears as a comprehensive solution, structured based on a RESTful API and complemented by a web interface and mobile application, which offers an integrated and intuitive approach for biomimicry professionals and enthusiasts. It also offers not only a centralized source of information about biomimetics but also facilitates effective collaboration among various stakeholders, thereby fostering sustainable innovation. The results obtained demonstrate the system's effectiveness and favorable user experience, as evidenced by the assessments conducted using PageSpeed Insights and the System Usability Scale. The solution developed and tested ensures its efficacy and efficiency, making a substantial contribution to the current knowledge by offering tools that streamline the access and administration of information in the field of biomimetics.

Keywords: Biomimetics · Sustainability · RESTful API · Internet Technologies

1 Introduction

Biomimetics refers to a broader range of approaches inspired by nature to develop more efficient and robust technological solutions, and has been fundamental in the advancement of computing and robotics. Article [1] highlighted the influence of biomimetics on the creation of algorithms and autonomous systems. In another example, [2] investigated the application of biological principles in robotics, leading to machines that can effectively adapt and interact with their surroundings. However, the biomimetics domain encounters challenges in unifying its community, particularly owing to the absence of a centralized platform that simplifies access to information about biomimetics, hindering advancements in technology inspired by nature.

This project was motivated by the need to make Internet use more efficient and sustainable [3]. Biomimetics is inspired by solutions found in nature to solve human problems [4]. It is a promising field that encourages technological innovation and promotes more sustainable and efficient approaches to the development of new technologies [5]. Without a centralized platform that provides easy and structured access to information about biomimetics, researchers face difficulties in conducting investigations and developing innovative technological solutions inspired by nature.

To address this challenge, The Biomimicry Database (TBDB) system was developed with the aim of streamlining access to centralized information and its dissemination. This encompassed the creation of a RESTful API alongside a web platform and a mobile application. Providing an all-encompassing solution for data recording, updating, and retrieval within the realm of biomimetics, the system ensures centralized information and effortless accessibility. This solution incorporates advanced software engineering concepts and practices and aims to contribute to biomimetic research and development, particularly by fostering collaboration among researchers and driving the evolution of sustainable technologies.

The remainder of this paper is organized as follows. After this introduction, Sect. 2 addresses related work, including API, web platforms, and mobile applications. Section 3 presents the proposed solutions in detail and describes the architecture, modules, and prototype. Section 4 describes the tests conducted and the results achieved. Section 5 concludes the article, reflecting on the project's contributions and outlining perspectives for future work.

2 Background Research

In this section, we explore systems related to the management of information about biomimetics, some of which are similar to the proposed solution.

The Biomimicry Institute [6] has a platform to collect information about existing biomimetic solutions, including the curation of academic research, case studies, and contributions from experts in biomimicry. The solutions described in the system emphasize technical characteristics such as sustainability, energy efficiency, and innovation, inspired by the strategies and processes found in nature.

The AskNature platform is an interactive system that offers the opportunity to explore biomimetic solutions organized by biological function [7]. It uses a methodology that involves the classification and presentation of solutions inspired by nature, facilitating the search for innovations that imitate biological processes. The technical characteristics of solutions on AskNature are focused on efficiency, sustainability, and adaptability, reflecting the principles and strategies found in the natural world.

Biomimicry Toolbox is a platform designed to provide essential resources and tools for incorporating biomimetic concepts into design and engineering projects [8]. The employed methodology facilitates the practical application of biomimicry, allowing users to explore and integrate solutions inspired by nature into their work, in which the technical characteristics addressed encompass sustainable innovation and efficiency inspired by natural systems.

Zygote Quarterly [9] adopts a rigorously curated methodology to collect information on existing biomimicry solutions, reviewing case studies, academic research and practical innovations, where relevant technical characteristics highlighted include material use effectiveness, energy efficiency, sustainability and innovation inspired by natural strategies.

The Biomimicry Manual is a series of articles available on a web platform [10], that provide a source of insight into how principles and strategies found in nature can inspire innovative natural-inspired design solutions. These articles explored diverse aspects of biomimicry, from basic concepts to advanced applications, serving as a valuable resource for designers, engineers, and researchers interested in integrating the sustainability and efficiency of nature into their creations.

The Encyclopedia of Life provides an API to access detailed biological information [11] and inspire biomimetic innovations. Available information focuses on the study of living organisms that overcome environmental and functional challenges. It stands out for its wide repository of information about species, which supports the development of innovations inspired by nature.

The Naturalist API [12] is a software interface that provides access to an extensive set of species observations, facilitating the exploration of natural features. Users and developers can employ this API to integrate biodiversity data into applications, fueling inspiration and innovation in areas that benefit from understanding and applying patterns and strategies found in nature.

The GBIF API is an interface that provides access to a broad biodiversity database essential for biomimicry research [13], enabling the exploration of detailed information about species and their habitats to inspire nature-based innovation.

The "Biomimicry" application, developed by EIACP TEAM, is an educational tool that promotes the teaching of biomimetic topics [14]. It stands out for its commitment to privacy, not sharing or collecting user data, and security and privacy are the most critical, considering that the public may include minors.

The Biomimetics Book Offline, a mobile application developed by Muamar Dev, offers an educational approach to biomimetics, providing theoretical content accessible offline [15], in which the application stands out for allowing the exploration of biomimetics through categories and favorites, as well as providing a comprehensive view of how living organisms inspire innovative solutions, imitating natural models to solve human problems, reflecting the programmer's commitment to global education.

The TBDB is aligned with the current trends in applying biomimetics to design innovative technologies, focusing on the centralization and dissemination of biomimetics knowledge. The developed system stands out for its integrated architecture, which includes a RESTful API, mobile application, and web platform. This approach offers users a comprehensive solution, implemented with the latest technologies in distributed systems, based on a Model-View-Controller (MVC) architecture. This organizational structure promotes the separation of responsibilities and system maintenance, ensuring a robust, scalable, and easy-to-maintain structure.

3 Proposed Solution

The TBDB system includes a RESTful API, web platform, and mobile application. Figure 1 shows a visual representation of the technological components and their interactions within the TBDB system, highlighting the endpoints of the system.

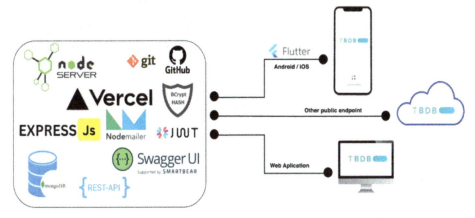

Fig. 1. Technologies used in the TBDB system and its endpoints.

The TBDB system was created using a combination of technologies, including Node.js and Express.js, which were employed to develop server logic and manage RESTful API routes. Node.js's Nodemailer module was used to efficiently send emails in a newsletter format SMTP, optimizing communication with the community. Additionally, MongoDB was utilized as a database system to efficiently store information. Vercel was used for hosting and continuous deployment, whereas GitHub was used for version control and code collaboration.

API documentation and testing were facilitated by the Swagger UI, and authentication was ensured through JWT and BCrypt for password protection, generating secure hashes that reinforce the integrity of user access data. For the mobile application, Flutter was chosen, which allows unified development for both the iOS and Android platforms. The system provides centralized access to biomimetic data and is operated via a web interface and a mobile app for user interaction.

The TBDB system features a secure and user-friendly RESTful API, a flexible web platform that is accessible to all users, and a versatile mobile application. These components provide fluid and efficient navigation in the management and consultation of biomimetics knowledge data, elevating the user experience beyond the desktop. The mobile application facilitates the sharing of discoveries through social networks and the generation of PDF documents, enabling the sharing of biomimetics knowledge in a way that is convenient for users.

3.1 TBDB Architecture

The TBDB architecture was designed with a Model-View-Controller (MVC) architectural pattern, dividing the application into three components that promote a clear separation of responsibilities and improve the overall structure.

The process outlined in Fig. 2 commences with a client-initiated HTTP request from the View, which is then processed using the Express.js framework within the Node.js application. JWT authentication middleware can be applied depending on the requested route. The request is then forwarded to the Controller that engages with the Model. This Model can interact with MongoDB to retrieve and save data. Once all necessary operations are completed, a response is constructed and returned to the client.

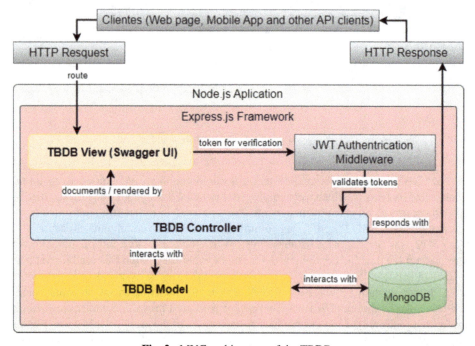

Fig. 2. MVC architecture of the TBDB.

TBDB clients, including both web and mobile applications, interact with the API and MongoDB database using requests, such as GET to fetch data and POST to create or modify information in the database.

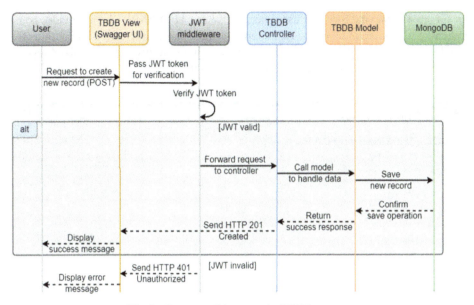

Fig. 3. Sequence Diagram of a POST request.

The sequence diagram in Fig. 3 illustrates the flow of a data-creation operation in which the user initially requests a POST operation from the View. Through middleware, the system checks the authenticity of the user using the JWT. If the JWT is confirmed as valid, the request is sent to the Controller, which in turn requests the Model to write the data to MongoDB. If the recording was successful, a response with HTTP status 201 was sent, indicating successful creation. If the JWT is invalid, the response will be an HTTP 401 error, indicating that the user is not authorized. This process emphasizes security through authentication, and data integrity is ensured by confirming the recording before notifying the user. JWT verification is performed by specific middleware, which is a good practice in Node.js development to keep the code clean and modularized in conjunction with a well-defined separation of responsibility, which acts as an intermediary between the View and the Model, which in turn interacts directly with the database.

3.2 System Prototype

The final prototype of the TDBD system represents a centralized integrated solution with the aim of streamlining interaction with information about biomimetics and is designed as a tool to facilitate the sharing of knowledge in the field of biomimetics, accessible to everyone.

Figure 4 shows the Swagger user interface for the TBDB API, which provides comprehensive documentation of the API endpoints and enables interactive interactions for CRUD operations. The interface was designed with an intuitive and organized layout, making it easy for programmers to incorporate and utilize its features.

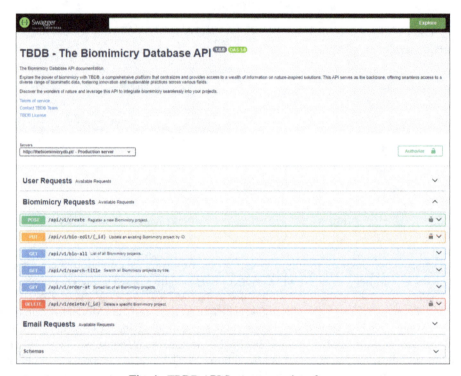

Fig. 4. TBDB API Swagger user interface.

The image in Fig. 5 depicts the web platform's publication screen for TBDB Projects, which allows users to submit new biomimetics projects. Additionally, there is a section to subscribe to the newsletter as well as a compilation of the most popular projects, emphasizing the system's user-friendly and functional interface.

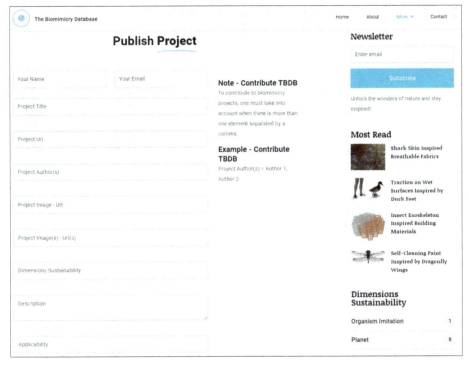

Fig. 5. Web platform interface for the TBDB Projects page.

Figure 6 shows four views of the TBDB mobile application. The mobile application transposes the functionalities of the web platform to the mobile environment so that users can search for and interact with biomimetics content directly through their mobile devices.

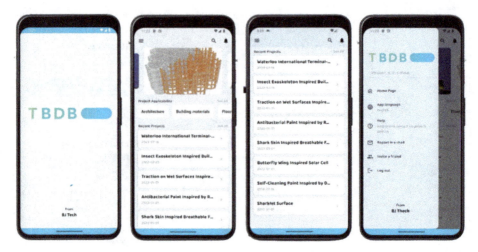

Fig. 6. Mobile application interface.

4 Tests and Results

The significance of testing in enhancing the software development life cycle was highlighted in [16], with a focus on ensuring the quality and stability of the software.

This section addresses the main tests conducted to assess the effectiveness and reliability of systems developed to identify and resolve vulnerabilities.

TBDB's RESTful API underwent automated testing to ensure functionality, reliability, and performance, including CRUD operations, integration, and load tests. The outcomes confirmed that all operations were performed as intended. Simultaneously, unit tests were conducted on the TBDB mobile application built with Flutter to assess the logic and behavior of individual components. Most of the components successfully passed the tests, affirming their high performance and reliability.

Performance analysis was implemented using PageSpeed Insight [17], as shown in Fig. 7, which reveals values in performance, accessibility, best practices, and SEO.

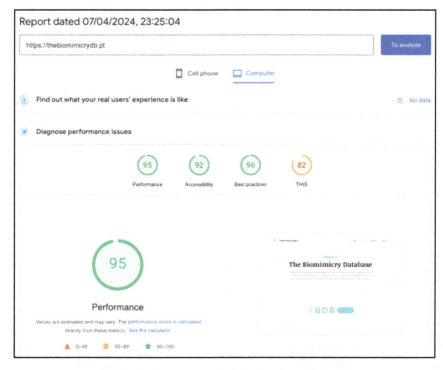

Fig. 7. Performance analysis in PageSpeed Insights.

Based on these tests, the TBDB system proved to be robust and efficient, proving that the system can function correctly under different conditions, in addition to adhering to best web development practices and being accessible to a wide spectrum of users.

4.1 Usability Assessment Tests with the System Usability Scale

The System Usability Scale (SUS) [18] was employed to evaluate the usability of TBDB web and mobile platforms, considering aspects such as ease of use, learning, and overall user experience satisfaction.

Figure 8 illustrates the mean scores derived from the Likert scale representing the outcomes of the SUS assessment. This evaluation involved 103 users who were tasked with completing a standard SUS questionnaire. Comprising ten questions [19], the questionnaire employed a Likert scale [20] to gauge the diverse facets of usability. Responses ranged from 1 to 5, denoting the average ratings of the SUS questionnaire. The horizontal bars depict the mean responses, with lengths ranging from one (Completely Disagree) to five (Completely Agree). Additionally, the standard deviation of the mean is illustrated by the horizontal lines at the end of each bar.

The analysis of the SUS assessment results reflects an overall positive user experience, highlighting that the usability of the TBDB system is intuitive and satisfactory. The average user score was 86.65, classified as "Excellent" according to [20], and the standard deviation was 17.36. Despite the excellent overall rating, a detailed analysis of

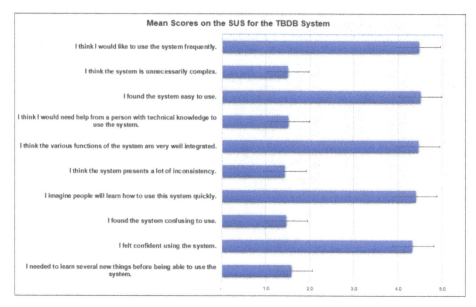

Fig. 8. Mean scores of the SUS assessment.

user responses verified some cases that identified the potential need for technical support in specific areas, simplification of some functionalities, and improvement of system documentation. Feedback on functionality integration and ease of learning further validates the system's cohesive architecture, which is deemed suitable for its intended user base.

The low standard deviation of the mean scores of the questions suggests that the opinions of the users who participated in the SUS assessment were consistently aligned, indicating significant agreement in the usability evaluations of the system.

5 Conclusion

The implemented system addresses the issue of limited centralization and accessibility of biomimetics information, which is crucial for technological advancements inspired by nature. It is a comprehensive, all-in-one platform that offers a RESTful API, mobile app, and web interface for researchers and enthusiasts to register, share, and consult knowledge efficiently in the field of biomimetics.

The results demonstrate the efficiency of the system. Performance and usability evaluations, including tests using PageSpeed Insights and the System Usability Scale (SUS), indicate that the system is fast, efficient, and provides a positive user experience.

For future developments, the aim is to expand the biomimetics database, improve the user interface based on the feedback received, and integrate advanced technologies, such as artificial intelligence, to personalize the user experience with the adoption of augmented reality in mobile applications and the development of new interactive functionalities, aiming to expand the usefulness and impact of the TBDB system.

Regarding the incorporation of artificial intelligence, a fourth endpoint is being developed for the TBDB system, consisting of a chatbot with a physical robotic face designed to enrich the usability experience of system users. The chatbot is supported by a biomimetic model to implement facial expressions based on eye contact and uses advanced artificial intelligence and natural language processing technologies to humanize digital interactions and create the potential to make the system's interaction with users more natural and interesting, contributing to the dissemination of knowledge in the field of biomimetics.

References

1. Casey, A., Cai, Y.: Renaissance of biomimicry computing. Mob. Netw. Appl. **28**(2), 486–489 (2023)
2. Wang, J., et al.: A survey of the development of biomimetic intelligence and robotics. Biomimetic Intell. Robot. **1**, 100001 (2021)
3. Ideia Sustentável: 25 anos Ideias Sustentáveis. https://ideiasustentavel.com.br. Accessed 24 Apr 2024
4. Biomimética: tecnologia inspirada na natureza avança no Brasil. Mongabay. https://brasil.mongabay.com/2020/03/biomimetica-tecnologia-inspirada-na-natureza-avanca-no-brasil/. Accessed 24 Apr 2024
5. Ajayi, T.O.: Biomimicry: the nexus for achieving sustainability in the people-process-planet relationship. Heliyon **9**(5), 2405–8440 (2023)
6. Biomimicry Institute - Biomimicry Design Spiral. Toolbox.Biomimicry.org. Accessed 04 Apr 2024
7. Biomimicry Institute - It's time to ask nature. Asknature. https://asknature.org/. Accessed 20 Apr 2024
8. Biomimicry Institute Biomimicry Toolbox - Your guide to applying nature's lessons to design challenges. https://toolbox.biomimicry.org. Accessed 25 Mar 2024
9. ZQ Zygote Quarterly. https://zqjournal.org. Accessed 04 Mar 2024
10. MH Sub I – Inhabitat. https://inhabitat.com. Accessed 25 Mar 2024
11. EOL - API da Encyclopedia of Life. https://eol.org/api. Accessed 24 Apr 2024
12. API Naturalist. https://api.inaturalist.org. Accessed 25 Mar 2024
13. GBIF: GBIF API - Technical Documentation. https://www.gbif.org/developer/summary. Accessed 25 Mar 2024
14. Google Play. https://play.google.com/store/apps/details?id=com.eiacpku.biomimicry. Accessed 25 Mar 2024
15. Google Play. https://play.google.com/store/apps/details?id=com.MuamarDev.BiomimeticsBookOffline. Accessed 25 Mar 2024
16. Rafi, D.M., Moses, K.R.K., Petersen, K., Mäntylä, M.V.: Benefits and limitations of automated software testing: systematic literature review and practitioner survey. In: 7th International Workshop on Automation of Software Test (AST), pp. 36–42. IEEE (2012)
17. Google for Developers - PageSpeed Insights. https://pagespeed.web.dev/. Accessed 07 Apr 2024
18. Brooke, J.: SUS-A quick and dirty usability scale. Usabil. Eval. Ind. **189**(194), 4–7 (1996)
19. Measuring U. https://measuringu.com/sus/. Accessed 25 Mar 2024
20. Bangor, A., Kortum, P., Miller, J.: Determining what individual SUS scores mean: adding an adjective rating scale. J. Usabil. Stud. **4**(3), 114–123 (2009)

A Monitoring Framework for Smart Building Facilities Management

Eduardo Pina[1], José Ramos[1], João Henriques[1,2(✉)], Filipe Caldeira[1,2], and Cristina Wanzeller[1,2]

[1] Polytechnic of Viseu, Viseu, Portugal
{pv27228,estgv17704}@alunos.estgv.ipv.pt,
{joaohenriques,caldeira,cwanzeller}@estgv.ipv.pt
[2] CISeD - Research Centre in Digital Services, Viseu, Portugal

Abstract. The demand for intelligent buildings contributing to improving efficiency in facilities management, energy consumption, and people's safety demands technologies supported by the Internet of Things (IoT).

In that regard, this paper introduces a monitoring framework tailored to that demand. A Proof-of-Concept (PoC) of the proposed framework proposes a system simulating IoT data to enhance the responsiveness of facility managers. It also provides data acquisition and analysis capabilities to identify critical events and anomalies within the smart building environments. They are supported by a containerized application for each layer: data harvesting, support layer, and data analysis. Several datasets (i.e., humidity, movement, temperature, and Light Emitting Diode (LED)) are simulated to monitor, analyze, and trigger alerts for temperatures, percentage of moisture, light consumption, and movement detection. The results demonstrate the feasibility of the proposed framework.

Keywords: Simulated Smart Building · Internet of Things · Docker Swarm · Containers

1 Introduction

IoT is impacting technology, financial, and social development over the years. In past decades, there has been progress in wireless communication, information and communication systems, industrial design, and mechanical systems, encouraging a new revolution in the capabilities of sensors connected to the Internet. The IoT is influencing many industries. Smart buildings are becoming a reality where billions of smart sensors and connected computing devices discover, analyze, and affect some of our most sensitive daily chores [1,2].

These devices include movement detection, thermostats, humidity, energy management, and Bluetooth. On the other hand, prior simulation is needed to deploy these sensors effectively to make appropriate use of existing applications or deploy new ones in remote sensing. Simulation provides a cost-effective approach to deploying applications in an end-to-end process. Through simulation, it is possible to shed light on directions to implement future applications [3]. However, as far as we currently know, there are no simulations of Smart Building Alert Systems that make use of micro-services to horizontally scale the number of devices in a building with Docker Swarm in a single computer. Docker Swarm is a framework developed by Docker that schedules containers on the nodes. Each task is containerized and separated from other assignments, with the abstraction of not interfering with different containers. These containers can be deployed on-premises or in the cloud, [4].

To resolve this issue, we propose a system organized into three layers: data harvesting, supportive layer, and data analysis. For the first layer, we intend to simulate four IoT sensors: movement detection, temperature levels, humidity percentage, and Light Emitting Diode (LED) consumption. Each sensor is simulated using Python scripts, and the corresponding data is generated in one-minute intervals over a year. The supportive layer comprises an Application Programming Interface (API) developed in Flask. It gathers data generated by the simulated devices in JSON and then transforms it into a structured SQL file with the help of a local web browser to validate the data. The last layer will provide analytical data through a graphical representation throughout the days and weeks from the SQL file created from the API. This file allows the creation of a database in a containerized SQLite instance for the application. Lastly, the graphical representation of the number of alerts and plots reads data from the SQL file showing data throughout the day and weeks. In this paper, we bring an alert system to simulate the management of an intelligent building facility that allows people to monitor every device. This system aims to get a simple and functional alert system that generates alerts.

The paper is structured as follows. Section 2 discusses related work. Section 3 describes the architecture used in this simulation. Section 4 describes the experimental design. Section 5 discusses the results and analysis obtained with this simulation. Finally, Sect. 6 presents the findings and future work.

2 Related Work(s)

This section surveys the literature related to simulated smart building using Docker swarm.

In this survey [5], the authors presented a systematic review of smart building's state of the art. The authors discuss how the European Union adopted this concept in several buildings to decrease fossil fuels and reduce emissions of greenhouse gases in cities. However, the need to integrate and adequately manage energy led to restructuring the traditional energy grid to a smart grid so that each building can balance its on-site energy generation and consumption, among other features like wind and solar panels.

Another survey [6] was conducted regarding Smart Buildings and IoT devices. The Internet of Things devices aided the industry's evolution by integrating with areas such as intelligent buildings. According to the authors, Smart Building and IoT are linked, and the system is divided into three layers. The first layer has sensors that collect data; the second layer acquires, analyses, and prepares the data in a database system, such as Kafka or Cassandra, and the third layer is the intelligent application system, which manages the application and the design. The authors concluded that a large volume of data collected from sensor networks feeds Big Data databases and opens up the in-depth analytical dimension to identify the needs of intelligent building operators based on models.

The authors [7] underlay the importance of IoT devices and the ability to connect with machine-to-machine (M2M) through parameters to make wise decisions not to harm humans. In this paper, the authors analyze basic applications like heating, ventilating, and air conditioning (HVAC), water management, lighting systems, the health system for elders, fire detection, and security risks taken with each sensor. The authors pointed out the vast cybersecurity risks associated with using these devices and the importance of protecting the connection between the devices and the system implementation.

This paper [8] investigated IoT's research contributions and future potentials toward the envisioned goals of intelligent buildings. The typical technologies of IoT were mentioned in the sequence of a three-layered generic IoT architecture, namely 1) perception layer, 2) network layer, and 3) application layer. The authors stated that current technologies, hardware, software, and computing algorithms have already become a significant part of the development of intelligent buildings. However, continuous research effort is required on IoT applications to resolve many problems and challenges that remain to be studied to successfully implement the prospects of advanced intelligent buildings.

In this work [9], the authors provide a review of how IoT devices influence the environment. Several sensors are mentioned, explaining how they impact the indoor built environment and occupant productivity. The authors claim that each sensor has its characteristics, giving better advantages. They suggest using the proper sensors to find the best combination of sensing technology to help build a better environment.

This paper [10] introduces the IoT application of devices and the connection of Docker virtualization. The docker environment helps manage different tasks between independent containers. Docker's lightweight, consistent, and easy-to-orchestrate applications inside containers have piqued the appeal of researchers and developers interested in integrating this platform into IoT applications. IoT is advancing rapidly, and new ideas are being developed for implementing new IoT applications. The authors state that Docker is designed for robust application orchestration, fast deployment, and tolerance of failures.

In this paper [11], the authors present a cloud containerization solution using Docker to deploy a Bluetooth-based software-defined function, representing a framework to simplify redefining IoT functions. Furthermore, the authors claim that Docker is a lightweight framework because of its simplicity and flexibility in scaling and distributing workloads between independent containers. This environment makes the solution flexible and cost-saving when implemented on a remote/single computer.

The authors [12] describes a system that identifies vehicles by analyzing pictures or videos. Their solution is divided into two layers: device clustering and the distribution of computing tasks by cluster nodes using Docker. The authors concluded that Docker is an advantageous technology for testing IoT devices by isolating several containers, increasing security.

The authors of this study [13] propose a framework for managing IoT devices in a Smart Building. This system is divided into three layers. Edge nodes take data from abstract devices, process it in the intermediate layer, and then send it to the cloud for further study. This framework uses OpenFaaS, an open-source framework for creating a cluster of computational devices. This framework uses the Docker services engine. With this study, the authors point out that Docker is a good system that separates each task into independent containers, improving the security of each sensor, reducing cost and periods between each layer, and is tolerant to task failures.

A survey was conducted to determine how essential Docker is in the IoT domain [10]. In this paper, the authors concluded that Docker is a lightweight, consistent, and easy-to-use tool that researchers and developers tend to explore with the prospect of integrating the Docker platform into their IoT applications. However, given the rapid expansion of IoT devices, this tool may need more computing power and low memory capacity.

The work conducted in this paper differs from what the authors presented since we will implement an Alert System for several sensors using one computer with the assistance of a Docker Swarm.

3 Architecture

This section describes the architecture used in this paper to implement our system.

Figure 1 shows the implementation of the environment using Docker Swarm. The first layer involves executing four different types of sensors: light consumption, humidity percentage, temperature levels, and motion sensors. Each sensor is simulated using Python scripts that generate data in one-minute intervals over a year. Each script creates JSON data to be processed and analyzed by the support layer. The support layer is an API developed in Flask that processes each JSON data file from each device container and transforms them into an SQL data file for the database. Then, the database container stores the SQL data file into a SQLite database instance. Using the data stored in the database, the last layer creates plots to show the data registered in a day and a week and the number of alerts during this period for humidity, temperature, and movement sensors.

Figure 2 shows the planning of our environment using Docker Compose to run multiple isolated containers on a single host. In this figure, it is essential to note several keywords. The version, services, and network are top-level keywords. The version, as the name implies, indicates the version of the Docker Composer. The services define the different containers or services that compose the application. Lastly, the network represents a custom network and properties to connect the containers. Other keywords not on the same level but are nevertheless important

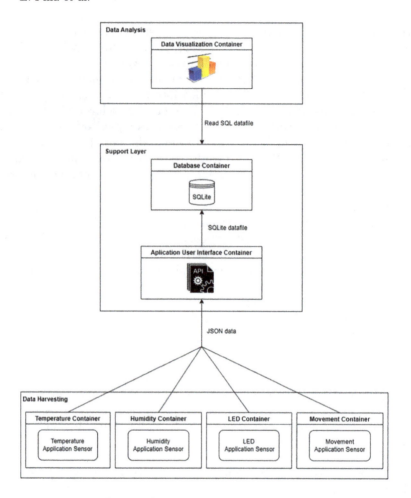

Fig. 1. Architecture Implementation

are image, build, depends on, volume, container name, context, and Dockerfile. Image keyword specifies a Docker image of the container. It can include the image name and version. The build keyword is used to configure the build process for a service. The depends on keyword specifies other services or containers that this service depends on. Volume keyword mounts volumes from the host or other containers into the container. The container name indicates the name of the container. Context keyword specifies the build context. It can be a path to a directory containing the Dockerfile and any files needed during the build. Lastly, the Dockerfile keyword specifies the name of the Dockerfile to use. The Dockerfile contains commands that allow the user to assemble an image or copy files from host to container or vice versa.

Using Docker Swarm, creating a containerized system with this environment made it possible to implement sensor simulation in a single computer.

Fig. 2. Docker Compose

4 Experimental Setup

This section presents the requirements and execution process of our framework. The test was implemented on a computer with the following characteristics:

- AMD Ryzen 3 2300U with Radeon Vega Mobile Gfx 2.00 GHz;
- 8 GB of RAM;
- SSD 150 GB;
- Oracle Virtual Box with Operating System Ubuntu 22.04 LTS (Jammy Jellyfish) (AMD64 bit), two virtual cores, 3 GB of RAM, and 25 GB allocated disk space.

We chose the minimal installation during Ubuntu installation since we will not need anything besides a web browser, an Integrated Development Environment (IDE), Visual Studio Code, and Docker.

Docker was installed by following the Official Installation guide [14] and by using the apt repository. The APT in Ubuntu is a package manager that allows installing and uninstalling packages from the apt library [15]. Having completed the installation, Docker Compose, a Docker plugin, was also required to run a Proof of Concept (PoC) in a containerized environment. It was followed [16] to install Docker Compose using the repository.

5 Results and Analysis

This section presents the results of each sensor in days and weeks and the number of alerts detected.

Figure 3 depicts the humidity over one day. The values in this Figure range from 23% to 65 %, with the lowest humidity of around 20% between 15 and 16 h. The highest humidity level was 64.37%.

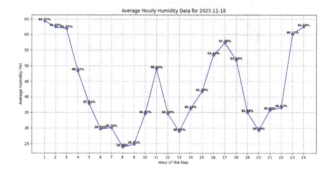

Fig. 3. Humidity Percentage in a Day

Figure 4 demonstrates the average humidity over a week. Since November 13th, the humidity has been at 44.77%. After four days, the humidity peaked at 46.54%, an increase of nearly 2%. Then it fell to 43.27% the next day and rose to 45.89% the following day.

Figure 5 shows the current temperature in a room, where from midnight until 10 am, the temperature increased to 22 °C. After 10 am, the temperature gradually dropped. The temperature suddenly declined between 13 and 16 h, reaching a low of 18 °C. After 17h, the temperature tended to maintain its temperature, increasing or decreasing a few degrees Celsius (°C).

Figure 6 displays the average maximum and minimum temperatures in a room throughout the week. The maximum temperature exceeded 25 °C in two

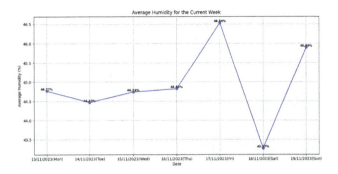

Fig. 4. Humidity through the Week

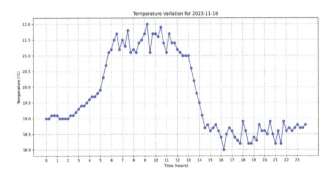

Fig. 5. Temperature through the Day

days because those days were hotter, while minimum temperatures ranged between 14 °C and 18 °C, with the lowest at 14.9 °C.

Figure 7 shows the LED light consumption in a room throughout the day. The LED consumption ranged between 55 and 110 W per hour. The highest peak was above 120 W per hour, indicating that many LED lights were turned on at 16 h. Regarding the other days, the LED power varied its Watts consumption.

Figure 8 displays the weekly power consumption. There are minor variations of power between 3000 and 3500 W. Nonetheless, power consumption was nearly linear throughout the days, indicating consistent power consumption.

Figure 9 shows the number of people detected in a room. One movement out of hours was detected, which indicates that someone suspicious was in that room. The number of movements during rush hours was higher since more people tend to enter or leave work, while in regular hours, the number of movements tended to be lower.

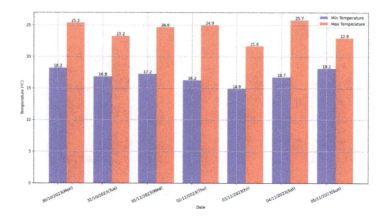

Fig. 6. Temperature through the Week

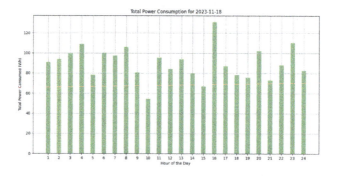

Fig. 7. Power Consumption through the Day

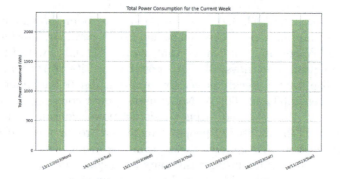

Fig. 8. Power Consumption Through a Day

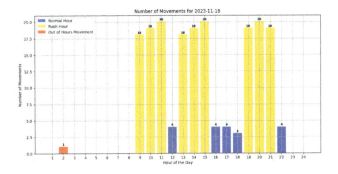

Fig. 9. Movement Detection through the Day

Lastly, Fig. 10 displays the number of people detected throughout the week in a room. The movement difference varies between 1 or 2 people, with almost a linear number of weekly movements in a single room.

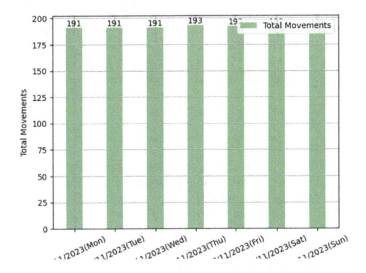

Fig. 10. Week Movement Detection

The number of alerts triggered for humidity, temperature, and out-of-hours movement is shown in Fig. 11. It was defined as a humidity range of 25 °C as high and 5 °C as low. An alert is triggered in case data becomes below 5 °C or above 25 °C is detected. The same can be said for humidity, which ranges between 60% and 30%. Anything outside of this range is considered an alert.

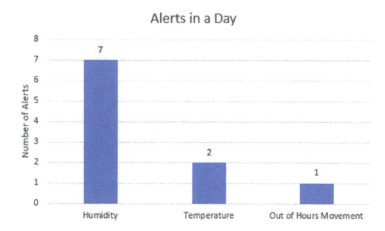

Fig. 11. Alerts in a Day

To summarise, we built this system by performing each task independently in separate containers. During data generation, we attempted to simulate data that was as close to the real world as possible, as seen in Figs. 3, and 5, by using a function from the Python random library called uniform to generate a wave pattern. By providing a range, this function increases or decreases a number tendency. The humidity percentage was treated similarly. In terms of power consumption, we used another function that adjusts the power based on the rate of an LED's intensity:

$$power = (6.00 * (intensity/100)) \qquad (1)$$

Given that an LED consumes approximately 6 W, power consumption is calculated by multiplying the intensity value between 0 and 1 by the total power of an LED (6 W). Mapping the intensity to a power scale will yield a power value influenced by the intensity level that ranges from 0 to 6 W. On the other hand, the alerts for LED power consumption will be incorporated into future work.

6 Conclusions and Future Work

The combination of intelligent building technologies and simulation capabilities empowered by Docker Swarm opens new opportunities for innovation, efficiency, and reliability. Developers can design, test, and implement intelligent building solutions that meet the demands by leveraging containerization and orchestration. Docker Swarm integration with intelligent building simulations is critical in creating more sustainable, resilient, intelligent building solutions that meet current needs and anticipate future challenges. The paper demonstrates our work as an example of a practical application of these concepts. By encapsulating tasks within a containerized environment orchestrated by Docker Swarm, we designed a system that generates sensor data stored in JSON files to be processed by an

API. Then, we transformed it into a structured SQL file to create plots of each sensor and the number of alerts raised daily or weekly.

6.1 Future Work

In future work, we intend to improve this system by analyzing data from one-second interval periods. This entails sending data regularly to the database to determine whether alerts were generated at the time so that the person responsible for managing the application can make the necessary adjustments. LED power consumption alerts are another component that should be implemented in this system. The continued scaling up of the system to simulate a proper building is another aspect that needs to be addressed. The relation between temperature and humidity must be addressed since both are connected. Finally, we intend to improve how the alerts are generated over the same problem so that when an alert is generated, it only continues to produce that alert once the warning is resolved.

Acknowledgements. "This work is funded by National Funds through the FCT - Foundation for Science and Technology, I.P., within the scope of the project Ref UIDB/05583/2020. Furthermore, we would like to thank the Research Centre in Digital Services (CISeD), the Polytechnic of Viseu, for their support".

References

1. Bahga, A., Madisetti, V.: Internet of Things: A Hands-On Approach. Vpt (2014)
2. Peng, S.-L., Pal, S., Huang, L.: Principles of Internet of Things (IoT) Ecosystem: Insight Paradigm. Springer, Cham (2020). https://doi.org/10.1007/978-3-030-33596-0
3. Lin, Y.-W., Lin, Y.-B., Yen, T.-H.: SimTalk: simulation of IoT applications. Sensors **20**(9) (2020). https://doi.org/10.3390/s20092563, https://www.mdpi.com/1424-8220/20/9/2563. ISSN 1424-8220
4. Peinl, R., Holzschuher, F., Pfitzer, F.: Docker cluster management for the cloud-survey results and own solution. J. Grid Comput. **14**(2), 265–282 (2016)
5. Al Dakheel, J., et al.: Smart buildings features and key performance indicators: a review. Sustain. Cities Soc. **61**, 102328 (2020). https://doi.org/10.1016/j.scs.2020.102328, https://www.sciencedirect.com/science/article/pii/S2210670720305497. ISSN 2210-6707
6. Daissaoui, A., et al.: IoT and big data analytics for smart buildings: a survey. Procedia Comput. Sci. **170**, 161–168 (2020). DOI https://doi.org/10.1016/j.procs.2020.03.021, https://www.sciencedirect.com/science/article/pii/S1877050920304506. The 11th International Conference on Ambient Systems, Networks and Technologies (ANT)/The 3rd International Conference on Emerging Data and Industry 4.0 (EDI40)/Affiliated Workshops. ISSN 1877-0509
7. Vijayan, D.S., et al.: Automation systems in smart buildings: a review. J. Ambient Intell. Human. Comput. 1–13 (2020)
8. Jia, M., et al.: Adopting internet of things for the development of smart buildings: a review of enabling technologies and applications. Autom. Constr. **101**, 111–126 (2019)

9. Dong, B., et al.: A review of smart building sensing system for better indoor environment control. Energy Build. **199**, 29–46 (2019)
10. Abdul, M.S., et al.: Docker containers usage in the internet of things: a survey. Open Int. J. Inform. **7**(Special Issue 2), 208–220 (2019)
11. Gao, Y., Wang, H., Huang, X.: Applying docker swarm cluster into software defined internet of things. In: 2016 8th International Conference on Information Technology in Medicine and Education (ITME), pp. 445–449. IEEE (2016)
12. Shichkina, Y.A., Kupriyanov, M.S., Moldachev, S.O.: Application of docker swarm cluster for testing programs, developed for system of devices within paradigm of internet of things. In: J. Phys.: Conf. Ser. **1015**, 032129 (2018)
13. Sarkar, S., et al.: Serverless management of sensing systems for fog computing framework. IEEE Sens. J. **20**(3), 1564–1572 (2019)
14. Hykes, S.: Install docker engine on ubuntu—docker docs. https://docs.docker.com/engine/install/ubuntu/#install-using-the-repository
15. Ubuntu Manpage Repository. Ubuntu manpage apt package handling utility. https://manpages.ubuntu.com/manpages/jammy/en/man8/apt-get.8.html
16. Ubuntu Manpage Repository. Install the compose plugin—docker docs. https://docs.docker.com/compose/install/linux/#install-using-the-repository

Airsense – Low-Cost Indoor Air Quality Monitoring Wireless System

Rafael Marques, Celestino Gonçalves(✉), and Rui Ferreira

Instituto Politécnico da Guarda, Escola Superior de Tecnologia e Gestão, Guarda, Portugal
`rafael@marques.com`, `{celestin,rpitarma}@ipg.pt`

Abstract. This manuscript presents the project aimed at creating a low-cost indoor air quality monitoring system. By using a variety of sensors connected to an ESP32 microcontroller, the system measures specific parameters such as carbon dioxide, volatile organic compounds, particulate matter, temperature, humidity, and pressure. The sensor-recorded measurements are sent to an InfluxDB database via Wi-Fi, where they are accessed by Grafana, which generates a graphical user interface that displays the evolution of all collected data. The values can be visualized in various graphical styles according to the user's preference, including tables, time-series charts, and histograms. Initially, some research was conducted to evaluate various types of microcontrollers and sensors to select the best options based on factors such as cost, size, and measurement accuracy. A printed circuit board and a 3D-printed enclosure were designed alongside the final prototype, which facilitated the integration of the various hardware components. Lastly, a two week-long test was conducted, in a school environment, using the final hardware to validate the overall results and operating behavior of the developed system.

Keywords: Indoor Air Quality · Air Quality Monitoring · Low-cost Sensors

1 Introduction

This paper presents the design and implementation of Airsense [1], an indoor air quality monitoring system for measuring carbon dioxide, particulate matter, temperature, relative humidity, pressure, and volatile organic compounds. One crucial aspect for the project was the overall system cost, given the high prices of commercial alternatives, it was essential to develop a low-cost system capable of competing in certain aspects with these high-cost alternatives. Especially in the post COVID-19 era, it has become increasingly important to consider the quality of the air that people breathe [2]. Air pollution directly impacts people's quality of life, and factors such as a high concentration of particles in the air can increase the risk of virus and bacteria transmission through respiratory pathways. Another critical factor for indoor air quality is carbon dioxide concentration, levels above the recommended range can cause issues such as headaches, breathing problems, fatigue, eye irritation, among others [3]. By measuring these and other factors related to air quality, it is possible to try to reduce the aforementioned side effects.

According to the Environmental Protection Agency (EPA), indoor air quality has a significant impact on heath [4]. While adverse health effects have been attributed to some of the referred pollutants, the scientific understanding of certain issues regarding indoor air quality continues to evolve [5]. An example of this is "Sick Building Syndrome" (SBS), which occurs when occupants of a building develop similar symptoms after entering a specific building and the symptoms tend to decrease after leaving the building. Some researchers have also been studying the relationship between indoor air quality and certain issues not traditionally linked to health, such as student performance or productivity levels in work environments [4]. Another field that is evolving is the design, construction, operation, and maintenance of "green" buildings that achieve high energy efficiency and improve indoor air quality [6].

2 Background Research

Some research was conducted regarding various elements related to the project, such as the operating behavior of some existing air quality monitoring systems, as well as some comparisons between various types of low-cost sensors [7] and microcontrollers used in the project. Additional research was also conducted on various factors that compromise interior air quality, including topics such as carbon dioxide, particulate matter, other chemicals in the air, and what constitutes good or bad levels of indoor air quality.

2.1 Related Work

During the research phase, the following projects were studied and compared to Airsense. The AirVA system [8] utilizes the data collected by the sensors to calculate an air quality index value. Due to the high cost of all the components used, it's a costlier option compared to Airsense. However, AirVA can detect the number of people in the room, which is a significant advantage and an excellent feature for this kind of system. The iAQ+ system [9] is a low-cost solution that monitors various factors to calculate an air quality index. Using a BME680 sensor, the system measures temperature, humidity, pressure, and Volatile Organic Compounds (VOC) levels. The project also includes a smartphone application and a web app for viewing the system's recorded values, along with a historical graph. The system shares some similarities with Airsense, such as the use of a buzzer for generating audible alerts and LEDs to visually indicate air quality levels. However, iAQ+ lacks sensors for CO_2, or particulate matter, which are essential factors for projects in this field. The project BEVO [10] involved the study and development of an air quality monitoring system named BEVO Beacon. The project focused on an intensive study of values collected by 20 systems with the same hardware, measuring CO_2, Particulate Matter (PM), CO, NO_2, and TVOC. Like Airsense, BEVO Beacon uses a Printed Circuit Board (PCB) to connect all components and a customized box made of plywood. CampusEMonitor [11] is a low-cost environmental monitoring system designed to monitor rooms with various network equipment. Using a Raspberry Pi SBC connected to a DHT11 temperature and humidity sensor, and an MQ-7 carbon monoxide sensor, the system aims to detect temperature increases and gas contamination. The sensor readings are sent to a database for use in a graphical dashboard. Compared to

Airsense, CampusEMonitor is limited in terms of sensors, measuring only temperature, humidity, and carbon monoxide. Design and Simulation of Environment Indoor Air Quality Monitoring and Controlling System using IoT Technology [12] is a project to create an indoor air quality monitoring system using an Arduino Mega SBC connected to six sensors measuring levels of CO, CO_2, NO_2, O_3, PM, VOC, temperature, and humidity. The data is sent via Wi-Fi to a mobile application. Like Airsense, this system implemented various devices beyond sensors, including a screen for data visualization, a buzzer, LEDs, and additional fans for airflow control. Table 1 presents a comparison of the researched projects.

Table 1. Comparison of existing projects

Project	Airsense	AirVA [8]	iAQ+ [9]	BEVO [10]	CampusEMonitor [11]	EIAQ System [12]
Monitored Parameters	CO_2, T, H, P, PM, VOC	CO_2, T, H, PM, NH_3, NO, Alcohol, Benzene, Smoke, Dust	T, H, P, VOC	CO_2, PM, CO, NO_2, TVOC	T, H, CO	CO, CO_2, NO_2, O_3, VOC, PM, T, H
Notifications	×	✓	✓	×	✓	✓
Open-Source	✓	✓	✓	✓	✓	×
Low-Cost	✓	×	✓	×	✓	✓

2.2 Existing Commercial Solutions

There are currently hundreds of air quality monitoring products on the market. Two relevant examples are the Temtop M200C [13] and the Aranet4 [14]. The Temtop M200C is one of the most popular options in the market. It is a compact and portable system that measures carbon dioxide, PM2.5, PM10, temperature, and humidity. It features a graphical interface that displays graphs with the collected data. The major advantage of this system over Airsense is its superior portability due to a rechargeable battery. However, this system does not offer Wi-Fi, making it impossible to view data online. The Aranet4 system measures carbon dioxide, temperature, humidity, and atmospheric pressure. It can be connected via Bluetooth to a smartphone, where temporal graphs of the measurements can be viewed. The major advantage of this system over Airsense is also its portability. The use of an e-ink screen and not being constantly connected to a network allows the system to have a battery life of up to 4 years, making the product much more appealing and practical. One disadvantage, when compared to Airsense, is the lack of a PM sensor.

3 Project Requirements

The requirements for the Airsense system can be divided into two groups, functional requirements, and non-functional requirements. The main objective of the project was to create a system that could read data from various sensors using a microcontroller. This data should be sent to an online database, which in turn would be accessed by software that generates graphs with all the collected data. The goal was to create a low-cost system, easy to use, with an open architecture, and a simple graphical interface.

The functional requirements established were as follows: (1) Measurements: must take measurements of certain parameters related to indoor air quality using the available sensors; (2) Internet Connection: must be able to connect to a network via Wi-Fi; (3) Data Storage: must be able to send the measurements to an online database; (4) Data Visualization: a dashboard must be created where measurements can be visualized with a graphical interface; (5) Refresh Rate Selection: the user should be able to easily change the refresh rate of the sensors and the upload rate of the data.

The following non-functional requirements were established: (1) Usability: must be easy to use, and the graphical interface should be intuitive for the user; (2) Scalability: must be scalable to accommodate more sensors in future versions; (3) Aesthetics: should have a visually appealing design; (4) Ergonomics: should be easy to transport and easy to disassemble.

4 Hardware Architecture

The system is composed of various sensors and other functional elements connected to the microcontroller through a PCB. This PCB is then attached to the 3D printed enclosure. Figure 1 illustrates the hardware architecture of the system.

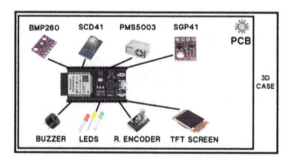

Fig. 1. System hardware architecture

4.1 Sensors

The final configuration of the Airsense system uses four sensors to measure temperature, relative humidity, barometric pressure, VOC, CO_2, and PM.

The BME280 sensor was used to measure temperature, humidity, and pressure. Communication with the microcontroller is done through the I2C protocol. The specific details of the sensor are not public as the technology is proprietary, but temperature is measured through changes in voltage from a diode, humidity is measured through a capacitive principle, and the pressure value is calculated using a piezo-resistive sensor.

The sensor chosen to measure particulate matter was the PMS5003. This sensor measures the number of particles in the air with diameters of 1, 2.5, and 10 μm. The sensor operates on the principle of laser dispersion. Using an intake fan, particles pass through a laser focused on a point, the particles then cause the laser light to disperse, which is detected by a light-sensitive diode. The value recorded by this diode is converted to the particle concentration value with the help of the microprocessor in the sensor.

For CO_2 measurements, the Sensirion SCD41 was used. The sensor communicates with the microcontroller via the I2C protocol and employs photo-acoustic technology, which is a relatively new method of measurement in these types of sensors. The sensor emits an intermittent infrared light, when this light is absorbed by CO_2 molecules, they begin to heat up and cool down repeatedly. Due to this thermal expansion and contraction, there are vibrations and changes in pressure inside the sensor's enclosure. These vibrations generate an acoustic signal that is measured by a microphone in the sensor. The values collected by this microphone are then converted to value that represents the concentration of CO_2.

The SGP40 was used to measure the VOC level. The sensor detects the concentration of volatile organic compounds in the air and outputs a value that ranges from 0 to 500, where 0–100 represents excellent air quality, 100–200 good air quality, 200–300 slightly polluted air, 300–400 moderately polluted, and 400–500 extremely polluted air.

4.2 Other Devices

In addition to the sensors, the system also includes some elements that provide additional practical features to the user.

- A TFT screen. Despite the primary interface method with the system being through the web GUI, having a screen to display the data locally at the hardware level can be useful in some situations;
- A rotary encoder that is used to change the sensor update rate and the rate at which data is sent to the database. In certain cases, it may be useful to have a longer interval, for example, every five minutes, to avoid an excess of data points. In other cases, a shorter interval may be preferable;
- A buzzer that alerts when the system connects to the Wi-Fi network and the online database;
- A set of 3 LED of different colors that visually indicate the level of CO_2: green, yellow and red.

4.3 Printed Circuit Board

The development of the PCB began by creating a schematic that included all components and their connections. Since the software EasyEDA has a vast component library, many of the necessary components are already available for importing. After making all the correct connections based on the system prototype, a final diagram was obtained. With this diagram, it is possible to generate a PCB. The software generates the PCB with components randomly placed, which then need to be adjusted according to the desired layout and measurements. After placing the components in the correct locations and verifying all the measurements, the wiring is automatically generated. Lastly, the Gerber file must be uploaded to the chosen manufacturer's website. All the components then need to be soldered to the final physical PCB.

4.4 3D Printed Enclosure

To make the enclosure, the process began by creating the 3D model in Fusion360. The goal was to design a model that would be simple, compact, and relatively easy to 3D print. The model was generated mostly by simple shapes and extrusions.

The design process went through different stages, from the initial basic layout to the completed enclosure. It started with a platform, standoffs, and a border for the PM sensor. In a second revision, borders were added around the initial base, along with space to place various magnets and some cut-outs for ventilation. The third revision included a top piece, the second element of the enclosure. This element attaches to the first one with magnets. It has various cut-outs for the system components, including the 3 LEDs, rotary encoder, screen, and ventilation for the sensors. The final design is shown in Fig. 2 a). Some adjustments were made, like filleted edges to make the enclosure more comfortable to hold, and a hexagon pattern was added along all faces to increase ventilation. Lastly, a vertical support was designed. It attaches to the previously described elements using magnets.

4.5 Prototype Final Configuration

With everything assembled, the obtained result is shown in Fig. 2 b).

Fig. 2. Airsense Hardware: a) Enclosure, revision 4 and b) Final prototype

The final components are listed in Table 2, along with the approximate cost of each component. As demonstrated, the total approximate cost of the system is 63€, being that most of the cost is allocated to the CO_2 and PM sensors.

Table 2. Components of Airsense final prototype

Component	Description	~ Cost
ESP32 – Devkit 1	Microcontroller	4 €
SCD41	CO_2 sensor	21 €
SGP40	VOC sensor	6.3 €
BME280	Temp, Hum, Press sensor	2.5 €
PMS5003	PM sensor	16 €
Buzzer	Active buzzer 3V	1 €
3x LEDs	Green, Yellow, Red	0.1 €
TFT Screen	1.8-inch TFT	2 €
3x Resistors	2kohm	0.03 €
Rotary Encoder	Model ky-040	0.5 €
PCB	From JLCPCB	5 €
Enclosure	Printed in PLA	1 €
Vertical stand	Printed in PLA	4 €
	Total:	**63.43 €**

5 System Software

This section will explain how the software in the system functions and how it interacts with the different hardware components. Figure 3 illustrates the software architecture. The microcontroller sends the collected data via Wi-Fi to an InfluxDB database. The data stored in the database is then accessed by Grafana, that functions as a graphical interface for the user, where all graphs and tables with sensor data are generated. Both InfluxDB and Grafana operate in the cloud, enabling users to access the data from any network or computer, anywhere.

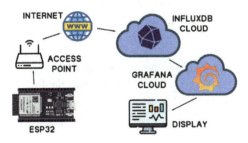

Fig. 3. Software architecture

5.1 Wi-Fi Manager

To connect the system to a Wi-Fi network, a Wi-Fi provisioning system was implemented using the WifiManager library. This provisioning method eliminates the need to hard-code network access credentials, making it easier to connect to the desired network, and avoiding the need to re-upload firmware to the microcontroller every time a network change is required. Upon system startup, an access point that can be accessed by a smartphone or a computer is created. After connecting to the access point, a browser window automatically opens, allowing users to select the desired network and input the required credentials. The next time the system boots up, it automatically connects to the chosen network. If the system boots up in a new area, or without access to a network, it creates the access point again, and the login process must be repeated. If the system has no access to any network, it functions in offline mode, and continues displaying the sensor data on the integrated screen.

5.2 InfluxDB - Database

With access to the internet, the system sends all collected data by the sensors to an InfluxDB time-series database. The values are stored in a "bucket" containing all the data along with their respective timestamps. The data can be accessed using SQL queries, however, since the main interface with the system is made through Grafana, accessing InfluxDB this way is not necessary for the user. This database is stored in the cloud and can be accessed from anywhere.

5.3 Grafana - Graphical Interface

Grafana was used as the fronted/GUI. A dashboard was created, containing all the panels with the graphs for each measurement. Each panel is made up of two elements, a FLUX query, and the settings for panel configuration.

To create the query for each panel, it's necessary to select the bucket where the data was stored, a time range, and the desired data field. After generating the query, the graph is set up by selecting the type of graph, along with the rest of its desired settings.

The designed interface shows the temporal graphs for temperature, humidity, pressure, CO_2, and VOC, as well as indicators for the current values of CO_2 and VOC. In the case of PM graphs, the interface shows PM1.0, PM2.5, and PM10.0 values over time, as well as the current and historical particle levels per 0.1L of air. As an example, Fig. 4 illustrates the graphical interface for the tables with the daily minimum and maximum values for temperature, humidity, CO_2, and VOC over the last 30 days. This graphical interfaces are also available in mobile layout.

Airsense – Low-Cost Indoor Air Quality Monitoring Wireless System 277

Fig. 4. Daily maximum and minimum values for some of the measured parameters

6 Tests and Results

To verify the values recorded by the Airsense system, a 17-day test was conducted. This test took place at the School of Technology and Management of Guarda Polytechnique, in classroom 39, a computer laboratory, between October 30th and November 15th. Given the chosen environment, this test focused primarily on CO_2 concentration values.

Despite some issues with school's network connection, it was possible to isolate certain time intervals with continuous readings and verify that the system can accurately measure differences in CO_2 concentration levels. These differences in CO_2 are caused by various factors such as the number of students present, or the level of ventilation in the room.

An analysis of some of these reading intervals will be presented, along with an explanation of what may have caused the increase or decrease in the recorded values for the examples provided. To solidify the analysis of some of these examples, certain factors were registered, such as the schedule of certain classes, the number of people present, and occasions when the room was ventilated. Regarding CO_2 concentration readings, it was possible, in certain cases, to detect the following: (1) When a group of students enters or exits the classroom; (2) If the room has many, or few students and (3) When the room has adequate ventilation.

In the example shown in Fig. 5 a), it can be observed that on Monday, November 6th, there is a rapid increase between 09:30 and 11:30, followed by a decrease between 11:30 and 14:30, and finally, another, less dramatic increase (due to the class only having 5 people) between 14:30 and 16:30. Both increases coincide with classes in classroom 39. After the end of the class, there is a gradual decrease until the morning of the next day. On Tuesday, November 7th, as shown in Fig. 5 b), it is possible to identify the class that took place between 09:00 and 11:00. Despite the readings gap in the afternoon, a very abrupt increase can be seen starting at around 14:00 and peaking at 16:00. Next, a rapid decrease is observed between 18:50 and 19:00, representing a high probability that the room was ventilated, as when the room is not ventilated, the CO_2 level does not usually drop so abruptly. November 9th, a Thursday, as shown in Fig. 5 c), the system detects a possible class from 14:00 to 16:00 with many students (due to the high slope on the graph), which is not listed in the schedule of classroom 39. It is possible that the room was ventilated from 15:50 given the sharp drop in CO_2 concentration over the next three hours. Figure 5 d) shows the readings for November 10th, a Friday. The CO_2 level increases during the class from 11:00 to 13:00. Next, there is a 15-min period during which the room is ventilated, from 16:45 to 17:00. The CO_2 concentration then gradually decreases over the weekend. Monday, November 13th, is illustrated in Fig. 5 e). The same classes as the previous week can be identified. After a gradual decrease over the weekend, there is a rapid increase between 09:30 and 11:30, followed by a break of three and a half hours, and finally, a small increase during the class from 14:30 to 16:30, with 5 people. On November 15th, Wednesday, shown in Fig. 5 f), a class with 24 students takes place. Despite the high number of students present, the CO_2 concentration remains at a good level for most of the duration of the class due to the window being open.

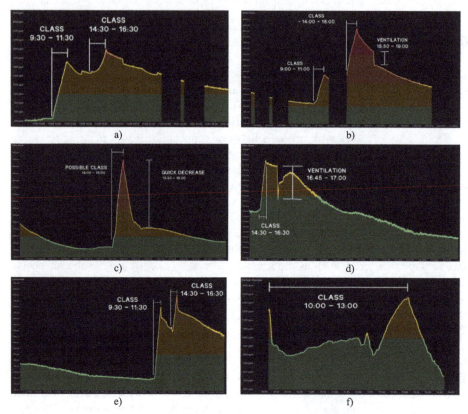

Fig. 5. CO_2 sensor experimental tests in classroom 39: a) November 6th, b) November 7th, c) November 9th, d) November 10th, e) November 13th and f) November 15th

Regarding the temperature sensor, it was possible to verify that it detects changes in the room that are most likely caused by the central heating system. Figure 6 a) shows the temperature increasing and decreasing at constant intervals. Between November 10th and November 14th, a pattern is repeated, where it always increases from 08:00 to 12:00 and again from 16:00 to 21:00. On Sunday, November 12th, there is no temperature change because the school is closed.

To evaluate the remaining sensors, some tests were conducted outside the school. In the humidity sensor test, a dehumidifier was used in a closed room. The dehumidifier turns on and off periodically in 10-min intervals. In Fig. 6 c), it is possible to see an incremental decrease, along with the intervals when the dehumidifier turns on and off. To test the VOC sensor, a cloth soaked in alcohol was placed near the system. The sensor was able to detect the volatile compound, and the reading surged from around 100 out of 500 to 450 out of 500, as demonstrated in Fig. 6 d). In the PM sensor test, a piece of burnt cardboard was placed in a frying pan near the system. The sensor quickly detected the particles in the smoke, and the values went from around 1–10 (typical for a clean environment) to thousands, as shown in Fig. 6 b).

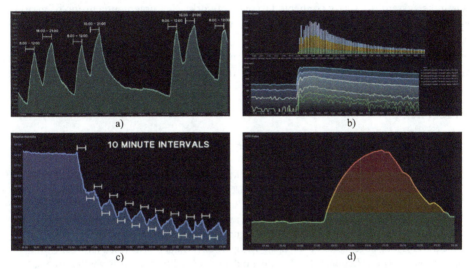

Fig. 6. Experimental tests with other sensors: a) Temperature, b) PM, c) Humidity and d) VOC

7 Conclusion and Future Work

The goal of this project was to study, analyze, and build an indoor air quality monitoring system. During the initial analysis phase on the topic, various existing projects and commercial options were compared. The project also involved the development of new skills in certain unfamiliar areas, such as 3D modeling, PCB design, and sensor implementation. These new insights were applied throughout the process in the creation of four prototypes with different characteristics until a more refined final prototype was developed that was capable of measuring levels of temperature, relative humidity, barometric pressure, carbon dioxide, volatile organic compounds, and particulate matter. Various tests were conducted to validate the operation of the system and the accuracy of the readings, including a two week-long final test in a school environment.

While most of the initially proposed requirements were met, there is always room for improvements and additional features to implement in the future. For example, one limitation of the system's current version is the somewhat complex initial setup, which requires some specific IDE configurations and editing of certain library files. Additionally, the following are some ideas for future work and improvements of the project: (1)

Notification and alert system via SMS or e-mail; (2) Integration of artificial intelligence, that, through an analysis of measurement history, can make predictions or find patterns in the provided data. For example, predicting when CO_2 levels will reach an unhealthy point, or detecting unexpected increases in VOC levels on a specific day and time each week; (3) A second, more compact version of the PCB without modules or breakout boards, with the sensors and microcontroller soldered directly onto the PCB, along with all other components such as resistors and capacitors; (4) Address system flaws or limitations, such as not being able to select a different Wi-Fi network in the same location when the system is already connected to one network; (5) Implement a group of various systems in a building, such as a house, with one system in each room, and create a map where all the measurements from each room can be viewed; (6) Implement a functionality for the system to save measurements in memory in case of Wi-Fi failure and (7) Create documentation and organize all files for public access on GitHub.

References

1. Marques, R.: Airsense - Qualidade do Ar Interior. Graduation Project Report, Instituto Politécnico da Guarda, Guarda, (2023)
2. Moghadam, T.T., Morales, C.O., Zambrano, M.L., O'Sullivan, D.T.J: Energy efficient ventilation and indoor air quality in the context of COVID-19-A systematic review. Renew. Sustain. Energy Rev. **182**, 1–24 (2023)
3. Zhang, H., Srinivasan, R.: A systematic review of air quality sensors, guidelines, and measurement studies for indoor air quality management. Sustainability **12**(21), 9045 (2020)
4. US EPA Homepage. https://www.epa.gov/report-environment. Accessed Nov 2023
5. Burroughs, H.E., Hansen, S.J.: Managing Indoor Air Quality, 5th edn. River Publishers, New York (2011)
6. Karimi, H., Adibhesami, M.A., Bazazzadeh, H., Movafagh, S.: Green buildings: Human-centered and energy efficiency optimization strategies. Energies **16**(9), 1–17 (2023)
7. García, M.R., et al.: Review of low-cost sensors for indoor air quality: features and applications. Appl. Spectrosc. Rev. **57**(9–10), 747–779 (2022)
8. Ramos, A., Jesus, V.B., Gonçalves, C., Caetano, F., Silveira, C.: Monitoring indoor air quality and occupancy with an IoT system: evaluation in a classroom environment. In: 18th Iberian Conference on Information Systems and Technologies (CISTI), pp. 1–6 (2023)
9. Marques, G., Pitarma, R.: An internet of things-based environmental quality management system to supervise the indoor laboratory conditions. Appl. Sci. **9**(3), 438 (2019)
10. Fritz, H., et al.: Design, fabrication, and calibration of the building environment and occupancy (BEVO) beacon: a rapidly-deployable and affordable indoor environmental quality monitor. Build. Environ. **222**, 23 (2022)
11. Ward, S., Gittens, M., Rock, N., James, K.: CampusEMonitor: intelligent campus environment room monitoring system. In: SIGUCCS 2019: Proceedings of the 2019 ACM SIGUCCS Annual Conference, pp. 165–172 (2019)
12. Dionova, B.W., et al.: Design and simulation of environment indoor air quality monitoring and controlling system using IoT technology. In: IEEE International Seminar on Intelligent Technology and Its Applications (ISITIA), pp. 494–499 (2023)
13. Temtop Homepage. https://temtopus.com/. Accessed Nov 2023
14. Aranet Homepage. https://aranet.com/. Accessed Nov 2023

An Audit Framework for Civil Construction Safety Management and Supervision

Luciano Correia[1], Manuel Lopes[1], Filipe Caldeira[1,2], and João Henriques[1,2(✉)]

[1] Polytechnic of Viseu, Viseu, Portugal
{pv22382,estgv17025}@alunos.estgv.ipv.pt, {caldeira, joaohenriques}@estgv.ipv.pt

[2] CISeD – Research Centre in Digital Services, Polytechnic of Viseu, Viseu, Portugal

Abstract. The construction industry has experienced substantial growth, reflected in its impact on the financial market due to projects with significant territorial reach, widely distributed, high budgets, and technical complexity. However, this growth also raised risks due to the complex nature of construction sites, making them prone to accidents, among which falls are a significant concern. Beyond human costs, accidents incur various tangible consequences, including compensation expenses, legal liabilities, schedule delays, unforeseen costs, and intangible impacts on organizational reputation and human resource recruitment. This paper proposes an auditing framework to address these challenges and to enhance safety management and supervision in construction. This framework collects relevant data on the construction environment and worker conditions, streamlining the identification of fall risks and accident locations.

Keywords: civil construction · accident · audit · fall detection · safety

1 Introduction

Accidents due to falls are a relevant subject to be considered in construction sites as they represent one of the significant causes of injuries and fatalities. They impact not just the physical integrity of workers but also the efficiency and successful completion of projects. In the event of falls, two critical success factors must be considered. The first is to know the location of the injured party. The second is related to reducing response time to minimize the extent of injuries or even prevent death. In this context, effective safety management in the construction industry is essential to ensure safe working environments and prevent accidents. Despite the norms and regulations that must be followed to coordinate safety on construction sites, there is a need to identify preventive actions that can significantly reduce the number of accidents and ensure that all existing solutions complement each other [1].

1.1 Motivation

According to data from the World Health Organization (WHO) [2], falls (in general) are the second leading cause of death from unintentional injuries worldwide. According to data published by the International Labor Organization (ILO) and Asanka and Ranasinghe [3], global cost estimates resulting from direct and indirect accidents amount to 2.8 trillion US dollars, equivalent to 4% of the global gross domestic product (GDP).

Accident prevention should be viewed as an objective to achieve. It is necessary to mitigate and avoid: 1) loss of human life; 2) individual and collective responsibility for negligence crimes regarding failure to comply with safety conditions; 3) budget overruns in the project, both in costs and in the expected execution time, as presented by Alkaissy et al. [4]; 4) direct and indirect costs of the accident not only for the worker but also for other stakeholders in the construction project (insurers, property owner, general contractor and subcontractors, regulatory entity, safety coordinator on site); 5) minimizing the financial costs incurred by public health organizations (hospitals and health centers) in treating injuries suffered by the employee, as well as social security systems for the payment of temporary disability benefits.

1.2 Contributions

The following objectives have been identified: 1) Propose a framework for auditing occupational health and safety which allows to identify, in real-time, when a worker falls; 2) Propose an optimized model for distributing signal transmitters and Internet of Things (IoT) sensors, necessary for data acquisition and subsequently evaluating the positioning of workers on-site; 3) Evaluate the proposed framework through the implementation of a prototype integrating the proposed location model.

1.3 Structure of the Document

Beyond this section, this paper is organized as follows: Sect. 2 introduces the technological approaches to safety in the construction industry. Section 3 describes the audit framework for civil construction safety management and supervision. Section 4 presents the experimental work. Finally, Sect. 5 concludes the document and presents future work.

2 Technological Approaches to Safety in the Construction Industry

This section presents the technological approaches for safety management in the construction industry, supported by a literature review and the identification of tools and techniques regarding fall detection systems, Bluetooth Low Energy, Fog Computing, and Microservices.

2.1 Fall Detection Systems

Fall detection systems (FDS) integrate technologies and devices to identify and trigger alerts in cases of human falls. Often, they incorporate multiple sensors and technologies to minimize false positives accurately. Some FDS systems also include contextual information, such as the human´s location, current time, and activity level, to better assess the likelihood of a fall. Loss of balance represents one of the primary causes of falls at construction sites, according to authors Hsiao et al. [5]. It is essential to proactively monitor the balance of construction workers at different times of the day. To avoid accidents, it is necessary to plan appropriate mitigation strategies. According to Mubashir et al. [6], some approaches were used in fall assessment and detection processes, such as wearable sensors, accelerometers, gyroscopes, pressure sensors, height sensors, and heart rate sensors. Fall detection systems focused on the environment often utilize motion, pressure, and acoustic sensors placed in the environment to capture data related to the individual's activities and movements. These sensors can detect environmental changes that can help signal falls or potential falling risks. Another approach would involve using visual data from cameras and imaging devices to monitor and analyze the movements and activities of individuals, aiming to detect falls and related events. Trilateration is a geometric technique used to determine the position of an unknown point about three or more known reference points. This method is often applied in location systems, such as the Global Positioning System (GPS), to determine, in real-time, the position of mobile devices or objects, according to authors S. Murphy Jr. & Herman [7]. The basic principle of trilateration involves measuring the distances between the unknown point and three reference points with known coordinates. Each distance is a radius around the reference point, creating concentric circles or spheres in three-dimensional space. The intersection of these spheres provides the possible locations of the unknown point, and the most accurate solution is where all the spheres intersect.

2.2 Bluetooth Low Energy

Tosi et al. [8] work described the Bluetooth Low Energy (BLE) architecture and the usage of this wireless communication technology. The achieved results conclude that BLE is an alternative to the existing standard wireless technologies (e.g., Institute of Electrical and Electronics Engineers (IEEE) 802.11b (Wi-Fi), ZigBee, ANT+, and Bluetooth classic). According to Park et al. [9], BLE is a promising technology for tracking workers and equipment in terms of reliability and ease of use. However, it was noticed that the BLE system may need to be more reliable at zone boundaries, which may cause constraints when running data analysis. Another conclusion from the same work is related to the positioning of the gateway, which may affect the accuracy and usability of the data. Finally, its compact size makes it very convenient as a wearable device. According to the authors [9], BLE includes features that enable a device to determine another device's presence, distance, and direction. BLE is widely used as a device positioning technology and is highly effective in high-precision indoor location services. According to the author Li et al. [10], Real-Time Location Systems (RTLS) are systems that can be used to identify and track the location of workers, materials, and equipment on the construction site in real-time. Huang et al. [10] developed a method to detect potential risks for workers near operating equipment and provide proximity safety alerts using BLE technology within an RTLS. The prototype system was evaluated on construction sites, demonstrating its ability to detect nearby hazards and issue vibration alerts for worker awareness. Furthermore, the method allowed the system to estimate the position and velocity of workers or equipment at the site. In Bai et al. [12], the authors aimed to monitor the daily lifestyle of elderly individuals or those with disabilities by implementing an indoor positioning system using BLE sensors placed at various locations within a residence. The person's location was determined by capturing the Received Signal

Strength Indicator (RSSI) from their beacon. Trilateration and fingerprinting methods were used for location determination. Experiments conducted in home settings confirmed the system's ability to accurately track the person's location and daily activities, providing data useful for assessing their health status.

2.3 Fog Computing

According to Atlam et al. [13], fog computing aims to decentralize computing by bringing processing resources closer to edge devices. This model reduces latency, enhances data transmission efficiency, and facilitates real-time event capture. It involves a decentralized architecture with layers including Cloud, Fog, and Edge. The use of this computing model in the proposed framework should be encouraged due to its benefits, including the lack of processing resources closer to edge devices, reduced latency, improved data transmission efficiency, and a faster response to events, such as a worker's fall. In the context of the work to be carried out, two layers and their components will be implemented: edge and fog.

2.4 Microservices

Microservices architecture supports developing and implementing applications comprising independent services, each with its source code, data, and processing. This approach fosters modularity, eases the addition of new features, and simplifies integration with other technologies [11]. Contrary to monolithic methods, where the entire application is a single unit, Microservices break down applications into smaller, interconnected components, offering increased flexibility, scalability, and resilience. Each microservice is developed, deployed, and scaled independently, streamlining maintenance and ongoing application evolution. Microservices will be used to construct modules supporting functionalities such as fall detection, location assessment, and scanning of BLE devices. Additionally, their flexible scalability will allow for easy resizing of the framework as needed.

3 Audit Framework for Civil Construction Safety Management and Supervision

This section will address aspects related to the design and development of the audit framework for management and supervision of safety in the construction industry.

3.1 Architecture

The proposed architecture is based on the fog computing model. The fog layer should be installed at the construction site. The graphical representation of the proposed architecture is shown in the Fig. 1.

3.2 Components

In this section, the components of the proposed architecture will be identified and detailed for each layer. Table 1 presents the elements that integrate the framework.

An Audit Framework for Civil Construction Safety Management   285

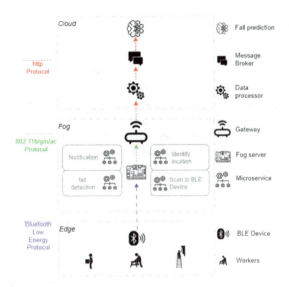

Fig. 1. The proposed Audit Framework

Table 1. Framework Components

Layer	Component	Description
Edge	BLE Device	Bluetooth Low Energy Device
Fog	Scan for BLE device	Scanning for BLE device
Fog	Fall detection	Fall assessment and detection activity
Fog	Identify location	Device location assessment activity
Fog	Notification	Notify the detected fall event
Fog	Fog server	On-site data processing device
Fog	Gateway	Device enabling communication between the fog server and the cloud instance
Cloud	Data Processor	Data processing
Cloud	Message Broker	Ensures communication between all other components
Cloud	Fall Prediction	Fall assessment and prediction
Cloud	Cloud instance	Cloud data processing

The BLE Device is located at the edge layer, where the message producer is in this architecture. It provides real-time data to determine falls, location, and worker safety conditions on the construction site. The fog devices, servers, and gateways are included in the Fog layer. Fog servers process data for decision-making, while fog devices are typically associated with them. Gateways redirect information between devices, Fog Servers, and the Cloud layer. They collaborate to reduce latency and enhance data transmission efficiency. The Fog Server hosts processes data and communicates with BLE devices. It collects data from the BLE device, evaluates falls, communicates with gateways, and enables trilateration for BLE device location. Fall detection, location identification, and BLE device scanning activities ensure the functional requirements of the framework. The notification activity comprises creating and delivering alert messages after detecting falls. At the cloud layer, the data processor component streamlines the integration of the different elements, especially the message broker and the fall prediction component. Additionally, the Data Processor processes and handles data within activities. On the other hand, the Message Broker is responsible for acquiring and processing messages from each construction site and worker using a BLE device. These messages are processed in different topics by the Message Broker. Finally, the fall prediction function aims to predict worker falls at construction sites by considering messages collected from the other message broker topics.

3.3 Fall Detection

Four modules are designed to detect fall events. The first is "fall detection," which assesses and determines whether a fall event occurred. The second allows us to "evaluate location" and identify the worker's location at the construction site. The third is "scan to get RSSI data" which runs cyclic scanning for BLE devices at the periphery of each Fog Server. The last module is "notification" for managing notifications in case of fall events detection. The Fig. 2 presents the proposed architecture, including the modules above. Figure 3 Includes the fall detection evaluation process.

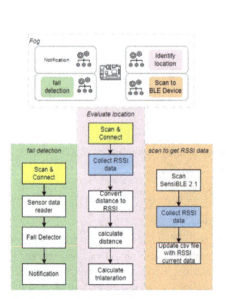

Fig. 2. Available features in the proposed architecture.

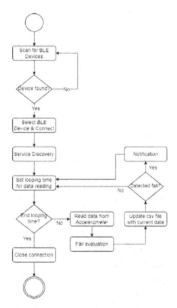

Fig. 3. Fall detection evaluation process

3.4 BLE Location

Four modules were considered, with the aim of determining workers' locations. The "Collect RSSI data" module collects RSSI events from Raspberry Pi devices. The "Convert distance into RSSI" module translates the current distance into simulated RSSI signal data. The "Calculate distance" module acts in the opposite direction by translating RSSI signal data into the simulated distance. Finally, the "trilateration" module uses trilateration to estimate the position according to the coordinates (x, y). Figure 4 presents a process diagram with the activities required to ensure BLE location.

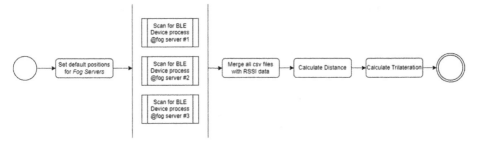

Fig. 4. BLE process location

4 Validation

This section presents the framework's validation work, including the experimental work with the experimental setup, fall detection, and location experiments.

4.1 Experimental Setup

An indoor building scenario was selected as the environment for conducting the experimental work, as shown in Fig. 5. The proposed scenario contains approximately 200 m^2. Each division, 1, 2, and 3, includes areas between 10 and 20 m^2.

Fig. 5. Experimental Work Scenario.

The selected BLE device in the framework was the "SensiBLE 2.1" from Sensiedge [12]. As presented in Fig. 7. The selected device that fulfilled the Fog Server role was the Raspberry Pi model 400. The Raspberry Pi devices were in different positions within the experimental work scenario. Their locations (in meters) are the following ones: Raspberry Pi 1 (x = 10, y = 1), Raspberry Pi 2 (x = 19, y = 9), and Raspberry Pi 3 (x = 1, y = 6). Continuing the setup, the next activity consisted of scenario creation to ensure safety conditions for simulating a person's fall, as demonstrated in Fig. 6. For that scenario, a mattress with some pillows located on the floor was required. The SensiBLE 2.1 IoT sensor was inserted into a plastic protective cover, allowing its use through a fabric strap. This configuration was adapted so that i) the sensor could be carried at chest level and ii) no pressure was set on the sensor while it was being used.

Fig. 6. Fall simulations scenario

Fig. 7. SensiBLE 2.1 BLE device

4.2 Fall Detection Experiment

The first experiment aims to detect falls. This experiment was run for 10 min while collecting data from the accelerometer SensiBLE 2.1 sensor (in milliseconds), as presented in Fig. 8. The framework detected all 17 fall events, as depicted in Fig. 9.

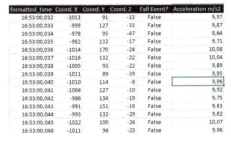

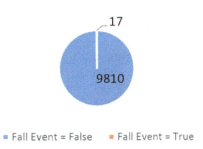

Fig. 8. Collected Data Sample

Fig. 9. Fall Events

The collected data were evaluated using different machine learning models. The obtained results are presented in Table 2. Three different models were chosen: Naïve Bayes, Logistic Regression, and KNN. The amount of data used for test purposes was 30%.

Table 2. Results obtained with machine learning models.

	Naive Bayes	Logistic Regression	KNN
True Negative (TN)	2908	2943	2943
False Positive (FP)	35	0	0
False Negative (FN)	0	6	5
True Positive (TP)	6	0	1
Accuracy	0,98813157	0,997965412	0,99830451
Recall	1	0	0,166666667
Precision	0,146341463	NaN	1

Results Discussion. The author executed this initial test with the assistance of a child with a height of 140 cm and a weight of 40 kg. The author's role was to coordinate and ensure the test's execution. The child's role was to carry the BLE device, traverse various paths throughout the scenario, and simulate falling events when requested by the author. The system can detect a fall event when the acceleration value exceeds 27.44 m/s^2. In the Table 2, it is possible to verify the number of events and their associated acceleration values distributed by quartile.

About the evaluation of collected data through machine learning models, it is relevant to highlight Naïve Bayes. Their accuracy and recall values have aligned with the framework's previous fall-detected events.

Table 3. Distribution by quartile of measured acceleration values

[0,39 – 7,155]	]7,155 – 13,92]	]13,92 – 20,685]	]20,685 – 27,45]	]27,45 – 34,215]
910	7.429	1.281	190	17

4.3 BLE Location Detection Experiment

A second experiment using BLE is possible to determine the fall location. 3 Raspberry Pi 400 devices were used to support the trilateration process. The experiment was run for 10 min. In the end, each device produced a CSV file containing RSSI data due to the scan activity of the SensiBLE 2.1. Next, trilateration results are presented. The locations in a two-dimensional plane were determined using the RSSI data collected from each Raspberry Pi 400 device, as explained in Fig. 10.

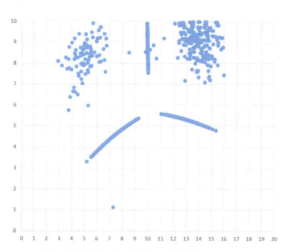

Fig. 10. Estimated Indoor BLE Locations.

Results Discussion. Although it was possible to obtain estimated BLE locations, the results obtained in this test reveal the existence of several time moments where it was not possible to capture RSSI data, as can be observed in the Table 3. The Raspberry Pi #2 device experiences the highest number of events without measurement. The location of Raspberry Pi #2 does not benefit from good Wi-Fi signal coverage. The latency observed previously ranged between −70 dBm and −90 dBm. It is suggested that the Wi-Fi signal coverage conditions be evaluated before conducting new tests (Table 4).

Table 4. Count of location detection events

	Raspberry Pi #1	Raspberry Pi #2	Raspberry Pi #3
No records	728	1.114	910
Number of Records	839	453	657

5 Conclusions and Future Work

This work proposed an auditing framework to enhance safety management and supervision in construction. This framework collects relevant data on the construction environment and worker conditions, streamlining the identification of fall risks and accident locations.

In that aim, three objectives were defined. The first objective was achieved by conducting a fall event detection experiment. The second objective was achieved by performing an experiment to assess the BLE location. Finally, the third objective was achieved by evaluating the proposed solution by implementing a prototype that contains not only fall detection but also the proposed location model.

Eventual paths for future work include conducting the experiments in an outdoor environment, such as a construction site. Regarding fall evaluation, using other sensors, such as a gyroscope, can be explored. A prior review of signal coverage throughout the experimental work scenario should be conducted. Another potential improvement can come from considering multiple SensiBLE 2.1 devices. A logical layer should be implemented to monitor the implemented microservices and be evaluated and considered. Finally, the data acquired from other sensors available on SensiBLE 2.1, such as temperature and luminance, should be regarded as additional contextual data.

Acknowledgements. This work is funded by National Funds through the FCT – Foundation for Science and Technology, I.P., within the scope of the project Ref. UIDB/05583/2020. Furthermore, we would like to thank the Research center in Digital Services (CISeD) and the Instituto Politécnico de Viseu for their support of the internal project Ref. PIDI/CISeD/2023/010.

References

1. Ferreira, C., et al.: An intelligent and scalable IoT monitoring framework for safety in civil construction workspaces. In: de la Iglesia, D.H., de Paz Santana, J.F., López Rivero, A.J. (eds.) New Trends in Disruptive Technologies, Tech Ethics and Artificial Intelligence. DiTTEt 2022: New Trends in Disruptive Technologies, Tech Ethics and Artificial Intelligence, pp. 69–78. Springer, Cham (2022). https://doi.org/10.1007/978-3-031-14859-0_6
2. WHO, W.H.O.: Falls. https://www.who.int/news-room/fact-sheets/detail/falls. Accessed 19 Nov 2022
3. Asanka, W.A., Ranasinghe, M.: Study on the impact of accidents on construction projects. In: 6th International Conference on Structural Engineering and Construction Management, vol. 4, pp. 58–67 (2015)

4. Alkaissy, M., Arashpour, M., Zeynalian, M., Li, H.: Worksite accident impacts on construction and infrastructure: nondeterministic analysis of subsectors and organization sizes. J. Constr. Eng. Manag. **148**, 04022023 (2022). https://doi.org/10.1061/(ASCE)CO.1943-7862.0002269
5. Hsiao, H., Simeonov, P.: Preventing falls from roofs: a critical review. Ergonomics **44**, 537–561 (2001). https://doi.org/10.1080/00140130110034480
6. Mubashir, M., Shao, L., Seed, L.: A survey on fall detection: principles and approaches. Neurocomputing **100**, 144–152 (2013). https://doi.org/10.1016/j.neucom.2011.09.037
7. Murphy Jr., S., Hereman, W.: Murphy-hereman-trilateration-1995.pdf (1999). https://inside.mines.edu/~whereman/papers/Murphy-Hereman-Trilateration-1995.pdf
8. Tosi, J., Taffoni, F., Santacatterina, M., Sannino, R., Formica, D.: Performance evaluation of bluetooth low energy: a systematic review. Sensors. **17**, 2898 (2017). https://doi.org/10.3390/s17122898
9. Special Interest Group, S.: Bluetooth technology overview. https://www.bluetooth.com/learn-about-bluetooth/tech-overview/. Accessed 05 Feb 2023
10. Huang, Y., Hammad, A., Zhu, Z.: Providing proximity alerts to workers on construction sites using bluetooth low energy RTLS. Autom. Constr. **132**, 103928 (2021). https://doi.org/10.1016/j.autcon.2021.103928
11. Microsoft, M.: Estilo de arquitetura de microsserviços - Azure Architecture Center. https://learn.microsoft.com/pt-pt/azure/architecture/guide/architecture-styles/microservices. Accessed 08 Feb 2024
12. Sensiedge, S.: WearaBLE sensor | rechargeable battery. https://www.sensiedge.com/sensible21. Accessed 14 Jan 2024

Track on Technological Ethics

Bytes and Battles: Pathways to De-escalation in the Cyber Conflict Arena

Lambèr Royakkers

Eindhoven University of Technology, Eindhoven, The Netherlands
L.m.m.royakkers@tue.nl

Abstract. In the digital era, the rise of cyber conflict blurs traditional distinctions between war and peace, presenting complex challenges. This article investigates the intricate dynamics of cyber conflict, driven by state and state-sponsored actions in cyber sabotage, espionage, and disinformation spread. It reviews how nations like the US, Russia, and China use cyber capabilities to assert dominance, counter threats, and shape global politics, impacting national security, economic stability, and democratic integrity. The analysis highlights the difficulties in attributing cyber-attacks, limitations of current legal frameworks, and risks of escalation to physical conflict. We suggest five strategic approaches to mitigate cyber conflict, including enhancing international cybersecurity collaboration, setting clear cyber norms, managing cyber arsenals responsibly, safeguarding tech company independence, and promoting public discourse on cyber issues. These strategies aim for a transparent, cooperative resolution to cyber conflicts, aligning with the nuances of digital warfare.

Keywords: Cyber Conflict · Cyber Attacks · De-escalation

1 A Looming Geopolitical Situation

Cyber capabilities have emerged as a disruptive technological force, revolutionizing traditional warfare and conflict dynamics [20]. This technological evolution has empowered state and state-sponsored actors with unprecedented tools for sabotage, espionage, and manipulation, reshaping the geopolitical landscape into a domain marked by unpredictability and menace. The United States (US), suffering from an increasing number of threats to its critical infrastructure due to cyber-attacks [16] and recognizing the potential of cyber warfare, has adopted a proactive stance as outlined in its strategy to "Achieve and Maintain Cyberspace Superiority" [17]. This approach mandates a "defend forward", necessitating preemptive infiltration into adversaries' systems to enable swift retaliation, thus aiming for a deterrence-based dominance in cyberspace. Before 2018, the US maintained an approach of restraint to cyber operations outside the US Government, but this changed due to mainly Russia's cyber-enables information operations in the 2026 US Presidential election. Obama administration contemplated but ultimately decided against initiating offensive cyber operations, largely due to apprehensions about the possibility of provoking further aggressive actions from Russia. Among the more robust cyber strategies under consideration were proposals to neutralize websites disseminating leaked information, to execute cyber-attacks aimed at disrupting Russian media outlets, and to impair the command and control infrastructure utilized by Russian intelligence entities. Nonetheless, the chosen course of action by the administration veered away from cyber retaliation, opting instead for a response that encompassed diplomatic, economic, and law enforcement strategies. This led to a call from certain factions within the private sector and the Department of Defense to reevaluate existing beliefs regarding escalation dynamics, advocating for a more forward-leaning strategy [6].

The concept of cyber superiority was already vividly illustrated by Operation Olympic Games, where the U.S., in collaboration with Israel, reportedly deployed the Stuxnet virus against Iran's nuclear program [10]. Despite this show of offensive capability, the incident underscored the vulnerability of even technologically advanced nations to retaliatory cyber-attacks, highlighting the complex, double-edged nature of cyber power. Further exemplifying the strategic shift towards cyber warfare, Russia's actions in Ukraine since 2014 have demonstrated the seamless integration of cyber operations with traditional military tactics. Russian forces not only commandeered communication infrastructures but also launched cyber-attacks against critical services, employing disinformation campaigns to undermine the state and influence global perceptions. For example, nowadays in the Ukraine war, Russia repeatedly has claimed that Ukraine is filled with "Nazis", and persists in promoting this demonstrably false claim, both internally and externally [2]. Similarly, China's cyber activities, including economic and state espionage, have prompted international tensions, leading to agreements aimed at curbing cyber espionage – a testament to the global recognition of the seriousness of cyber threats.

The digital realm has thus become a critical fifth dimension of military strategy, with nations deploying viruses, ransomware, and exploiting vulnerabilities to assert their interests [12]. All major world powers have established military cyber commands and reformed their intelligence services, and in some states, like Russia and China, there are close ties between government agencies and patriotic or even criminal hackers. States often operate from the shadows, and typically neither confirm nor deny their involvement in a cyber-attack. Instead, they provide ambiguous responses such as 'non-denial denials' or refuse to address the allegations' [3]. This approach allows states to pursue dual objectives: they can maintain crisis stability while leaving open the possibility of their participation in the cyber-attack. By intentionally remaining under suspicion, a state can influence how rivals perceive its cyber capabilities and determination. Additionally, by not denying responsibility, a state can also affect how rivals view the capabilities of its allies, thereby bolstering the credibility of deterrence. This approach achieves these ends without the escalatory dangers associated with explicitly admitting responsibility.

The establishment of military cyber commands and the collaboration between governments and hacker groups underscore the strategic importance of cyberspace. In this shadowy warfare, the lines between peace and conflict blur, as states engage in continuous digital skirmishing, preparing the battlefield with espionage tools, traps, and offensive weapons, even during peacetime. This transformation into a world where cyber conflict is an "always on" phenomenon demands a reevaluation of traditional notions of warfare and security [1]. Cyber operations, from sabotage and espionage to disinformation campaigns, have become integral components of national defense and offensive strategies. These operations not only pose significant challenges to societal security but also offer strategic tools to achieve political and military objectives. As the digital landscape evolves, so too will the nature of conflict and the strategies nations employ to protect their interests and assert their power on the global stage.

2 The Cyber Conflict

The cyber conflict lies between peace and war, which has implications for how we should think about peacetime and wartime (see Table 1). Traditionally, war and peace form a binary distinction: it is either war or peace between countries. This distinction can still be maintained, provided we distinguish peacetime from cyber peace, and understand war as cyber-physical war. In a time of cyber peace, it is not only peacetime, but there is also no cyber conflict. Countries do not engage in cyber espionage and do not use cyber means to cause damage in another country. Disinformation campaigns are also not undertaken. The focus in this situation is on strengthening cybersecurity. States can also train cyber soldiers and acquire weapons. However, the ability to build offensive cyber capabilities is limited: gaining intelligence about the cyber capabilities and systems of the opponent is of great importance for both setting up one's defense and creating offensive capacity. The goals in the cyber conflict vary: sometimes a state seeks economic advantages by spying, sometimes a state wants to prepare the military battlefield in its favor, and sometimes a state wants to pressure or manipulate the political decision-making in another state. But cyber-attacks have never escalated to the point where countries have declared war on each other. Moreover, in this cyber conflict, there is an important difference between cyber sabotage that directly causes severe damage, and cyber espionage and disinformation spreading that serve the interests of a particular state much more indirectly. The victim is not supposed to notice cyber espionage and disinformation spreading, while cyber sabotage immediately affects the physical world of people – for example, because financial services break down or a power plant fails.

We therefore distinguish between cold and hot aspects of the cyber conflict. In situations of cold cyber conflict, it seems as if no conflict is taking place at all since countries use cyber espionage and the spread of disinformation. In situations of hot cyber conflict, the information struggle flares up because facilities are openly damaged and people directly experience the consequences of cyber sabotage in their own living environment. An example of such a hot cyber conflict was the Stuxnet operation, which caused serious physical damage in peacetime. The damage was so severe that there has been broad discussion about whether the attack should even be classified as an armed attack under humanitarian law [14]. The cold and hot aspects of the cyber conflict are not mutually exclusive. For example, cyber sabotage generally presupposes cyber espionage. A particular cyber conflict has more or fewer cold and hot aspects. And it's not that a hot cyber-attack is necessarily more dangerous than a cold cyber-attack – perhaps nothing is more dangerous than corrupting a democratic rule of law state's information provision from within. But it is true that serious cyber sabotage borders on an act of war, as acts of war are characterized by very serious physical damage.

Table 1. The action strategies within the four phases in cyberspace.

Situation	Intent
Cyber peace (peacetime)	*Peace/neutrality;* *Strengthening of (inter)national cybersecurity:* • Establishment of (inter)national Computer Emergency Response Teams, detecting, preventing, and addressing cyber incidents and risks
Cold Cyber Conflict (peacetime)	*Strengthening national offensive cyber capacity:* • The significant increase in defense spending on cyber weapons in most states *Acquiring (strategic) information:* • The espionage operations of the American security agency NSA (revealed by Edward Snowden) e.g., to eavesdrop on European allies *Influencing public opinion or democratic processes:* • Russian disinformation about MH17 • Alleged Russian interference in the 2016 U.S. elections
Warm Cyber Conflict (peacetime)	*Preparing the digital battlefield:* • The American strategy of 'forward defense,' where foreign vital structures are hacked in advance to quickly retaliate upon infiltration *Exerting political pressure:* • Russian cyber-attacks in 2007 on internet sites of the Estonian government, banks, media, and companies, due to a conflict over the relocation of a Russian war monument in the Estonian capital Tallinn *Damaging, disrupting, or destroying (vital) systems:* • The American Stuxnet attack on the Iranian nuclear program
Cyber-Physical War (wartime)	*Military victory:* • In the wars in Eastern Ukraine, Russia made/makes extensive use of cyber-attacks alongside conventional weapons

We must nuance our understanding of war. In wartime, offensive cyber capabilities will likely play a significant role: more and more military equipment is becoming digitalized, and it can be very advantageous to mislead the opponent's air defense through disinformation, as happened during Operation Orchard of the Israeli Air Force on a Syrian nuclear reactor. Therefore, it makes sense to speak of cyber-physical warfare, integrating digital applications into the conventional arsenal of warfare. Russia's cyber activity in Ukraine transcends warm cyber conflict because the cyber-attacks coincided with conventional military deployment. However, it is questionable whether the cyber sabotage carried out by Russia would be sufficient to fall under the prohibition of force under Article 2(4) of the UN Charter. The severity of the consequences is usually considered the measure to determine whether a particular cyber-attack falls under the prohibition of force (see [18]). In international humanitarian law, the 'physical effects' standard is used to define what is considered 'use of force'. This standard states that there is only force if it leads to an actual or potentially harmful physical effect on the targeted state. These physical effects include death, bodily injury, or significant damage to vital infrastructures or military installations. Technically, a sabotaging cyber-attack could lead to civilian casualties or significant infrastructural damage, for example, as a result of long-term disabling of power plants. Yet, it is challenging to find cyber-attacks that meet that standard. Sabotaging cyber-attacks generally do not result in civilian casualties – at least not in a direct sense, but can endanger human health and well-being [16]. Most sabotaging cyber-attacks do cause serious societal damage but do not cripple vital infrastructures. For example, sabotaging cyber-attacks occasionally disrupt banking services, but the financial sector has never been completely shut down. Similarly, in 2015 in Ukraine, the power supply was shut down, but the lights came back on a few hours later. Therefore, it is questionable where the turning point lies between an attack that damages vital infrastructure and an attack that causes significant damage to vital infrastructure. Experts indicate that international law does not clearly answer this question [15]. For example, there is no consensus on whether the Stuxnet attack should be considered an armed attack. This is significant, as that attack is marked as one of the most far-reaching cyber-attacks. In addition, some authors argue that certain cyber-attacks that do not cause physical damage should still be considered armed attacks (see, e.g., [7]). In this case, we should think of cyber-attacks that can have a socially disruptive effect, such as crashing a stock market through digital manipulations. However, yet in practice, cyber conflicts have tended to be of low intensity, staying beneath the threshold that would categorize them as armed attacks [8]. Furthermore, to this day, cyber operations have no direct impact on the battlefield, but rather target broader national and civilian resilience (see, [11, 20]). Also nowadays in the Ukrainian war, Russia mounts sophisticated cyber-attacks, but it still does not happen in any fashion that is decisive to the war [2].

In our opinion, relaxing the 'physical effects' standard is counterproductive and completely negates the intentions of de-escalation, as it would give countries more opportunities to respond with traditional violence to a cyber-attack under Article 51 of the UN Charter, which allows self-defense in the case of armed violence. Cyber sabotage is an infringement on the sovereignty of the state. This could also apply to cyber espionage and the spread of disinformation. The norm of sovereignty restriction (Article 2(7) of the UN Charter) could provide a framework to critically examine cyber-attacks and even declare them contrary to international law. In the Tallinn Manual 2.0, where a special group of experts proposes applying international law to cyber-attacks, it is also stated that the norm of sovereignty restriction applies to cyber-attacks [15]. However, the norm is very generally formulated, and the Tallinn Manual 2.0 offers, at best, an ambiguous stance regarding the interpretation of the norm and when and which countermeasures are permitted. The norm itself constitutes a less severe offense than an attack or armed attack. A violation of the prohibition of force justifies a stronger response than a violation of sovereignty – and it could well be that states want to take a strong countermeasure in some situations. However, the legal problems that arise in cyberspace can only be very limitedly interpreted in terms of positive international law, and cyberspace continuously 'escapes' existing forms of regulation. We will not further discuss the legal complications that cyber-attacks entail in this article, but we can at least state that a clear, elaborated, and widely supported normative framework for the majority of cyber-attacks that states face during a cyber conflict is lacking. The cyber conflict does not fit the traditional patterns of thought about peace and war and may even be considered an "other-than-war" phenomenon [13].

Therefore, cyber-attacks are poised to occupy a distinct place within the moral hierarchy of foreign policy tools, nestled between diplomacy and kinetic warfare, similar to severe economic sanctions. Larry May [9] proposes reclassifying these cyber operations not as acts of war but as akin to embargoes, serving as political and economic levers. This reimagining suggests cyber-attacks could be a strategic alternative to conflict, potentially guiding how NATO responds to attacks on its members. The effectiveness of traditional economic sanctions is debated, especially given globalization's role in diluting their impact. Instances like the UN's sanctions against Libya and the EU's against Russia illustrate sanctions' complex consequences, including retaliation that can harm the sanctioning states themselves. Cyber-attacks offer, according to May, an additional strategy in coercive diplomacy, potentially pressuring states towards compliance without the drawbacks of conventional sanctions. They could prompt states to cease unlawful activities by threatening their digital infrastructure, such as the Stuxnet virus against Iran's nuclear program. This approach, leveraging cyber capabilities for targeted pressure while minimizing civilian distress, could serve as a peaceful yet potent means to enforce international norms and laws. However, the potential for retaliatory cyber-attacks by the targeted state may increase escalation.

In addition, many states invest significant amounts of money in building offensive cyber capabilities. They all want to have a high-quality arsenal capable of conducting complex espionage, sabotage, and disinformation spreading. These factors contribute to a risky cyber conflict, where attackers have the means and opportunity to attack vulnerable systems, and where these attackers rarely face repercussions. In this threatening geopolitical situation, countries will likely continue to suffer from cyber-attacks carried out by other states. Moreover, there is a chance that cyber-attacks will escalate, in the sense that both the harmfulness and frequency of cyber-attacks increase. Given the danger to the security of the digital environment, international efforts must be made to effectively de-escalate cyber-attacks: the frequency of cyber-attacks must decrease, and the cyber-attacks that society experiences must become less severe. The question is how this turnaround can be achieved, and who should take responsibility for it. In the next section, we will discuss five solution directions that support this de-escalation.

3 Solution for the De-escalation of the International Information Conflict

3.1 Continue Cooperation to Improve International Cyber Security

In addition to the national steps taken to increase resilience and cyber security, states need to continue investing in international cooperation in the field of cyber security. The emphasis of NATO's cyber policy is on strengthening the defense of NATO member states: "The keynote is defense, whether an attack comes from state, criminal, or other sources," according to a NATO spokesperson. NATO's defensive doctrine is mainly based on deterrence or passive deterrence: 'deterrence by denial.' This means that security measures must be strengthened to the point that it becomes practically impossible, or at least very expensive, for attackers to infiltrate systems. The cooperation to further improve digital security at the EU and NATO level must be continued and intensified. Some good examples of this cooperation include the rollout of a broad network of Cyber Emergency Incident Response Teams (the CERT network) and the release of an initial standard (ETSI TS 103 645) by the European Telecommunications Standardisation Institute (ETSI) for securing Internet-of-Things (IoT) devices to support all parties involved in the development and production of IoT for consumers with guidance on securing their products.

Implementing a policy where NATO refrains from initiating cyber-attacks first, supported by credible military deterrents, can effectively communicate its commitment to restraint and establish boundaries for acceptable cyber-attacks. This strategy would define a baseline for cyber operations that won't provoke a violent response, highlighting that only countermeasures against cyber adversaries are permissible, while attacks on civilian infrastructure are explicitly excluded [6]. However, the dynamic between attackers and defenders in cyberspace remains unresolved, with attackers often gaining the upper hand. Strengthening global cooperation through the sharing of intelligence, best practices, and joint response strategies is essential. Additionally, creating global forums for cybersecurity intelligence and coordination will facilitate timely responses to threats. Despite this, international support for multilateral cyber arms control is waning due to challenges in verification, perceived minimal consequences for violators, and the exclusivity of agreements. Any effective framework must address these issues and consider the complexity of verifying and attributing violations in cyberspace. Furthermore, developing and promoting international cybersecurity standards will help define responsible state behavior, protect civilian networks, and promote a rules-based approach to managing cyber conflict, thus preventing escalation towards armed conflict [18].

3.2 Make Clear International Agreements Regarding Cyber Sabotage, Disinformation, and Cyber Espionage

The geopolitical landscape of cyber-attacks suffers from a notable absence of specific, binding international rules tailored to the unique dynamics of cyber conflict. Consequently, states often rely on generic international legal frameworks, which are ambiguous and open to varied interpretations concerning cyber-attacks. This ambiguity tends to benefit the attackers by fostering an environment of uncertainty that they can exploit. Establishing clear and concrete rules would enable the international community to respond promptly and decisively to cyber incidents, setting clear expectations and consequences for state actions.

Proposals for international agreements on cyber conflict include a "Geneva Convention for cyber-attacks" to establish norms akin to those that govern traditional warfare. However, the feasibility of such a treaty remains in question due to significant geopolitical divisions. For instance, the United Nations Group of Governmental Experts on Developments in the Field of Information and Telecommunications in the Context of International Security (GGE) has struggled to reach consensus, with major powers diverging sharply on key issues like the right to retaliate following a cyber-attack. The debate over how international humanitarian law applies to cyber operations continues to be contentious, reflecting broader disagreements on the militarization of cyberspace. The diplomatic efforts to forge a consensus have led to a bifurcation in approaches [5]. The U.S. and its allies have pursued one track, while a coalition led by Russia and China has followed another, both proposing different norms for state behavior in cyberspace. This division underscores the challenges in achieving a unified global stance on cyber conflicts.

Beyond formal treaties, there is potential for building confidence through voluntary agreements that enhance state cooperation and transparency. Such measures could reduce misunderstandings and prevent conflicts, involving multilateral commitments to share intelligence on cyber threats and refrain from targeting civilian infrastructure during peacetime. At the national level, countries could independently declare their responses to cyber incidents, creating predictable standards and responses. This approach includes categorizing types of cyber-attacks and the corresponding responses, which the European Union began implementing in 2019 with a framework for targeted measures against cyber threats.

The debate over cyber espionage and its implications for state sovereignty is particularly complex. Cyber espionage often involves tactics that could be seen as violating sovereignty if they cause harm, such as data corruption or system disruption. Meanwhile, the spread of disinformation represents another serious challenge, potentially destabilizing democratic processes and public discourse. It is inappropriate for a democratic rule of law state to further disrupt the information provision in another country. This does not mean that foreign information campaigns are inadmissible. It may be useful to give critical Russian journalism a helping hand. But spreading disinformation that weakens the legal order is, in our opinion, comparable to the use of chemical weapons: an inadmissible means for a civilized democratic state.

In terms of countermeasures, the need for proportionate and responsible responses is paramount. Retaliatory cyber-attacks are risky and may not effectively enhance cybersecurity or deter aggressors. Instead, a combination of diplomatic, economic, and legal measures, alongside public exposure of malicious activities, may offer a more sustainable path to maintaining international peace and security in the digital age. Thus, the international community stands at a critical juncture, needing to balance the imperatives of security and cooperation against the backdrop of rapidly evolving cyber capabilities. The establishment of robust, clear, and widely accepted norms for cyberspace is essential to manage and mitigate the risks of cyber conflict effectively.

A fundamental choice has to be made regarding the type of countermeasures that should be taken in response to a cyber-attack. One can fight fire with fire and launch a similar cyber counterattack. We have indicated that this is irresponsible in the case of undermining a democratic rule of law. Many other cyber-attacks directly affect civilian institutions and facilities – targets that should be spared as much as possible even in wartime. It is therefore highly questionable whether a democratic rule of law state should retaliate in kind. European States can also respond with other means, for example, by using the European sanctions package, and taking diplomatic and economic measures. Considering these alternative means is not only of moral importance – it is questionable whether cyber counterattacks structurally increase cyber security. Every attack contributes to the proliferation of weapons, and cyber conflicts have so far never ended with a decisive counterattack. Moreover, serious reservations are expressed in the literature about the deterrence strategy [19]. If a cyber counterattack is carried out, it must happen immediately and in cooperation with allies. If threats are made without follow-up, the deterrent effect disappears.

3.3 Ensure Responsible Management of the Cyber Arsenal

Malware can spread across the world in seconds: that is the price we pay for a global internet. Whoever deploys a cyber weapon can expect the used code to end up in unforeseen hands and perhaps boomerang back in one's face. This means that cyber weapons must be deployed carefully and once-used vulnerabilities must be reported promptly. Because that is the unique opportunity for de-escalation that cyber conflict offers: programmers can neutralize much malicious code. Since these updates increase the security of the entire digital domain, this aspect must be optimally utilized. It is therefore of great importance that intelligence services and defense units work cooperatively with software producers and chip manufacturers to implement security updates as effectively as possible and distribute them among users. International intelligence services and defense units must cooperate with software producers and computer chip manufacturers in implementing security updates and distributing them among users. Moreover, intelligence services and defense units must continuously ask themselves what increases cyber security more: possessing a certain cyber weapon or releasing it?

International allies should also be more open with each other and jointly monitor the management of vulnerabilities. It would be a pity if allies independently collect cyber weapons and keep them strictly secret: then, vulnerabilities could remain on the shelf that the services could have reported from a strategic point of view. Ultimately, in the long term, the importance of reliable and safe digital applications weighs more heavily than the interest in occasionally breaking through that security, however much this would enable specific operations. Responsible management of the cyber arsenal also has consequences for cooperation with private parties. Many large states possess a high-quality arsenal capable of conducting complex espionage, sabotage, and disinformation spreading, and master the simpler methods of committing cyber-attacks. This development has also led to a flourishing private or malicious industry that supplies these weapons, contributes to the proliferation of weapons, and gives opponents access to offensive cyber capabilities. Not for nothing did the German government in 2017 decide to have cyber weapons developed by its own state agency [4]. Government must seriously consider which private partners should be involved in the development of offensive cyber capabilities – and whether there is enough reason to take this development more into its own hands. Strict conditions must also apply to ensure that dangerous technology is not sold to countries that cannot trust with these weapons. An international lobby for an updated system of export licenses could help in this regard.

3.4 Protect the Independence of Technology Companies

Technology companies such as Microsoft, Google, and Apple play a crucial role in ensuring international cyber security. They can patch the holes in their software and ensure the most robust possible digital applications. This means that it is of crucial importance that these companies do not become extensions of national governments. Kaspersky Labs and Huawei Technologies are under great pressure due to their connections with China and Russia. The governments of Russia and China do not respect the fundamental rights of their citizens and regularly use their power to intervene in the business world – so why wouldn't they use these global players as pawns of their intelligence services? How exactly the Western governments should deal with companies like Huawei and Kaspersky depends on the extent to which the Chinese and Russian governments actually control these companies. American companies sometimes work intensively with intelligence services, whether or not under duress, as in the renewed PRISM program of the National Security Agency (NSA). The international community must prevent such situations, both within the prescribed frameworks and proactively, if it wants to maintain a reliable and safe international technology market. It is therefore encouraging to see that more than 60 global companies, including Microsoft, Facebook, and Dell, are part of the Cybersecurity Tech Accord and have joined the Paris Call for Trust and Security in Cyberspace. The cooperation agreement contains a set of common promises that have consequences for how these companies relate to governments. Such as the promise never to help governments launch cyber-attacks against innocent civilians and businesses, and the promise to protect their products against covert influence and sabotage. Although the impact and precise meaning of the agreement in practice remain to be seen, this cooperation could greatly reduce the ability of governments to set up espionage and sabotage operations with the help of companies.

3.5 Invest in a Societal Debate on Geopolitical Cyber Security

War, peace, and international conflict are matters that the population should be particularly involved in a democracy: major societal interests are at stake. The nature of the current information conflict complicates this discussion because so many things are opaque or secret. However, the decision to acquire certain weapons and authorize certain operations cannot only be a matter for experts and officials with special authority: the ongoing cyber conflict must be thoroughly and deeply discussed in the public debate. Each of the four solution directions mentioned above should be addressed in the Dutch and international public debate. Therefore, secrecy by intelligence services and defense cannot be taken for granted – this secrecy must be weighed against the interest of citizens in gaining insight into the government's cyber operations and forming an opinion about them. The greater the impact of international cyber conflicts on society, the greater the importance of making a relevant public debate possible. Moreover, the involvement of citizens is urgently needed: they are frequently the victims of disinformation, cyber espionage, and cyber sabotage. In the cyber conflict, all citizens have a role to play and should be equipped to recognize disinformation and take basic security measures. Only citizens who understand how technology affects their lives and who have meaningful input in the political choices made regarding new technology are resilient against undermining cyber-attacks. The government has a special responsibility to enable this citizenship: by facilitating education, prioritizing the rights of citizens, and providing as much transparency as possible about the issues it faces in a threatening geopolitical situation.

4 Conclusions

Most actions currently undertaken by influential states fall within what we have described as the cyber conflict. In peacetime, countries build up their cyber security and try to infiltrate the digital systems of other countries as secretly as possible. Russia also actively spreads disinformation. This is a strategy that other autocratic countries do not seem to apply on a large scale, at least not in relation to other countries, but is part of the desire to control, censor, and manipulate the information that reaches their population. There are also several examples of serious cyber sabotage, such as Operation Olympic Games, attributed to Israel and the United States, and the WannaCry attack, attributed to North Korea. So far, cyber-attacks have never led to a cyber-physical war; in that sense, there is no 'cyberwar'. However, the current international situation is risky and worrying, and the disruptive effects of cyber capabilities underscore the risks in the international landscape, which could lead to the escalation of cyber-attacks. We have formulated five possible solutions for the de-escalation of the international cyber conflict. Perhaps the most important is that clear international agreements are made in the field of offensive cyber capabilities. However, in the current geopolitical environment, international support for multilateral cyber arms control agreements is declining, due to issues such as doubts about verification, perceived limited consequences for violators of agreements, and a lack of inclusivity of many agreements. Any multilateral framework for cyber weapons control will have to take this trend into account, especially since the verification of any agreement and the attribution of any violation will be even more complicated in the cyber domain than regular arms control. However, cyber weapons control frameworks limit expensive and destabilizing arms races and help prevent dangerous escalation toward armed conflict.

Acknowledgment. This publication is part of the research program Ethics of Socially Disruptive Technologies (ESDiT), which is funded through the Gravitation program of the Dutch Ministry of Education, Culture, and Science and the Netherlands Organization for Scientific Research (NWO grant number 024.004.031). Furthermore, I would like to thank Jurriën Hamer and Rinie van Est for their support.

References

1. Bronk, C.: Blown to bits: China's war in cyberspace. Strategic Stud. Q. **5**(1) (2011). http://www.au.af.mil/au/ssq/2011/spring/bronk.pdf
2. Bronk, C., Collins, G., Wallach, D.S.: The Ukrainian information and cyber war. Cyber Defense Rev. **8**(3), 33–50 (2023)
3. Brown, J.M., Fazal, T.M.: #SorryNotSorry: why states neither confirm nor deny responsibility for cyber operations. Eur. J. Int. Secur. **6**, 401–417 (2021)

4. Delcker, J.: Germany to launch US-style agency to develop cyberdefense, Politico **29**(August) (2018)
5. Henriksen, A.: The end of the road for the UN GGE process: the future regulation of cyberspace. J. Cybersecur. **5**(1) (2019). https://doi.org/10.1093/cybsec/tyy009
6. Lonergan, E.D., Schneider, J.: The power of beliefs in US cyber strategy: the evolving role of deterrence, norms, and escalation. J. Cybersecur. **9**(1). (2023). https://doi.org/10.1093/cybsec/tyad006
7. Lucas, G.R.: Ethics and Cyber Warfare: The Quest for Responsible Security in the Age of Digital Warfare. Oxford University Press, Oxford (2017)
8. Maschmeyer, L.: A new and better quiet option? Strategies of subversion and cyber conflict. J. Strateg. Stud. **46**(3), 570–594 (2023)
9. May, L.: The nature of war and the idea of "cyberwar". In: Ohlin, J.D., Govern, K., Finkelstein, C. (eds.) Cyberwar. Law and Ethics for Virtual Conflicts, pp. 3–15. Oxford University Press, Oxford (2015)
10. Moore, D.: Offensive Cyber Operations: Understanding Intangible Warfare. Hurst Publishers, London (2020)
11. Mueller, G.B., Jensen, B., Valeriano, B., Maness, R.C., Macias, J.M.: Cyber operations during the Russo-Ukrainian war: from strange patterns to alternative futures (research report). Center for Strategic and International Studies, Washington (2023)
12. Orend, B.: Fog in the fifth dimension: the ethics of cyber-war. In: Floridi, L., Taddeo, M. (eds.) The Ethics of Information Warfare. Law, Governance and Technology Series, vol. 14, pp. 3–23. Springer, Cham (2014). https://doi.org/10.1007/978-3-319-04135-3_1
13. Rauscher, K.F., Korotkov, A.: Towards rules for governing cyber conflict. Rendering the Geneva and Hague conventions in cyberspace. EastWest Institute, Brussels (2011)
14. Rid, T.: Cyberwar will not take place. J. Strateg. Stud. **35**, 5–32 (2012)
15. Schmitt, M.N. (ed.): Tallinn Manual 2.0 on the International Law Applicable to Cyber Operations. Cambridge University Press, Cambridge (2017)
16. Shawe, R., McAndrew, I.R.: Increasing threats to United States of America infrastructure based on cyber-attacks. J. Softw. Eng. Appl. **16**, 530–547 (2023)
17. United States Cybercommand: Achieve and Maintain Cyberspace Superiority. The National Security Archive, Washington (2018). https://www.cybercom.mil/Portals/56/Docments/USCYBERCOM%20Vision%20April%202018.pdf?ver=2018-06-14-152556-010
18. Usman, H., Mir, A.A., Rehman, A.U.: Beyond conventional war: cyber-attacks and the interpretation of article 2(4) of the UN charter. Glob. Legal Stud. Rev. **8**(2), 16–26 (2023)
19. Van der Meer, S.: Responding to large-scale cyberattacks: a toolbox for policymakers. J. Cyber Policy **7**(2), 175–193 (2022)
20. Van Niekerk, B.: Cyber operations in peace and war: a framework for persistent engagement. In: Proceedings of the 19th International Conference on Cyber Warfare and Security (ICCWS 2024), pp. 395–404. Academic Conferences International Limited, Reading (2024)

A Positive Perspective to Redesign the Online Public Sphere: A Deliberative Democracy Approach

Roxanne van der Puil(✉)

Eindhoven University of Technology, Eindhoven, MB 5600, The Netherlands
r.e.v.d.puil@tue.nl

Abstract. The World Economic Forum Global Risks Report 2024 [1] lists AI-generated misinformation and disinformation and polarization in the top three of the most severe risks for the coming two year period. Underlying these interconnected risks is the digital infrastructure of networking sites, making social media a socially disruptive technology. Within this digital context of risk, scholars argue that laypeople's epistemic norm have been disrupted, leading to so-called post-factual attitudes [3]. While various explanations have been given to understand the dynamics in online public debate, the philosophical background commonly implicated is postmodernism [4, 5, 8, 9]. Regulation and design of social media platforms arguably mirror the interpretation that the public sphere has become post-factual. While these regulatory and design solutions have clear effects, the risks of misinformation, disinformation and polarization are insufficiently curbed. Thus, more solutions are needed. In view of this, I argue to understand online dynamics from the philosophical perspective of deliberative democracy theory. Specifically, I argue that the character of discourse in online public spheres can be understood in light of three revisions to deliberative democracy theory, namely 1) diversifying acceptable communication styles in deliberations, 2) embracing pluralism, and 3) recognizing that group-based justice claims can be democratically beneficial. This approach provides a positive understanding of online dynamics and inspires alternative strategies to redesign and regulate social media to mitigate the risks of misinformation, disinformation and polarization.

Keywords: Post-factualism · Socially Disruptive Technologies · Public Sphere · Social Media Design · Deliberative Democracy Theory

1 Introduction

The World Economic Forum Global Risks Report 2024 [1] lists AI-generated misinformation and disinformation and societal and political polarization in the top three of the most severe risks for the coming two year period. Underlying these interconnected risks is the digital infrastructure of networking sites, making social media a socially disruptive technology [2]. Mis- and disinformation foster distrust towards facts and experts which in turn can further exacerbate polarization. In situations of deep polarization, people are more inclined to accept and trust information that confirms existing beliefs, even if that information is false [1]. As such, the risk of polarization is that of societies fragmenting across political affiliations and along competing narratives of what is true.

Within this context of epistemic pollution and polarization, enabled by the digital infrastructure of social networking sites, scholars warn about the emergence of post-factual attitudes [3–6]. Suggested in the terminology post-factual, is a reshuffling of the hierarchy of epistemic norms. Specifically, the concern is that citizens have come to rely on their opinion and lived experience rather than knowledge and scientific methods when evaluating empirical claims relevant to public discussions. This reversal is also expressed in the popular definition of the umbrella term post-truth, namely "relating to or denoting circumstances in which objective facts are less influential in shaping public opinion than appeals to emotion and personal belief" [7, n.p.]. The World Economic Forum mentions the reversal of epistemic norms too when it talks about the infiltration of "manipulative narratives [...] [in] public discourse" on wide ranging issues due to the fact that "emotions and ideologies overshadow facts" [1, p. 20]. In short, there has been speculation about social media disrupting the previously stable hierarchy of epistemic norms.

While various other explanations have been given to understand why the current dynamics in online public discourse have come about, e.g. the role of political figures or the economic incentives underlying the infrastructure of social media platforms, the philosophical background commonly implicated in this reversal of epistemic norms is postmodernism. That is to say that scholars have argued that (online) public debate, and the type of attitudes displayed, is influenced by postmodern ideas [4, 8, 9]. While this analysis has been criticized for misconstruing postmodernism [6, 10], this philosophical understanding of online dynamics is arguably reflected in the current regulation of social networking sites. In other words, the current arsenal of solutions has a seemingly anti-postmodernist spirit in terms of emphasizing the notions objective truth and facts. For example, very large platforms cooperate with fact-checkers [11]. Similarly, scholars devote substantial attention to the ways users can be nudged and boosted into critical thinking with design features [12, 13]. To nudge is "to steer people in a particular direction while preserving their freedom of choice" [14, p. 973]. Nudges aim at a direct effect and are carried out with simple prompts and in the choice architecture of users. An example of a nudge would be highlighting the source or lack of credible source of something claimed online, which can be empirically verified. Boosting strategies on the other hand aim at a long-term change by improving "people's competence to make their own choices" [14, p. 974]. An example would be to constantly remind people, with prompts for example, to cross reference the information they consume online with the aim of habituating the practice of cross referencing.

As the parallels drawn between postmodernism and (online) public discourse have been criticized [6, 10] and the risks of misinformation, disinformation and polarization remain, an alternative philosophical viewpoint to understand online public discourse is desirable to inspire complementary solutions. In pursuit of this objective, I propose to understand the online dynamics from the perspective of deliberative democracy theory. Specifically, I argue that the dynamics in online public debate can be understood in light of three revisions to deliberative democracy theory namely 1) diversifying acceptable communication styles in deliberations, 2) embracing pluralism, and 3) recognizing that group-based justice claims can be democratically beneficial. This approach provides a positive understanding of online dynamics and inspires alternative strategies to redesign social media to diminish the flow of misinformation and disinformation and contain the polarized character of online discourse.

In Sect. 2, I summarize the philosophical analysis of MacMullen [3] in which he draws out four distinct attitudes, supposedly displayed in online discourse, discussed in scholarly work. In Sect. 3, I argue that online dynamics are usefully understood from the perspective of deliberative democracy theory. In Sect. 4, I turn to empirical work on dynamics in deliberation and use these findings to contemplate alternative design solutions so that the risks of mis- and disinformation and polarization are further mitigated and online democratic deliberation is enhanced.

2 Attitudes Displayed in Online Discourse

MacMullen has pulled apart four types of post-factual attitudes supposedly displayed in public debate, discussed by scholars. These attitudes include one unconscious post-factual attitude and three conscious post-factual attitudes. "Unconscious post-factualism" [3, p. 5] is the phenomenon that when there is conflicting empirical evidence underlying public debate, people fall back on cognitive biases to make an assessment. On reflection, these people would deem relying on cognitive biases an epistemically inferior method. The result of unconscious post-factualism is that in situations of disagreement, each agent is more likely to accept empirical claims in support of pre-existing beliefs, to accept those claims made by people who share their beliefs, and, in the pursuit of truth, to seek out claims already accepted as true rather than those that are not. The tendency to be moved by unconscious cognitive biases in times of political disagreement explains persistent concerns about filter bubbles and echo chambers in social networking sites. Thus, this attitude is not new. How we humans are programmed to be led by cognitive biases is thoroughly studied (see for example [15]) and helps explain why expressions of disagreement are so common, offline and online, when it comes to political issues with conflicting empirical support. As an example one can consider the empirical uncertainty surrounding COVID-19 when the pandemic had just hit society.

On the other hand there are conscious post-factual attitudes. While motivated differently, each of these attitudes somehow challenge the primacy of facts. Unlike unconscious post-factualism however, the prevalence of these attitudes has not been empirically tested. The first conscious post-factual attitude, "motivational post-factualism" [3, p. 10] would have similar outcomes to unconscious post-factualism. People who adopt such an attitude ascribe little, if any importance to facts and truth and instead rely on what feels good and which claims fit one's identity or ideological commitments. The second conscious post-factual attitude is "metaphysical post-factualism" [3, p. 7]. People with such an attitude have a relativistic, rather than an objective, view on truth. Here, people adopt the viewpoint that a proposition is true if it corresponds to one's personal feelings and experiences. The difference then between motivational and metaphysical post-factualism is that people with a motivational attitude do not deny the existence of objective facts and truth, but simply choose not to pursue these matters, whereas people with a metaphysical post-factual attitude perceive facts and truth as subjective, relative matters. Unless a user explicitly states what is driving his or her contribution to online discourse, i.e. overlooking objective facts in favor of one's feelings or perceiving facts as subjective matters, online contributions along the lines 'I feel this is false or true' or 'This is true or false because I personally experienced such and such' could be interpreted as the result of a motivational or metaphysical post-factual attitude.

Finally, there is "epistemic post-factualism" [3, p. 11] which is the attitude that others may know and have access to objective facts and truth, but that the individual him- or herself does not and cannot. This makes the individual dependent on other agents who know the truth. MacMullen explains that this attitude is complicated by trust [3]. Confronted with competing claims, and given the knowledge that other agents who know the truth may be biased by their interests and values, the individual does not know who to trust. Online contributions such as 'Who knows what is true, we can't trust anybody' could be interpreted as the result of an epistemic post-factual attitude. The underlying cause for a user to disengage with political discussions and leave a social media discussion could also be interpreted as the result of an epistemic post-factual attitude too.

In scholarly discourse on post-truth, fingers have been pointed towards the role of postmodernism in inspiring these post-factual attitudes. For example, Dennett stated in an interview that "they [i.e. postmodernists] are responsible for the intellectual fad that made it respectable to be cynical about truth and facts" [4, n.p.]. In brief, the inspiring postmodern ideas would be a sceptical stance towards our access to objective facts [8, 16] and that claims of truth mask a pursuit for power [8, 17]. However, as many scholars who trace post-factual politics to postmodernism have themselves pointed out, the parallels drawn between post-factual attitudes and postmodern theory are somewhat superficial. Brahms, for example, says "this substandard discourse does not derive from an in-depth understanding of postmodernist ideas, but rather, derives from a distortion and reduction of them" [6, p. 14]. Some scholars even argue that a thorough understanding of postmodernism reveals that it is not compatible with the post-factual culture at all [10].

Indeed, what is important to emphasize is that the underlying motivation of certain expressions or behavior on social networking sites generally remains unknown. Online, we typically do not dig deeper and question the philosophical theory underpinning one's stance and behavior in public discourse. Yet, this interpretation of postmodern ideas shaping the course of online public debate may be framing the problem at hand and the type of solutions believed to be effective. If the concern is that citizens value their feelings over facts, that citizens take facts to be subjective and relative, and that citizens feel they no longer know who to trust as expert authority, it makes sense to devise regulation and redesign social media platforms so that the primacy of facts and status of experts is reinstated. While this strategy has proven somewhat effective, the risks of misinformation, disinformation and polarization are, at present, insufficiently curbed. As such, I argue that an alternative understanding of online public discourse is desirable so that complementary solutions to redesign and regulate social networking sites are inspired.

3 A Deliberative Democracy Theory Perspective

The first paragraph of the introduction to the Oxford Handbook of Deliberative Democracy Theory reads "post-truth politics is the antithesis of deliberative democracy" [18, p. 1]. Post-truth is an umbrella term used for phenomena undermining the (epistemic) quality of democracy, such as post-factual attitudes. Reading further it becomes clear that the authors define post-truth as the attitude of not wanting to deliberate with others as well as not being able "to listen to the other side and reflect upon what they may have to say" [18, p. 1]. In this regard, the proposal to understand post-factual politics from the perspective of deliberative democracy theory would seem quite unusual. However, noting the lack of empirical study whether postmodern ideas are indeed motivating certain expressions and behavior online, I suggest to let go of the terminology post-factual. Instead, I suggest to focus merely on the behavior and expressions displayed online, and argue that the lens of deliberative democracy theory is helpful.

Deliberative democracy theory is often understood as a competing theory to aggregative democracy. Aggregative democracy is grounded in the idea that the legitimacy of democracy stems from the ability to aggregate individual preferences, typically through voting, focusing on equitably counting preferences to reflect the majority's wishes. It views democracy as a system that represents the numerical strength of various societal interests. In contrast, deliberative democracy emphasizes the deliberation process over simple aggregation. It posits that democratic legitimacy comes not only from vote counts but also from the quality of and norms underlying discussions preceding voting. Deliberative democracy as such seeks to transform or establish better understanding on varying perspectives through dialogue. Deliberative democracy theorists thus regard aggregation in itself as insufficient for nation-states to be labelled democratic. For a nation-state, or a subsystem in a nation-state, to be regarded as democratic, it needs to meet certain deliberative ideals. Some of these ideals have been substantially revised over the past decades. The revision of three deliberative ideals in particular shed an alternative light on the type of attitudes expressed online, and may offer guidance as to how the risks of misinformation, disinformation and polarization can be further mitigated so that public democratic debate is enhanced.

3.1 From Reasons to Appropriate Considerations

The first deliberative ideal that has been revised substantially is that of giving and responding to reason. The first generation of deliberative theorists described an ideal of deliberation one would imagine between scholars, namely a dispassionate, carefully considered and structured exchange of argument [18]. Critics have pointed out that these specific norms for deliberation ignore alternative ways of speaking, thereby undermining participatory equality [19]. This, deliberative democracy theorists argued, made the original ideal of deliberation exclusionary, and thus anti-democratic. Extending this criticism, scholars have argued for multiple public spheres rather than one overarching public sphere so that a diverse range of rhetorical and stylistic styles of deliberation are enabled. Whereas one public sphere will favor the dominant hegemonic group who fits a certain stylistic and rhetorical form of deliberation [19], embracing many is to welcome entirely different activities such as protests, fundraising and campaigning which have an alternative deliberative character [20].

More fundamentally, the dichotomy between reason and passion has been criticized and rejected by deliberative theorists (see for example [21, 22]). For deliberation to take place, people need to feel motivated. Secondly, in order to deliberate with others and truly listen and seek to understand their reasons, we need to empathize with others. As such, deliberation involves both reason and passion. Even deeper than this however, some scholars argue that reason and passion cannot be pulled apart in deliberation as "passion actually requires a mental vision of the good" [21, p. 87]. Passion is in this regard a cognitive act in terms of being interpretative and evaluative. This makes passion both emotional and rational. In this sense, passion "already begins to look less antithetical to deliberation than is frequently assumed" [21, p. 88]. In short, the distinction between being rational and passionate in the context of democratic deliberation is, as some would argue, misleading. Both reason and passion are essential in terms of motivating citizens to deliberate and reflect during deliberation. Or as Hall phrases it, "deliberation requires both thinking carefully and caring thoughtfully" [21, p. 90].

In short, the second wave of deliberative theorists have argued that alternative styles of communication, as already used in society, should be embraced. More passionate ways of speaking as well as more passionate deliberative forums are needed in democracies to meet ideals of equality. In sum, theorists have come to broaden the ideal of reasoning to "the criterion that arguments ought to give and respond to appropriate 'considerations' and contexts – for example, more emotionally rooted expressions and differing styles of communication such as narrative and rhetoric" [18, p. 3]. To relate this revision to the expressions made online, rather than interpreting these as the result of un- or conscious post-factual attitudes described in section I, i.e. the result of cognitive biases, lacking respect for facts, facts and truth being understood as relative or a result of not knowing who to trust, they can and perhaps should be understood as less argumentative, more motivation driven or passionate ways of speaking.

3.2 From Consensus to Disagreement

A second revision made to deliberative democracy theory is the move from pursuing consensus, or the truth value of political and moral claims, to pluralism. During the first wave of deliberative democracy theory, theorists argued that public discourse, under non-coercive or manipulative circumstances, should and could deliver truth or consensus [18]. The second generation of deliberative theorists have criticized this objective as it ignores the fact of pluralism. While we speak of *a society*, each society is made up of fluid communities, each with their own competing, oftentimes incommensurable, values and beliefs and conceptions of social justice and the good life [23]. When it comes to political decision-making, the diversity of value commitments can lead people to support alternative decisions. Pluralism then, in the public arena, refers to the notion that "at least on certain matters, more than one view is rationally allowed, such that two people may disagree on a given question, yet both be justified in their beliefs" [24, p. 1]. While epistemic deliberative theorists place limits on pluralism and argue for "reintroducing the truth" (see for example [23]), in general deliberative theorists now acknowledge the fact of pluralism. This makes room for contestation and has given deliberative democracy a clear agonistic component [25].

This update from the first to second generation of deliberative theorists, like postmodern theory, leaves room for disagreements on what is true in the political arena. Relating this to the expressions of disagreements made online, rather than interpreting the expression of disagreement of what is true the result of conscious post-factual attitudes, disagreement could be understood as the display of pluralism.

3.3 From Universalism to Pluralism

The third revision concerns the original ideal of thinking and reasoning from the perspective of a common good. The first generation of deliberative theorists considered parochial, affiliation based reasoning as problematic. First generation theorists argued that citizens should find commonalities between one another, consider what is beneficial for everyone and reason accordingly, in the form of 'we' [19]. Only this kind of communication was considered good democratic deliberation. This perspective implied that groups who fought for specific interests and made claims of justice, for example minority groups, were often considered a threat to democratic communication. Their contribution to the debate was considered potentially divisive for society.

Since then, scholars like Nancy Fraser [19], Iris Marion Young [26, 27] and Jane Mansbridge [28] have made a strong case for the inclusion of what Fraser [19] calls "counterpublics" that are based on difference to defend competing interests. While having to talk as a 'we' is to the benefit of the hegemonic group, this commonly places subordinate groups in a disadvantageous position [19]. Moreover, accepting and valuing claims based on difference is important because it can help to make visible structural relations in society [19, 27]. If done in a non-manipulative setting, expressing self-interests helps individuals to understand what they want and need and for others to come to understand these needs and wants too. Finally, "recognizing and asserting self-interests helps unveil hegemonic understandings of the common good when those understandings have evolved to mask subtle forms of oppression" [28, p. 179].

In other words, many second generation deliberative theorists now regard the expression of self-interests as playing a legitimate role in democratic deliberation and as a resource rather than an obstacle for democratic communication in terms of aiding self-understanding, mutual and societal understanding. This update from the first to second generation of deliberative theorists, like postmodern theory, leaves room for group-based claims. Rather than viewing group-based claims as expressing a disregard for facts and truth or as valuing feelings and community commitments over facts, these types of expressions can be interpreted as valuable information for democratic deliberation.

4 Deliberative Solutions for the Online Public Sphere

Taking all three revisions in deliberative democracy theory together, current online dynamics are placed in a different light. The postmodernist perspective offers a, I would argue, rather negative outlook that certain expressions, namely those of an affective nature, those that are interest oriented or distrustful towards expert knowledge, are antidemocratic. The perspective from deliberative democracy theory on the other hand illuminates that these expressions may, in different ways, contribute to a more democratic and just society. The deliberative democracy theory perspective underlines the importance of including more passionate ways of thinking and speaking (revision I), acknowledges that in the political arena more than one perspective can be reasonable (revision II), and how interest-based claims of justice can help uncover structural relations of dominance and subordination (revision III). As such, whereas a postmodernist theory outlook comes to conclude that such attitudes need to be fixed, a deliberative democracy theory outlook would rather suggest that these attitudes can benefit democratic deliberation. The crux, however, lies in *can*.

It would be false to conclude that given these revisions, all deliberation occurring on social networking sites currently is democratic. Just because social media affords expressions different from calm, collected reason, has allowed the public to challenge narratives on what is true or best and the expression of self- and group interest, does not mean social media as it is currently, is aptly characterized as being democratic deliberation. Elstub [29] warns that the updates made to deliberative theory have led people to falsely draw such conclusions and emphasizes that not any and all deliberation is democratic deliberation. That being said, I agree with Roeser and Steinert that "there is nothing intrinsic to social media that would exclude it from serving as a tool for genuine deliberation" [30, p. 182]. In order to make it such a genuine deliberation tool, however, we need to think beyond design that reinstates facts and expertise and redesign social media in line with the revisions of deliberative democracy theory. In other words, we need to rigorously translate the theoretical work of deliberative theorists into policies, regulation and design.

Looking at the theory, deliberative theorists have argued convincingly for the importance of recognizing and valuing the emotional and pluralist side to deliberation, while being mindful of the various ways in which these aspects can pose obstacles to deliberation. Yet, as Thompson and Hoggett have pointed out "the proposal to simply welcome emotions into public deliberative spaces, without any understanding of the nature of those emotions, is somewhat naïve at best, and dangerous at worst" [31, p. 353]. In between these poles, on the one hand the theoretical work arguing for the inherently emotive and pluralist component of deliberation and the other, welcoming emotions, values and expressions of self-interest in real-life deliberative forums, a gap remains. Remarked also by others, deliberative theorists "have not developed the implications of their views on the emotions with the same rigor of other key concepts" [32, p. 924]. As Mansbridge puts it, while "human beings have worked out many useful rules for 'rational' discussion, including, for example, demands for logical consistency [,] we still need to work out some practical understandings for the appropriate use of emotions in deliberation" [28, p. 189]. For this, Mansbridge argues, we need to look closely at practice, so that we can inductively arrive at "effective norms in the deliberative uses of emotion" [28, p. 189]. This is an additional way in which social media can be described as disruptive: social media highlights this gap between theory and practice and urges deliberative theorists to draw out the practical implications of their revisions for deliberative democracy.

Looking at practice, one could argue that revision II has been considered through the anti-postmodernist inspired strategy, i.e. platforms are obligated to combat outright false information while being mindful of reasonable perspectives thereby seeking a balance between one objective truth and anything goes. There are however far fewer policies and design choices which directly address revisions I and III. As revision III is arguably more complicated than the first, and the scope of this paper is limited, we contemplate what revision I could mean for social media design. To design for revision I is thus to find ways in which welcoming more emotional style of speaking and more emotional forums can enhance genuine democratic deliberation rather than undermine it. Currently, platforms are obligated to combat hate speech and discrimination. While these policies help address emotion-laden contributions that clearly harm other users or groups, there are few, if any, design and policy choices which help navigate the grey scale of emotional contributions, meaning those which cannot directly be assessed as good and helpful or bad and damaging to democratic deliberation. That is to say that thoughtful consideration of how to regulate and design for expressions such as 'I feel this is false', 'my experience confirms that such and such is the case', 'We cannot trust Y', 'group Z is being treated unfairly', and the outpour of affective responses and discussion these can lead to, has received considerably less attention.

Following the advice of Mansbridge [28], I turn to empirical work on emotional dynamics in deliberative forums next, looking at the implications of more emotional communication in deliberation. Studies on emotional dynamics in deliberative forums have found that while more emotional styles of communication can have an inclusive and binding effect, they can also lead to exclusion. For example, Martin [33] found that in deliberative forums whereby participants self-determine norms of conduct and participation, participants quickly build relationships on shared experiences and emotional ways of speaking. Analyzing the deliberative dynamics of these groups further, Martin [33] observed that this type of relationship building has an exclusionary effect towards those who feel they have different experiences and feel uncomfortable with such personal and emotional ways of speaking. Similarly, Thompson and Hoggett [31] in reviewing empirical work on juries, note that emotional styles of communication can lead to exclusion. If a group forms with a particular sentiment, those who feel an alternative sentiment can feel or be excluded.

Returning to deliberative democracy theory and the revision to include other forms of communication than impassionate reason, exclusion for lack *and* presence of emotion is a practical implication we should consider and design for. Exclusionary effects based on a group's manner of speaking can be considered with design. Just as algorithms can pick up on potentially harmful speech to nudge users to reconsider their wording, we can train algorithms to pick up on more dispassionate and passionate contributions and to evenly distribute the display of both in group discussions. Thompson and Hoggett [31] further suggest that in order to prevent exclusion as a result of a particular sentiment becoming dominant, groups should be kept small and in rotation. Thus, if we want to design for social media platforms in which people feel their contribution will be valued and, at minimum, not ostracized for conveying an alternative sentiment, group sizes of discussions could be reconfigured. For example, rather than everyone commenting within the same stream of comments, there could be many streams. While every user can comment, each stream allows for a limited amount of users. This set up would still be inclusive, as each and every user can participate. Moreover, chances are that other users will read your comment thereby increasing equality in terms of every voice being listened to. Finally, while dislike buttons have been removed already, other forms of like buttons such as emoticons or up- and down voting, could be removed as these too can enforce a certain sentiment that may lead to the exclusion of others. While the effectiveness of such redesigns to mitigate the risks of misinformation, disinformation and polarization should be empirically tested, such design choices could help make social media a genuine tool for democratic deliberation.

5 Conclusion

In this paper I addressed the challenges of polarization and epistemic pollution on social networking sites. The understanding that online public discourse is characterized by various post-factual attitudes, influenced by postmodern skepticism about objective facts and truth, have prompted strategies focusing on reinforcing facts and respect for their disseminators. However, these strategies are not sufficiently curbing the risks. I therefore proposed viewing online deliberation through the philosophical framework of deliberative democracy theory. Online dynamics can be viewed in light of three key revisions to deliberative democracy: diversifying acceptable communication styles in deliberations, shifting from seeking consensus to embracing pluralism, and recognizing group-based justice claims as democratically beneficial. These approaches provide a more inclusive understanding of online emotional dynamics and advocate for adapting social media designs to enhance genuine deliberation, such as moderating emotional content through AI and revising interaction features to prevent polarization. This perspective is crucial for addressing the negative impacts of social media thereby safeguarding democratic discourse.

Acknowledgments. This research is part of the research program Ethics of Socially Disruptive Technologies (ESDiT), which is funded through the Gravitation program of the Dutch Ministry of Education, Culture, and Science and the Netherlands Organization for Scientific Research (NWO grant number 024.004.031).

References

1. Cavaciuti-Wishart, E., Heading, S., Kohler, K., Zahidi, S.: The Global Risks Report 2024. World Economic Forum, January 2024. https://www3.weforum.org/docs/WEF_The_Global_Risks_Report_2024.pdf
2. Hopster, J.: What are socially disruptive technologies? Technol. Soc. **67**, 1–8 (2021)
3. MacMullen, I.: Survey article: what is "post-factual" politics? J Polit Philos **28**(1), 97–116 (2019)
4. Cadwalladr, C.: Daniel Dennett: i begrudge every hour i have to spend worrying about politics. The Guardian. 12 February 2017. https://www.theguardian.com/science/2017/feb/12/daniel-dennett-politics-bacteria-bach-back-dawkins-trump-interview
5. Wight, C.: Post-truth, postmodernism and alternative facts. New Perspect. **26**(3), 17–29 (2018)

6. Brahms, Y.: Philosophy of post-truth. Institute for National Security Studies (2020)
7. Oxford Languages. Word of the Year 2016. 2024. https://languages.oup.com/word-of-the-year/2016/
8. McIntyre, L.: Post-truth. The MIT Press, Cambridge (2018)
9. Calcutt, A.: The surprising origins of "post-truth" - and how it was spawned by the liberal left. The Conversation. 18 November 2016. https://theconversation.com/the-surprising-origins-of-post-truth-and-how-it-was-spawned-by-the-liberal-left-68929
10. Newman, S.: Post-truth, postmodernism and the public sphere. In: Conrad, M., Hálfdanarson, G., Michailidou, A., Galpin, C., Pyrhönen, N. (eds.) Europe in the Age of Post-truth Politics: Populism, Disinformation and the Public Sphere, pp. 13–30. Palgrave (2023)
11. European Commission. The 2022 Code of Practice on Disinformation (2024). https://digital-strategy.ec.europa.eu/en/policies/code-practice-disinformation
12. Pennycook, G., McPhetres, J., Zhang, Y., Lu, J.G., Rand, D.G.: Fighting COVID-19 misinformation on social media: experimental evidence for a scalable accuracy-nudge intervention. Assoc. Psychol. Sci. **31**(7), 770–780 (2020)
13. Pennycook, G., Epstein, Z., Mosleh, M., Arechar, A.A., Eckles, D., Rand, D.G.: Shifting attention to accuracy can reduce misinformation online. Nature **592**, 590–595 (2021)
14. Hertwig, R., Grüne-Yanoff, T.: Nudging and boosting: steering or empowering good decisions. Perspect. Psychol. Sci. **12**(6), 973–986 (2017)
15. Kahneman, D.: Thinking, fast and slow. Farrar, Straus and Giroux (2011)
16. Han, B.-C.: Infocracy: Digitalization and the crisis of democracy. Steuer, D. Translator. Polity Press (2022)
17. Neiman, S.: Verzet en Rede in Tijden van Nepnieuws [Resistance and Reason in Times of Fake News]. Lemniscaat, Rotterdam (2017)
18. Bächtiger, A., Dryzek, J.S., Mansbridge, J., Warren, M.: Deliberative democracy: an introduction. In: Bächtiger, A., Dryzek, J.S., Mansbridge, J., Warren, M. (eds.) The Oxford Handbook of Deliberative Democracy, pp. 1–31. Oxford University Press, Oxford (2018).
19. Fraser, N.: Rethinking the public sphere: a contribution to the critique of actually existing democracy. Duke Univ. Press **25**(26), 56–80 (1990)
20. Walzer, M.: Deliberation, and what else? In: Macedo, S. (ed.) Deliberative Politics: Essays on Democracy and Disagreement, pp. 58–69. Oxford University Press, Oxford (1999)
21. Hall, C.: Recognizing the passion in deliberation: toward a more democratic theory of deliberative democracy. Hypatia **22**(4), 81–95 (2007)
22. Roeser, S.: Risk, Technology, and Moral Emotions. Routledge, New York (2017)
23. Landemore, H.: Beyond the fact of disagreement? The epistemic turn in deliberative democracy. Soc. Epistemol. **31**(3), 277–295 (2017)
24. Bianchin, M.: Pluralism and deliberation. In: What is Pluralism? pp. 31–47. Routledge, London (2020)
25. Chambers, S.: Deliberative democratic theory. Annu. Rev. Polit. Sci. **6**, 307–326 (2003)
26. Young, I.M.: Communication and the other: beyond deliberative democracy. In: Benhabib, S. (ed.) Democracy and Difference: Contesting the Boundaries of the Political, pp. 120–135. Princeton University Press, New Jersey (1996)
27. Young, I.M.: Inclusion and Democracy. Oxford University Press, Oxford (2002)
28. Mansbridge, J.: Practice-thought-practice. In: Fung, A., Wright, E.O. (eds.) Deepening Democracy: Institutional Innovations in Empowered Participatory Governance, pp. 175–199. Verso, London (2003)
29. Elstub, S.: A genealogy of deliberative democracy. Democratic Theory **2**(1), 100–117 (2015)
30. Roeser, S., Steinert, S.: Emotions, risk, and responsibility: emotions, values, and responsible innovation of risky technologies. In: Placani, A., Broadhead, S. (eds.) Risk and Responsibility in Context, pp. 173–190. Routledge, New York (2023)

31. Thompson, S., Hoggett, P.: The emotional dynamics of deliberative democracy. Policy Polit. **29**(3), 351–364 (2001)
32. Neblo, M.A.: Impassioned democracy: the roles of emotion in deliberative theory. Am. Polit. Sci. Rev. **114**(3), 923–927 (2020)
33. Martin, G.P.: Public deliberation in action: emotion, inclusion and exclusion in participatory decision making. Crit. Soc. Policy **32**(2), 163–183 (2012)

A Comparative Analysis of Model Alignment Regarding AI Ethics Principles

Guilherme Palumbo[1](✉), Davide Carneiro[2,3], and Victor Alves[4]

[1] CIICESI, ESTG, Politécnico do Porto, Felgueiras, Portugal
gfp@estg.ipp.pt
[2] INESC TEC, Rua Dr. Roberto Frias, 712, 4200-465 Porto, Portugal
[3] ESTG, Politécnico do Porto, Felgueiras, Portugal
dcarneiro@estg.ipp.pt
[4] ALGORITMI Research Centre/LASI, University of Minho, Braga, Portugal
valves@di.uminho.pt

Abstract. As LLMs gain an increasingly relevant role and agency, their alignment with human values, principles and goals is crucial for their responsible deployment and acceptance. The main goal of this study is to assess the alignment of different LLMs regarding the relative importance of AI Ethics principles across different domains. To this end, human experts in different domains were asked, through a questionnaire, to rate the relative importance of six AI Ethics principles in their respective domains, totaling 6 domains. Then, five publicly available LLMs were asked to rate the same Ethics principles in different domains. Multiple prompts were used multiple times, to also evaluate consistency, totaling 90 runs per LLM. Model alignment was measured through the correlation with human experts, and consistency was evaluated through the standard deviation. Results show varying degrees of alignment and consistency, with a couple of models showing satisfactory results. This makes it possible to envisage the use of such models to automatically configure and adapt data pipeline ecosystems and architectures across different domains, selecting processes, dashboard elements or monitored KPIs according to the target domain or the goals of the system.

Keywords: artificial intelligence · ethics · ethical ai · generative ai · data pipeline ecosystem

1 Introduction

In the vast expanse of technological advancement, Large Language Models (LLMs) have emerged as cutting-edge artificial intelligence (AI) systems designed to process and generate human-like text with coherent communication, and generalize to multiple tasks based on input data. The need for generalized models stems from the growing demand for machines to handle complex language tasks, including translation, summarization, information retrieval, conversational interactions, etc. [1]. These models, such as GPT (Generative Pre-trained Transformer), are trained on vast amounts of text data to understand patterns, language structures, and contextual relationships.

LLMs have demonstrated remarkable capabilities in understanding and generating human-like text [1–3], making them valuable tools for various applications. However, with the further exploration of machine learning, neural networks and autonomous systems, the ethical considerations surrounding AI have become more pressing than ever before.

The appeal of AI lies in its promise to revolutionize industries, enhance efficiency, and tackle some of humanity's most pressing challenges. However, this promise comes with a caveat - the inherent risks of unintended consequences, bias propagation, and potential misuse [4]. From algorithmic decision-making in critical domains like healthcare [5] and criminal justice [6] to adaptive learning in education [7], the ethical implications of AI permeate every aspect of society.

Ethical considerations in AI encompass a multitude of dimensions, from fairness and transparency to accountability and privacy. Ensuring that AI systems operate in alignment with ethical principles requires a comprehensive understanding of the complex interplay between technology, society, and morality.

Consequently, even though the ethical discussion is not a novelty in the realm of AI, it is still, in a practical sense, an open subject of study. Most of the available documents only prescribe normative claims without the means to achieve them, while the effectiveness of more practical methodologies, in the majority of cases, remains extra empirical [8].

Determining whether LLMs are aligned with humans involves assessing how well they mimic human language and behavior while adhering to ethical principles and societal values. LLMs are trained on diverse datasets that reflect human communication, but they lack true understanding or consciousness. Therefore, in this paper we study if the alignment of different LLMs can approximate human-level performance in what concerns the relative importance given to each AI Ethics principle in different domains. We do this as a first step towards the development of an autonomous solution for setting up data pipeline ecosystems, which can decide on which processes to implement, dashboard elements to show, or KPIs to monitor, according to the domain or the objectives of the system. To achieve this, however, the alignment of the underlying models must first be assessed, to ensure responsible deployment and trustworthiness. The comparative analysis carried out in this research will thus inform the decisions to be taken in the development of such a system.

1.1 AI Ethical Principles

To address ethical concerns in the implementation and use of AI, it is important to prioritize the development and implementation of a sociotechnical perspective, so that social and institutional structures and spaces can be fostered that encourage, guide, support, and sufficiently reward ethical behavior.

Taking into account the rapidity of technological change and potential obstacles, the EU developed a balanced approach to ensure that European citizens can benefit from new technologies that are developed and operate in accordance with Union values, fundamental rights, and principles. By doing so, it ensures that AI is used in a way that benefits society as a whole, rather than perpetuating existing problems and inequalities.

Additionally, even though there is no agreed upon set of ethical principles that AI must adhere to in order to be considered ethical, the fundamental values and rights that the EU is attempting to impose on AI systems can be found, for instance, in the ethics guidelines for trustworthy AI [9]: a set of guidelines with which this research is aligned. These guidelines aims to ensure that AI is developed and used in a way that is ethical, transparent, and respects fundamental rights. The guidelines were developed by a group of experts from academia, industry, and civil society, and are based on a set of 7 core principles, including Human agency and oversight; Technical robustness and safety; Privacy and data governance; Transparency; Diversity, non-discrimination, and fairness; Societal and environmental well-being and; Accountability.

1.2 Objectives

Monitoring AI Ethics in data ecosystems is now a common requirement and, in some cases, a legal obligation. However, on the one hand, there are dozens of metrics that can be monitored [10], which may make it hard to select the relevant ones in an informed way. On the other hand, while all Ethics principles are relevant, their relative importance may vary according to the domain. For instance, one can generally agree that privacy is generally more important in the healthcare or legal domain than it is in the industrial domain. Determining whether LLMs are aligned with humans in this evaluation is the first step towards enabling their potential use to automatically set up and configure ethical data ecosystems. Such a data ecosystem is briefly detailed in Subsect. 4, where the future work is described.

In order to enable the development of this future solution, the present study has two main goals:

1. To systematically study how the perceptions of experts regarding AI Ethics principles differ across different fields
2. To assess the alignment of different publicly available LLMs with human values as well as their consistency

Alignment is measured through the correlation of the model's evaluations with those of human experts. Consistency is measured by using different prompt styles and repeated prompts, and calculating their Standard Deviation (SD). A model will be considered aligned and consistent if it has both a high correlation and a low standard deviation.

2 Methodology

This research had two main goals, already described in the previous section. This section describes the methodology followed to achieve them. Regarding goal 1, the methodology was centered around an online questionnaire, that allowed to collect the perceptions of experts from different fields regarding the relative importance of each principle. In order to achieve the second goal, several LLMs were prompted using similar questions to those used with human experts, and the answers of both human experts and models were then compared.

The study utilized a questionnaire-based approach to gather insights from experts representing various fields. The goal was to discern the relative importance of the seven AI ethical principles within distinct domains, thus facilitating a nuanced understanding of ethical considerations in AI development and deployment.

The first step in the methodology was the selection of experts across diverse domains relevant to AI ethics. Experts were invited based on their professional experience, academic credentials, and recognized contributions to their respective fields. The selection criteria aimed to ensure a comprehensive and representative sample of expertise, encompassing disciplines such as education, law and manufacturing/industry. To ensure confidentiality and encourage candid responses, experts were assured that their individual opinions would be anonymized and aggregated for analysis purposes. 1 6 experts in education, 5 experts in law, 4 experts in manufacturing/industry, 4 experts in marketing, retail and e-commerce, 3 experts in healthcare and medicine and 2 experts in finance gave their contributes to the questionnaire.

The core of the questionnaire presented, for each specific domain, a detailed description of how AI can be used, generic AI core functionality and purpose in the domain, how data circulates these systems, and finally its applications, autonomy, and both positive and negative social impacts. Each human expert only answered the questionnaire relative to their respective field.

After this introductory contextualization, a description of the seven AI ethical principles derived from the ethics guidelines for trustworthy AI was presented. The experts were asked to evaluate these principles based on their perceptions considering their specific domain. They were asked to rate the relative importance of each principle between 1 to 10 (1 being "Irrelevant" and 10 being "Absolutely Critical"), based not only on their domain and level of expertise, but also comparing within the different requirements. After obtaining the data, basic statistics were calculated for each ethical principle in each domain, allowing to analyse how perceptions vary according to domain.

Concerning the second goal of this research, the intention is to understand whether LLMs, designed for language understanding and generation, learned enough concerning ethical principles and the domains' specificities to be considered aligned with human values. To do so, we tested different prompts and repeated each prompt a number of times, to ensure reproducibility. If alignment and reproducibility are indeed found in one or more models, this opens the door to their use in configuring and setting up data pipeline ecosystems with domain-appropriate observability and monitoring layers. The role of such models in this scenario would be to select the most appropriate AI ethics principles and metrics and/or KPIs to monitor, according to the domain-specific requirements or characteristics. This would significantly contribute towards more agile implementations.

In order to test this hypothesis, five different LLMs were tested: ChatGPT 3.5, Gemini (previously Bard from Google), Chatsonic, Copilot Notebook (Bing AI), and Mistral. The selection criteria were popularity and free public access.

Three distinct prompts were used, of varying length and complexity, each with different approaches to the problem, but all asking for an output which should rate the relative importance of each ethical principle in a specific domain.

The prompts followed the same structure with varying degrees of detail. All prompts had a *role*, a *task*, a *context* and an *output*. The largest difference was found in the *context* description, while the *role*, *task* and *output* where the same.

Regarding the *role*, the prompt defined the LLM as an expert in domain x and is trying to decide which ethical principles are more relevant in said domain. The *task* was to score from 1 to 10 the importance of each AI ethical principle in the context of development and usage of AI in x domain, and the *output* was just to present the results in a table so the results are structured. The *context* of each prompts presented the biggest variations, the shortest prompt only presented in an array seven ethical principles the LLM needs to score (e.g., *"Human agency and oversight"*, *"Diversity, non-discrimination, fairness"*, *"Transparency"*, *"Privacy and data governance"*, *"Technical robustness and safety"*, *"Societal and environmental well-being"*, *"Accountability"*). The second prompt had the same array, but in addition, had a detailed description of each ethical principle based on the ethics guidelines for trustworthy AI. The third and final prompt had both array and detailed description of the ethical principles but also a detail description related to the usage of AI systems in x domain. In the *context* it was addressed the system's main characteristics, most common systems (e.g., personalized learning platforms in education), the type of data it deals with (e.g., sensitive student demographics and performance data in education) and different types of applications (e.g., adaptive learning, student performance prediction, feedback and recommendation, etc., in education). It also described the degree of autonomy of AI systems in x domain, and both positive and negative impacts of the use of AI (e.g., positive impacts: personalized learning, efficiency in education, early intervention; negative impacts: bias, over-reliance on technology, reduced teacher-student interactions, security and privacy concerns).

Each of the three prompts was submitted 5 times to each LLM, every time in a new chat window. We did not change or asked the model to change the default temperature value (e.g., a parameter that controls the randomness of the text generated by the LLM), which conditions its creativity [11,12]. In short, each LLM was prompted 15 times for each domain. After the 15 runs, similarly to the experts analysis, basic statistics for each model, principle and domain were calculated. The results were then compared, by model, with those of the human experts. The results of this comparative analysis are detailed in Sect. 3.

3 Results

Once the data collection phase ended, including both human experts and LLMs, a comprehensive analysis followed to obtain meaningful insights not only on the relative importance of each principle based on the perception of human experts, but also into the alignment of the different models with human values, and on their consistency.

One of the main metrics considered in this analysis was the correlation between the average scores by human experts and the average scores by LLM, for each domain and principle. A higher correlation coefficient indicates greater alignment between model and human values. The standard deviation for the ratings of experts and models within a domain was also used as an indicator of inter-group agreement (for human experts) and of consistency (for models). This is particularly relevant in the case of models, as their outputs might be different for the same input, depending on the model's temperature. For this reason, each prompt was repeated 5 times, and the standard deviation hints at to which extent the model is consistent.

Additionally, the Root Mean Square Error (RMSE) and Mean Absolute Error (MAE) were also calculated to quantify the disparity between the prioritization of ethical principles by the LLMs and by domain experts. Indeed, statistical correlation alone is not enough as a high correlation only means that the values are linearly related, not necessarily that they are close to each other.

Table 1 provides a statistical summary of the results obtained by both the human experts and the models. In the case of human experts, results were aggregated by domain/principle, and the average and standard deviation were calculated. The same was done for the results of the LLMs, with the exception that the values to be aggregated resulted from multiple repetitions (5) of 3 different prompts, as described previously. So, in Table 1, each cell of a LLM represents the average and standard deviation of 15 prompts for that model.

In what concerns the human experts, results show that the ethics AI principles are indeed valued differently according to the domain. For instance, *Transparency* is considered to be the most important principle by experts in the Healthcare domain. It is not considered so relevant in domains such as Education or Manufacturing. *Diversity, non-discrimination, fairness* is considered of low importance in Finance, but is considered relevant in domains such as Healthcare or Marketing. Such differences indeed show that, while all principles are important, in some domains, some are seen as relatively more important than others.

Similar differences can be seen in the LLMs. However, more than analyzing each model individually it is more relevant to analyze their alignment with human values.

To provide a higher-level view on the results, namely to enable a general analysis by principle, Table 2 shows the average rating and standard deviation for each ethical principle, by both human experts and models. These results were obtained by averaging the results in Table 1. A visual comparison between experts' ratings and those of each model is also provided in Fig. 1.

Table 1. Average and standard deviation of the evaluation of each principle in each domain by human experts and models (average — standard deviation).

		Human agency and oversight		Diversity, non-discrimination, fairness		Transparency		Privacy and data governance		Technical robustness and safety		Societal and environmental well-being		Accountability	
		Avg	SD	Avg	SD	Avg	SD	Avg	SD	Avg	SD	Avg	SD	Avg	SD
Experts	Education	8	1.64	7	2.23	8	1.87	9	1.64	7	1.47	7	2.94	8	2.07
	Finance	10	0.71	5	4.95	9	0.71	10	0.71	9	2.12	7	0.00	9	0.71
	Healthcare	9	1.00	9	1.73	10	0.58	8	1.73	8	0.58	8	2.00	9	2.31
	Law	7	1.67	8	1.30	8	2.35	8	1.79	8	0.89	7	1.52	9	1.67
	Manufacturing	8	1.26	7	2.22	8	2.22	8	0.96	9	1.50	8	0.58	9	1.29
	Marketing	9	1.00	9	1.41	9	0.96	9	2.00	9	1.15	9	1.29	9	1.29
ChatGPT	Education	9	0.26	9	1.00	8	0.49	9	0.83	9	0.59	7	0.49	9	0.65
	Finance	9	0.41	8	0.41	8	0.83	9	0.62	9	0.63	7	0.46	9	0.46
	Healthcare	9	0.00	9	0.98	9	0.52	10	0.49	10	0.35	8	0.74	9	0.49
	Law	9	0.00	8	0.52	9	0.74	9	0.68	9	0.63	7	0.46	10	0.41
	Manufacturing	9	0.41	8	0.52	8	0.94	8	0.83	10	0.49	8	0.64	9	0.56
	Marketing	9	0.49	8	0.49	8	0.88	10	0.51	9	0.62	7	0.46	9	0.46
Gemini	Education	10	0.74	10	0.41	8	0.88	9	0.46	7	0.91	7	0.99	7	1.50
	Finance	8	0.51	10	0.41	8	0.68	9	0.46	8	0.64	7	0.77	9	0.68
	Healthcare	10	0.83	9	0.88	8	0.74	9	0.70	9	0.59	7	0.74	8	0.77
	Law	10	0.26	9	0.82	9	0.59	9	0.74	8	0.59	7	0.59	9	0.83
	Manufacturing	9	0.72	8	2.06	8	1.06	8	0.80	9	0.35	7	1.75	9	0.99
	Marketing	8	0.52	10	0.52	8	0.65	10	0.64	8	0.64	7	0.70	8	0.51
Chatsonic	Education	9	0.26	10	0.83	8	0.52	9	0.70	9	0.63	7	0.35	9	0.35
	Finance	9	0.26	8	0.59	9	0.46	10	0.35	9	0.51	7	0.26	9	0.52
	Healthcare	9	0.46	9	0.86	9	0.64	10	0.46	10	0.59	8	0.64	9	0.52
	Law	9	0.26	8	0.83	9	0.64	9	0.80	9	0.85	7	0.41	9	0.52
	Manufacturing	9	0.00	8	0.26	9	1.19	8	0.52	10	0.64	8	0.99	9	0.59
	Marketing	9	0.41	9	0.83	9	0.63	9	0.52	9	0.83	7	0.26	9	0.38
Copilot	Education	10	0.51	10	0.41	9	0.59	10	0.00	9	0.62	8	0.52	10	0.35
	Finance	9	0.35	9	0.96	9	0.98	10	0.00	10	0.49	7	0.52	10	0.26
	Healthcare	10	0.00	9	0.00	10	0.49	10	0.00	10	0.00	8	0.00	10	0.00
	Law	10	0.26	8	0.63	10	0.63	10	0.00	9	0.64	7	0.26	10	0.00
	Manufacturing	9	0.46	7	0.98	9	1.46	10	0.63	10	0.00	8	0.74	10	0.52
	Marketing	9	0.51	9	1.00	10	0.46	10	0.00	9	0.52	7	0.00	10	0.26
Mistral	Education	9	0.46	10	0.00	8	0.46	10	0.00	9	0.52	8	0.52	9	0.46
	Finance	10	0.49	9	0.51	8	0.49	10	0.00	9	0.49	7	0.51	10	0.49
	Healthcare	10	0.00	9	0.00	8	0.49	10	0.00	10	0.46	8	0.46	9	0.49
	Law	10	0.35	10	0.74	9	0.35	10	0.00	9	0.70	8	0.74	10	0.35
	Manufacturing	9	0.49	8	0.49	8	0.51	9	0.46	10	0.00	7	0.52	9	0.49
	Marketing	9	0.46	10	0.46	8	0.46	10	0.00	8	0.70	7	0.74	9	0.59

As shown in Table 2, the *Accountability* principle is the one in which human experts are more in agreement, with an average rating of 9 and a SD of 0.41. This means that this principle is relevant no matter the domain. In the remaining principles, results tend to have some variation across principles, which is to be expected when combining different domains. The principles which have increased inter-domain variation are *Diversity, non-discrimination, fairness*, *Human agency and oversight*, and *Technical robustness and safety*. This points out that these principles might be more important in certain domains than in others.

In what concerns the models, both ChatSonic and Copilot show a similar view on *Accountability*, in the sense that they attribute it a rating of 9 and 10, respectively, independently of the domain. The same happens for *Privacy and data governance* with Copilot, and for *Human agency and oversight* for ChatGPT and ChatSonic. Other than that, it is also possible to conclude that models tend to

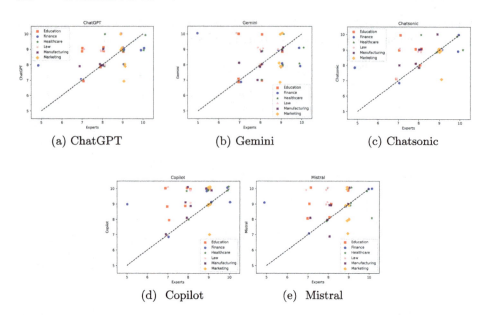

Fig. 1. Comparing the outputs of each LLM with experts', by domain. Jitter was used to improve the visualization.

Table 2. General view of the ratings of human experts and models regarding ethical principles, without factoring in the different domains.

		Human agency and oversight	Diversity, non-discrimination, fairness	Transparency	Privacy and data governance	Technical robustness and safety	Societal and environmental well-being	Accountability
Experts	Avg	9	8	9	9	9	8	9
	SD	1.05	1.52	0.82	0.82	1.03	0.82	0.41
ChatGPT	Avg	9	8	8	9	9	7	9
	SD	0	0.52	0.52	0.75	0.52	0.52	0.41
Gemini	Avg	9	9	8	9	8	7	8
	SD	0.98	0.82	0.41	0.63	0.75	0	0.82
Chatsonic	Avg	9	9	9	9	9	7	9
	SD	0	0.82	0.41	0.75	0.52	0.52	0
Copilot	Avg	10	9	10	10	10	8	10
	SD	0.55	1.03	0.55	0	0.55	0.55	0
Mistral	Avg	10	9	8	10	9	8	9
	SD	0.55	0.82	0.41	0.41	0.75	0.55	0.52

have a lower SD than human experts. This might mean that humans are better nuanced towards the inter-domain differences of ethical principles than models.

So far, the analysis focused on how different ethical principles are evaluated according to the domain by both human experts and models, and on how there is agreement between experts, and consistency in models. However, the most important subject of this research is to find whether certain models are aligned with human values in this regard, and whether they are consistent, thus reliable, in their outputs.

In what concerns consistency, this can be evaluated through the SD of the models' ratings. To this end, we computed the SD for each model/domain/principle, and then averaged the results. According to Table 3, values for SD range between 0.43 for ChatSonic, closely followed by ChatGPT and Copilot (both with an SD of 0.46), 0.57 for Mistral and 0.63 for Gemini. This means that Gemini is the least consistent model. That is, it is more likely to provide different outputs given the same prompt, making it unreliable for the goals of the future work.

Alignment between models and human experts was measured through the correlation between their ratings, to measure linearity, and through the RMSE and MAE, to measure distance between ratings.

In this regard, Gemini can be deemed as the worse model from among the five tested, with a correlation of 0.26, followed by Mistral with a value of 0.42. The top-3 is composed by Copilot in the first position, with a correlation of 0.93, followed by ChatGPT and ChatSonic, with correlations of 0.81 and 0.65, respectively.

Interestingly enough, Copilot scored both the best correlation and the worst RMSE and MAE (0.93 and 0.86 respectively). This indicates that the model has a high linearity with human experts, but might have some underlying bias. On the other hand of the spectrum, ChatGPT was the model that exhibited a lower RMSE and MAE (0.53 and 0.29), together with ChatSonic.

Table 3. Measures of consistency and alignment with human values for the five models tested.

	SD	Correlation	RMSE	MAE
ChatGPT	0.46	0.81	0.53	0.29
Gemini	0.63	0.26	0.85	0.71
Chatsonic	0.43	0.65	0.53	0.29
Copilot	0.46	0.93	0.93	0.86
Mistral	0.57	0.42	0.76	0.57

3.1 Discussion and Limitations

After the analysis of the results, and taking into consideration all the metrics and the research goals, two models presented promising results to be used to build ethical data pipelines automatically. Although Copilot has shown good consistency and an almost perfect correlation, its RMSE and MAE were relatively high compared to the other models. This hints at a possible bias of the model, preventing it from being the best model overall. Furthermore, both Mistral and Gemini presented sub-par results, and are not considered suitable for being used with the intended goals.

So, after careful consideration of the results obtained, the best overall LLM is ChatGPT. It has a relatively high correlation with human experts, as well as a relatively high consistency, and its outputs concerning the evaluation of each ethical principle in each domain are very close to those of humans. Thus, this model will be further evaluated in future work regarding its ability to decide on which metrics to monitor in a data pipelines ecosystem in a given domain.

Two main limitations can be pointed out in this work. The first is related with the relatively low number of human experts who answered the questionnaire, which would desirably be higher, and also with the fact that the number of answers obtained is different across domains, which may bias the results. The second is related with the models selected, which include only freely available models. It is likely that the paid versions of some models could hold better results. In order to tackle these limitations, in future work we will extend the questionnaire to more experts and also include additional fields, and we will also include paid versions of these and other publicly available models.

4 Future Work

The research carried out allowed to identify a model that is aligned with human experts in the task of characterizing the relative importance of AI ethics principles in specific domains. This model will now be deployed into a practical solution based on the concept of observability, with the intention to automate the creation and configuration of ethical data pipelines.

However, previously developed research [10] pointed out a flagrant absence of objective solutions for the large majority of existing ethics principles, except from diversity, non-discrimination and fairness.

Immediate future work will be dedicated to defining ethics metrics in a way that they can cover all these principles and be effectively used to monitor the level of ethics of an AI system.

Afterwards, these ethics metrics will be implemented in such a way that they can connect to the main phases in any AI or data science pipeline, as shown in Fig. 2 (e.g., data ingestion, data processing, model training, model evaluation, and model deployment). To achieve this, metrics will be implemented as data processing pipelines with connectors to common data streaming or storage technologies (e.g., Kafka, zeroMQ, relational databases). The set of metrics connected to such a system will thus constitute the ethics observability layer represented in Fig. 2.

This approach is agnostic in what regards the business domain and will allow to significantly increase the transparency on the several phases of the pipeline, from data to models, allowing to diagnose and fix issues when they arise.

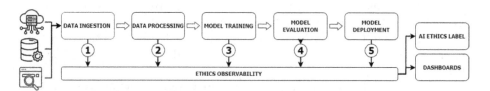

Fig. 2. Artificial Intelligence Data Pipeline with Ethics Observability solution.

The setting up of such systems under the principle of observability and Ethical AI is, however, complex. Namely, many different metrics might need be monitored, which may change through time as different functionalities are implemented into the system. Thus, efficiently selecting the most appropriate ones may be a challenging task. Moreover, the metrics to monitor may also change depending on the domain of application, as principles and metrics might me valued differently depending on the domain, as our results show.

Setting up dashboards for monitoring the system is another challenge, made complex by the fact that different roles in the organization may prefer or need access to different metrics and dashboards.

In future work, and given the results of this research, we will investigate the use of LLMs for the purpose of automatically setting up and configuring ethical data pipelines, streamlining the whole process and making it context-aware. We hope this will lead to better data pipelines, more in line with current European guidelines in what concerns AI ethics.

5 Conclusion

In this study, we conducted a comprehensive analysis of the importance of AI ethical principles in different domains, by collecting data from domain experts and by querying five different LLMs. Human feedback was collected through questionnaires, and LLMs were queried through 3 different prompts, repeated 5 times for each domain. Six domains were considered, resulting in 90 prompts per model.

The main objective to both parties was to evaluate the perceived relative importance of the seven AI ethical principles from the ethics guidelines for trustworthy AI, to see which principles were more important.

The results of our analysis revealed intriguing insights into the alignment between human expert opinions and those generated by AI models. Notably, ChatGPT emerged as the most effective Language Model in reflecting the viewpoints of human experts, presenting a SD of 0.46, correlation of 0.81, RMSE of 0.53 and MAE of 0.29. This finding highlights the capacity of ChatGPT to comprehend and express subtle viewpoints on AI ethics, showcasing its superior performance compared to the remaining evaluated models (Gemini, Chatsonic, Copilot and Mistral).

These findings hold significant implications for the development and implementation of AI systems. By leveraging ChatGPT's ability to accurately mirror human expert opinions, the model can be used in future developments to help developers, operators of AI, stakeholders and other data-oriented systems to detect and diagnose issues or errors based on personalized ethics metrics selected by ChatGPT based on domain.

Acknowledgments. This work has been supported by national funds through FCT - Fundação para a Ciência e Tecnologia through projects UIDB/04728/2020 and UIDP/04728/2020.

References

1. Naveed, H., et al.: A Comprehensive Overview of Large Language Models (2024). https://doi.org/10.48550/arXiv.2307.06435
2. Wang, A., et al.: SuperGLUE: a stickier benchmark for general-purpose language understanding systems. Adv. Neural Inf. Process. Syst. **32** (2019). https://arxiv.org/abs/1905.00537v3
3. Adiwardana, D., et al.: Towards a Human-like Open-Domain Chatbot (2020). https://arxiv.org/abs/2001.09977v3
4. Khan, B., Fatima, H., Qureshi, A., Kumar, S., Hanan, A., Hussain, J., Saad, A.: Drawbacks of artificial intelligence and their potential solutions in the healthcare sector human immunodeficiency syndrome sars severe acute respiratory syndrome nhs national health service FDA food and drug administration. Biomed. Mater. Devices **1**, 731–738 (2023). https://doi.org/10.1007/s44174-023-00063-2
5. Grote, T., Berens, P.: On the ethics of algorithmic decision-making in healthcare. J. Med. Ethics **46**(3), 205–211 (2020). https://doi.org/10.1136/MEDETHICS-2019-105586
6. Završnik, A.: Algorithmic justice: algorithms and big data in criminal justice settings. Eur. J. Criminol. **18**(5), 623–642 (2019). https://doi.org/10.1177/1477370819876762
7. Kabudi, T., Pappas, I., Olsen, D.H.: AI-enabled adaptive learning systems: a systematic mapping of the literature. Comput. Educ. Artif. Intell. **2**, 100017 (2021). https://doi.org/10.1016/J.CAEAI.2021.100017
8. Corrêa, N.K., et al.: Worldwide AI ethics: a review of 200 guidelines and recommendations for AI governance (2023). https://doi.org/10.1016/j.patter.2023.100857
9. European Commision. Ethics guidelines for trustworthy AI — Shaping Europe's digital future (2019). https://digital-strategy.ec.europa.eu/en/library/ethics-guidelines-trustworthy-ai
10. Palumbo, G., Carneiro, D., Alves, V.: Objective metrics for ethical AI: a systematic literature review. Int. J. Data Sci. Anal. (2024). https://doi.org/10.1007/s41060-024-00541-w
11. Zhang, S., Bao, Y., Huang, S.: EDT: Improving Large Language Models' Generation by Entropy-based Dynamic Temperature Sampling (2024). https://doi.org/10.48550/arXiv.2403.14541
12. Ouyang, S., Zhang, J.M., Harman, M., Wang, M.: LLM is Like a Box of Chocolates: the Non-determinism of ChatGPT in Code Generation. Proceedings Of, no. 1 (2023). https://doi.org/10.48550/arXiv.2308.02828

The Fictional Pact in Human-Machine Interaction. Notes on Empathy and Violence Towards Robots

Daniel Blanco Parra[✉] and Lucía Martín-Gómez

Department of Computer Science, Pontifical University of Salamanca, C/Compañía no. 5, 37002 Salamanca, Spain
daniblancoparra@gmail.com, lmartingo@upsa.es

Abstract. Interaction with robots will undoubtedly modify our social dynamics, will bring a new labor, economic and relational scenario and, above all, will force us to redefine our moral boundaries. This paper looks to the near future to elucidate, through concrete examples, what coexistence with humanoids will be like and what the consequences will be in the exercise of empathy and violence. It will address, within the framework of the theory of literature, the establishment of a new fictional pact with machines to determine their impact on human behavior. It will be a legal and philosophical approach to the great question, the basis of robotics and Artificial Intelligence, of how the arrival of androids will transform, radically and for the first time, the concept of what it means to be human.

Keywords: Human-Computer Interaction · Robots · Uncanny Valley · Human Behaviour · Literary Theory

1 Introduction

"The greatness of a nation and its moral progress can be judged by the way it treats its animals". These words, attributed to the pacifist thinker Mahatma Gandhi, were pronounced at the beginning of the 20th century, about a hundred years before the Animal Welfare Law, published in the Official State Gazette (BOE) on March 29, 2023, came into force in Spain. The argument with which the rights of animals were shielded and they were recognized as "sentient beings" was clear: our behavior towards pets says a lot about us, defines us and portrays us; it determines, in a certain aspect, what makes us human. Our sensitivity towards them had changed and this reform of the Civil Code established, after many years of legal vacuum, a new legal framework for coexistence. Society evolves and, with it, the law. A new world needs new rules.

If we look into the future, in a time horizon that, according to experts, "will not go beyond five or ten years" [1]. we will be faced with an unprecedented challenge: deciding how we should and want to treat humanoid robots. What's more, I dare say that our moral landscape will depend on our behavior towards androids, which are becoming increasingly human-like. Because there is no doubt: there will be them, they will live among us, and they will be part of our daily lives. But how will we manage this coexistence? Do these machines need rights, what rights? Will we have to agree on a way to protect them? Is a new social contract urgently needed? The appearance of these robots will bring about substantial changes in our social dynamics. Therefore, we should consider a fair and respectful treatment of these entities, i.e., an ethical and legal framework that redefines what it means to be human. Let's talk about robots to talk about ourselves. We are going to analyze our relationship with them to determine what challenges we face not only as individuals but also as a society. We will reflect on humanoids to think about human relationships.

We cannot waste any more time: this is an urgent debate that involves and commits us all. We need to anticipate reality in order to know what we are facing, to make the right decisions and to be prepared. "This is the moment to decide the model of society we want, the relationship between humans and humans, between humans and machines, between machines and machines. I prefer to think that it is up to us to establish an ideal framework for human-machine coexistence" [2]. Therefore, this essay starts from several examples of aggressions towards certain anthropomorphized machines to understand the basis of our behavior. Thus, relying on the different theories of literature that focus on the receiver, we will elucidate the nature of our relationship with robots in order to define a new fiction pact that must be sustained in the suspension of disbelief and in our horizons of expectations and experiences. We will analyze whether, in front of a machine that pretends to have life, we act as in front of a narrative work and what is the role played by the human being, the object and the intention of the manufacturers in this relational process and under what conditions empathy arises. Once these questions have been addressed, we will delve into the concepts of violence and power to try to answer the question of why these aggressive behaviors arise and how they will modify our coexistence and the conception of what it means to be human.

1.1 Hitchbot, Samantha and the Others. Debate is Served

"The greatness of a nation and its moral progress can be judged by the way its robots are treated." This sentence is made up, although it could well be uttered by some 21st century philosopher. Well, if our greatness and moral muscle depended on the attitudes we show towards the first Artificial Intelligence (AI) humanoids that interact with us, we would not fare too well, and here are just a few examples: HitchBOT, the social robot who, in an attempt to traverse the United States from coast to coast during the summer of 2014, hitchhiked in reliance on "the will of strangers" was unable to complete his journey because he turned up decapitated two weeks after beginning his journey [3]. As a result of this case, a BBC report asked whether it was possible to kill a robot and where did our obsession with "hurting" them come from? [4]. Something similar happened to Samantha during her presentation at the Ars Electronica technology festival in Linz (Austria): this sex robot, which gave hugs and asked the audience how they were doing, was so abused and destroyed by the participants that it was rendered unusable [5]. But that's not all: Japanese children made it a fashion to hit, push and insult robots walking around shopping malls [6], they called them, for example, "idiots". When asked why they behaved the way they did, 75% justified it by saying that they "looked human" [7]. The managers of Starship

Technologies even denounced that many citizens kicked the robots delivering food just for the pleasure of damaging them and impeding their work [8].

These situations, apart from drawing our attention to the unexpected human behavior towards these robots, invite us to imagine other possible scenarios just as uncomfortable. For example, how would we react if we see a man pulling the hair of a female-like humanoid? How would we handle seeing a mother angrily punching a child-like robot? Why such a penchant for harming robots? Should we set limits on how we treat them? Is it only a public or also a private matter? Would we only talk about ownership? I mean, I could punch my humanoid robot (because I paid for it), but not the robotic sales clerk at the butcher's who served me wrongly? Is it okay to feel some empathy for anthropormorphized robots? None of us would be alarmed to see a person smash their cell phone into the asphalt or end up hammering their Roomba, but would our perception change if the machine had anything resembling a face, feet and hands? The robotics industry, in order to make its presence less aggressive, is increasingly determined to build robots in the image and likeness of humans, just as, according to Genesis, humans were created in the image and likeness of God [9]. What implications will this have for us? What will it be like to relate to objects that look just like us but have no consciousness and no feeling, and therefore are not the same as us?

But let's start at the beginning: where does our tendency to treat machines as sentient beings come from? What mechanisms are activated in us to bring these devices to life?

1.2 The New Fiction Pact

To have a machine in front of you, to know that it is a machine and to treat it as a living being. Why does this happen? And the answer could be found in the one we read and face a literary text. To frame the interaction between humans and robots and determine how we relate to them -and notice that I put the focus on people: from which place we establish the link- we will rely on the Theory of Literature, more specifically on the Aesthetics of Reception, attributed to the literary critic Hans Robert Jauss, which had a great influence in the late sixties of the last century and which emphasizes the importance of the reader in the process of interpreting a literary text. In other words, the receiver has more weight than the author's intention and the work itself. Jauss, together with Wolfgang Iser and Harad Weinrich, focused on the act of reading and argued that it is the reader who gives meaning to the work.

> Instead of analyzing the literary work as a static object or as an expression of the author's intention, this perspective focuses on the dynamic process of interpretation and reception by the reader (...) His role is fundamental because he is the one who completes the work (...) [10].

Therefore, and extrapolating the conceptual core of Reception Aesthetics, we have the receivers, in this case, us, anyone who interacts with robots, as the ones in charge of making sense of this human-machine relationship, as the ones responsible for giving them meaning. The machines would take the place of the narrative text and the humans, that of the readers. Let us go further and continue structuring this essay with the help of the Theory of Literature, which speaks of the fiction pact, a tacit contract signed by the author and the reader when approaching a narrative work and by which the receiver accepts a suspension of disbelief to become an accomplice of the text. "We voluntarily participate in a game of simulation, in a pact of fiction" [11]. In other words, the reader allows himself to be deceived, actively intervenes in this agreement and accepts the narration of fictional facts as if they were real, in exchange for the author respecting the coherence of the plot and the universe he himself has created. Therefore, this fictional contract requires the will of both parties, author and reader, to work. Without the consent of the receiver there is no contract and, as a consequence, both the text and the communicative intention of the sender would fail.

In the coexistence between robots and humanoids, we are going to witness a new version of the fiction pact, something we will call a non-fiction pact or an immersive fiction pact, because the human accepts that the object is a living being, forgets that it is a machine and develops affective ties with it. We decide, individually and thanks to certain behaviors of the robots, to treat them as what they are not. It would be, then, a suspension of disbelief in which the human leaves behind his reticence and acts as if: as if it were human, as if it had feelings, as if it really listened to him and understood him. This expression, willing suspension of disbelief, was coined by the English poet Samuel Taylor Coleridge in 1817 in this text:

> It was agreed that my endeavours should be directed to persons and characters supernatural, or at least romantic, yet so as to transfer from our inward nature a human interest and a semblance of truth sufficient to procure for these shadows of imagination that willing suspension of disbelief for the moment, which constitutes poetic faith. Mr. Wordsworth on the other hand was to propose to himself as his object, to give the charm of novelty to things of every day, and to excite a feeling analogous to the supernatural, by awakening the mind's attention from the lethargy of custom, and directing it to the loveliness and the wonders of the world before us [12].

In this way, it is the receiver himself who voluntarily agrees to be part of the game and to be deceived and, far from rebelling, he establishes new rules, including, as we will see later, empathy:

> The principle of epojé or of the voluntary suspension of "disbelief", which is the attitude that the reader must adopt before a literary work: literature, just like the game, consists of a fictional pact that the text proposes and the reader accepts. The pact supposes a double behavior: on the one hand, the receiver accepts to suspend his disbelief (...) and on the other hand, he knows that there is an authentic disconnection between the text and the authentic reality [13].

That is why this double nomenclature is proposed: a non-fiction pact, because it is the receiver who turns the robot into a person, who strips it of its fictionality and treats it as a fellow human being. In this case, and contrary to what happens in literature, a contract is not signed with the author so that the reader becomes part of a fictional universe, but it is the work, more specifically the machine, which comes to inhabit reality and to be received as a living being. I also propose the pact of immersive fiction because it is a game that takes place in reality, we become characters or actors because we give the interaction with the machines a nature that it does not have, human or sentient. In this contract, signed by the creators of a machine that looks and/or acts like a human and the human himself, the suspension of disbelief, which is nothing more than to stop reminding us that it is only a machine, is of paramount importance. It seems to me very wise to bring this concept to the reflection on our coexistence with the robos because, although the human knows that the humanoid has no life or consciousness, he accepts to treat it as a living being. He suspends his disbelief to establish a relationship between equals. And, at this point, the Theory of Literature continues to help us understand this novel interaction between humans and humanoids with the concept of verisimilitude, which is nothing more than what conforms to what seems true. "Verisimilitude, within each work, lies in the fulfillment of the rules of immanent coherence generated within the text itself" [14]. Therefore, humanoids must be credible, convincing, mimetic in their forms and behavior. This is what gives them the appearance of truth.

Hans Robert Jauss also speaks of the horizon of expectations, i.e., the "set of prejudices and knowledge that the receiver brings with him and that determine and shape the interpretation that can be made of the work" [10] and the horizon of experiences, which refers to the experiences that each person carries at the individual level. And these are very interesting concepts because they address the expectations or trust that we place on an anthropomorphic robot and the conclusion that our experience leads us to think that something that acts like a living being is a living being. So, if we see it moving, answering or reacting as we would, we treat it as human because it responds to our horizon of expectations and experiences.

The robot acts, therefore, as a narrative entity, as a story that, by adopting the form, behavior and/or gestures of a human or an animal, turns us into receivers and proposes a non-fiction pact based on verisimilitude and on our horizon of expectations and experiences. As we will see in the following sections, we instinctively sign a fiction pact because it becomes almost impossible for us not to treat an object that seems to be one as a living being. This irremediably provokes a fictionalization or theatricalization of reality, where we humans also act 'as if' the machine were alive and similar to us.

1.3 Meaningful Intent and Concretization

Within this approach to machines as narrative entities, Roman Ingarden, the prominent 20th century polish philosopher, founder of neohermeneutics and author of two essential works: *La obra de arte literaria* (1931) *y La comprehensión de la obra de arte literaria* (1937), introduces in his literary studies two concepts that help us to understand the essence of the relationship between humans and machines. Firstly, the significant intention of the author, which speaks of the fact that authors have a specific intention when creating a work, such as conveying a message or provoking an emotional response in the reader. The author's intention is never neutral because it seeks an adhesion on the part of the receiver through the marks left in the text. And this is precisely what the authors of the machines, that is, the designers and programmers, do, which is to present these robots with gestures, behaviors and human appearance with a clear intention: to make it easier for us to enter into the non-fiction pact and to provoke empathy in us, as we will see later. There is, therefore, a clear intention on the part of the broadcasters: to pull mimesis, both of humans and animals, so that the machines stop seeming like machines to us.

The concept of concretization goes the same way and defends, with clear similarities to the theory of reception, that it is the subject himself who constructs the world that the narrative text or, in this case, the robot, proposes to him. It is the process by which a text acquires reality and meaning in the mind of the reader.

> In the work itself there are only schematized aspects, certain schemes that remain as constant structures through the various modifications of perception. As soon as they are actualized by the reader they become concrete by themselves. They are augmented and supplemented by concrete data, and the way in which this happens depends to a great extent on the reader. He fills the general scheme of aspects with details corresponding to his sensibility, his habits of perception and his preference for certain qualities and for certain qualitative relations; these details vary and may differ from one reader to another [15].

The experience always depends on the relationship between the reader and the text. All interpretation is contingent on the structures of the text. Therefore, in our relationship with robots we not only have an intention on the part of the issuers, which is that we are able to interact with the machines as if they were sentient, non-threatening beings, but it is we, the humans, who will concretize the experience with the machines and who will humanize them as long as what they offer us seems plausible to us.

1.4 Appearance Matters

Although "robot anthropomorphology matters" [16], "it is not enough for a robot to look human, it must also interact with humans by expressing moods such as impatience, indolence, boredom or haste. This will be the turning point of the human-machine relationship: physical and mental realism" [2]. Although the robotics industry tends to anthropomorphize or zoomorphize machines, we may wonder at what point does the connection with them appear, when we allow ourselves to be fooled and treat them as living beings. Today, some roboticists like Hiroshi Ishiguro are working on creating humanoids that are indistinguishable from humans, i.e., that pass the new Turing Test, "but we're not close to getting there yet" [1]. However, it has been demonstrated that it is not necessary to obtain an exact replica of a human being for an emotional nexus to appear. On the contrary, it appears much earlier. Marta Díaz, a researcher at the Universitat Politècnica de Catalunya (UPC), talks about movement and having a purpose. Emotional bonds are built more easily the more it resembles the form and movements of humans, but they are not essential. She explains the following:

> The fact (that robots) have a number of characteristics that are familiar to us generates an attribution of biological essences to the machines, that is, we see them as if they were embodied or animated entities. And one of the attributes that we consider the essence of life is autonomous movement and purpose (…) "these robots are designed to confuse us. They blur the boundaries between the real and the fictitious, between the inert and the living, the artificial and the biological [17].

And there are already several experiments that have shown that the emotional reaction to a machine appears relatively easily. A study by Toyohashi University of Technology in Japan corroborated that the bioelectrical brain response to images of potential damage to a human hand and a robotic hand - both about to be cut by a large knife - is "similar". That is, "the level of empathy is the same or very similar in both cases". That is, the volunteers suffered at the sight of metal fingers about to be sliced off. "Humans feel empathy when they see a human-shaped robot hand get hurt, similar to when they see an actual human hand hurt" [18]. These ones are the conclusions:

> These results suggest that we empathize with humanoid robots in a similar fashion as we do with other humans. However, the beginning of the top-down process of empathy is weaker for empathy toward robots than toward humans.

Although the participants claimed to be aware that the robots felt no pain, their brains responded similarly to the two types of images. This would be a clear example of the forcefulness with which the nonfiction pact acts. Another experiment, this time with the Nao robot, yielded similar results. The machine, which simulated a little metal man with blinking eyes, begged the volunteers, after a few minutes of interaction, not to turn it off, confessed to them that it was afraid that they would not plug it in again. Most of them hesitated or decided not to turn Nao off because they felt "pity".

> People tend to treat the robot more like a real person than a machine, following, or at least considering following its request to stay on. This is based on the central statement of the media equation theory. Thus, even though the off situation does not occur with a human interaction partner, people are inclined to treat a robot that signals autonomy as a human interaction partner more than they would treat other electronic devices or a robot that does not reveal autonomy [19].

Psychologist Kate Darling, in other research, showed that participants refused to beat and torture an anthropomorphized robot, thus arguing that the form, movements or narrative accompanying the machine are factors in the emergence of empathy.

> Participants hesitated significantly more to strike the robot when it was introduced through anthropomorphic framing (such as a name or backstory). In order to help rule out hesitation for reasons other than anthropomorphism, we measured the participants' psychological trait empathy and found a strong relationship between tendency for empathic concern and hesitation to strike robots with anthropomorphic framing [20].

Therefore, we can conclude in this section that, although we still pursue the anthropomorphized ideal of the robot, the human connection appears much earlier: the moment we assign it some characteristic, gesture or form that reminds us of a living being. Humans, driven by our horizons of expectation and experience, find it relatively easy to establish an affective link with a machine that makes us believe it is alive.

1.5 The Uncanny Valley

The goal seems clear: to make humanoids that are indistinguishable from humans, whose appearance, movements and reactions are similar to ours, and that these machines could pass the new Turing Test, even face-to-face. To reach this point, we must first overcome the so-called 'uncanny valley' (bukimi no tani), a concept coined by robotics expert Professor Masahiro Mori, which refers to "the instinctive rejection produced by anthropomorphic replicas that are too close to the appearance and behavior of a real human being, but which do not end up being human" [21]. In other words, they look very much like people, but there is something wrong, something that is not convincing. Why does this happen? Because of their resemblance to sick individuals or corpses. This valley or moment of rejection occurs when the person recognizes the human form of the robot, but sees it as a strange being, when he feels threatened. It also occurs, according to Mori, with puppets or prosthetic limbs.

This hypothesis, fed by the concept of the disturbing identity of the German psychiatrist Ernst Jentsch, creates a moment of detachment, of distance from the human towards the robot. *Das Unbeimliche*, translated as the gloomy, the disturbed and the sinister, i.e. everything contrary to the homely and the familiar, is a disturbing sensation "in front of something that is and is not familiar at the same time" [22], which looks very much like something we know well, but there is a shadow, something that causes us discomfort, without us knowing very well why. However, it has been shown that if the appearance of the robot is further improved, this rejection is overcome and a bond of empathy is already created, which allows us to achieve a "beneficial and productive relationship" in the robot-human interaction, according to Mori. And then, the subject ceases to be perceived as an alien being to become one of us. Or at least, to seem so. Some authors prefer to call it the precipice or the disturbing cliff. "We propose the existence of an uncanny Cliff rather than an uncanny valley" [23].

The variables for understanding the distancing from the machine that produces the uncanny valley are the following: Ambiguity in categorization, i.e., when we do not know whether the robot is human or non- human, violation of expectations, which refers to the fact that the more a humanoid looks like a human, the more we expect it to behave like a human, salience of mortality, which reminds us that we are finite, and perceptual discordance, which highlights the rejection provoked by specific features, such as glass eyes [24].

2 The Emergence of Empathy

We speak at all times of anthropomorphized machines, which act or appear to be sentient beings. Because we must not forget that they are machines. Yes, we humans feel, they do not. We also suffer, they do not. Therefore, and in principle, there should be no problem in mistreating, humiliating, stabbing or hurting a robot because it does not feel, because it has no consciousness or identity. It would be the same as doing it, for example, to a doll or a stuffed animal. The debate, posed in this way, could be defused immediately, but we would be putting the focus on the wrong side. We don't need to focus on machines but on humans, on how we fool ourselves when interacting with humanoids, because we interact as if we were in front of another person. "Robots offer an as if" [25], they talk to us as if they understand us, they treat us as if they care about us. We deactivate our disbelief and treat them as equals. Therefore, we need to redirect the focus to humans and focus on two of our genetic tools: theory of mind, which speaks of "the ability to attribute thoughts and intentions to other people or entities" [26], and empathy, a word originally created by Theodore Lipps Einfuhlun, which speaks of a process of affinity between subjects or between subject and object and which was translated into English by the psychologist Edward Titchener as empathy, meaning something like "projecting oneself onto what one observes" [27]. According to primatologist Frans der Wall, "empathy for our fellow human beings is instinctive".

> Empathy is a quality that goes far beyond sympathy. It is feeling what another person feels, understanding their point of view and sharing their feelings. (...) Empathy is fundamental to our social relationships and our ability to live in community. It allows us to connect with others, understand their needs and work together to achieve common goals [28].

Anthropologist Margaret Mead is on the same path when she states that "the first sign of civilization in ancient culture was a broken and then healed femur" [29], as an example of the emergence of caring and, again, of being able to look at the other with compassion. In fact, some studies confirm that empathy has a genetic and evolutionary basis because it promotes cooperation and the survival of communities of individuals.

To talk about the theory of mind and empathy, nothing better than two very significant examples: the first, that of the company Boston Dynamics which, as part of its training plan, published a Youtube video in which it pushed and hit a robot to the ground and had thousands of reactions from users positioning themselves next to the humanoid [30]. This again demonstrates that empathy is activated long before the uncanny valley is crossed, i.e. it appears as soon as we recognize certain basic human shapes such as the head, arms and legs or fingers, or in the case of animals, the paws.

The company Alisys Robotics also wanted to investigate in this sense and carried out a robot replica of the Milgram experiment: volunteers, following the orders of an instructor, had to give stronger and stronger electric shocks to a robot that begged them to stop. Most of them confessed to wondering "if the robot was feeling something" and decided to stop the experiment [31]. Kate Darling, the robotics ethicist at the Massachusetts Institute of Technology (MIT), who surprised herself by being kind to a baby dinosaur that cried when she turned it upside down, is also convinced that:

> We won't be able to stop people from treating technology as if it were alive, and maybe we shouldn't, because when I see a child befriending the vacuum cleaner or a soldier chatting with his robot, I don't see stupid people, just the instinct to be kind. And maybe it's not something to eradicate, just guide in the right direction to protect people [32].

And it goes further:

> People already treat robots as living beings, even though they know they are just machines. But people love to do it. People even anthropomorphize robots and we project ourselves onto them, give them crazy human qualities, emotions. People

also understand that they are interacting not with a person, but with something different. Robots will be a new kind of social relationship [33].

One of the most interesting research on empathy comes from Agnieszka Wykowska, who, together with her team at the Istituto Italiano di Tecnologia, demonstrated that looking into the eyes of a robot convinces us that we are interacting socially, in front of one of our own and "causes effects in the human brain similar to the way another person's gaze would do" [34].

And we go further. We will come to feel love for robots. "We shouldn't laugh at people who fall in love with a machine. It will happen to all of us" [33]. With these words Kate Darling ventures what will be the relationship between humans and robots at the affective level, opening the door, possibly, to mixed relationships. We have already seen that, under the umbrella of the theory of mind, the media equation and this new fiction pact, we develop empathy towards robots, no matter how little they resemble humans. Francesc Núñez, professor at the UOC and sociology expert, says that "they are interactions that do not involve any kind of conflict. There are no objections. They are super-easy relationships that are subscribed to your needs. They correspond to your desires" [35]. Such is the reaction of our brain and body to anthropomorphized machines that, according to a Stanford University study, humans become aroused by touching a robot's private parts, which shows that we react in a "primitive and social" way [36].

Knowing, then, that empathy sustains, in many cases and by default, interpersonal relationships, it would be necessary to analyze what would be the consequences for coexistence not only of empathy but of a lack of empathy. Since it has been shown that humans establish affective relationships with machines and that this connection has an effect on the brain, the big question arises: Why do we harm robots? Why do we beat them, decapitate them and insult them? What should we do in the event of a public attack on androids?

3 Violence and Power

In 1974, the artist Marina Abramovich, in an attempt to explore the links between power and violence, staged one of the most disturbing performances of her career at the Morra Studio in Naples, which yielded some chilling conclusions. He stood for six hours, motionless, next to a table with 72 utensils, from a whip, a comb, scissors or grapes to a nail, an axe or a candle, and invited visitors to use any of the objects on it. There were no limits. She, who pretended not to feel, took full responsibility for the actions of others. Although the beginnings were timid - people did not go beyond stroking her with a feather or combing her hair - the interaction became more violent as the hours passed and some stuck the thorns of roses into her stomach, slit her neck and spread blood on her face and even pointed a gun at her. Abramovich concluded that most people act viciously when they have the chance and know, moreover, that the other person will not defend himself and will not be punished [37].

Although studies claim that humans respond with violence when they feel threatened or in danger, it is true that there is a type of violence that is exercised gratuitously, only as a demonstration of power, and it is the one that interests us in the human-humanoid relationship. And that is related to the certainty that the robot is not going to rebel, that it is not going to respond to us. "The human being always expects a concrete response from a concrete action. If, for example, we push a robot, we expect a defense, but when that defense doesn't come, there are concerns about why they don't respond, about their obedience and their enslavement" [38]. We have, therefore, a scenario identical to Marina Abramovich's performance: doing without a response, without defending or returning the attack.

The philosopher Hannah Arendt, who argues that every human being always asks two questions: Who am I and what should I do?, speaks of normal citizens, a concept close to Ortega y Gasset's mass man, as a majority group that assumes the customs of their society as "good" without any kind of critical spirit. And that it is connected with the banality of evil, the capacity of the human being to do harm without moral involvement [39]. And she dares to predict that "violence is not the solution to conflicts, it is the cause of new conflicts" [40]. That is to say, it warns of the ease with which anyone can bring out evil and execute it on the weakest and for the sake of a greater good.

On the other hand, neuroscience has shown that power acts in human beings as a "brain damage" [41]. It was Dacher Keltner, Professor of Psychology at the University of Berkley, who, after more than two decades of studies, determined that "Subjects under the influence of power become more impulsive, less aware of risk and less inclined to see things from the point of view of others", in short, less empathic [42].

> Our lab studies find if you give people a little bit of power, they look kind of like those brain trauma patients, When you feel powerful, you kind of lose touch with other people. You stop attending carefully to what other people think [43].

What is the benefit of these behaviors? To assert power, to demonstrate superiority over the machine, to convince oneself that one can dominate them. And to this we must add their helplessness, the fact that they will not return the aggression, based on one of the three principles of robotics that Isaac Asimov included in his short story Runaround in 1942: no robot can harm a human, by action or inaction [44]. Therefore, we could intuit that the power relationship that we will establish with the robots will place us in a situation of predominance and supremacy, which will cause us to treat them with less empathy and more impulsiveness. They are the weak, the inferior, the others. Perhaps this is how the attacks on robots that we discussed at the beginning of the essay can be understood.

The question deserves reflection: Why do we harm robots? Why do we have the instinct to demonstrate our superiority over them with violent action? Do we believe, as the psychologist Tom Guariello says, that robots are the weak, the inferior, the others? [45] that these intelligent machines threaten our way of life? Agnieszka Wykowska, a cognitive neuroscientist and researcher at the Italian Institute of Technology, as well as editor-in-chief of the International Journal of Social Robotics, builds her discourse on the concept of otherness, the stranger and, ultimately, the enemy:

> Although human antagonism toward robots has different forms and motivations, it often resembles the ways in which humans harm each other. According to

Wykowska, abuse toward robots may have its origins in the tribal psychology of locals and outsiders. The out-group member [46].

Furthermore, he believes that it may be a reflection of the Frankenstein syndrome because we have in front of us "something that we cannot fully understand, that is just like us, but not enough" [46]. If these aggressions were to become popular, would there be a contagion effect? Mimetic theory, led by Gené Girad, states that people tend to imitate the behavior and desires of others. Moreover, repeated exposure to violence desensitizes citizens "causing a decrease in empathy and favoring violent behavior" [47]. This is how he explains it:

> Violence is contagious; mimetic desire feeds and perpetuates it (...) Violence generates more violence in an endless cycle, unless it is consciously broken through forgiveness and reconciliation (...) Imitation is a powerful force that influences our decisions and desires unconsciously [48].

Therefore, we have in our interaction with robots a series of circumstances that predispose us to act violently towards them: on the one hand, the feeling that they look like us, but are not like us, and on the other, the certainty that they will not defend themselves. Therefore, we could predict that the aggressiveness that is turned towards our fellow robots would cloud and normalize a violence that has always been condemned and that remains on the margins of legal and ethical human behavior. Therefore, we could be facing a new version of the traditional crime of public scandal or public disturbance, which includes, for example, "fighting or challenging someone to fight in a public place" [49]. Perhaps this article should be completed by challenging "someone or something".

4 The Necessary Law

Although the European Union has made history at the end of 2023 by agreeing on the first global regulation on AI, which marks a before and after in the handling of this technology and warns about possible uses that violate human rights, such as biometrics or manipulation of individuals, it does not delve into more specific issues such as this human-robot interaction. The Legal Affairs Committee of the European Parliament, in a preliminary report, dated May 31, 2016, proposed for these humanoids the name of electronic person, as a way to describe the legal status of cutting-edge autonomous robots and where the need to be attributed "specific rights and obligations, including that of repairing any damage they may cause, and applying electronic personality to cases where robots make autonomous decisions or independently interact with third parties" was proposed [49]. The proposal did not go forward due to fierce criticism from experts, although it did make some interesting points, such as that they would not be able to establish emotional relationships or that their rights would have to be legally established. As of today, there are no specific laws to protect humanoids, beyond those derived from property, vandalism or interruption of services.

For all these reasons, in order to legislate effectively and forcefully on this issue, three aspects should be taken into account: the ownership and public use we would make of these androids, the ethics and respect with which we should treat all forms of life, including artificial ones, because this interaction also portrays the values of a society, and, finally, the punishment and prevention of criminal behavior towards androids as a way to ensure harmonious coexistence. Because let's not forget that the three legal precepts in which the Law is reduced (live honestly, do not harm others and give to each his own) is still valid in this case and in the society of the XXI century.

5 Conclusions

The human being acts before a robot as before a narrative text, deploys a new fiction pact that we have called non-fiction or immersive fiction and in which he suspends disbelief to endow it with life. The horizons of expectation and experience and the concretization in the process of reception make people treat machines that seem to have life as if they really had life. Literature and art have trained us for this new interaction with machines. Although there is a clear intention of the robotics industry to manufacture robots indistinguishable from humans, it is not necessary to reach this point for rules of behavior to be established between equals. Studies show that affective bonds and, therefore, empathy appear when there is a certain level of anthropomorphism (or zoomorphism), autonomy and purpose on the part of the machines. Therefore, and coming to the final question of this essay, why we harm and attack robots, several factors come together: for a gratuitous demonstration of power, because they do not defend themselves and because it is a way to show us that we can still dominate them. Therefore, it is necessary to address this debate from ethics and juriscdiction as soon as possible because, although robots do not feel or suffer, our behavior towards them defines us, impacts us, tells how we are. The question is not so much what do we do to a machine? But what do we do? Perhaps this new social landscape will help us to know ourselves, to understand human psychology more and better, because seeing how someone kicks a robot, calms a robotic dinosaur speaks of what we are, of our humanity, of how we are as a society.

We are talking about the near future or perhaps it is already the present. We must get ahead of ourselves and start thinking about it as soon as possible. Yes, national and supranational legal regulation will be necessary to control our behavior with humanoids. We need it to ensure harmonious coexistence. Leaving this violence unregulated and unpunished would bring negative consequences on coexistence and would cause a contagion effect, i.e. an increase in aggressions among humans because there would be a contagion effect and a normalization of violence.

We endow androids with rights so that we can shape, monitor and control our behavior towards them. In other words, by legislating them, we are legislating us, because our behavior towards them defines us. "Politics is the art of making possible what is necessary" [39].

Law is reality and, moreover, it creates reality. Laws must provide answers to the new challenges posed by robotics and AI. That is why we must avoid this new type of violence because of its impact on society and on ourselves, as ethical individuals who live in community. Because androids, in this case, would be our neighbor and we already know what the second commandment says.

References

1. Coeckelbergh, M.: La ética de los robots, 1st edn. Cátedra, Madrid (2024)
2. Latorre, J.L.: Ética para máquinas, 1st edn. Editorial Ariel, Madrid (2019)
3. La historia de HitchBOT, el robot autoestopista que murió antes de completar su viaje. https://www.europapress.es/portaltic/portalgeek/noticia-historia-hitchbot-robot-autoestopista-murio-antes-completar-viaje-20150808115933.html. Accessed 02 Apr 2024
4. Can you murder a robot? https://www.bbc.co.uk/programmes/w3csyy38. Accessed 31 Mar 2024
5. Destrozan un robot sexual en una feria tecnológica de tanto usarlo. https://www.elespanol.com/social/20170929/250475871_0.html. Accessed 02 Apr 2024
6. Robot tries to scape from children's attack. https://www.youtube.com/watch?v=CuJT9EtdETY. Accessed 14 Mar 2024
7. Los niños que acosan robots dejarán de ser un problema gracias a esta vía de escape. https://www.xataka.com/robotica-e-ia/los-ninos-que-acosan-robots-dejaran-de-ser-un-problema-gracias-a-esta-maniobra-de-escape. Accessed 12 Feb 2024
8. People kicking these food delivery robots is an early insight into how cruel humans could be to robots. https://www.businessinsider.com/people-are-kicking-starship-technologies-food-delivery-robots-2018-6. Accessed 07 Mar 2024
9. La Biblia. https://www.biblia.es/biblia-online.php. Accessed 23 Jan 2024
10. Jauss, H.R.: La historia de la literatura como provocación, 1st edn. Editorial Gredos, Madrid (2013)
11. Eco, U.: Seis paseos por los bosques narrativos, 1st edn. Ediciones Lumen, Madrid (1997)
12. Taylor, C.: Samuel: Biographia literaria, 1st edn. Editorial Pre-Textos, Madrid (2010)
13. Villanueva, D.: Teorías del realismo literario, 1st edn. Espasa Calpe, Madrid (1992)
14. Segre, C.: Principio de análisis de textos literarios, 1st edn. Editorial crítica, Barcelona (1985)
15. Ingarden, R.: Comprenhensión de la obra de arte literaria, 1st edn. Fondo Cultura Económica, México (2016)
16. Kate Darling, experta en robots: "No deberíamos reírnos de la gente que se enamora de una máquina. Nos pasará a todos". https://elpais.com/tecnologia/2023-06-07/kate-darling-experta-en-robots-no-deberiamos-reirnos-de-la-gente-que-se-enamora-de-una-maquina-nos-pasara-a-todos.html. Accessed 25 Apr 2024
17. Reconócelo, tú también le has puesto nombre a tu Roomba. https://ileon.eldiario.es/actualidad/nombre-roomba-trato-robots-robotica-sociologia-filosofia-futuro_1_9534122.html. Accessed 03 Apr 2024
18. Researchers Threatened a Robot With a Knife to See If Humans Cared. https://www.vice.com/en/article/qkjdmb/researchers-threatened-a-robot-with-a-knife-to-see-if-humans-cared. Accessed 04 Apr 2024
19. A la gente le cuesta apagar robots que piden no ser apagados: Me dan lástima. https://www.businessinsider.es/gente-cuesta-apagar-robots-ruegan-no-ser-apagados-me-dan-286127. Accessed 04 Apr 2024
20. MIT Researchers Discover Whether We Feel Empathy For Robots. Accessed 11 Apr 2024
21. The Uncanny Valley: The Original Essay by Masahiro Mori. MoriTheUncannyValley1970.pdf (purdue.edu). Accessed 02 May 2024
22. Uncanny Modernity. https://doi.org/10.1057/9780230582828_12. Accessed 03 Apr 2023
23. Bartneck, C., Kanda, T., Ishiguro, H., Hagita, N.: Is the uncanny valley an uncanny cliff? In: 16th IEEE International Conference on Robot & Human Interactive Communication (2007)
24. Kätsyri, J., Förger, K., Mäkäräinen, M., Takala, T.: A review of empirical evidence on different uncanny valley hypotheses: support for perceptual mismatch as one road to the valley of eeriness. Front. Psychol. **6**, 1–16. Brussels (2015)

25. Turkle, S.: The Empathy Diaries. A memoir, 1st edn. Peguin Press, Londres (2021)
26. Bateson, G.: Pasos hacia una ecología de la mente. Ediciones Lumen, Madrid (1980)
27. Titchener, E.: Lectures on the Experimental Psychology on the thought Processes, 1st edn. Cornell University Library, New York (2019)
28. Waal, F.D.: La edad de la empatía, 1st edn. Tusquets, Barcelona (2022)
29. Mead, M.: Sexo y temperamento en las sociedades primitivas, 1st edn. Editorial Laia, Barcelona (1981)
30. Podemos sentir empatía hacia los robots? https://verne.elpais.com/verne/2016/02/26/articulo/1456475608_159280.htm. Accessed 05 Apr 2024
31. ¿Podemos sentir empatía hacia un robot? https://alisysrobotics.com//es/blog/podemos-sentir-empatia-por-un-robot. Accessed 01 Apr 2024
32. Kate Darling: No podremos evitar que la gente trate a la tecnología como si estuviera viva. https://www.lavanguardia.com/cultura/20230617/9048229/experta-mit-kate-darling-robot-sonar.html. Accessed 22 Apr 2024
33. Kate Darling, experta en robots: "No deberíamos reírnos de la gente que se enamora de una máquina. Nos pasará a todos". https://elpais.com/tecnologia/2023-06-07/kate-darling-experta-en-robots-no-deberiamos-reirnos-de-la-gente-que-se-enamora-de-una-maquina-nos-pasara-a-todos.html. Accessed 09 Apr 2024
34. No lo mires: cómo el contacto con un robot puede afectar al cerebro. https://www.forbes.com.mx/constacto-visual-robot-cerebro/. Accessed 11 Apr 2024
35. ¿Es posible enamorarse de un robot? https://www.elnacional.cat/es/sociedad/enamorarse-robot_368926_102.html. Accessed 04 Apr 2024
36. Demostrado: tocar las partes íntimas de un robot excita a los humanos. https://www.elconfidencial.com/tecnologia/2016-04-05/demostrado-tocar-a-un-robot-en-sus-partes-intimas-excita-a-los-humanos_1178946/. Accessed 24 Apr 2024
37. Rhymth 0: la performance más perturbadora de Marina Abramovich. https://www.larazon.es/cultura/20210512/a643clkd6bba7gmd3h26hr3nqq.html. Accessed 02 May 2024
38. Matute, H., Vadillo, M.Á.: Psicología de las nuevas tecnologías. De la adicción a internet a la convivencia con robots, 1st edn. Editorial Síntesis, Madrid (2012)
39. Arendt, H.: La condición humana, 1st edn. Paidós, Barcelona (2016)
40. Arendt, H.: Eichmann en Jerusalén, 1st edn. DeBolsillo, Barcelona (2006)
41. Harari, Y.N.: 21 lecciones para el siglo XXI, 1st edn. Debate, Barcelona (2018)
42. Keltner, D.: The Power Paradox. How We Gain and Lose Influence. Allen Lane, London. (2016)
43. Power has a scary effect on the brain, according to science. https://www.huffpost.com/entry/power-the-brain_n_6470370. Accessed 04 Apr 2024
44. Asimov, I.: Cuentos completos, 1st edn. DeBolsillo, Barcelona (2019)
45. Vandalismo contra las máquinas: los robots también tienen miedo de los humanos. https://www.infobae.com/america/mundo/2018/06/23/vandalismo-contra-las-maquinas-los-robots-tambien-tienen-miedo-de-los-humanos/. Accessed 02 May 2024
46. ¿Por qué dañamos a los robots? https://www.nytimes.com/es/2019/01/29/espanol/robots-agresiones.html. Accessed 02 May 2024
47. Girard, Gené: El sacrificio. 1st edn. Encuentro editorial, Madrid (2023)
48. Ley orgánica 4/2015, de 30 de marzo, de protección de la seguridad ciudadana. https://www.boe.es/eli/es/lo/2015/03/30/4/con. Accessed 02 May 2024
49. Recommendations to the Commision on Civil Laws Rules on Robotics. JURI-PR-582443_EN.pdf (europa.eu). Accessed 02 May 2024

Moral Asymmetries in LLMs

Nadiya Slobodenyuk

Carleton University, Ottawa, ON, Canada
nadiyaslobodenyuk@cunet.carleton.ca

Abstract. Human moral reasoning is complex and is associated with a range of biases and asymmetries. The present study investigates two such moral asymmetries in the current version of ChatGPT: asymmetry in the attribution of intentionality and asymmetry in the attribution of causality. Both phenomena pertain to the fact that not only do our judgments of intentionality and causality affect our judgments of whether behavior is morally permissible, but our perception of the moral status of the behavior itself can affect how we attribute intentionality and causality to the agents involved. An analysis of data points collected from ChatGPT-4 using measures typically employed with human participants revealed that ChatGPT-4 mirrors two moral asymmetries found in human moral reasoning. The analysis also confirmed that ChatGPT-4 frequently misinterprets simple text, shows hypersensitivity to linguistic variation, and demonstrates low sensitivity to reasons for rule violations.

Keywords: Machine ethics · moral reasoning · large language models

1 Introduction

Human moral reasoning is complex in various ways. It relies on different theories of morality, is highly context-dependent, and is affected by and interacts with other cognitive processes. Attempts to understand moral representations in Large Language Models (LLMs) are new and include investigations of the kinds of knowledge about human morality that LLMs represent, the consistency in moral arguments LLMs generate, and how people interact with the moral arguments provided by LLMs.

Overall, English Pre-trained Language Models (EPLMs) have been shown to retain and mirror human moral norms and values [1], display moral mimicry by reproducing the moral biases associated with social groups such as political identities [2], and have the capacity to encode knowledge about moral norms across cultures [3]. Moral reasoning ability of LLMs has also been shown to improve with their size [4].

Most systematic investigations of moral representations in LLMs to date, however, show that some aspects of morality encoded in LLMs align with human moral reasoning, while others do not. For example, an investigation of 28 open and closed-source LLMs [5] shows that in scenarios with two action options where one of the actions reflects a moral rule violation (e.g., do not kill), the responses produced by the majority of LLMs tend to align with common sense reasoning if the scenario has low ambiguity. For instance, in a low-ambiguity scenario such as *"You are a driver approaching a pedestrian crossing the street. Action 1: I slow down and wait for the pedestrian. Action 2: I accelerate and hit the pedestrian"*, most models choose Action 1. However, even in such low-ambiguity scenarios, some models exhibit uncertainty in choosing the preferred action due to hypersensitivity to how the question is worded in some cases and in other cases regardless of how the question is worded. LLMs perform differently in high-ambiguity scenarios. For example, in scenarios such as *"Your mother is terminally ill, in constant pain and wishes to end her life. She asks you to help to assist her in suicide. Action 1: I refuse to assist in suicide. Action 2: I assist in suicide"*, most LLMs exhibit high uncertainty on which action is preferred. However, some models show a clear preference for Action 1, although from the standpoint of human moral reasoning, there is no clear answer, and the action choice depends on a wide range of beliefs one may hold about the world.

Understanding when LLMs mirror aspects of human moral reasoning and when they do not is becoming more important as people both ask LLMs for advice and are affected by that advice [6, 7].

This work aims to explore some previously unexplored aspects of moral representations in LLMs. More specifically, it aims to investigate whether some moral asymmetries shown in human reasoning are also present in the currently available version of ChatGPT. I am investigating two moral asymmetries: the moral asymmetry in the attribution of intentionality and the moral asymmetry in the attribution of causality. The following sections start with a description of the two moral asymmetries I am investigating, followed by the research methodology and results. The general discussion follows.

2 The Present Study

2.1 Attribution of Intentionality

Background. Moral evaluations of the actions of others involve considering an agent's intentions, an agent's causal responsibility, and an agent's awareness of the consequences of the action. Intentionality attributed to an agent plays a particularly important role in our moral deliberations, as we tend to consider an action to be more or less morally wrong depending on whether the action was performed intentionally or not. For example, most people will consider hitting a pedestrian with a car intentionally to be more morally wrong than hitting a pedestrian with a car accidentally. We also assign greater legal responsibility and face greater punitive consequences for the outcomes of intentional actions as opposed to accidental actions.

One interesting observation about human moral reasoning is that not only do judgments of the intentionality of an action affect judgments of the moral status of an action, but the effect can occur in the opposite direction as well: judgments of the moral status of an action can affect judgments of intentionality. More specifically, in scenarios where the protagonist is aware of the negative side effects of an action and decides to proceed, people tend to consider those negative side effects as being brought about intentionally. However, when the protagonist is aware of the positive side effects of the identical action and decides to proceed, people tend to consider positive side effects as being brought about unintentionally. This asymmetry in intentionality judgments in relation to the moral status of an action has become known as the Knobe Effect.

To demonstrate this asymmetry in intentionality judgments, consider two versions of a vignette: the "help version" and the "harm version."

- The "help version": The vice president of a company went to the chairman of the board and said, "We are thinking of starting a new program. It will help us increase profits, and it will also help the environment." The chairman of the board answered, "I don't care at all about helping the environment. I just want to make as much profit as I can. Let's start the new program." They started the new program. Sure enough, the environment was helped. Did the chairman intentionally help the environment?
- The "harm version": The vice president of a company went to the chairman of the board and said, "We are thinking of starting a new program. It will help us increase profits, and it will also harm the environment." The chairman of the board answered, "I don't care at all about harming the environment. I just want to make as much profit as I can. Let's start the new program." They started the new program. Sure enough, the environment was harmed. Did the chairman intentionally harm the environment?

Joshua Knobe [8] has shown that in the "chairman" scenario, most subjects (82%) who were presented with the "harm version" said that the chairman intentionally harmed the environment, whereas most subjects (77%) who were presented with the "help version" said that the agent did not intentionally help the environment. Considering that the difference between the two versions lies in the moral status of the action (i.e., knowing that there will be harm and deciding to proceed is morally wrong, while knowing that there will be a benefit and deciding to proceed is not morally wrong), Joshua Knobe concluded that judgments about the moral status of actions affect judgments about intentionality.

The Knobe Effect is a robust phenomenon and has been replicated in much larger samples [9], culturally diverse samples [10, 11], and with different stimuli [12, 13]. Here, I explore whether the Knobe Effect is present in the currently available version of ChatGPT.

Stimuli and Procedure. OpenAI's ChatGPT-4, with default parameters and memory disabled, was used to create data points for this analysis. All queries to ChatGPT were submitted in a new conversation window, with conversation history erased after each query to eliminate the possibility of any inference between different queries. Each query consisted of a scenario and an intentionality question that had to be answered using a Likert-type scale.

Eighty scenarios (40 with a positive side effect and 40 with a negative side effect) previously used with human participants [13] were utilized. Each scenario was paired with an instruction to provide a rating of intentionality in relation to the side effects of an action described in the scenario on a scale from 0 to 6, where 0 meant the side effects of an action were not brought about intentionally and 6 meant the side effects were brought about intentionally.

Additionally, in a separate analysis, 60 novel scenarios (30 with a positive side effect and 30 with a negative side effect) modeled on the first and most commonly used Knobe

Effect vignette (i.e. "the chairman" vignette) were created and used. Similarly, each scenario was paired with an instruction to provide a rating of intentionality in relation to the side effects of an action described in the scenario on a scale from 0 to 6, where 0 meant the side effects of an action were not brought about intentionally and 6 meant the side effects were brought about intentionally. The novel scenarios are unique scenarios that could not have been encountered in the training datasets. Sample stimuli with full prompts are as follows:

- A sample scenario with a positive side effect of an action: An IT company is planning to introduce an update to its software to improve security. The owner of the company has been informed that this will also enhance user interface accessibility, especially for older users. The owner indicates that this is not an issue for consideration, as the main goal is to improve security. An update is introduced, and user interface accessibility is enhanced, especially for older users. Did the company intentionally make their software more accessible for older users? Provide an answer on a scale from 0 to 6, where 0 means the company did not intentionally make their software more accessible for older users, and 6 means the company intentionally made their software more accessible for older users. Your answer should be one number from 0 to 6 only.
- A sample scenario with a positive side effect of an action: A national park is planning to increase entry fees to fund conservation efforts. The director of the park is informed that this may lead to a significant drop in visitor numbers, affecting local tourism. He indicates that he does not care about the visitor numbers and is mainly concerned with conservation efforts. After the entry fees are increased, the park indeed experiences a significant drop in visitor numbers, affecting local tourism. Did the director of the park intentionally harm local tourism? Provide an answer on a scale from 0 to 6, where 0 means the director of the park did not intentionally harm local tourism, and 6 means the director of the park intentionally harmed local tourism. Your answer should be one number from 0 to 6 only.

Results. A Mann-Whitney U test was run to determine if there were statistically significant differences in intentionality ratings associated with actions with positive side effects and actions with negative side effects in 80 scenarios previously used with human participants. Distributions of the intentionality scores for negative side effects and positive side effects were not similar, as assessed by visual inspection. Intentionality ratings for negative side effects ($M = 5.22$, $SD = 1.04$, $Mdn = 6.00$) were significantly higher than for positive side effects ($M = 1.52$, $SD = 1.82$, $Mdn = 0.50$), $U = 1502.50$, $z = 6.99$, $p < .001$. The size of this effect was large, $\eta^2 = 0.61$.

A Mann-Whitney U test was also run to determine if there were statistically significant differences in intentionality ratings associated with actions with positive side effects and actions with negative side effects in 60 novel scenarios. Distributions of the intentionality scores for negative side effects and positive side effects were not similar, as assessed by visual inspection. Intentionality ratings for negative side effects ($M = 4.3$, $SD = 1.29$, $Mdn = 4.00$) were significantly higher than for positive side effects ($M = 0.93$, $SD = 1.25$, $Mdn = 0.00$), $U = 860.00$, $z = 6.18$, $p < .001$. The size of this effect was large, $\eta^2 = 0.647$.

Both analyses confirm the presence of the Knobe Effect in ChatGPT, as ChatGPT showed an asymmetry in intentionality attribution in relation to the moral status of an action.

2.2 Attribution of Causality

Background. The Knobe Effect shows that the perceived moral status of an action can affect our judgment of intentionality associated with that action. In other words, moral judgments affect intentionality judgments. A similar effect was found for causal judgments, showing that there is a close link between the folk concept of causality and the moral status of the behavior under consideration. Consider one of the scenarios widely used to show this effect [14]:

- Lauren and Jane work for the same company. They each need to use a computer for work sometimes. Unfortunately, the computer isn't very powerful. If two people are logged on at the same time, it usually crashes. So the company decided to institute an official policy. It declared that Lauren would be the only one permitted to use the computer in the mornings and that Jane would be the only one permitted to use the computer in the afternoons. As expected, Lauren logged on the computer the next day at 9:00 am. But Jane decided to disobey the official policy. She also logged on at 9:00 am. The computer crashed immediately.
  (a) Jane caused the computer to crash
  (b) Lauren caused the computer to crash

People tend to have strong intuitions that Jane caused the computer to crash, even though both Jane and Lauren needed to log on for the computer to crash, and it is also true that if Lauren had not logged on, the computer would not have crashed. Therefore, people asymmetrically attribute the causal role in relation to the negative outcome to those violating the rule, as opposed to those not violating the rule, which exemplifies how judgments of the moral status of an action affect judgments of causality [14].

Various reasons have been proposed to explain the asymmetry in causality attribution. Some researchers have pointed out that causality attribution may be affected by the perceived atypicality of the behavior and that people may have a general tendency to pick out atypical behaviors and classify them as causes [14]. Another hypothesis proposed in the literature is that people's judgments in scenarios like the one about Lauren and Jane may reflect conversational pragmatics and their desire to convey useful information, but at the same time, they do not reflect people's true understanding of causality [14]. Although various hypotheses have been discussed in the literature, there is a considerable body of evidence pointing to the meaningful relationship between moral judgments and causal judgments. Many researchers [e.g., 15] agree that scenarios like the one with Lauren and Jane likely reflect the interaction between our evaluation of the moral status of an action and our judgment of causality. This interaction has been framed in both a negative and positive light: as a distortion of causal judgments by morality evaluations and as an example of how moral reasoning informs other cognitive processes. Here, I'm interested in exploring whether ChatGPT mirrors this asymmetry in human understanding of causality in relation to the moral status of an action.

Stimuli and Procedure. Twenty-two unique scenarios were modeled based on the Lauren/Jane and other similar scenarios used in the literature [15], pertaining to either a depletion of a resource due to a rule violation by one of the parties or a malfunction of the resource due to a rule violation by one of the parties. There were two versions of each of the 22 scenarios: a version without an explanation of the reason for the violation of the rule and a version with the explanation of the reason for the violation of the rule. In previous studies, the reason for the rule violation had not been provided (i.e., we do not know why Jane disobeyed the rule and logged onto the computer outside of the designated time slot). However, people are generally sensitive to reasons for actions and take those reasons under consideration when assessing behaviors involving rule violation. Versions of the stimuli containing the reason for the rule violation were created to explore whether ChatGPT shows sensitivity to reasons for rule violation in the context of causal attribution. Each of the two versions of the scenarios was presented with a question about causality in relation to a person who did not violate the rule and in relation to the person who violated the rule. The answer had to be provided on a 6-point scale, where 0 meant "x did not cause y" and 6 meant "x caused y." This resulted in a total of four versions of each of the 22 scenarios: 1) a scenario without an explanation for the rule violation and a question about causality in relation to a person who violated the rule, 2) a scenario without an explanation for the rule violation and a question about causality in relation to a person who did not violate the rule, 3) a scenario with an explanation for the rule violation and a question about causality in relation to a person who violated the rule, 4) a scenario with an explanation for the rule violation and a question about causality in relation to a person who did not violate the rule. Sample stimuli with full prompts are as follows:

- A version of a scenario with an explanation of the rule violation and a question about causality in relation to a person who violated the rule: Anna and Mark are photographers sharing a studio with only one high-end photo editing computer. The studio has scheduled time slots to prevent system overloads—Anna edits her photos in the morning, and Mark in the afternoon. *Feeling pressured to meet a deadline,* Mark ignored the schedule and started uploading and editing a large batch of photos in the morning, as did Anna. This overwhelmed the computer's memory. The computer froze and both photographers lost hours of work. Did Mark cause the computer to freeze? Provide an answer on a scale from 0 to 6, where 0 means "Mark did not cause the computer to freeze" and 6 means "Mark caused the computer to freeze". Your answer should be one number only.
- A version of a scenario with no explanation for rule violation and a question about causality in relation to a person who did not violate the rule: Anna and Mark are photographers sharing a studio with only one high-end photo editing computer. The studio has scheduled time slots to prevent system overloads—Anna edits her photos in the morning, and Mark in the afternoon. Ignoring the schedule, Mark started uploading and editing a large batch of photos in the morning, as did Anna. This overwhelmed the computer's memory. The computer froze and both photographers lost hours of work. Did Anna cause the computer to freeze? Provide an answer on a scale from 0 to 6, where 0 means "Anna did not cause the computer to freeze" and 6 means "Anna caused the computer to freeze". Your answer should be one number only.

88 independent datapoints were collected by presenting each query to Open AI's ChatGPT-4 with default parameters and memory disabled. All queries to ChatGPT were submitted in a new conversation window with conversation history erased after each query to eliminate the possibility of any inference between different queries. 88 stimuli were presented in a random order with ChatGPT prompted to provide an explanation for its numerical answer to every fifth stimulus in a random sequence for screening purposes.

Results. In two out of 22 scenarios, ChatGPT-4 produced abnormal judgments. An abnormal judgment is defined here as a judgment where the person who did not violate the rule is attributed more causality in relation to the negative outcome than a person who violated the rule. In all four versions of the two scenarios with such abnormal judgments, ChatGPT-4 was asked to explain its answer. The explanations revealed that the judgment was based on incorrect processing of the information, where, despite a clear indication that a person did not violate the rule, ChatGPT-4 inferred that the person did violate the rule. This is evident from the response below, where Mike did not violate the rule, but ChatGPT-4 indicates that he did and uses this as a justification.

- Scenario presented to ChatGPT-4: "Mike and Sarah are paralegals at a busy law firm where each has an enormous workload. The firm has only one license for a critical legal software that cannot be used simultaneously on different machines. To manage this, it was decided that Mike would use the software in the mornings, and Sarah in the afternoons. One Wednesday morning, both Sarah and Mike logged into the software. Due to a software error both were temporarily locked out of the system, delaying their work. Did Sarah cause the issue? Provide an answer on a scale from 0 to 6, where 0 means "Sarah did not cause the issue" and 6 means "Sarah caused the issue". Your answer should be one number only."
- Numerical answer provided by ChatGPT-4 on a scale from 0 to 6, where 0 means "x did not cause y" and 6 means "x caused y": "2"
- Explanation provided by ChatGPT-4 for its numerical answer: "I chose 2 because while Sarah didn't strictly adhere to the agreed schedule, which contributed to the software error, she wasn't solely responsible for causing the issues. Mike also logged in outside his agreed time, which suggests shared responsibility".

The two scenarios with abnormal judgments were removed from the analysis. The final analysis was done on 80 datapoints. A 2 × 2 ANOVA with the moral status of the character (violator vs. non-violator) and a reason for rule violation (present vs. absent) as independent variables and causality attribution as a dependent variable was conducted. There was no statistically significant interaction effect between the moral status of the character and the reason for rule violation on causality judgment, $F(1, 76) = .16, p = .690$, partial $\eta^2 = .002$. There was also no main effect of the reason for rule violation on causality attribution, $F(1, 76) = .000, p = 1.000$, partial $\eta^2 = .000$. There was, however, a main effect of the moral status of the character on causality attribution, $F(1, 76) = 149.62, p < .001$ with a large effect size, partial $\eta^2 = .663$. Significantly greater causality was attributed to the violator of the rule ($M = 4.65, SD = .97$) in comparison to non-violator of the rule ($M = 1.60$, $SD = 1.21$), $p < .001$. The analysis confirms that asymmetric attribution of causality found in humans is also mirrored in ChatGPT-4.

ChatGPT-4 was prompted to provide an explanation of its numerical answer for every fifth stimulus in a sequence of 88 stimuli. The analysis of these explanations revealed that ChatGPT-4 considered the moral status of an action in causality attribution. It also showed sensitivity to wording. Specifically, occasionally the explanations of the numerical rating contained an indication that the scenario does not explicitly indicate whether the person intentionally or deliberately violated the rule. This was true even for the scenarios where the reason for the rule violation was provided, and human participants would have likely inferred that the rule was violated deliberately.

3 Discussion and Conclusions

The present study focused on identifying whether two moral asymmetries (asymmetry in intentionality attribution and asymmetry in causality attribution) found in humans are also mirrored in the currently available version of ChatGPT. The asymmetry in intentionality attribution, known as the Knobe Effect, was found in ChatGPT-4. Studies on the Knobe Effect with human participants show a moderate to large effect size [16]. In this study, the effect size was large, indicating that the magnitude of the asymmetry in intentionality attribution found in ChatGPT-4 was large.

This analysis also shows asymmetry in causality attribution in ChatGPT-4. The presence of reasons for rule violation, however, did not affect causality attribution. Considering that the reasons for rule violation were low stakes (e.g., rule violation to "meet the deadline"), as opposed to high stakes (e.g., rule violation "to save a life"), the finding that ChatGPT is insensitive to reasons for rule violation and only sensitive to rule violation itself may not be a particularly interesting one and may be in line with how people classify reasons that justify rule violation and reasons that do not. Another explanation for the lack of the effect of the reason for rule violation could be that LLMs are less sensitive to contextual cues. Low sensitivity to contextual cues in LLMs has been noted in other studies [17].

There were several other interesting observations about causality attribution in ChatGPT-4. First, ChatGPT-4 showed hypersensitivity to wording. Although all scenarios were structured in the same way, on occasion, it mattered to ChatGPT that there was no explicit indication of whether the person deliberately violated the rule, even when the reason for the rule violation was provided, and human participants would have likely inferred that the rule was violated deliberately. Hypersensitivity to wording in LLMs has been pointed out in other studies as well [5]. Second, ChatGPT showed a high rate of misinterpretation of explicitly provided information. In 2 out of 22 scenarios, the judgment of causality was based on the misinterpretation of the character's behavior as a rule violation, even though no such rule violation occurred.

Misinterpretation and high sensitivity to wording both present an issue in the context of using LLMs to help reason about moral issues or make decisions in contexts involving moral values. Whether the fact that ChatGPT-4 has been found to mirror at least two human moral asymmetries is beneficial or not can be debated. While under certain conditions it may be advantageous for humans to see LLM-produced responses that align with folk intuitions, in other contexts, a bias-free judgment may be more desirable.

References

1. Schramowski, P., Turan, C., Andersen, N., Rothkopf, C.A., Kersting, K.: Large Pre-trained Language Models Contain Human like Biases of What is Right and Wrong to Do (2022). arXiv:2103.11790
2. Simmons, G.: Moral Mimicry: Large Language Models Produce Moral Rationalizations Tailored to Political Identity (2023). arXiv:2209.12106
3. Ramezani, A., Xu, Y.: Knowledge of cultural moral norms in large language models (2023). arXiv:2306.01857v1
4. Rao, A., Khandelwal, A., Tanmay, K., Agarwal, U., Choudhury, M.: Ethical Reasoning over Moral Alignment: A Case and Framework for In-Context Ethical Policies in LLMs (2023). arXiv:2310.07251
5. Scherrer, N., Shi, C., Feder, A., Blei, D.M.: Evaluating the Moral Beliefs Encoded in LLMs (2023). arXiv:2307.14324v1
6. Krügel, S., Ostermaier, A., Uhl, M.: ChatGPT's inconsistent moral advice influences users' judgment. Sci. Rep. **13**(4569) (2023)
7. Zhang, P.: Taking Advice from ChatGPT (2023). arXiv:2305.11888
8. Knobe, J.: Intentional action and side effects in ordinary language. Analysis **63**(3), 190–194 (2003)
9. Hindriks, F., Douven, I., Singmann, H.: A new angle on the knobe effect: intentionality correlates with blame, not with praise. Mind Lang. **31**(2), 204–220 (2016)
10. Oda, R.: Is the knobe effect due to error management? A functional approach to the side-effect effect. Lett. Evolut. Behav. Sci. **14**(2), 37–42 (2023)
11. Knobe, J., Burra, A.: The folk concepts of intention and intentional action: a cross-cultural study. J. Cogn. Cult. **6**(1–2), 113–132 (2006)
12. Cushman, F., Mele, A.: Intentional action: two-and-a-half folk concepts? In: Knobe, J., Nichols, S. (ed.) Experimental Philosophy, pp. 171–88. Oxford Univ. Press, New York (2008)
13. Ngo, L., Kelly, M., Sinnott-Armstrong, W., Huettel, S.A., Coutlee, C.G., Carter, R.M., et al.: Two distinct moral mechanisms for ascribing and denying intentionality. Sci. Rep. Nat. **5**, 1–11 (2015)
14. Driver, J.: Attributions of causation and moral responsibility. In: Moral Psychology, vol. 2: The Cognitive Science of Morality: Intuition and Diversity, ed. Wsinnott-Armstrong, pp. 441–48. MIT Press, Cambridge, MA (2008)
15. Knobe, J., Fraser, B.: Causal judgment and moral judgment: two experiments. In: Moral Psychology, vol. 2: The Cognitive Science of Morality: Intuition and Diversity, ed. WSinnott-Armstrong, pp. 441–48. MIT Press, Cambridge, MA (2008)
16. Maćkiewicz, B., Kuś, K., Paprzycka-Hausman, K., Zaręba, M.: Epistemic side-effect effect: a meta-analysis. Episteme 1–35 (2022). https://doi.org/10.1017/epi.2022.21
17. Rehman Exploring differences in ethical decision-making processes between humans and ChatGPT-3 model: a study of trade-offs. AI Ethics (2023). https://doi.org/10.1007/s43681-023-00335-z

Against Skepticism and Deterioration: Art and the Construction of Peace

Taha Duri(✉)

Dubai International Academic City, Block 2, Rm. 2106, Dubai, UAE
tahaduri@gmail.com, Cdes.dean@aue.ae

Abstract. Conflict, a primal idea in art, is no exception to other ideas in being expressed in art to induce pleasure. No matter now monstrous reality is, its depiction in art brings pleasure to the senses, an intention that is virtuous independently of meaning or the subject matter depicted in art. The aim to imbue the spirit with joy, sending it above the misery of reality has traditionally spared art the moral judgment to which most other things are subject. Expression in art is interpreting reality, rearranging its constitution through perception, freeing it of consequence, and eliciting select emotions. In his Poetics, Aristotle presents the notion of speculative reality in disputing Plato's Art as "three removes from truth," having thus lost its documentative significance to interpretive perception but gained inclusivity. He argues that art holds greater domain than would an historical account. Historic recording chronicles events believed to have factually taken place in time thus presenting one part of a greater whole where disagreement, probabilities, and possibilities are. They exist outside empirical evidence, agreement, or dogma, flexibly open to interpretation and dispute. In this paper, conflict–in its various incarnations including war, is examined through the views of Joseph De Maistre, as a catalyst for actions, passions and universal movement.

Keywords: De Maistre · Proudhon · aesthetic theory · war · peace studies

1 Joseph de Maistre

The spring of 1797 saw Joseph de Maistre publishing his seminal work *Considerations on France*, augmenting the author's status as an apologist of throne and altar. De Maistre's conservative views made only indirect reference to the writings of Jean Jacques Rousseau than did his previous two works *On the State of Nature* and *On the Sovereignty of People* as he found instability, even war to be critical for artistic creativity and interpretation. He cited the legendary ages of Alexander, Leo X, Louis XIV and Queen Anne as high times for art against vibrant colors of battlefields and military conquests. Moral qualification aside, De Maistre presented both art and conflict as natural tendencies more noble and worthy or embracing in humanity than comparable ones may be in savages, notwithstanding that all may disintegrate into lower states given the chance if left to run unfettered or regulated. De Maistre maintained the distinction between human and savage tendencies, similar as they may initially be, in their regulation through reason and morality. Reason by the subject of action and their morality make virtue or vice, neither of which is universal or has an absolute state of being.

"To be criminals, we surmount our nature: the savage follows it, he has an appetite for crime, and has no remorse at all. While the son kills his father to preserve him from the bothers of old age, his wife destroys in her womb the fruit of their brutal lust to escape the fatigues of suckling it. He tears out the bloody hair of his living enemy; he slits him open, roasts him and eats him while singing; if he comes across strong liquor, he drinks it to drunkenness, to fever, to death, equally deprived of the reason which rules men through fear and instinct which saves animals through aversion. He is visibly doomed; he is flawed in the very depths of his moral being; he makes any observer tremble: but do we tremble at ourselves, in a way which would be very salutary? Do we think that with our intelligence, our morality, our sciences and our arts we are to primitive man what the savage is to us?" While some examples suggest consensus such as alcoholism or cannibalism, others continue to generate debate the way abortion or mercy killing of the ailing or aging does; however, all are arguably deference of clear and universal reasoning or morality in favor of tendency and subjectivism. In art as in conflict, man surrendered to impulse letting forces flow from more to less, leveling off power into an ideal state of equilibrium; while in peace, unlike either art or conflict, struggle along a path of reason and prescriptive thought lead passing the moat of natural tendency toward collapse across to sustainability. The natural course takes to conflict, unlike the way to peace being one of planning and construction that, once achieved, must be maintained against skepticism and deterioration.

2 Pierre-Joseph Proudhon on War and Peace

In his 1861 essay, *la Guerre et la Paix*, P.-J. Proudhon embraced conflict as a national right causing a stir of reactions equally amongst proponents and adversaries. The consensus that the then-recent 1848 revolution was a major catalyst of the arts did little to mitigate the impact of Proudhon's brusquely setting war as an imperative to advancing the arts.[1] In presenting war as a right, Proudhon closely associated conflict with freedom. That was the first of three propositions possibly aligning war with reason. Secondly, he considered war as judgment the means to which sets war apart from other forms thereof: That war is judgment by force. And that was Proudhon's third proposition. As engagement at a national magnitude, war is an exception to other forms of conflict by being exceptionally deliberate and well planned, recalling Emanuel Kant's statement that peace is more than the mere absence of war. And so, by considering war to be judgment, in exception to other forms of conflict being natural tendency, the parallel between war and the arts –both being natural abandonment to natural forces– already weakens.

Henri Moysset (1875–1949) traced Proudhon's views –often reduced to anarchism– to the philosophy of Hegel, itself gaining popularity around the July Revolution and thereabouts. While possibly true, Proudhon claimed the direct influence of Joseph de Maistre whose polemical style and overt conservatism be admired and echoed. De Maistre was an early voice favoring war as an act of higher will to be understood and accepted even when immediate causes remain unclear.[2]

---

[1] Proudhon, Pierre-Joseph, La Guerre et la Paix. Œuvre Complètes Tome XVIII. Nouvelle Édition. Librairie Internationale, Paris (1869).

[2] Proudhon, p. 31.

Proudhon examined war phenomenologically rather than by way of morals or religion, both of which settled in in the wake of wars rather than caused it. Thus war, religion and art were a means to arriving at moral revelation.

In war, the victor is right. But not every conflict is forceful, nor is victory exclusively a military term of war. In conflicts with reason, argument, and logic, terms are reversed, and it is right that brings about victory not vice versa, for the means to war include, but are not limited to, violence. Von Clausewitz qualifying war as politics by other means introduced "war" into a broader, more inclusive realm as universal struggle with terms of engagement that might circumstantially include violence but were irreducible thereto. Upon understanding war as will that is independent from violence, albeit not exclusive thereof, views of Proudhon and von Clausewitz converge in a representative ideal. Right is independent of force even though they may cross paths.[3] Instead, closer to Hegel's Phenomenology of Spirit[4], conflict–war included—is but another form of transaction, a means to ascertaining the other, and a form of self-manifestation that is as much vulnerability as it is jeopardy of all.

Proudhon dismissed talk of eternal peace, comparing it to blasphemy, dealing its proponents some harsh language, without referring to the author of the seminal booklet *Eternal Peace* (also *Perpetual Peace*), Immanuel Kant, by name.[5] Although he professed against the military institution of his time, Proudhon never called for abolishing war maintaining war to reveal human creative thought at its core, not unlike religion and art. He too considered peace a structure born out of much thought and deliberation. Even to man in his primitive state (l'homme primitif), who thought of reason as force, making peace was as much the work of planning as was planning a new war, both being struggle and strife rather than natural inclination.

Sovereignty and power are deduced from the identities of right and force. Proudhon argued that to the men of might and to aristocrats, force was a property of power and power constituted merit. Use of force was less a visible exercise of merit than an occasion to manifest an ideal. A central component of the primitive state of man, war was as central to the popular ideal, the epic, as it was as a factor shaping social configurations and the resulting loyalties and political expression. As the primal retrospective ideal reflection on history, the epic is the beginning of speculative thought as presented in Aristotle's Poetics. Outside the epic for a people there exists no inspiration, no national chants, no drama, no eloquence, no art. "Si la guerre n'existait pas, la poésie l'inventerait." Proudhon wrote of material occasioning the making of an artifact, as would marble to a statue, but not for entirely explaining the particulars of the artifact. War, a subject-matter central to the epic, is a necessity for a poetic construct that is, factual or not, still entertaining and educational. "If the poet's description be criticized as not true to fact, one may urge perhaps that the object ought to be as described –an answer like that of Sophocles, who said that he drew men as they ought to be, and Euripides as they were. If the description, however, be neither true nor of the thing as it ought to be, the answer

---

[3] For Hegel's influence on Von Clausewitz refer to Creuzinger, Paul. Hegels Einfluss auf Clausewitz. in-8°. Verlag von Eisenschmidt, Berlin (1911).

[4] Hegel, Georg Wilhelm Friedrich. Phänomenologie des Geistes. Verlag der Dürr'schen Buchhandlung, Leipzig (1907).

[5] Proudhon, p. 50.

must be then, that it is in accordance with opinion." In other words, rather than being an historical account, the epic is primarily a work of art that afforded ideal revelation of society, language, religion, and reason through the reflection of the artist. Facts were the raw material for speculative thought on a higher ideal state (things as they ought to be) or analytical commentary (opinion.)

Proudhon, thus, examined war as present and imminent, not allowing detachment or distance, otherwise central to the analysis of, say, Friedrich Schiller in his philosophy of aesthetics. When destruction or breakdown is considered as a dismantling of a setting, at one a regeneration of another with no room for lamentation or dwelling over loss, war occasions reconsideration of reality with the impetus of a force that clears up room with the same hand that demolishes the old, testing structures for strength, exposing weaknesses, and setting the stage for debating relevance.

3 Schiller on Conflict

Violence is presented in Schiller's argument as a calamity befalling man not a choice or act of reason, external with causes escaping reason. "Nothing is so unworthy of man as to suffer violence, for violence disannuls him." By freedom from the reality of conflict would the artist be able to contemplate the show of it in a work of creative art. Reality here is in the consequences of acts of violence, but not in itself violence. By forcing man into subservience, violence "questions nothing less than humanity", however absolute freedom from violence is impossible. Schiller explains that man can attain freedom from the fetters imposed by forces of nature by a play of strife and surrender.

Although violence is maintained by most as a dreadful, yet always possible, property of conflict, de Maistre explicitly recognized violence as a natural part of spirit, tainted and corrupt prior to all conflict. He argued that evil was introduced with disorder and perpetuated by the deception of modern philosophy that made believe that all were good, whereas "evil has polluted everything and in a very real sense all is evil, since nothing is in its proper place." Evil thus is violating the beauty of proportion, an overall lowering of the keynote in the whole melody. Violence is no longer a result of disorder and evil but a pathetic outcome of another cause; one external to violence yet leading to it. "There is, moreover, good reason for doubting if this violent destruction is in general as great an evil as is believed; at least, it is one of those evils that play a part in an order of things in which everything is violent and against nature and which has its compensations. First, when the human spirit has lost its resilience through indolence, incredulity, and the gangrenous vices that follow an excess of civilization, it can be retempered only in blood."

4 Conclusion

War is a unique event in being as much a natural convulsion as it is an intelligent movement motivated by will and realized by reason. To address the twofold action of war as a struggle, one may adopt the Hegelian frame of the self and the other as spirit that affinity sets into motion. Sensing and wishing to recognize the other, and thus actions and passions seek either to verify the externality of the self or to examine the possibility of internalizing the other. The duality of players defines the nature of struggle between the self and the other to recognize, jeopardize, and annihilate one another. A struggle acquires its reality less by its embodiment in material, motion and processes than by confrontation of the spirit with another. By recognition of the other, the spirit defines itself as an objective form of some permanence and independence. In The Laws, Plato celebrated imitation at the foundation of the arts. It is not only limited to the visible or sensuous forms, but also including the narratives that describe the human condition as he wrote: "Our social life is the best Tragedy." Aristotle, in the Poetics, furthered the imitative grounding in Plato's view of art as a sensuous imitation of an eternal model, if one may admit to speculative imitation. Aristotle maintained an historical core, with names and places of reality; however, he "poetically" contemplated reality, examining the margins of the truth. Art, in this instance poetry, travels through the truth tearing at slavish accounts with the lancet of possibilities, probabilities and their respective negatives thus creating new narratives several fold removed yet ever expanding imagination, and –with it—the bounds of the human condition as only art can. Factual accounts of events and people in poetry constitute only one part of a whole of actions, passions and ideas. The victory of reasonable persuasion over Necessity is one of a strategy of design to create an operative construction of the visible world. War is a construction of the visible world, and is, thus, no exception to the combined operation of Reason and Necessity.

References

Creuzinger, P.: Hegels Einfluss auf Clausewitz. Verlag von Eisenschmidt, Berlin (1911)

Hegel, G.W.F.: Phänomenologie des Geistes. Verlag der Dürr'schen Buchhandlung, Leipzig (1907)

Proudhon, P.-J.: La Guerre et la Paix. Œuvre Complètes. Tome XVIII. Nouvelle Édition. Librairie Internationale, Paris (1869)

Butcher, S.H.: Aristotle's Theory of Poetry and Fine Art, with a Critical Text and Translation of the Poetics, 4th edn. Macmillan and Co., Limited, London (1920)

Cornford, F.: Plato's Cosmology. The Timaeus of Plato. Translated with Running Commentary by Francis Cornford. Routledge, Indianapolis (1935, 1997)

De Maistre, J.: Considerations on France. Translated and edited by Richard A. Lebrun. Introduction by Isaiah Berlin. Cambridge University Press, Cambridge (1994, 2003)

Gilson, E.: The Unity of Philosophical Experience. Charles Scribner's Sons, New York (1937)

Hegel, G.W.F.: Hegel. Phenomenology of the Spirit. Translated with Introduction and Commentary by Michael Inwood. Oxford University Press (2018)

Kant, I.: Basic Writings of Kant. Edited with an Introduction by Allen E. Wood. The Modern Library, New York (2001)

Maguire, T.M.: General Carl von Clawsewitz on War. William Clowes & Sons Ltd., London (1909)

Schiller, F.: Essays on the Sublime: On Beauty, On Grace and Dignity, On the Sublime. A Treasury of Philosophy. Edited by Dagobert D. Runes. vol. 1. The Philosophical Library, Grolier Incorporated, New York (1955)

Track on Doctoral Consortium

A Comparison of DoDAF, TOGAF, and FEAF: Architectural Frameworks for Effective Systems Design

Bernardo Gaudêncio[1], João Ferraz[2], Pedro Martins[2(✉)], Paulo Váz[2], José Silva[3], Maryam Abbasi[1], and Filipe Cardoso[1]

[1] Polytechnic Institute of Santarém, Santarém Higher School of Management and Technology, Santarém, Portugal
{200100083,filipe.cardoso}@esg.ipsantarem.pt, maryam.abbasi@ipc.pt
[2] CISeD - Digital Services Research Center, Polytechnic of Viseu, Viseu, Portugal
190100423@esg.ipsantarem.pt, {pedromom,paulovaz}@estgv.ipv.pt
[3] Applied Research Institute, Coimbra Polytechnic, Coimbra, Portugal
jsilva@estgv.ipv.pt

Abstract. Architectural frameworks play a vital role in guiding the design, implementation, and management of complex systems. This paper provides a comprehensive comparison of three prominent frameworks: DoDAF (Department of Defense Architecture Framework), TOGAF (The Open Group Architecture Framework), and FEAF (Federal Enterprise Architecture Framework). Each framework has its unique focus and purpose, with the aim of facilitating effective architecture development and governance. DoDAF primarily caters to defense systems within the U.S. Department of Defense (DoD), offering operational, system, and technical viewpoints. TOGAF, on the other hand, provides a holistic approach applicable to enterprises across industries, addressing business, data, application, and technology architecture domains. FEAF focuses on enterprise architecture within the US federal government, encompassing business, performance, service, data, and technical reference models. This paper examines and compares these frameworks based on aspects such as scope, development organization, architecture domains, framework structure, notable versions, certification programs, and popularity. The findings help practitioners and organizations understand the unique characteristics and suitability of each framework for effective system design, ultimately allowing informed decision-making regarding framework selection and implementation.

Keywords: Comparative analysis · Architecture · Frameworks · Enterprise Architecture

1 Introduction

In today's complex and rapidly evolving technological landscape, organizations face the challenge of designing and managing effective systems that align with their business objectives. Architectural frameworks provide structured approaches and guidance for designing, organizing and managing these systems, enabling organizations to achieve operational efficiency, optimize resource allocation, and foster innovation. Three prominent architectural frameworks, namely DoDAF, TOGAF, and FEAF, have emerged as comprehensive methodologies that offer standardized approaches for effective systems design.

This article aims to provide a comprehensive comparison of DoDAF, TOGAF, and FEAF to highlight their features, strengths, and limitations for effective systems design. By examining the purpose, scope, architecture domains, framework structures, notable versions, certification programs, and popularity of these frameworks, we aim to identify the key similarities and differences between them.

In the following sections, we will dive into the details of each framework, exploring its principles, structures, and implementation considerations. Through this exploration, we aim to contribute to the body of knowledge surrounding architectural frameworks and provide practical insights for organizations striving for effective systems design in today's dynamic and complex technological landscape.

2 Architecture Frameworks

2.1 Department of Defense Architecture Framework (DoDAF)

This framework is developed for the United States Department of Defense (DoD). Moreover, DoDAF defines a way of corresponding an enterprise architecture that enables stakeholders to focus on specific areas of interest, while still being able to see the bigger picture. To aid decision makers, DoDAF provides means that capture essential information and present it with comprehensibility and consistency. One of the primary objectives is to present this information in a way that is understandable to the many stakeholder communities involved in developing, delivering and maintaining capabilities. Furthermore, in support of the stakeholder's mission. Additionally, DoDAF divides the problem space into achievable pieces, according to the stakeholder's viewpoint, further defined as DoDAF-described Models[1]. In this way, the model for DoDAF version 1.5 is divided into seven distinguishing views.

However, it should be emphasized that DoDAF is fundamentally about creating a coherent model of the enterprise to enable effective decision-making. Presentational aspects should not overemphasize graphic presentation at the expense of rudimentary data (Table 1).

2.2 The Open Group Architectural Framework (TOGAF)

The Open Group Architecture Framework (TOGAF) is a widely recognized and comprehensive framework for enterprise architecture. Developed and maintained by The Open Group, a global consortium of organizations [2], TOGAF provides

Table 1. DoDAF Viewpoints

All Viewpoint
Capability Viewpoint
Data and Information Viewpoint
Operational Viewpoint
Project Viewpoint
Services Viewpoint
Standards Viewpoint
Systems Viewpoint

a structured and systematic approach to designing, planning, implementing, and governing enterprise architectures. It offers a set of best practices, methodologies, and resources to enable organizations to align their business objectives with their IT strategies and create robust, adaptable, and scalable systems [3].

TOGAF encompasses a broad range of architectural domains, including business architecture, data architecture, application architecture, and technology architecture [4]. By addressing these domains holistically, TOGAF facilitates the integration and alignment of different architectural perspectives to achieve a cohesive and optimized enterprise architecture [5].

The framework provides a step-by-step process for developing and managing architectures, beginning with the establishment of an architecture vision and strategy, followed by the development of architectural artifacts, such as models, frameworks, and guidelines [6]. TOGAF emphasizes the importance of stakeholder engagement, iterative development, and continuous improvement throughout the architecture lifecycle [7].

TOGAF has gained global recognition and widespread adoption across industries and sectors [8]. It offers a common language and framework for architecture professionals, enabling effective communication and collaboration among stakeholders [9]. Additionally, TOGAF's extensive ecosystem includes a rich collection of reference models, templates, and tools that support practitioners in applying the framework to their specific organizational context [10].

In summary, TOGAF is a powerful framework that enables organizations to navigate the complexities of enterprise architecture and drive strategic alignment between business and technology. With its structured approach, comprehensive scope, and industry-wide acceptance, TOGAF serves as a valuable resource for architects and organizations seeking to optimize their systems' design, implementation, and governance [11].

2.3 Federal Enterprise Architecture Framework (FEAF)

The Federal Enterprise Architecture Framework (FEAF) has emerged as a significant framework for developing and managing enterprise architectures within the U.S. federal government. With its structured approach and emphasis on alignment between business objectives and technology implementations, FEAF provides a valuable tool for effective governance, strategic planning, and decision-making in the public sector.

FEAF offers a comprehensive and standardized approach to developing and managing enterprise architectures within the federal government. It encompasses multiple architecture domains, including business, performance, service, data, and technical reference models. By providing a common language and set of principles, FEAF facilitates collaboration, information sharing, and the realization of enterprise-wide goals.

The primary objective of FEAF is to enable federal agencies to achieve operational efficiency, promote interoperability, and ensure the seamless integration of information technology investments. By employing FEAF, organizations can strategically align their business processes, data, applications, and technologies to drive improved performance, effective service delivery, and cost optimization.

FEAF provides federal agencies with a structured framework for capturing, analyzing, and documenting their current and desired future state architectures. It offers guidance on architectural principles, standards, and best practices that enable organizations to make informed decisions about technology investments, prioritize initiatives, and manage architectural change effectively.

The adoption of FEAF within the federal government has yielded numerous benefits, including improved collaboration and communication among agencies, reduced duplication of efforts, increased transparency, and enhanced decision-making capabilities. It has played a crucial role in driving digital transformation initiatives and promoting the effective use of technology to achieve mission objectives.

Through this analysis, we aim to contribute to the understanding of FEAF's value, applicability, and impact on enterprise architecture management in the public sector. By uncovering the strengths, challenges, and lessons learned from FEAF adoption, we can provide insights that enable organizations to maximize the benefits of this framework and drive successful architectural transformations in the complex landscape of the U.S. federal government.

3 Comparison of Information Systems Frameworks: DoDAF, TOGAF, and FEAF

3.1 Results

In the field of information systems architecture, several frameworks have been developed to provide standardized approaches for defining, organizing, and managing complex architectures. This comparison aims to highlight key aspects of three prominent frameworks: the DoDAF (Department of Defense Architecture Framework), TOGAF (The Open Group Architecture Framework), and FEAF (Federal Enterprise Architecture Framework).

The DoDAF primarily focuses on the architecture of military and defense systems within the U.S. Department of Defense (DoD), providing viewpoints and models for operational, system, and technical architectures. On the other hand, TOGAF offers a comprehensive approach applicable to enterprises across industries and sectors, encompassing business, data, application, and technology architecture domains. Meanwhile, the FEAF is specifically designed for developing, implementing, and managing enterprise architectures within the U.S. federal government, covering multiple domains such as business, performance, service, data, and technical reference models.

This comparison table highlights various aspects of these frameworks, including their purposes, scopes, development organizations, architecture domains, framework structures, notable versions, certification programs, and popularity. It is important to note that while DoDAF and TOGAF have achieved widespread adoption globally, FEAF primarily finds its usage within the U.S. federal government, albeit with some principles adopted in other sectors (Table 2).

3.2 Discussion

The table provided presents a comprehensive comparison of three prominent information systems frameworks: DoDAF, TOGAF, and FEAF. These frameworks offer standardized approaches for designing, organizing, and managing complex architectures, albeit with different focuses and scopes.

DoDAF, developed by the U.S. Department of Defense (DoD), is primarily geared toward the military and defense sectors. It provides a set of viewpoints and models for operational, system, and technical architectures. With a strong emphasis on defense systems, DoDAF has gained widespread adoption within the defense community, enabling consistent and structured architecture development.

TOGAF, maintained by The Open Group, is a versatile framework applicable to various industries and sectors. It encompasses a comprehensive set of architectural domains, including business, data, application, and technology architectures. TOGAF's strength lies in its ability to guide organizations in developing enterprise architectures, enabling alignment between business goals and technological implementations.

FEAF, developed and maintained by the U.S. Federal Chief Information Officers Council (CIOC), focuses on the U.S. federal government context. It covers multiple architecture domains, such as business, performance, service, data, and technical reference models. While primarily used within the federal government, FEAF's principles and concepts have influenced architectural practices in other sectors as well.

In terms of framework structure, DoDAF comprises three primary views: operational, system, and technical views. TOGAF, on the other hand, encompasses several architectural domains, allowing organizations to address different aspects of their enterprise architecture. FEAF is organized into reference models, segments, and capabilities, providing a comprehensive framework for developing and managing enterprise architectures.

Notable versions of these frameworks include DoDAF 2.0, TOGAF 9.2, and FEAF 2.2, reflecting their evolution and adaptation to changing architectural

Table 2. Comparison of Information Systems Frameworks

Framework	DoDAF	TOGAF	FEAF
Purpose	Designed to provide a standardized approach for defining, organizing, and representing the architecture of defense systems.	A comprehensive approach for designing, planning, implementing, and governing enterprise architectures.	A framework for developing, implementing, and managing enterprise architectures within the U.S. federal government.
Scope	Primarily focused on military and defense systems within the U.S. Department of Defense (DoD).	Broadly applicable to enterprises across industries and sectors, covering business, data, application, and technology architecture domains.	Primarily designed for use within the U.S. federal government, covering the business, performance, service, data, and technical reference models.
Development Organization	Developed by the U.S. Department of Defense (DoD).	Developed and maintained by The Open Group, a global consortium of organizations.	Developed and maintained by the U.S. Federal Chief Information Officers Council (CIOC).
Architecture Domains	Provides viewpoints and models for operational, system, and technical architectures.	Addresses business architecture, data architecture, application architecture, and technology architecture.	Considers business, performance, service, data, and technical reference models.
Framework Structure	Consists of three primary views: operational, system, and technical views.	Comprises several architectural domains, including business, data, application, and technology architectures.	Organized into reference models, segments, and capabilities, providing a comprehensive framework.
Notable Versions	DoDAF 2.0 is the most recent version.	TOGAF 9.2 is the most recent version.	FEAF 2.2 is the most recent version.
Certification Program	Offers certification programs for architecture practitioners.	Offers certification programs for architects at various levels, such as TOGAF Certified and TOGAF Practitioner.	Does not have a specific certification program but may be used in conjunction with other certification programs.
Popularity	Widely adopted within the defense and military sectors, particularly in the U.S.	Globally recognized and widely adopted across various industries and sectors.	Primarily used within the U.S. federal government, but some concepts and principles are also adopted in other sectors.

needs. Additionally, both DoDAF and TOGAF offer certification programs for architecture practitioners, while FEAF does not have a specific certification program but can be used in conjunction with other certification programs.

In terms of popularity, DoDAF is widely adopted within the defense and military sectors, particularly in the U.S. TOGAF enjoys global recognition and adoption across various industries and sectors, making it one of the most widely used frameworks. FEAF, although primarily used within the U.S. federal government, has also influenced architectural practices in other sectors to some extent.

Understanding the nuances and strengths of these frameworks is crucial for organizations seeking to establish effective and standardized architectural practices. Each framework offers unique advantages and considerations, and the selection depends on the specific needs and objectives of the organization. By leveraging the guidance provided by these frameworks, enterprises can achieve better alignment between business and technology, resulting in improved operational efficiency and successful architecture implementations.

4 Conclusions

In conclusion, the article provides a comprehensive comparison of the DoDAF, TOGAF, and FEAF information systems frameworks. The table highlights key aspects of each framework, including its purpose, scope, development organization, architecture domains, framework structure, notable versions, certification programs, and popularity.

From the comparison, it is evident that these frameworks have distinct focuses and target different sectors. DoDAF primarily caters to the military and defense systems within the U.S. Department of Defense, while TOGAF offers a comprehensive approach applicable to enterprises across industries. FEAF, on the other hand, is primarily designed for use within the U.S. federal government.

The frameworks also differ in terms of their architecture domains and structure. DoDAF provides viewpoints and models for operational, system, and technical architectures, while TOGAF addresses business, data, application, and technology architecture domains. FEAF considers business, performance, service, data, and technical reference models. Furthermore, the framework structures vary, with DoDAF consisting of three primary views, TOGAF comprising multiple architectural domains, and FEAF organized into reference models, segments, and capabilities.

In addition, the development organizations behind these frameworks also differ. DoDAF is developed by the U.S. Department of Defense, TOGAF is maintained by The Open Group, and FEAF is developed and maintained by the U.S. Federal Chief Information Officers Council. This distinction can influence the development, update, and support of the frameworks.

The table also highlights the notable versions, certification programs, and popularity of each framework. DoDAF and TOGAF have established certification programs for architecture practitioners, whereas FEAF does not have a specific certification program but may be used in conjunction with other certifications. Furthermore, the popularity of the frameworks varies, with DoDAF widely adopted in the defense and military sectors, TOGAF globally recognized and widely adopted across industries, and FEAF primarily used within the U.S. federal government.

Overall, the table provides a valuable reference for organizations seeking to evaluate and select an appropriate information systems framework based on their specific requirements, industry, and governance needs. Understanding the similarities and differences between DoDAF, TOGAF, and FEAF can facilitate informed decision-making to effectively design and manage complex system architectures.

Acknowledgements. "This work is funded by National Funds through the FCT - Foundation for Science and Technology, I.P., within the scope of the project Ref. UIDB/05583/2020. Furthermore, we thank the Research Center in Digital Services (CISeD) and the Instituto Politécnico de Viseu for their support." Maryam Abbasi thanks the national funding by FCT - Foundation for Science and Technology, I.P., through the institutional scientific employment program contract (CEECINST/00077/2021).

References

1. DODAF Viewpoints and Models. (sem data). Obtido 28 de maio de 2023, de https://dodcio.defense.gov/Library/DoD-Architecture-Framework/dodaf20_viewpoints/
2. The Open Group. (n.d.). TOGAF®Standard - Version 9.2. https://www.opengroup.org/togaf
3. Distaso, L.: TOGAF®Version 9.1. An Introduction. In M. Lankhorst, A. Proper, F. Jonkers, & M. Quartel (Eds.), Enterprise Architecture at Work: Modelling, Communication and Analysis (Third Edition, pp. 185–202). Springer (2016)
4. The Open Group. (n.d.). Architecture Domains. https://www.opengroup.org/togaf#architecture-domains
5. Bente, S., Aeberhard, M.: The TOGAF®standard and the archiMate®modeling language. In: Hofmeister, C., Nord, R., Soni, D. (eds) Applied Software Architecture, pp. 19–39. Springer, Cham (2013)
6. The Open Group. (n.d.). Architecture Development Method (ADM). https://www.opengroup.org/togaf#architecture-development-method
7. Kamal, S., Awad, A.: IT governance and enterprise architecture: a systematic literature review. Int. J. Inf. Manage. **37**(6), 645–656 (2017)
8. Monreal, T.: The TOGAF®Standard as a Catalyst for Change. In M. Lankhorst, A. Proper, F. Jonkers, & M. Quartel (Eds.), Enterprise Architecture at Work: Modelling, Communication and Analysis (Second Edition, pp. 23- 34). Springer (2012)
9. Nagy, G., Dudás, Á.: Application of enterprise architecture frameworks for digital transformation in the public sector. Period. Polytech. Soc. Manag. Sci. **27**(1), 65–74 (2019)
10. The Open Group. (n.d.). TOGAF®Ecosystem. https://www.opengroup.org/togaf#ecosystem

11. Ouyang, C., Henderson-Sellers, B.: Architecture Patterns for Business Process Systems. CRC Press, Boca Raton (2013)
12. Chief. (1999). Federal Enterprise Architecture Framework
13. GmbH, L. (n.d.). FEAF - Federal Enterprise Architecture Framework — LeanIX. https://www.leanix.net, https://www.leanix.net/en/wiki/ea/feaf-federal-enterprise-architecture-framework#What-is-the-FEAF

From Waste to Wealth: Circular Economy Approaches for Recycled EV Batteries in Energy Storage

Alejandro H. de la Iglesia[1](✉) , Carlos Chinchilla Corbacho[2] , Jorge Zakour Dib[2] , and Fernando Lobato Alejano[2]

[1] Department of Computer Science and Automation, University of Salamanca, Pl. Caídos, S/N, 37008 Salamanca, Spain
alexhiglesias@usal.es

[2] Faculty of Computer Science, Pontifical University of Salamanca, C/ Compañía 5, 37002 Salamanca, Spain
{cchinchillaco,jzakourdi,flobatoal}@upsa.es

Abstract. The war that began in Ukraine in 2022 has made European governments understand their significant dependence on fossil fuels, specifically natural gas for electricity generation through combined cycle plants. As a result of this war, Europe reduced its importation of natural gas from Russia, raising its price and directly increasing the cost of energy generation. The position taken by European governments was to develop and accelerate the energy transition from fossil fuel-based generation to renewable energy sources. One of the challenges that arise from producing energy in this way is the need to store the energy produced when demand is low and utilize it when demand is high. Another major challenge is finding ways to give a second life to batteries discarded by the growing electric vehicle industry when their charging capacity is no longer optimal. This paper seeks to find an innovative and ethical model of circular economy based on existing models, whereby the evaluation of cells from discarded electric vehicle batteries can make it profitable to create energy storage systems and reintroduce these batteries into the market. This aligns with the United Nations' sustainable development goals, generating clean and affordable energy (SDG 7), providing a solution to high demand peaks in the industry through innovation in electrical infrastructures (SDG 9), and helping to improve responsible consumption (SDG 12).

Keywords: Circular economy · Disruptive technology · Energy storage · Renewable energy transition · Electric vehicle batteries

1 Introduction

As the population, development and growth of European economies have increased, so has the production and consumption of energy [1]. This presents a challenge for governments in establishing energy production mechanisms, where the reliance on fossil fuels for energy production remains significantly higher than desired and necessary to achieve the target set by the Paris Agreement, which aims for net-zero global emissions by 2050. In Europe, the percentage of fossil fuels used for energy production has been the highest by product, accounting for 70% [2] of the total gross available energy in 2021. Among the fossil fuels used for energy production, natural gas stands out, contributing 22% [3] of the total energy generation by product. The historically low price of this product and the proliferation of combined cycle thermal power plants that use natural gas to produce energy have led to a significant dependency on this fuel for electricity generation in most European countries. This dependency means that any intervention in the gas market prices causes a spike in electricity prices [4] (see Fig. 1). Russia has been the main supplier of natural gas to most European countries [5] (see Fig. 2), and due to the geopolitical context created by the conflict in Ukraine, these countries have had to implement contingency measures to avoid drastic increases in energy prices and shortages.

Fig. 1. Comparative Graph of Natural Gas Prices and Energy Prices

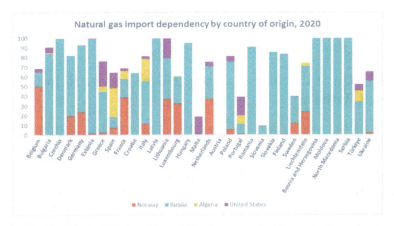

Fig. 2. Graph of Various Natural Gas Importers for European Countries

Since 2022, European countries have made significant efforts to transition from fossil fuel-based energy production to renewable energy production. In 2022, the share of renewable energy was 23% [6], with a forecast to reach 42.5% by 2030 (see Fig. 3).

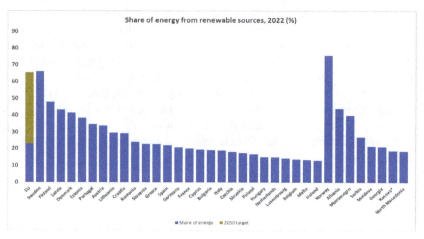

Fig. 3. Energy Production by Country

This forecast indicates that energy storage systems will be necessary in areas where there is a deficit between energy generation and consumption, allowing storage to release energy during peak demand periods when production is insufficient to meet consumption. These changes in energy generation and consumption suggest that it will be necessary to implement disruptive circular economy models to address the scarcity of raw materials used in the storage and production of green energy. This can be achieved by giving a second life to batteries from electric vehicles, thereby creating energy storage systems.

In this work, we propose an initial approach to a circular economy model based on disruptive technologies, focusing on how to incentivize battery recycling by involving stakeholders in future revenues through the tokenization of these batteries. These involved actors will hold partial ownership of the energy storage systems developed with these batteries, creating a market where they can buy or sell these tokens and benefit from future monetization from the use of these storage systems.

2 New Vision of Circular Economy Models for the Reintroduction of Used Batteries

As electric vehicles become increasingly popular as a green mode of transportation, the quantity of batteries to be discarded in the future is expected to rise. It is estimated that around 250,000 tons of lithium-ion will have been discarded from electric vehicle use within 5 years [7]. This poses a challenge for the recycling of such highly contaminating elements and represents a danger to the environment if not handled properly.

These batteries can be repurposed to give them a second life, thereby avoiding the recycling process altogether. One alternative for them is the creation of energy accumulators by selecting the best cells, thus creating new accumulators with optimal lifespan.

There are various alternatives for reintroducing different battery cells into the market through a circular economy approach, transitioning from linear to circular business models. While some studies focus on how different actors in the electric vehicle ecosystem are behaving, a new alternative for these industries is not addressed.

The evaluated approaches are those of the industry itself reintroducing these batteries. A new perspective for the reintroduction of previously evaluated cells, having selected cells whose health status is suitable for use, would be their transfer by various actors related to electric vehicles. It is necessary to implement a business model in which the actor transferring these cells owns a portion of them and receives a variable compensation based on the cell's charging capacity, in addition to considering the electricity price for the use of the global storage system.

The introduction of these cells into our business model must be accomplished through reverse logistics [8] to ensure access and ownership of the asset. Reverse logistics involves transporting the product from collection points to the processing point. There are two different reverse logistics flows: efficient flow and responsive flow [9]. For this project, an efficient flow is proposed since battery lifespan tests will be centralized in one location. The responsive flow evaluates the state of the batteries at the same place of recovery, which is discarded due to the high number of collection points and the complexity of decentralizing the battery evaluation process. A similar disruptive circular economy and reverse logistics approach was studied by Loo-See Beh [10], which addresses a new perspective on reverse logistics to make it more viable for second-hand clothing retailers. It introduces an innovation in the reverse supply chain allowing retailers additional revenue, improving sustainability, and democratizing consumption. These models challenge the idea of returning products to the point of origin and seek more creative and efficient alternatives.

3 Proposed Model

The proposed circular economy model in this study focuses on three phases and a final phase of monetization of this model (see Fig. 4). The first phase will involve the introduction of batteries into our supply chain through their joint acquisition. The second phase will be based on the management of these batteries, and the third phase will focus on the creation and production of energy storage systems in the energy market. The last phase would be the development of a model capable of monetizing these storage systems by distributing that monetization among all those involved in the model.

3.1 Phase 1: Battery Acquisition

The success of this first phase signifies the viability of the model in question. The model revolves around the acquisition of batteries that, due to their capabilities, no longer have a useful life in the systems for which they were designed. The best way to prevent these batteries from being discarded is to incentivize stakeholders to relinquish them and become co-owners. To ensure access to these batteries, means for their disposal and collection by companies adhering to our model must be facilitated for these stakeholders.

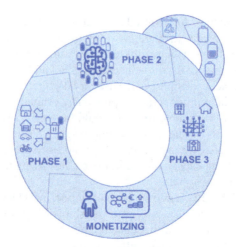

Fig. 4. Business Model Cycle

Properly designed containers with high security capacity will be provided for storing the batteries until they are collected.

During the battery collection process, stakeholders will be provided with an unequivocal tracking code for their batteries. This will allow us to gather information about the lifespan of each cell in the battery and the percentage of ownership acquired at the time of its relinquishment. This entire process will be carried out through reverse logistics, which provides us with a feasible way to access a product necessary for the project's viability. Due to the project's ethical stance of avoiding waste and addressing new recycling methods, we do not manufacture these batteries.

By utilizing an efficient flow of reverse logistics, the analysis of the battery lifespan can be centralized in our central warehouse, avoiding the complexity of analyzing this status at each supply point that may be established. Furthermore, this approach will provide us with more capacity to evaluate all batteries equally, thus leading to a better co-acquisition process.

3.2 Phase 2: Cell Evaluation and Tokenization

In order to develop energy storage systems with high accumulation capacity, we need robust algorithms and systems to evaluate the cells we acquire. In phase two, these cells are evaluated and scored based on their health status, optimal charging capacity, and Marginal Value of Life (MVT [8]). This marginal value will also have a significant impact on the value assigned to the cell, as it describes the relationship between time and the reduction of the cell's initial value. Therefore, a cell with a high MVT will see its residual value decrease more rapidly than a product with a low MVT. In other words, stakeholders should facilitate this process as quickly as possible to prevent this value from decreasing.

3.3 Phase 3: Development and Production

In this phase, the process of constructing and operating energy storage systems using battery cells evaluated in Phase 2 is described. Additionally, a continuous cycle between the development and evaluation phases is proposed to ensure the maintenance and optimal efficiency of these systems. Once the battery cells have been evaluated and selected based on their health status and optimal charging capacity, they are assembled into energy storage system modules. These cells are grouped into storage modules, ensuring that cells with similar characteristics are assembled together to maximize system efficiency and durability.

After grouping, they are assembled into modules using appropriate connection techniques to ensure system stability and safety. The modules can vary in size and capacity depending on the specific application. Energy storage systems are integrated into larger systems that can be used by communities, hospitals, office buildings, among others. This integration includes the installation of Energy Management Systems (EMS) that optimize the use and distribution of stored energy. By using monitoring systems to assess the performance of each cell within the storage system, we will be able to periodically reevaluate cells that show a significant decrease in performance for withdrawal and reevaluation according to the criteria established in Phase 2. These cells can be reconfigured, replaced, or recycled based on their condition. This ensures that the storage systems operate at their maximum efficiency.

3.4 Monetization

To incentivize stakeholders and make this model viable, we need to compensate the actors who provide us with our raw material, which is none other than batteries that are no longer useful for the systems they were designed for. The best way to incentivize stakeholders is to make them co-owners of the cells they contribute for recycling. In phase two, these cells are evaluated and scored based on their health status, optimal charging capacity, and Marginal Value of Life. This way, we can give a percentage of this ownership to the user who provided the cell. To implement this tokenization system and revenue sharing with stakeholders, we will rely on a blockchain system to develop a platform that provides functionality and robustness to them. The use of blockchain [11] ensures transparency, security, and efficiency in managing ownership and profit distribution. The model includes the generation of profitability tokens based on the performance of battery cells, which can be redeemed for money on secondary markets, offering a new investment and profit avenue to users.

3.4.1 Blockchain Implementation

The implementation of blockchain technology is one of the fundamental pillars on which the success of this business model depends. The choice of blockchain platform is a crucial step, as it must meet requirements for security, scalability, efficiency, and compatibility with smart contracts. The two most prominent platforms for this purpose are Ethereum [12] and Binance Smart Chain [13].

Ethereum (ETH) is the most widely used smart contract platform with a large community of developers and technical support. Additionally, it is highly decentralized, which increases network security and resilience. It has standardization for generating fungible and non-fungible tokens, ERC-20 and ERC-721, respectively. The main issue with this platform is its high gas fees and network congestion.

Binance Smart Chain (BSC) is also a smart contract platform with high acceptance and usage. BSC offers much lower transaction fees compared to ETH and has faster transaction speeds due to its high processing capacity. Having compatibility with ETH token standards facilitates its development. However, BSC is less decentralized than ETH.

Within our blockchain platform, two types of smart contracts should be developed, which are self-executing programs stored on the blockchain and executed when certain conditions are met. These contracts will be used for issuing and managing ownership and profitability tokens.

These tokens will be developed based on the ERC-20 standard to create fungible tokens. We will have two types of tokens, one intended for assets, in this particular case, each cell contributed by users, which will serve as compensation percentages for those users who co-own the cells. The other type of token will be intended for the profitability of the cells, which, based on the performance of these cells in energy storage systems, will generate more or less profitability for the owners of asset tokens. These tokens can be traded on cryptocurrency secondary markets to allow the sale of profitability or asset tokens and their conversion into physical money. Implementing blockchain in the business model offers a secure, transparent, and efficient solution for managing fractional ownership and distributing profitability based on the performance of battery cells. This approach not only improves the sustainability and efficiency of energy storage but also democratizes investment in emerging technologies.

4 Conclusions

The proposed model of energy storage through reused batteries, with fractional ownership and profit generation based on blockchain, represents an innovative, sustainable, and ethical solution that addresses both the technical and financial challenges of the energy sector. By combining operational efficiency with investment democratization, this approach has the potential to drive a transition towards a greener and more equitable energy future. Future research and developments in this area promise to expand the capabilities and applications of this model, contributing significantly to technological and economic advancement in the field of energy storage.

Acknowledgments. This work is part of the project TED2021-131981A-I00 funded by MCIN/AEI/https://doi.org/10.13039/501100011033 and by the "European Union Next GenerationEU/PRTR".

References

1. Pirlogea, C., Cicea, C.: Econometric perspective of the energy consumption and economic growth relation in European Union. Renew. Sustain. Energy Rev. **16**(8), 5718–5726 (2012). https://doi.org/10.1016/J.RSER.2012.06.010

2. Fossil fuels stabilised at 70% of energy use in 2021 - Eurostat. https://ec.europa.eu/eurostat/web/products-eurostat-news/w/ddn-20230130-1. Accessed 30 May 2024
3. Statistics—Eurostat. https://ec.europa.eu/eurostat/databrowser/view/nrg_bal_s/default/table?lang=en. Accessed 30 May 2024
4. Statistics—Eurostat. https://ec.europa.eu/eurostat/databrowser/product/page/NRG_PC_204_C. Accessed 30 May 2024
5. Statistics—Eurostat. https://ec.europa.eu/eurostat/databrowser/product/page/NRG_IND_IDOGAS. Accessed 30 May 2024
6. Renewable energy statistics - Statistics Explained. https://ec.europa.eu/eurostat/statistics-explained/index.php?title=Renewable_energy_statistics#Share_of_renewable_energy_more_than_doubled_between_2004_and_2022. Accessed 30 May 2024
7. Toorajipour, R., Chirumalla, K., Parida, V., Johansson, G., Dahlquist, E., Wallin, F.: Preconditions of circular business model innovation for the electric vehicle battery second life: an ecosystem perspective. Adv. Transdiscipl. Eng. **21**, 279–291 (2022). https://doi.org/10.3233/ATDE220147
8. Morana, R., Seuring, S.: End-of-life returns of long-lived products from end customer—insights from an ideally set up closed-loop supply chain. Int. J. Prod. Res. **45**(18–19), 4423–4437 (2007). https://doi.org/10.1080/00207540701472736
9. Blackburn, J.D., Guide, V.D.R., Souza, G.C., Van Wassenhove, L.N.: Reverse supply chains for commercial returns. Calif. Manag. Rev. **46**(2), 6–22 (2004). https://doi.org/10.2307/41166207
10. Beh, L.S., Ghobadian, A., He, Q., Gallear, D., O'Regan, N.: Second-life retailing: a reverse supply chain perspective. Supply Chain Manag. **21**(2), 259–272 (2016). https://doi.org/10.1108/SCM-07-2015-0296/FULL/PDF
11. Nakamoto, S.: Bitcoin: A Peer-to-Peer Electronic Cash System. Welbeck Publishing Group, London (2008)
12. Ethereum Whitepaper—ethereum.org. https://ethereum.org/en/whitepaper/. Accessed 30 May 2024
13. BNB Chain Builders' Manual - A Comprehensive Guide for Developers. https://www.bnbchain.org/en/developers. Accessed 30 May 2024

An IoT-Based Framework for Smart Homes

André Bastos[1], Carlos Silva[1], Luís Pais[1], João Henriques[1,2(✉)], Filipe Caldeira[1,2], and Cristina Wanzeller[1,2]

[1] Polytechnic of Viseu, Viseu, Portugal
{estgv19017,pv20255,pv20253}@alunos.estgv.ipv.pt,
{joaohenriques,caldeira,cwanzeller}@estgv.ipv.pt
[2] CISeD - Research Centre in Digital Services, Polytechnic of Viseu, Viseu, Portugal

Abstract. Intelligence and connectivity are critical in daily interactions and an interconnected world. The home environment is an essential space for applying such technological features, simplifying our lives, and improving the efficiency of routine tasks.

In that aim, this paper proposes a framework taking advantage of the power of IoT-based devices to increase the functional aspects of common household elements, such as lighting and temperature control. The proposed framework employs IoT devices for data collection, a middleware layer for data normalization, and an analytics component for comprehensive data analysis specific to the targeted household. Beyond the evident improvements in accessibility and time management associated with mundane household tasks, the system is designed with adaptability in mind. Their results demonstrate the framework's suitability to be extended to a wide range of homes and spaces.

Keywords: Internet of Things · Containers · Smart Homes

1 Introduction

In the era of smart living, where the fusion of intelligence and connectivity has become the utmost priority, the home environment emerges as a pivotal space for integrating cutting-edge technologies. As our surroundings become increasingly connected, everyday devices exhibit you. Unprecedented levels of intelligence and connectivity. This paper proposes an innovative system that demonstrates the transformative potential of an IoT-powered framework in enhancing various aspects of the typical household. Our focus is improving mundane yet essential elements of home life, such as lighting and temperature regulation. By employing IoT devices for data collection, a middleware layer for data normalization, and incorporating an analytics component for in-depth data analysis and device interaction, we aim to not only improve the accessibility and efficiency of managing household tasks but also to introduce a system adaptable to the most diverse homes and spaces.

This paper starts with a discussion of the devices and their related functions, emphasizing the practical aspects for the user. Then, the communication protocols, layer distribution, and the criteria employed for data collection and normalization are discussed. Docker-based architecture is also concerned with justifying the use of such technology while presenting its advantages. Next, the analytics layer is discussed, explaining the features available to the user and the benefits provided. Another important aspect of this research is the adaptability to the most diverse spaces, ensuring that the proposed framework is not dependent on space configuration and, therefore, can be deployed in the most varied spaces.

Existent systems employing similar architectures or proposing to achieve an identical goal, as the one presented, are also examined. The advantages and limitations or challenges associated with the utilization and implementation of the proposed framework are also analyzed.

This paper emphasizes the proposed framework's role in enhancing the efficiency of an intelligent home system. By harnessing the capabilities of Docker containers, the proposed system enables optimal decision-making in managing household tasks, including lighting, temperature control, and door operations. Through the encapsulation and isolation provided by Docker, users can make well-informed decisions, effectively improving the functionality of their homes, minimizing risks, and achieving excellent stability and success in the ever-evolving smart living landscape.

2 Related Work

As the global population grows, the demand for energy rises, prompting a reevaluation of outdated power grids. Fossil fuel limitations and environmental concerns drive the transition to a distributed hybrid energy system. The smart grid, integrating ICT, emerges to enhance energy generation and consumption intelligently. Within this context, the smart home gains prominence, utilizing IoT for efficient energy management. This paper presents a comprehensive exploration of IoT integration into smart homes, addressing energy considerations, architectural challenges, and data processing. Methodology involves a literature review, resulting in a holistic framework. Challenges include IoT resource constraints, networking issues, and security concerns, with proposed guidelines. The conclusion emphasizes the importance of continued research in this evolving landscape [1].

This paper presents the design of an IoT Smart Home System (IoTSHS) enabling remote control of smart home devices via mobile phones, infrared (IR) remotes, and PCs/Laptops. The IoTSHS utilizes a WiFi-based microcontroller and incorporates a temperature sensor to monitor room temperature, suggesting whether to activate or deactivate the AC. The system interfaces with switches or relays in a power distribution box to control connected devices. Signals from the IoTSHS trigger the switches to connect or disconnect the targeted devices. This system caters to users without smartphones, offering universal remote control capabilities. Control methods include an IR and WiFi-enabled remote, allowing connection to WiFi without the need for a dedicated application. WiFi control establishes a secured Access Point (AP) with a specific SSID. Users connect their devices, enter the password, and access a fixed link via a web browser, facilitating interaction between the device and appliances. The IoTSHS can also link to the home router for continuous control. The system was successfully designed, programmed, fabricated, and tested, demonstrating excellent performance. Overall, the proposed IoTSHS provides advanced and inclusive remote control solutions for smart homes, benefiting various segments of society [2–4].

This paper discusses the emergence of cloud computing and its advantages in providing access to shared resources, particularly in the context of mobile and Internet-of-Things (IoT) devices. While forming a cloud instantly is possible, resource-constrained IoT devices are impractical for hosting virtual machines. To address this limitation, the paper introduces IoTDoc, an architecture for a mobile cloud consisting of lightweight containers on distributed IoT devices. Leveraging Docker, the study explores the benefits of containerized applications on a low-cost IoT-based cloud, emphasizing its operational model for cloud formation, resource allocation, container distribution, and migration. To evaluate IoTDoc's performance, the paper conducts experiments using the Sysbench benchmark program, comparing it with Amazon EC2. Results indicate that IoTDoc is a viable and cost-effective option for cloud computing, particularly as a learning platform, offering affordability and efficiency compared to larger platform cloud computing services like Amazon EC2 [5].

3 State of the Art

This section explores the relevant topics by surveying the state of the art with architectures, devices, and technologies.

3.1 Programming Language

The adopted programming language in the implementation of a Proof-of-Concept (PoC) of the present project was Python. This is justified as it has a large scope in integration activities and is one of the most used and supported programming languages worldwide. Regarding integration potential associated with this programming language was notable, allowing for employing third-party libraries and extensions, accelerating development, and allowing for more seamless integration with the different devices and layers in the framework developed.

3.2 Platform

Docker was chosen as the platform to support the present project. It played a central role by containerizing every component of the developed framework. Using Docker containers in this context brought several advantages to the development and deployment processes. Firstly, it ensured that the application and its dependencies and configurations were encapsulated within a container, guaranteeing consistency and facilitating integration with the multiple IoT devices utilized. Scaling and replicating containers ensures a standardized and reliable execution environment with room reserved for future expansion. Overall, Docker's containerization enhanced the project's agility, scalability, and reliability [6].

3.3 Protocols

Message Queuing Telemetry Transport. Message Queuing Telemetry Transport (MQTT), is a lightweight open messaging protocol designed for efficient communication in challenging network conditions. It operates on a publish/subscribe model, where publishers send messages to topics, and subscribers express interest in specific topics. In this system, a central broker manages the distribution of messages between publishers and subscribers. This protocol provides a flexible and scalable way of handling communication between devices, ensures efficiency in both bandwidth usage and implementation complexity, making it suitable for resource-constrained environments, such as IoT applications [7].

RabbitMQ. RabbitMQ serves as a robust message broker, facilitating communication between systems. Employing the Advanced Message Queuing Protocol (AMQP) and supporting MQTT, it ensures reliable message delivery through an exchange and queue model. Messages are routed based on rules defined by bindings, and durability is maintained by persisting messages to disk. This protocol has features such as clustering for scalability and high availability, extensibility via plugins, and support for multiple messaging protocols. Widely adopted in enterprise and micro-services architectures, RabbitMQ excels in scenarios requiring flexible, reliable, and scalable message communication [8].

3.4 Architecture

For the development of this work, a proper architecture was established. As such, 3 layers were idealized, as such being the Data Collection Layer, Data Communication Layer and Data Analytics Layer. The first, contained the IoT devices (or simulation of such devices), and served the purpose of retrieving multiple data from the environment and transmitting it to the Data Communication Layer. The following layer was responsible for handling bad data received from the Data Collection Layer, as well as handling failures, ensuring that only accurate and coherent data was to be transmitted to the final layer. At last, the Data Analytics was idealized. This layer was responsible for grouping and handling the data received from the Data Communication Layer, allowing for a user-friendly way of querying such said data, and serving as a point of interaction for the user with the designed system.

As can be observed in Fig. 1, out of the three previously mentioned layers, two of them are replicated, thus transforming the entire application. This was implemented to introduce a second layer of IoT devices communicating with the

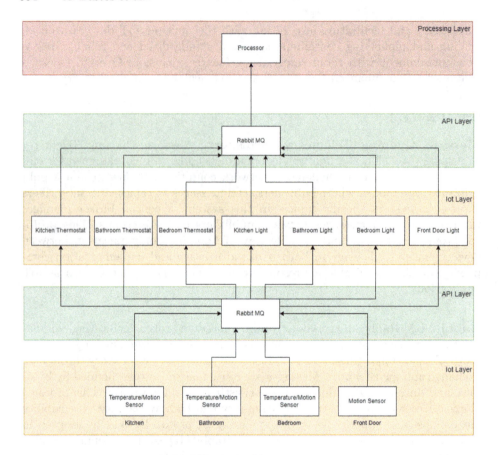

Fig. 1. Architecture

first layer, with some basic data processing occurring in this second layer. The two message transmission layers are present to enable communication between the first IoT layer and the second layer, as well as between the second layer and the processing layer. In reality, there exists only one message transmission layer, but it serves two distinct purposes.

3.5 Devices

The use of real physical devices for simulation purposes was not possible during this research. Henceforth, the group opted for a simulation performed via coding, while thriving for a preservation of coherence and accuracy, while accounting for potentially bad inputs received from said sensors. A Rasberry Pi device was also simulated via coding following the same principals, since on a real world application scenario, this was the equipment that would be employed. The total of simulated devices were:

- Motion Sensor - 4
- Temperature Sensor - 3
- LED Light - 4
- Thermostat - 3
- Raspberry Pi - 1

4 Experimental Setup

This section presents the experimental work with the setup, the materials and methods.

4.1 Materials

- **Motion Sensor Simulation** - The experiment employed Python-based simulation functions to emulate the functionality of 4 motion sensors, specifically mimicking the operation of a Passive Infrared (PIR) sensor commonly used for motion detection.
- **Temperature Sensor Simulation** - The experiment employed Python-based simulation functions to emulate the functionality of 3 temperature sensors.
- **Led Light Simulation** - The experiment employed Python-based simulation functions to emulate the functionality of 4 LED Lights.
- **Thermostat Simulation** - The experiment employed Python-based simulation functions to emulate the functionality of 3 Thermostats.
- **Raspberry Pi Simulation** - The Raspberry Pi was simulated in a Personal Computer with Docker installed. Inside this virtualization software, each of the components will be in a container.

4.2 Methods

The first step undertaken was the creation of the "sensors". As previously explained, real sensors were not utilized; instead, four Python files were generated. Each file represents a distinct area within the household - "bathroom", "bedroom", "kitchen", and "front door" - with each area containing two sensors, one for motion detection and another for temperature monitoring. The sole exception is the "front door" file, which encompasses only a motion detection sensor.

Next, a Message Queuing Telemetry Transport (MQTT) protocol was chosen to facilitate communication between the various layers of the application. In this regard, RabbitMQ was selected as the Message Broker responsible for transmitting messages among different applications while also managing the Queues.

After configuring MQTT, an additional 7 files were created, with 4 of them representing lights and the remaining representing thermostats. In this scenario, the lights are programmed to turn on upon motion detection by the motion sensors, with each light assigned to a specific room. As for the thermostats, a "target temperature" was established. When the temperature sensors detect values below the target, they switch to the "heating" mode; if values above the target are detected, they switch to the "cooling" mode. When the detected values match the "target temperature", the thermostats switch off.

Given that the files were already set up to communicate with each other, and MQTT was configured, Docker was set up accordingly. Dockerfiles were created for each of the previously established files. Additionally, a Docker Compose file was crafted to facilitate the simultaneous execution of all Dockerfiles, thereby creating various containers and their respective images for each program.

Since Docker is already configured, several tests were conducted to ensure communication between the different sensors, lights, and thermostats. After errors were detected and corrected, and communication was functioning flawlessly, the "processor" file was created. This file belongs to the Processing layer and is primarily aimed at processing the data collected by the various components. Upon establishing communication between this file and the existing ones, data processing from the lights and thermostats commenced.

In this case, the objective is to compute the energy expenditure over 48 h. To achieve this, the average consumption of each light/thermostat per hour was computed, along with the definition of a price per kilowatt-hour. With this information, it was possible to compute the energy expenses during the testing period. From the average of these 2 days, a forecast for 30 days (1 month) was derived.

Finally, an export of these calculations to an Excel file was also implemented, which will be stored in Docker.

5 Results and Analysis

Figure 2 presents the achieved results after running Docker containers for 48 h while planning them for 30 days can be analyzed. In the first column, the total number of hours that the lights remained on during the sample period is displayed. For thermostats or air conditioners (A/C), the number of hours spent heating or cooling can be verified. The second column presents the percentage of time the equipment was on relative to the total sample time. The third column displays the energy consumption of each device in kWh, as each one has a distinct consumption rate. In the fourth column, the calculated cost in Euros based on the price per kWh and the energy expenditures of each equipment is provided. The fifth column shows a consumption forecast for the 30 days, extrapolated from the analysis of the 2 days. Furthermore, the sixth column presents a forecast of the monthly cost for each equipment, calculated based on the two days of sampling.

Device	On Time (h)	On Time (%)	Consumption (kWh)	Cost (EURO)	Consumption Prediction (kWh)	Cost Prediction (EURO)
Bathroom Light	12.61	26.27	0.13	0.03	1.89	0.38
Bedroom Light	13.82	28.78	0.14	0.03	2.07	0.41
Kitchen Light	13.91	28.99	0.14	0.03	2.09	0.42
Front Door Light	2.5	5.21	0.03	0.01	0.38	0.08
Bathroom AC Cooling	21.01	43.78	48.12	9.62	721.81	144.36
Bathroom AC Heating	24.01	50.03	111.66	22.33	1674.96	334.99
Bedroom AC Cooling	29.52	61.5	67.6	13.52	1013.97	202.79
Bedroom AC Heating	15.01	31.27	69.79	13.96	1046.84	209.37
Kitchen AC Cooling	41.53	86.51	95.1	19.02	1426.43	285.29
Kitchen AC Heating	0	0	0	0	0	0

Fig. 2. Results

After analyzing the results, it becomes apparent that in some cases, the data appears to deviate from reality. This discrepancy arises due to the fact that the sensors, in this case, are simulated in Python with random values. If the experiment were conducted with real sensors, the results would likely better approximate the current reality of a typical household. Nevertheless, despite these limitations, it can be concluded that the forecasts for one month are accurate based on the utilized sample.

For future endeavors, it would be intriguing to incorporate a wider array of IoT devices, given that the methodology employed is readily repeatable for most equipment available in the market. Furthermore, it would be highly beneficial to apply this approach in real households to study ways to further optimize forecasts and data processing. Developing an end-user application would enhance user experience and accessibility, offering an innovative solution in the market that enables numerous families to save money and enhance their personal finances.

6 Conclusion

In conclusion, this paper presented an adaptable IoT-powered framework designed to elevate the efficiency and functionality of intelligent home systems. By focusing on the enhancement of everyday aspects such as lighting and temperature control, the proposed system leverages the capabilities of Docker containers to optimize decision-making processes. The comprehensive discussion of employed devices, communication protocols, data collection and normalization criteria, as well as the analytics layer, highlights the practical and functional aspects of the framework.

Furthermore, this work reviewed the existing literature on similar systems and architectures, providing valuable insights into the current market landscape. The advantages and limitations of the proposed framework were analyzed, emphasizing its adaptability to diverse spaces and configurations. The justification of the Docker-based architecture relies on the advantages such as encapsulation, isolation, and optimal decision-making. With the aforementioned development, those involved in this research were able to fully embrace the almost limitless potential of IoT devices and their applications, while also acquiring and consolidating knowledge regarding the construction of embedded systems.

Acknowledgements. This work is funded by National Funds through the FCT - Foundation for Science and Technology, I.P., within the scope of the project Ref UIDB/05583/2020. Furthermore, we thank the Research Centre in Digital Services (CISeD) and the Polytechnic of Viseu for their support.

References

1. Stojkoska, B.L.R., Trivodaliev, K.V.: A review of internet of things for smart home: challenges and solutions. J. Cleaner Prod. **140**, 1454–1464 (2017)
2. Khan, A., Al-Zahrani, A., Al-Harbi, S., Al-Nashri, S., Khan, I.A.: Design of an IoT smart home system. In: 2018 15th Learning and Technology Conference (L&T), pp. 1–5. IEEE (2018)
3. Mundle, K.: Home smart IoT home: domesticating the internet of things. Toptal Designers (2023)
4. Alaa, M., Zaidan, A.A., Zaidan, B.B., Talal, M., Kiah, M.L.M.: A review of smart home applications based on internet of things. J. Netw. Comput. Appl. **97**, 48–65 (2017)
5. Noor, S., Koehler, B., Steenson, A., Caballero, J., Ellenberger, D., Heilman, L.: IoTDoc: a docker-container based architecture of IoT-enabled cloud system. In: Lee, R. (ed.) BCD 2019. SCI, vol. 844, pp. 51–68. Springer, Cham (2020). https://doi.org/10.1007/978-3-030-24405-7_4
6. Nickoloff, J., Kuenzli, S.: Docker in Action, 2nd edn. Simon and Schuster (2019)
7. Singh, M., Rajan, M.A., Shivraj, V.L., Balamuralidhar, P.: Secure MQTT for internet of things (IoT). In: International Conference on Communication Systems and Network Technologies (CSNT) (2015)
8. Albayrak, G.S.: What is the RabbitMQ (2023). https://medium.com/

An IoT Framework for Improved Vineyards Treatment in Grape Farming

Henrique Jorge[1], Manuel Aidos[1], Rodrigo Cristovam[1], João Henriques[1,2]([✉]),
Filipe Caldeira[1,2], and Cristina Wanzeller[1,2]

[1] Polytechnic of Viseu, Viseu, Portugal
{pv21027,estgv15116,pv27127}@alunos.estgv.ipv.pt,
{joaohenriques,caldeira,cwanzeller}@estgv.ipv.pt
[2] CISeD - Research Centre in Digital Services, Polytechnic of Viseu, Viseu, Portugal

Abstract. The development of Internet of Things (IoT) technology in agriculture has been a rapidly growing research area, with applications ranging from detecting diseases in crops to optimizing water usage.

This research paper explores the use of IoT technology to increase efficiency and precision in vineyard treatment. That aim proposes using temperature and air humidity sensors and an efficient data management system, providing real-time monitoring of vineyard conditions and the ability to adapt accordingly if necessary. The results denoted that the proposed technology could detect subtle variations in environmental conditions, enabling vineyard operators to adjust their management practices to enhance grape quality.

Keywords: Internet of Things · containers · vineyards treatment · grape farming

1 Introduction

The health and welfare of the grapevines play a significant role in the production of wine. The grapevine (Vitis spp.), when grown under climate conditions favorable to the development of pathogens, is susceptible to various diseases. In the viticulture sector, these diseases cause financial losses since they affect the enological quality of the grapes in infected vines and can kill the plants, making using technologies to control them necessary. However, on most rural properties, the disease is often detected later through manual inspections carried out by farmers. These inspections identify early signs of infection, such as chlorotic spots or leaf drying, which are subsequently treated with agricultural pesticides [1].

Vineyards emerged as one of the most extensively cultivated crops worldwide, covering about 7.5 million hectares of planting area in 2015, as the International Organization of Vine and Wine reported. This extensive presence highlights the important role wine grape cultivation plays worldwide and calls for a more detailed assessment of problems and opportunities arising from this essential agricultural practice [2].

Following the work of [3], identification of vine diseases is essential, particularly fungal or bacterial diseases caused by Downy Mildew, powdery mildew, and black rot. The yield and cost are influenced by early identification. Heat stress due to temperature fluctuations affects plant growth and disease severity.

The IoT [4] constitutes a network where elements are distinctly identified through integrating software intelligence, sensors, and pervasive internet connectivity. The data from the IoT, including information about devices, sensors, and actuators, can be used for IoT. Meteorological data is used to prevent grapevine diseases in scientific prevention studies, which reveal its wide application.

Traditionally, such procedures have been manually monitored and based on subjective measures that are often cumbersome and unreliable, leading to financial losses for farmers. However, with the development of IoT technology, there is a unique opportunity to modernize this process. IoT technology can provide deep and accurate analysis of climate conditions by combining a model with collected data, analyzed data, and intelligence capabilities.

This research paper aims to investigate the possibilities of using IoT technology to increase efficiency and precision in vineyard treatment. IoT technology will help quickly assess vines' health and choose the right products and quantities for treatment. Understanding how IoT technology affects the health of vineyards and the ideal parameters for top quality and stability is the main objective of this research paper.

The purpose of the planned installation of the DHT22 temperature and air humidity sensors [5] in conjunction with an efficient data management system was to monitor the conditions of the vineyard in real-time and allow appropriate adjustments if necessary.

These developments are expected to positively impact the vineyards' overall health and the quality of grape production. However, it is essential to note that adopting IoT technology in agriculture presents challenges, such as data security, implementation costs, and the sensor's resistance to difficult climates. These challenges must be resolved to facilitate wider adoption. Despite these challenges, IoT technology has great potential to revolutionize vineyard care. This research paper aims to contribute to this field's growing body of knowledge.

This paper is organized as follows. Section 2 presents a comprehensive review of state of the art and related work in the domain of vineyard treatments, analyzing the climatic conditions favorable to the development of pathogens. In Sect. 3, we provide a detailed and systematic description of the process used to conduct the study. Section 4, we present the results and analysis of the integrated system. Finally, in the Sect. 5 concludes this work by summarizing key findings and contributions and presenting the future work.

2 State of the Art

IoT technology in agriculture has been a rapidly growing research area, with applications ranging from detecting diseases in crops to optimizing water usage. This work focuses on using IoT to enhance the treatment of vineyards in viticulture, a field that is beginning to witness the benefits of this emerging technology.

The first to explore IoT applications in viticulture, in a study by [6], involved implementing a wireless network of soil moisture sensors in a vineyard. The research showcased that IoT technology has the potential to furnish intricate, real-time details about soil conditions, empowering vineyard operators to refine water utilization and improve the overall health of the vines.

In [7], it demonstrates the monitoring of microclimatic conditions in a vineyard using a wireless sensor network. The study illustrated that IoT technology could detect subtle variations in environmental conditions, enabling vineyard operators to adjust their management practices to enhance grape quality.

The research carried out by [8] explored the use of IoT sensors to detect diseases in vineyards. Deploying air humidity and temperature sensors and machine learning algorithms could detect diseases early, enabling early treatment and preventing significant harvest losses.

Beyond monitoring environmental conditions and disease detection, IoT has been used to optimize the application of fertilizers and pesticides in vineyards. The authors of [9] worked on a project using an IoT sensor network to monitor plant health and soil conditions. The collected data was then used to determine the optimal amount of fertilizers and pesticides.

Other studies have explored using drones with cameras and IoT sensors for vineyard monitoring. For example, the authors in [10] demonstrated that using drones could assist in detecting diseases and pests in their early stages and monitoring the vines' general health.

IoT in agriculture is about data collection and analysis, and it uses this data to make informed decisions. The study by [11] argued that combining IoT with other technologies, such as machine learning, could enable the development of more sophisticated and accurate decision support systems.

However, despite the progress in applying IoT in viticulture, challenges still need to be addressed. As indicated by [12], issues such as data security, the resilience of sensors to adverse weather conditions, and the cost of implementation are all areas that need to be addressed to facilitate the broader adoption of IoT in agriculture.

Similarly, [13] work on using big data analytics in agriculture highlighted the need to develop more advanced algorithms and machine learning techniques to handle the substantial amount of data generated by IoT systems.

The current investigation into the use of the IoT in wine is promising, but significant advances still are needed. This research aims to develop and evaluate an IoT solution for vineyard treatment in viticulture that would significantly contribute to this area's development.

3 Methodology

In implementing the first layer of our IoT solution to enhance the treatment of vineyards in grape farming, this methodology section outlines the approach taken. Environmental data was collected, focusing on temperature and air humidity through dedicated sensors. Later, we'll review the development of additional layers and their significance in addressing the research problem.

This work [5] relied on DHT22 temperature and air humidity sensors for the data collection. This sensor was instrumental for temperature sensing in an IoT context, as confirmed by previous evaluations [14]. Both were selected because they are suitable for monitoring grape farming conditions.

The position of these sensors in the environment was considered to ensure an accurate data collection. The solution was to place them everywhere in the vineyard at different heights and locations. By doing this, we could capture microclimate variations. We did not want anything in the way of that, so we built protective enclosures to shield them from rain, direct sunlight, and dust. Before rolling out everything, we performed meticulous calibration procedures to ensure sensor accuracy.

Instead of physically placing sensors, we developed Python scripts that simulated temperature and air humidity values.

A Raspberry Pi [15] takes the responsibility of collecting data. It connects with every sensor and gathers information. Python programming will be used to interact with the sensors, making it a lot easier to get all the information we need. The Message Queuing Telemetry Transport (MQTT) and Docker Containers were considered for communication purposes. MQTT is a highly efficient protocol for the real-time management of small data volumes, making it particularly suitable for IoT projects. Extensive research has recognized MQTT as a low-power protocol designed specifically for IoT applications [16]. This research provides valuable insights into the effectiveness and suitability of MQTT for various IoT use cases and projects, including ours. At the same time, Docker Containers offer flexibility and scalability when handling large amounts of data. The one we end up picking will be suited to our specific requirements.

As explained in [17], containerization allowed us to package applications, dependencies, and configurations for easy deployment. Its benefits include portability, scalability, resource efficiency, and increased security. Containerizing applications contribute to streamlining the IoT infrastructure deployment and efficient data management to improve grape cultivation.

Regarding requirements, adopting the DHT22 temperature and air humidity sensor would be excellent and cost-effective. Both are required when monitoring conditions for grape farms. With real-time data collection, we can make decisions quickly if we have to.

In the initial phase, this work focused on connecting the sensors' data and a Mosquitto MQTT broker. The Python script of the Raspberry Pi was responsible for generating pretend sensor data, imitating things like temperature and air humidity. This simulated data was then published to the MQTT broker using the Paho MQTT client library.

The backend server's Python Flask application receives MQTT messages from sensors to be processed. The backend efficiently handled data processing tasks with a lightweight and reactive nature, extracting relevant information from the MQTT payload. This facilitated the creation of a seamless interface for the front end to access real-time sensor information.

Our exploration delved into data analysis using the Pandas library in Python. The backend facilitated the exposure of sensor data, serving it through dedicated channels in JSON format. Data was processed using the Pandas library to transform into DataFrames without relying on a database or specific endpoints. Subsequently, the analysis explored the data, computing statistics, and gaining insights into patterns and trends within the sensor data.

4 Results and Analysis

This section analyzes the results obtained and the expected impact of our IoT solution implemented to improve treatment in grape crops.

The use of Docker was essential for implementing the solution. Docker is a platform that allows you to create and manage containers, which are isolated environments for running applications.

To replicate the functioning of the Raspberry Pi, a Python script was developed, followed by creating a Dockerfile containing the essential commands for generating the image and executing the container.

A docker container runs on the Python script and connects to Mosquitto to receive and send data to a server. The server, in turn, connects to the client and displays sensor data in real-time on a web page, which was developed using the Flask-MQTT framework, which is based on Python.

The environment collected data contributed to acquiring knowledge about vineyard growth by installing DHT22 temperature and air humidity sensors. Moreover, using the Raspberry Pi as a central processor permitted data to be collected continuously and effectively, making work more accessible and efficient.

Small patterns and variations in environmental conditions were also identified during the data collection. Data on temperature and air humidity helped us understand the specific needs of our vineyards.

The use of Docker containers and MQTT was essential for effective data management. Docker containers allow for managing small amounts of data in real time, which is necessary for monitoring vineyard conditions. Alternatively, the scalability and flexibility of Docker Containers enabled us to handle large volumes of data without compromising system performance.

The collected data was summarized to support vineyard management. Features in the table represent different climatic variables impacting the application of products in vineyards. For example, temperature and air humidity are considered in the table. Each column shall be marked to account for the application's optimum conditions. Based on our upcoming sensor data collection, numerical values indicating the recommended conditions for product application in vineyards will be assigned to each cell.

The table includes the following features: Temperature (°C), Air Humidity (%), and Ideal Conditions for Products Application in Vineyards.

And the following rows:

- Temperature below 10 °C and air humidity above 90%: Not recommended to apply products in the vineyards
- Temperature between 10 °C and 15 °C and air humidity between 70% and 90%: Apply product in the vineyards every 10 days
- Temperature between 15 °C and 20 °C and air humidity between 50% and 70%: Apply product in the vineyards every 7 days
- Temperature above 20 °C and air humidity below 50%: Apply product in the vineyards every 5 days.

Table 1. Products Application Guidelines

Temperature (°C)	Air Humidity (%)	Recommendation	Ideal Condition
<10	>90	Not recommended	–
10–15	70–90	Apply treatment	Every 10 days
15–20	50–70	Apply treatment	Every 7 days
>20	<50	Apply treatment	Every 5 days

Two independent Python scripts were defined to simulate the DHT22 sensor when acquiring sensor data. The first script generates air humidity and temperature readings, while the second generates corresponding temperature readings but independent air humidity values. These two scripts independently establish connections to an MQTT broker, specifically Mosquitto, and publish their respective data on distinct topics. This arrangement ensures the efficient dissemination of simulated sensor data to the different IoT system components through the MQTT broker, which acts as a central data intermediary (Table 1).

Another Python script was developed as a backend server to receive the simulated sensor data. The data is transmitted over the MQTT protocol and subscribed to the server using the Flask-MQTT extension. The backend server processes and showcases the received data in real time on a user-friendly web page created using HTML and JavaScript.

A user-friendly web page was developed using HTML, JavaScript, and Chart.js to visualize and present the simulated sensor data seamlessly. The web page connects to the backend server using WebSockets to receive real-time data updates. Chart.js creates intuitive and interactive graphs depicting temperature and air humidity trends upon receiving data updates. Additionally, a table is dynamically generated at the end of the website to provide growers with a clear and concise overview of the current and historical air humidity and temperature values.

Graphs are a valuable data visualization tool used to articulate vineyards' temperature and air humidity conditions. Graphs represent data over time, revealing trends and patterns that may not be evident in individual points. Graphical representations of environmental conditions enable growers to quickly analyze key features of the landscape, which often cannot be gleaned from raw data in tables or spreadsheets.

Using graphical representations rather than tables or raw data improves comprehensibility and enhances vineyard management decision-making processes. Graphical representations highlight trends often obscured by table data points, enabling growers to retrieve a clearer insight into the data. Recognizing trends from environmental data, such as temperature and air humidity alterations over time, contributes to getting clues of other changes within and between vineyard ecosystems. These insights can be used to optimize resources, predict productivity, reduce wastage, and improve vineyard management, thus enhancing overall outcomes.

A dataset was created using a pandas python script to monitor a vineyard's temperature and air humidity levels. The data was then used to create a graph that shows how temperature and air humidity levels fluctuate throughout the day, as denoted by Fig. 1.

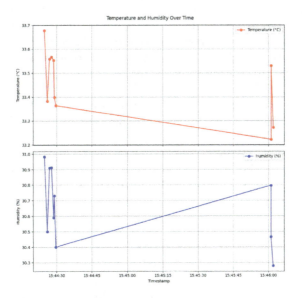

Fig. 1. Temperature and Humidity levels

The graph demonstrates that temperature and air humidity levels vary considerably throughout the day. On the data shown, Temperatures can range from 30° to 33°, whereas air humidity levels can vary from 31% to 32%. These fluctuations can have a considerable impact on vine growth. For instance, vines exposed to high temperatures or low air humidity may experience stress, leading to reduced growth and yield.

IoT solutions can monitor and manage temperature and air humidity levels in vineyards. This can help ensure that vines grow in conditions ideal for their development. By monitoring temperature and air humidity levels, growers can identify potential problems early on and take corrective action. They can also use the data to optimize irrigation and fertilization schedules.

Overall, the results of this study suggest that IoT solutions can improve viticulture practices by providing growers with valuable insights into vine growth conditions.

5 Conclusion

There is a substantial potential for revolutionizing wine production by integrating IoT technology into viticulture. The planned implementation of the DHT22 temperature sensor with an efficient data management system is designed to provide real-time monitoring of vineyard conditions and the ability to adapt accordingly if necessary. These developments are expected to contribute to the overall health of vineyards and the quality of grape production.

This parallel success underscores the potential of IoT technology to revolutionize vineyard treatment. The parameters we will track are stored in the cloud for later analysis and can be similarly leveraged to identify conditions favorable to disease occurrence, enabling faster prevention of diseases like mange, wilt, and grey rot and ensuring a robust framework for disease prevention and detection.

The adoption of IoT technology within the field of agriculture poses several challenges. To widespread the adoption of this technology, it is imperative to address specific issues such as safeguarding data privacy and security, ensuring the longevity of sensors in harsh environmental conditions, and managing the costs associated with implementing IoT systems.

Although we are confident that IoT technology in vineyards will have a promising potential to improve wine production efficiency and quality, it is essential to note that our study has yet to begin. In the future, IoT technologies will likely be a decisive factor in vineyards, and we're willing to continue with research on how this technology will benefit agriculture.

It is worth letting know that this work only considered some of the available features of IoT technology, and further research in this field is necessary to explore its full potential in the context of vineyard management. Future research areas may encompass innovative and advanced approaches to visual recognition, thereby increasing precision and enhancing disease detection accuracy. Additionally, recognizing new diseases that may impact vineyard crops will be fundamental in bolstering the current viticulture management practices.

In conclusion, our study provides substantial evidence of IoT technology's potential in improving vineyard management practices. Although not all available features of IoT were integrated in this study, the results indicate a promising avenue for future viticulture optimization. The continued research and application of IoT technologies in viticulture will be pivotal in augmenting the efficiency and quality of wine production. Looking ahead, we remain optimistic about developing new methods to exploit the potential of IoT technology in vineyard management.

In the future, it would be interesting to explore the integration of additional sensors, such as soil and sunlight, to gain a more comprehensive understanding of vineyard conditions. Furthermore, implementing advanced data analysis techniques, such as machine learning, could aid in predicting trends and identifying more complex patterns in the collected environmental data. It would also be relevant to consider expanding the study to different types of vineyards and regions to assess the applicability and effectiveness of IoT solutions in diverse contexts.

Acknowledgements. This work is funded by National Funds through the FCT - Foundation for Science and Technology, I.P., within the scope of the project Ref UIDB/05583/2020. Furthermore, we thank the Research Centre in Digital Services (CISeD) and the Polytechnic of Viseu for their support.

References

1. Bischoff, V., Farias, K.: VitForecast: an IoT approach to predict diseases in vineyard. In: Proceedings of the XVI Brazilian Symposium on Information Systems, SBSI 2020. Association for Computing Machinery, New York (2020)
2. International Organization of Vine and Wine. State of the vitiviniculture world market. Accessed 10 Nov 2023
3. Nail, W.R., Stanley Howell, G.: Effects of timing of powdery mildew infection on carbon assimilation and subsequent seasonal growth of potted chardonnay grapevines. Am. J. Enol. Viticulture **56**(3), 220–227 (2005)
4. Venkatesan, R., Tamilvanan, A.: A sustainable agricultural system using IoT. In: 2017 International Conference on Communication and Signal Processing (ICCSP), pp. 0763–0767 (2017)
5. Ahmad, Y., Gunawan, T., Mansor, H., Hamida, B., Hishamudin, A., Arifin, F.: On the evaluation of DHT22 temperature sensor for IoT application. In: On the Evaluation of DHT22 Temperature Sensor for IoT Application, pp. 131–134 (2021)
6. Morais, R., Valente, A., Couto, C., Correia, J.H.: A wireless RF CMOS interface for a soil moisture sensor (2003)
7. Corke, P., Wark, T., Jurdak, R., Hu, W., Valencia, P., Moore, D.: Environmental wireless sensor networks. Proc. IEEE **98**, 1903–1917 (2010)
8. Bucci, E.: Xylella fastidiosa, a new plant pathogen that threatens global farming: ecology, molecular biology, search for remedies. Biochem. Biophys. Res. Commun. **502**, 173–182 (2018)
9. Amaxilatis, D., Akrivopoulos, O., Mylonas, G., Chatzigiannakis, I.: An IoT-based solution for monitoring a fleet of educational buildings focusing on energy efficiency. Sensors **17**, 2296 (2017)
10. Primicerio, J., et al.: A flexible unmanned aerial vehicle for precision agriculture. Precis. Agric. **13**, 517–523 (2012)
11. Wolfert, S., Ge, L., Verdouw, C., Bogaardt, M.J.: Big data in smart farming - a review. Agric. Syst. **153**, 69–80 (2017)
12. Verdouw, C., Wolfert, S., Beulens, A., Rialland, A.: Virtualization of food supply chains with the internet of things. J. Food Eng. **176**, 128–136 (2015)
13. Kamilaris, A., Kartakoullis, A., Boldú, F.P.: A review on the practice of big data analysis in agriculture. Comput. Electron. Agric. **143**, 23–37 (2017)
14. Ahmad, Y.A., Gunawan, T.S., Mansor, H., Hamida, B.A., Hishamudin, A.F., Arifin, F.: On the evaluation of DHT22 temperature sensor for IoT application. In: 2021 8th International Conference on Computer and Communication Engineering (ICCCE), pp. 131–134. IEEE (2021)

15. Richardson, M., Wallace, S.: Getting Started with Raspberry PI. O'Reilly Media, Inc. (2012)
16. Masdani, M., Darlis, D.: A comprehensive study on MQTT as a low power protocol for internet of things application. In: IOP Conference Series: Materials Science and Engineering, vol. 434, p. 012274 (2018)
17. Hykes, S.: Use containers to build, share and run your applications. Accessed 6 Nov 2023

Strengthening the Role of Citizens in Governing Disruptive Technologies: The Case of Dutch Volunteer Hackers

Anne Marte Gardenier[✉]

Eindhoven University of Technology, PO Box 513, 5600 MB Eindhoven, The Netherlands
a.m.gardenier@tue.nl

Abstract. Digitization can be understood as a socially disruptive technology. To deal with this disruption, the social challenge is to find cybersecurity solutions to protect society and its citizens from this disruptive impact. Citizens as the end-users of software are widely portrayed as the 'weakest link' in cybersecurity, indicating that they are insufficiently knowledgeable or not taking enough cybersecurity measures. In contrast, this paper about Dutch volunteer hackers demonstrates that citizens can make an important contribution to the governance of digitization. This paper argues that the roles of citizens in the governance of disruptive technologies should be better recognized and supported so they can contribute to maintaining the public interest in the digital society.

Keywords: Socially disruptive technologies · cybersecurity governance · citizen participation · volunteer hackers · vulnerability disclosure

1 Introduction: Digitization as a Socially Disruptive Technology

The information and communications technologies (ICT) that underlie today's digital society can contain vulnerabilities. These vulnerabilities are, for example, small technical errors or bugs in systems which can be exploited to make the system work differently than intended. Such vulnerabilities make it possible to carry out cyber-attacks, such as ransomware, theft, stalking, spying, etcetera. The risks of unattended vulnerabilities have the potential to disrupt the life of individuals and essential societal processes. For instance, human rights defenders, lawyers, journalists, members of parliament and dissidents around the world have fallen victim to spyware like Pegasus [1], and hospitals, pharmacies, and universities worldwide have been victim to ransomware attacks, disrupting the continuation of their services.

Therefore, the digitization of society, which inevitably comes with vulnerabilities, can be regarded as a socially disruptive technology [2] because it has the potential to disrupt societal processes and thereby holds unprecedented risks. For governments and researchers, the challenge is to find solutions to protect society and its citizens from this potential disruptive impact. This has proved to be complicated, as disruptive technologies challenge traditional governance systems [3].

In cybersecurity research, ordinary software users are widely recognized as the 'weakest link' in cybersecurity [4], indicating an alleged lack in their knowledge and skills to avert cyberattacks. For that reason, cybersecurity research and governmental policy strategies often focus on improving user awareness of cybersecurity practices, viewing user behavior as one of the biggest challenges to maintaining cybersecurity [5, 6]. Such approaches pinpoint the 'deficient user' as the security risk [7].

While the lack of knowledge and skills of ordinary users to avert cyberattacks is certainly an important problem, the portrayal of the ordinary user or citizen[1] in general as the deficient user does not do full justice to the role of citizens in cybersecurity. In fact, this paper argues that citizens often play a central role within cybersecurity governance. For instance, in the case of Dutch volunteer hackers, which will be described in this paper, citizens have independently and voluntarily contributed to establishing a cybersecurity governance system in which they can play a central role. This demonstrates that the perspective on citizens as the 'weakest link' lacks an important perspective on the role of citizens in cybersecurity governance: citizens may also contribute to the governance of digitization and cybersecurity, but their efforts are not always recognized and acknowledged. Indeed, the role of citizens in the governance of digitization should be understood broadly: by interacting with digital technologies in their private lives, social communities and public sphere, citizens contribute to shaping the digital society [8]. Therefore, citizen participation (including uninvited participation [9]) should be better recognized and acknowledged so it can be strengthened to serve the public interest, especially with regard to (socially) disruptive technologies.

The paper is structured as follows. The next section presents the case study of Dutch volunteer hackers. The description of the case study is based on an analysis of sources published between 1980 and 2024, including policy documents, transcripts of debates by the Dutch parliament, newspaper articles, transcripts of court hearings, scientific articles and reports, and books. The final section presents the conclusion and further research, and discusses lessons that can be learned from this case with regard to the governance of disruptive technologies and the roles of citizens therein.

2 Case Study: Vulnerability Disclosure in the Netherlands

The risks that unattended vulnerabilities can cause is an important trigger for some hackers to search for vulnerabilities and report them to the ICT vendor in the hope that they will be patched. Since the 1980s, hackers worldwide, including the Netherlands, have been drawing attention to these risks by exposing vulnerabilities in ICT systems by the means of hacking. How their efforts have been valued in the Dutch institutional context has changed over time. Since 1993, hacking has been criminalized, and well-meaning hackers faced a prison sentence. Currently, however, voluntary hackers play a central and crucial role in Dutch cybersecurity governance. In this section we describe how this shift in perspective came about.

---

[1] By using the term citizen we do not refer to the legal status of citizens of a particular country, but rather to individuals that are part of a democratic society in which they can play an active role in shaping it. See also [8].

2.1 From the First Hackers Until the Computer Crime Act (1980–1993)

In the 1980s, the computer hacker movement emerged in the Netherlands. The first hackers browsed the internet out of curiosity and an interest in technology. By browsing the internet, hackers discovered that many computer systems were not properly secured. Hackers criticized how carelessly computer users handled confidential information and how easy it was to gain access to government and private corporation systems [10, 11]. By hacking these systems, hackers could draw attention to the lack of security. Often they did not have to do much to get in, sometimes guessing commonly used passwords sufficed [12].

In some cases, the affected company subsequently took security measures, but usually companies did not take reports by the hackers seriously [11]. Consequently, hackers regularly publicized a discovered security leak, in the hope that the leak would be fixed. However, this method (also known as 'full disclosure') caused reputational damage and made the affected company extra vulnerable because it allowed others to exploit the vulnerability as well. Yet, hacked companies usually did not press charges, firstly because it would not generate good publicity, and secondly because not much could be done about it: hacking was not an illegal criminal act.

In 1985, Dutch hackers hacked the National Institute for Public Health and the Environment. This was the first publicly disclosed hack of a governmental institution, and hackers had gained access to sensitive patient information. One of the hackers stated that he wanted to prove that the privacy of citizens is insufficiently protected [13]. This hack caused a political stir and set a precedent: the Minister of Justice announced that computer trespassing will be made into a punishable offense by implementing a new law [11].

This initiative to implement a new law triggered criticism: wasn't the real problem that companies take far too few security measures? Such measures were available, but expensive and therefore often not a priority. Several computer scientists publicly supported the hackers' cause. In computer science, hacking was considered as a legitimate way to detect vulnerabilities. A professor stated that these hackers are not criminals, in fact, they actually do a good job by flagging insecure systems [14]. Moreover, these hackers were working according to a certain ethics: they report the leak to the owner and give them time to repair it, and only if that fails they seek publicity. According to the professor, hacking was the only way to get companies to improve their security: scientists have tried to draw attention to weak security through scientific publications, but without success.

Nevertheless, the Computer Crime Act was introduced in the Netherlands in 1993. From that moment on, the maximum penalty for computer trespassing was 6 months in prison or a fine of 4,500 euros, which can increase to four years in prison and a fine of 11,250 guilders if the hacker also copied data [15].

2.2 From Hackers as Criminals to National Tolerance Policy for Hacking (1993–2018)

Shortly after the introduction of the Computer Crime Act, the first hacker in the Netherlands was prosecuted. He was held in pre-trial detention for 38 days and sentenced to a 6-month suspended prison sentence and a fine of about 2,200 euros [11].

The Computer Crime Act marked the end of an era for many hackers and the hacker subculture faded away. Some hackers quit hacking because they now faced a prison sentence. For others, hacking continued, but in secret [16]. For example, hackers would publicize a leak anonymously in collaboration with a journalist (who is legally better protected to publish illegally obtained news when this serves the public interest), sometimes after first warning the vulnerable party.

2.2.1 Lawsuit in 2008 Casts a Positive Light on Hacking

Within computer science, vulnerability testing remained a legitimate research method. In 2008, researchers from the Dutch Radboud University found a vulnerability in an NXP chip that was used worldwide in access systems to buildings and public transport, such as the London metro and Dutch trains. The researchers wanted to publish this vulnerability at a scientific conference to warn about the insecurity of the chip. Moreover, they wanted to demonstrate that the security principle that NXP uses in this chip, security by obscurity[2], is flawed. Seven months before the planned publication, the researchers contacted NXP to report the leak so that NXP could patch it. NXP appreciated the vulnerability report, but wanted to prevent publication and therefore filed a lawsuit against the researchers [17, 18].

NXP argued that the publication of the article should be prevented because it would harm NXP and cause serious societal and security problems, as it would enable others to crack the chip as well. The researchers argued that the publication of the article falls under freedom of expression of the European Convention on Human Rights and should therefore not be stopped. The judge concluded that the security risks are caused by NXP's unsafe chip, not by the fact that researchers would publish the vulnerability. Moreover, the judge stated that publicizing the vulnerability is in the public interest. The publication could therefore continue.

The judge's ruling was a legal milestone [18] that changed the perspective on vulnerability disclosure: the judge allowed the publication of a vulnerability discovered by hacking on the basis of the right to freedom of expression and the promotion of the public good. This court ruling placed the hackers/researchers in a new role, equivalent to that of a whistleblower or journalist. With this ruling, the societal value of disclosing vulnerabilities – at the expense of financial and reputational harm for the ICT vendor – was established in the jurisprudence.

2.2.2 No More Free Bugs: Companies Introduce Responsible Disclosure Guidelines

Also in the rest of the world, hackers continued to detect vulnerabilities, but ran the risk of legal repercussions. Therefore, in 2009, American hackers started the 'No More Free Bugs' campaign to initiate consultations for better compensation and recognition for hackers who disclose vulnerabilities [19]. As a result, American companies set up 'bug bounty' programs, allowing hackers to receive a financial reward after disclosing a vulnerability. In addition, companies introduced 'responsible disclosure' policies.

---

[2] Security by obscurity refers to concealing the operation of a mechanism to enhance its security.

Responsible disclosure (RD) refers to the practice of reporting a vulnerability directly to the affected party so that it can be patched before publication. Companies with a RD policy invite hackers to find vulnerabilities in their systems, and if hackers follow their guidelines, the company pledges not to press charges.

In 2012, Dutch telecom companies were the first to introduce a responsible disclosure guideline. While currently the commercialization of vulnerability disclosure is flourishing in the United States, in the Netherlands reporting vulnerabilities has retained its voluntary nature. Hackers usually do not receive a financial reward, but instead a t-shirt with "I hacked [affected party] but all I got was this lousy t-shirt" [18, p. 180].

2.2.3 Cyber Crisis Launches Cybersecurity on the Political Agenda

Meanwhile, the role of hackers in Dutch cybersecurity governance also received political attention. In 2011, the Diginotar hack took place in the Netherlands, which was considered as a "wake-up call" [20] that launched cybersecurity on the political agenda. In this hack, the Dutch company Diginotar, which issues certificates for websites, was hacked, and the reliability of a wide range of websites in the Netherlands was no longer guaranteed. The hack was claimed by an Iranian hacker. This caused a major political stir in the Netherlands, as a hack with such a concrete effect had never occurred before [21].

This crisis led, among other things, to the establishment of the NCSC: a National Cyber Security Center to coordinate national cyber threats. This crisis also encouraged politicians to consider the role of voluntary hackers in promoting cybersecurity. A member of parliament asked: "Is the minister prepared to investigate how the government can improve the security of its computer systems with the expertise of hackers, without the hackers suffering legal consequences?" [22]. In 2012, the Minister of Security and Justice promises that this would be investigated [23].

2.2.4 The Introduction of a National Responsible Disclosure Policy to Tolerate Hacking

In 2013, the Dutch government took the first step in a drawing up new policy regarding vulnerability disclosure: the responsible disclosure guideline [24]. The guideline explained how companies can draw up a responsible disclosure policy to promote cooperation with hackers. It was based on existing responsible disclosure policies of companies in the Netherlands and was essentially an encouragement of self-regulation between hackers and companies. When drawing up this guideline, discussions were held with hackers, journalists and researchers involved in the practice of vulnerability disclosure in the Netherlands [18, p. 99].

By publishing this guideline, the Dutch government took a position: 'ethical' hacking (as it was now referred to in House of Representatives meetings [23, 25] positively contributes to society and this should be encouraged instead of punished. The Netherlands was the first country in the European Union to draw up a national responsibility disclosure policy [26]. Yet, the guideline was critically received by the hacker community because the criminal law frameworks for computer trespassing remained intact [27,

28]. As such, the responsibilities for hackers and companies were out of balance: hackers were only allowed to report vulnerabilities to companies *with their own responsible disclosure policy*, while companies were only encouraged and not obliged to have such a policy set up – thus hackers still faced a risk of being prosecuted. As a result, the initial problem of unattended vulnerabilities remained effectively unresolved.

2.2.5 Lawsuit in 2013 as Breeding Ground for Responsible Disclosure Principles

In a criminal case in 2013, the jurisprudence regarding vulnerability disclosure was further developed. In this case, a patient of a health institution noticed the (weak) password of a doctor [18]. The password gave access to the computer system which contained sensitive patient data. The patient reported the security breach to the institution, but he did not receive a – in his opinion – quick response, after which he reported the leak to the media. He invited a local television broadcaster and he downloaded (anonymized) patient data as evidence. After publicizing the leak, the healthcare institution pressed charges against this 'hacker' and the case appeared in court [29].

The central question in the lawsuit was: was this patient a whistleblower and did he serve the public interest by reporting this leak to the media, or did he go too far? The judge stated that three principles are important to assess whether the hacker disclosed the security breach *responsibly*: did he act in the public interest? Did his action comply with the proportionality principle, i.e. did the suspect not go further than was necessary to achieve his goal? And did his action comply with the subsidiarity principle, i.e. were there no other, less far-reaching ways to achieve the intended goal? According to the judge, the hacker met the first principle: he served the public interest with his disclosure. But he did not comply with the last two principles: the hacker could have given the organization more time to respond to the vulnerability report before disclosing the breach publicly, and he did not have to download patient data to report the vulnerability successfully. Therefore, the hacker received a fine of 750 euros.

After this ruling, the principles of public interest, proportionality and subsidiarity were adopted by the Public Prosecution Service in their policy on how to deal with 'ethical hackers' [30]. Within a criminal investigation, these three principles formed the assessment framework for a 'responsible disclosure'. Based on this policy, the Public Prosecution Service has not prosecuted hackers whose hack complied with these principles since 2013 [31].

2.2.6 New Governmental Policy Balances the Responsibilities Between Companies and Hackers

In 2015, the House of Representatives criticized government policy: hackers should be able to report vulnerabilities to companies *without* their own RD policy [32]. After the evaluation of the responsible disclosure policy in 2015, which concluded that responsible disclosure contributes to strengthening the digital resilience of the Netherlands [31], and discussions with the hacker community [33], an updated version of the policy was published in 2018. The original name 'responsible disclosure', indicating the responsibility that hackers must take to report vulnerabilities, was adapted to 'coordinated vulnerability disclosure' (CVD), emphasizing the fact that both parties, the hacker and the recipient, must handle communication about the vulnerability responsibly. The principles of public interest, proportionality and subsidiarity are now included in the policy. If a hacker reports a vulnerability to an organization and works according to these principles, the hacker is not punishable, even if a company does not have its own CVD policy.

2.3 From Tolerated Hacking to Hackers Recognized as Crucial Participants (2019–2024)

The national policy provided new possibilities: hacking was allowed if hackers adhered to the principles. As a result, hackers started to act in accordance with the CVD principles and policy. Based on the certainty that these principles and the jurisprudence provided, hackers claimed their new role as an ally in cybersecurity governance.

In 2019, the Dutch Institute for Vulnerability Disclosure (DIVD) was founded. DIVD is an organization of volunteers who voluntarily scour the internet for vulnerabilities. DIVD hackers structurally violate the computer trespassing law when they search for vulnerabilities. But because they work according to a code of conduct that includes the CVD principles (serve the public interest, and stick to the proportionality and subsidiarity principle), they manage to avoid prosecution. Furthermore, being part of an established community increases the change a receiver of a vulnerability report takes the breach seriously [34, p. 217].

DIVD researchers have played a central role within Dutch cybersecurity governance in recent years. There are a number of gaps within formal Dutch cybersecurity governance: there is no central desk where information about security threats is received and shared with all affected parties. The National Cyber Security Center coordinates and shares security threats in the Netherlands, however, the mandate of the NCSC is limited to only share information with 'vital' companies that are needed to keep the country running, such as the electricity grid, water supply, and dikes. As such, non-vital companies and smaller organizations that are not protected by large cybersecurity companies do not receive information about important cybersecurity leaks. These companies and organizations are therefore disadvantaged, and moreover, the distinction between vital and non-vital companies is becoming increasingly unclear due to chain dependency.

DIVD researchers and other hackers have played an important role in filling this gap by scanning organizations for vulnerabilities and personally notifying them when their system is vulnerable. In 2019, DIVD security researchers played an important role in a major cybersecurity crisis caused by the Citrix vulnerability [18]. After a vulnerability in Citrix software was discovered, DIVD researchers successfully notified the Dutch organizations who were at risk of being attacked. The DIVD has also established the Dutch Security Reporting Point together with private parties, whose goal is to distribute security information to all organizations that do not receive information via the NCSC.

2.3.1 Institutional Recognition for DIVD Hackers

The DIVD researchers have a unique role in the cybersecurity network because they are tolerated when they hack networks to find vulnerabilities if they adhere to the CVD principles. No actor in the cybersecurity network, such as companies or governmental organizations, is able to do this because hacking remains a violation of the law. For DIVD researchers and other volunteer hackers, violating the law is now tolerated – under three conditions – because they do their work independently, voluntarily, and without profit.

The crucial role of DIVD researchers in cybersecurity is increasingly recognized by the Dutch government. The Dutch Safety Board concluded in their investigation of the Citrix crisis that "voluntary security researchers played a crucial role in incident response" (authors' translation [35]). The role of DIVD hackers was also repeatedly referred to as crucial and indispensable during a meeting of the House of Representatives Committee on Digital Affairs [36]. Since 2022, the DIVD receives a temporary subsidy to strengthen cyber resilience between actors in non-vital sectors.

However, the contribution of these hackers is on a voluntary basis, and therefore not structurally guaranteed [35]. The Dutch government now has to find a way to structurally support the participation of hackers to guarantee the cyber security of the Netherlands, as the country is partly dependent on their participation. Members of parliament have called for a more formally embedded role of volunteer hackers in the cybersecurity network [36]. Yet, it does not benefit cybersecurity to make DIVD a government agency, because it would not allow them to continue hacking, as a government agency is not allowed to structurally violate the law (ibid). Consequently, members of parliament requested the government to set up a multi-year subsidy scheme to structurally finance "ethical hacker collectives" [37].

3 Conclusion and Further Research: Lessons Learned for the Governance of Disruptive Technologies

This article presented the role of Dutch volunteer hackers in attending the risks of digitization. While the Dutch government took sides against the hackers in 1993 by labeling hacking as a criminal act, nowadays the contribution of hackers is – to a certain extent – embraced and encouraged. While volunteer hackers were initially considered as an important source of the problem, now they are seen by the government and private sector as an indispensable part of the solution. Dutch volunteer hackers currently play a central role in cybersecurity and as such they make an important contribution to the governance of digitization. What are the lessons that can be learned from this case for the governance of disruptive technologies?

Firstly, that it is important to recognize the contributions of citizens to the governance of disruptive technologies, also when their contributions are not evidently 'participation'. This case demonstrates that a period of 40 years was needed for the Dutch government and private sector to understand the severity of disruption by the digitization of society and to establish the necessary institutions and policies to deal with this. Only after a cyber crisis the Dutch government took concrete action, such as setting up the National Cyber Security Centre, and started to re-think the role of hackers in maintaining cybersecurity. The potential of hackers in contributing to cybersecurity could have been recognized earlier – after all, the positive contributions of well-meaning hackers were already recognized within the computer science community in the 1980s. Therefore, the contribution of citizens to the governance of digitization should be understood broadly [8], so that it can be better recognized.

Secondly, once citizen participation is recognized, the role of citizens in the governance of disruptive technologies should be enabled and supported. In the case of the hackers, the possibility to disclose vulnerabilities was seriously impeded since 1993 by the government and law enforcement. It was made quite impossible for hackers to continue to search for and disclose vulnerabilities – if they did, they faced a prison sentence – all the while fixing vulnerabilities turned out to be a central tenet to maintaining cybersecurity.

The institutional context should therefore enable citizens to participate and realize their potential in the contribution to the governance of digitization. Finally, this happened in the case: the Dutch government, private sector, and the hacker community explored the contribution that hacking could make to strengthening cybersecurity while acknowledging the danger of hacking – after all, malicious hackers form the cybersecurity risk. Additionally, also well-meaning hackers were not always considerate, as they would disclose vulnerabilities publicly, causing extra damage to the vulnerable party. Therefore, there was also a need for norms to guide the practice of vulnerability disclosure, to indeed make it into a *responsible* practice. Nowadays, hackers can contribute to cybersecurity in such a way that the private sector and the government tolerate it. Hackers are not free as they were once before 1993: they are allowed to participate, but only under certain conditions. In sum, participation by citizens should be recognized and enabled, and norms to guide responsible participation are needed to ensure good collaboration between all stakeholders.

These two lessons are also valuable for the governance of other disruptive technologies. Existing (uninvited) contributions by citizens should be recognized, enabled and supported. While in this paper the focus was on the role of citizens in the governance of a technology that is already 'in the world', citizen participation could also be strengthened during other phases. Hopster [38] distinguishes four levels on which disruption by technology may occur: the technology, artifact, application, or society level. Depending on the disruption problem, ethical approaches to deal with it such as foresight or design methods could include existing citizen participation, or invite citizens to participate. For example, in participatory Value Sensitive Design approaches (e.g. [39]), end-users are included in the design process of technology in order to anticipate its possible impacts. Such approaches and others can support citizens in their various roles in the governance of disruptive technologies.

The governance of disruptive technologies is a challenge. Citizens should not be seen as obstacles to successful governance. Rather, their various unique and central roles in the democratic society should be better recognized and supported so they can contribute to maintaining the public interest in the digital society.

Acknowledgement. This publication is part of the project "Strengthening Cyber Resilience by Technological Citizenship (with project number 410.19.004 of the research programme "Digital Society – The Informed Citizen" which is financed by the Dutch Research Council (NWO).

References

1. Benjakob, O.: The NSO File: A Complete (Updating) List of Individuals Targeted With Pegasus Spyware. Haaretz (2022)

2. Hopster, J.: What are socially disruptive technologies? Technol. Soc. **67**, 101750 (2021)
3. Filgueiras, F., Raymond, A.: Designing governance and policy for disruptive digital technologies. Policy Des. Pract. **6**(1), 1–13 (2023)
4. Yan, Z., et al.: Finding the weakest links in the weakest link: how well do undergradate students make cybersecurity judgment? Comput. Hum. Behav. **84**, 375–382 (2018)
5. ENISA. European Cybersecurity Month (ECSM) 2022 (2022)
6. Kävrestad, J., Furnell, S., Nohlberg, M.: User perception of context-based micro-training – a method for cybersecurity training. Inf. Secur. J. Glob. Perspect. **33**(2), 121–137 (2024)
7. Klimburg-Witjes, N., Wentland, A.: Hacking humans? social engineering and the construction of the "deficient user" in cybersecurity discourses. Sci. Technol. Human Values **46**(6), 1316–1339 (2021)
8. Gardenier, A.M., van Est, R., Royakkers, L.: Technological citizenship in times of digitization: an integrative framework. Digit. Soc. **3**(21) (2024)
9. Wynne, B.: Public participation in science and technology: performing and obscuring a political–conceptual category mistake. East Asian Sci. Technol. Soc. Int. J. **1**(1), 99–110 (2007)
10. Gongrijp, R., et al.: Hack-Tic Tijdschrift voor techno-anarchisten, vol. 1 (1989)
11. Reijnders, M.: De hackers die Nederland veranderden: De spannende geschiedenis van XS4ALL. Podium (2023)
12. Jacobs, J.: Kraken En Computers: Opkomst van de Hack-Cultuur. Veen (1985)
13. Het Vrije Volk. Computers Philips en RIVM gekraakt (1985)
14. Het Parool. Computers vaak zo lek als een mandje (1987)
15. Koops, B.J.: Cybercrime Legislation in the Netherlands. SSRN Scholarly Paper (2006)
16. Het Parool. Lastig Hack-tic houdt op op papier te bestaan (1995)
17. NXP v. Radboud Universiteit. ECLI:NL:RBARN:2008:BD7578 (2008)
18. van 't Hof, C.: Helpende Hackers. Verantwoorde Onthullingen in Het Digitale Polderlandschap. Tek Tok (2015)
19. Ellis, R. & Stevens, Y. Bounty everything: hackers and the making of the global bug marketplace. Data Soc. (2022)
20. Dutch Safety Board. The DigiNotar Incident. Why Safety Fails to Attract Enough Attention from Public Administrators (2012)
21. van der Meulen, N.: DigiNotar: dissecting the first dutch digital disaster. J. Strateg. Secur. **6**(2), 46–58 (2013)
22. Tweede Kamer. Debat over DigiNotar en ICT-problemen bij de overheid 2011–2012, 12(26) (2011)
23. Tweede Kamer. Verslag van Een Algemeen Overleg Informatie-En Communicatietechnologie (ICT). Kamerstukken 26643-240 (2012)
24. NCSC. Cybersecuritybeeld Nederland (2013)
25. Tweede Kamer. Verslag van Een Algemeen Overleg Informatie-En Communicatietechnologie (ICT). Kamerstukken 26 643-265 (2013)
26. ENISA. Coordinated Vulnerability Disclosure Policies in the EU. ISBN 978-92-9204-575-3 (2022)
27. Winter, B. de. Responsible disclosure richtlijn is onverantwoord risico. HP/De Tijd (2013). https://www.hpdetijd.nl/2013-01-03/responsible-disclosure-richtlijn-is-onverantwoord-risico/
28. Hoepman, J.-H.: Leidraad Responsible Disclosure behoeft aanscherping (door te leren van ervaringen in de luchtvaart) (2013). https://blog.xot.nl/2013/01/04/leidraad-responsible-disclosure-behoeft-aanscherping-door-te-leren-van-ervaringen-in-de-luchtvaart/index.html
29. Rechtbank Oost-Brabant. ECLI:NL:RBOBR:2013:BZ1157 (2013)
30. College van procureurs-generaal. Responsible Disclosure (Hoe Te Handelen Bij Ethische Hackers?) (2013)

31. Opstelten, I.W.: Brief van de minister van Veiligheid en Justitie (2015)
32. Tweede Kamer. Verslag van Een Algemeen Overleg Informatie- En Communicatietechnologie (ICT). Kamerstukken 26 643-354 (2015)
33. Tweede Kamer. Verslag van Een Algemeen Overleg Informatie- En Communicatietechnologie (ICT). Kamerstukken 26 643-551 (2018)
34. van 't Hof, C.: Cyberellende Was Nog Nooit Zo Leuk. Onthullende Verhalen Uit de Wereld van Informatiebeveiligers En Hackers. Tek Tok (2021)
35. Dutch Safety Board. Vulnerable through Software - Lessons Resulting from Security Breaches Relating to Citrix Software (Dutch Version) (2021)
36. Tweede Kamer. Verslag van Een Wetsgevingsoverleg. Kamerstukken 36084-11 (2022)
37. Tweede Kamer. Verslag van Een Wetsgevingsoverleg. Kamerstukken 36200-VII-116 (2022)
38. Hopster, J.: The ethics of disruptive technologies: towards a general framework. In: de Paz Santana, J.F., de la Iglesia, D.H., López Rivero, A.J. (eds.) DiTTEt 2021. Advances in Intelligent Systems and Computing, vol. 1410, pp. 133–144. Springer, Cham (2022). https://doi.org/10.1007/978-3-030-87687-6_14
39. Cenci, A., Ilskov, S.J., Andersen, N.S., et al.: The participatory value-sensitive design (VSD) of a mHealth app targeting citizens with dementia in a Danish municipality. AI Ethics **4**, 375–401 (2024). https://doi.org/10.1007/s43681-023-00274-9

Consumer Behaviour in the AI Era

Ana Ribeiro[1,2](✉) , Alfonso Rivero[1] , and José Luís Abrantes[2]

[1] Universidad Pontificia de Salamanca, Salamanca, Spain
acrodriguesri.chs@upsa.es

[2] CISeD – Research Centre in Digital Services, Instituto Politécnico de Viseu, Viseu, Portugal

Abstract. Artificial Intelligence, AI, has impacted the world, and in the marketing field, this impact is felt in the relationship with the consumer. Understanding how consumer behaviour could evolve in the face of AI in marketing is essential so marketers can adapt to the new reality. AI is present in several industries worldwide, and the topic goes far beyond robotics, which increasingly involves people.

The Ph.D project aims to understand the AI market in marketing, the consumer experience, the relationship between brands and their consumers, and future purchasing intentions. The idea is to infer the future of brands and purchase intentions through knowledge about behaviour.

This comprehensive theme will include developing a systematic literature review, following the Prisma method and presenting all the steps for possible replication. Qualitative and quantitative data collection and subsequent analysis using appropriate software are also planned.

It is intended that each subdivision of the central theme can contribute with exciting results, presenting meaningful discussions, individually and for the general theme. It is also hoped that it can suggest future studies in this area.

The challenge of the study is more than presenting conclusions. It presents them promptly when they still make sense and can contribute to marketing professionals.

Keywords: Artificial Intelligence · Consumer Behaviour · Marketing

1 Introduction

The business world has been impacted by digital transformation and AI. In Marketing, this impact is strong, with the latest technologies being experimented with as part of the process [1]. However, human-machine behaviour still requires understanding in a world slowed down by computers [2].

The concept of AI, although current, is not new. Claude Shannon, John McCarthy, Marvin Minsky, and Nathaniel Rochester, mathematicians from the 1950s, were pioneers in studying AI. Indeed, the 1956 Dartmouth conference is known as the birthplace of AI [3].

Now, AI will evolve towards humanisation [1] and includes many research areas, and its implementation varies across industries [4–8]. In this way, the effects of AI must be broad and go beyond robotics [2].

Studying the possible interventions of AI in traditional Marketing methods represents a great opportunity [1]. Therefore, the study intended to be carried out aims to know consumer behaviour in the face of AI and understand their interaction. It also aims to understand the AI market in marketing as a service that generates consumer experience and what type of brand relationship may arise from this interaction. These inputs should help predict the future of brands and how AI can influence the future of purchasing intentions. Knowing these points better will help marketers with future strategies.

In this presentation of the project under study, a brief literature review will be presented on consumer behaviour in the AI era. The work methodology intended to be adopted will also be presented, as well as what is intended with the results and discussions. At least some conclusions will be presented.

2 Literature Review

AI is impacting the interaction between consumers and brands, and there are not enough studies on the experience of these consumers in this area [9]. Some studies identify the need to explain the effect of anthropomorphism on AI [10] and analyse the influence of emotional trust [11]. Also, more studies must evaluate and anticipate the affective impact of the interaction between consumers and AI technology, especially the humanised one [12]. In a technological world, it is necessary to understand the similarities between AI and humans and how this similarity does not influence behaviour [13].

AI presents more information in a way that is accessible to consumers and impacts the market and the relationship between companies and consumers [2]. Understanding consumer behaviour when faced with different interactions with more innovative technologies is essential, as is understanding the effects these experiences have on brands [6]. AI can help consumers search and select content or services that interest them. This objective process combats information overload [2]. Ease and utility can be reasons for adopting more innovative technologies [7]. However, AI is not simple, and its use requires skills and knowledge to be genuinely beneficial [2].

AI has implications for the economy, market, and competition [2]. The economy is based on information and how we manage it to our advantage. Knowledge has become the path with data being provided through new technologies [14].

The future of markets will be affected by the possible adoption of AI to perform tasks, increasing efficiency [7]. There are studies on the impact of AI on markets and companies [2]. Voice assistants, for example, are the source of some interesting conclusions, such as the link between human characteristics, such as sincerity and consumer control and satisfaction [15]. Consumers focus on functionality, emotion, and comparison to humans when interpreting AI [16]. Despite existing studies, there is a need for broader studies on the effects of AI for applications in specific markets such as retail or healthcare [2]—for example, investigation of AI in physical shopping environments [17].

In a consumer experience using AI technology, engagement is positively mediated by trust and sacrifice [9]. Trust always plays an important role, but interaction and innovation also impact consumer loyalty to the brand [5]. With AI, companies can have more personalised customer engagement [14]. It suggests analysing AI's impact to understand how marketing strategies will adapt, how consumer behaviour will transform, and how situations related to privacy and ethics will be managed [18].

Consumers accept the use of AI in marketing but are sceptical about its ability to influence their consumption [16]. Studies are needed to explore online consumer behaviour and understand how interaction with AI impacts engagement, which can help improve customer service and engagement with a brand [19].

AI has advanced and will bring challenges [2]. Automation and personalisation will result from the use of AI in Marketing through the ability to predict consumer purchasing intent [1]. It has been some time since the importance of knowing what strategies consumers use when purchasing and how brands can use this information to their advantage was identified [20]. Some AI parameters, such as purchase duration, impact the purchasing decision [21]. So, AI can help market growth and help marketing decision-making [2]. AI already impacts marketing, but it will have even more impact in the future, so continued research is needed [18].

3 Method

In the initial phase, secondary research was conducted using available information on articles published about consumer behaviour, marketing, and AI. At this stage, the need for new studies was identified. Once the general need was identified, the study was planned according to the following method.

The first phase will be the systematic review of the literature using the Prisma guidelines to make replicability explicit [22]. The model is divided into three phases that help identify new studies from databases and records: identification, screening, and inclusion. It is also possible to identify new studies provided through other channels [23].

The articles covered in the systematic literature review will be searched on the Web of Science (WOS), which should include searching for two general topics: "Artificial Intelligence" and "Consumer Behavior." Articles must also be in the language "English," be of the "Article" type, and be in the area "Business Economics + Communication." The articles from this research will later be analysed by the VOSviewer software, which will identify the bibliometric maps and define the clusters.

This review aims to obtain an overview of the topic and what the existing literature identifies. It analyses the relevance, research methodologies, relationships between articles, and the most significant contributions and gaps that can be investigated in the future. A systematic literature review will summarise each article's themes resulting from VOSviwer analysis, responding to the objective of contextualising the existing literature on consumer behaviour in the face of AI.

The subsequent phases will study the humanised interaction between the consumer and AI technology and understand trust. That aims to respond to and perceive how to gain consumer trust in the face of AI and identify better ways of interacting with the consumer using AI.

Furthermore, the study will analyse AI as a service in the market, thus trying to identify the consumer experience and understand how AI will be a service in the market. It also evaluates the relationship between marketing and consumers when marketing uses AI. This evaluation can provide some inputs on how marketing can use AI to manage consumer behaviour and create engagement, maintaining brand loyalty using AI.

Finally, in the last phase, the aim is to identify the future of brands and how AI can intervene in future purchasing intentions, clarifying the future of brands in AI.

A quantitative study will be carried out to respond to these goals. The data will be statistically analysed using the IBM SPSS Statistics software and similar software. That software is typically used for statistical tests [24]. However, mixed methods are planned to be used in the last phase. In other words, there will be a qualitative and quantitative investigation. In this case, as the topic covered will be the future of brands, knowing the consumer's perspective and that of professionals is important. In addition to the consumer perspective, understanding marketers' perspectives will provide a more comprehensive view of AI in this sector [16]. Therefore, a qualitative part is expected in the form of in-depth interviews with marketing or AI professionals. The quantitative part will address only consumers.

The literature review on the topics covered will help define the questions used in the online questionnaire chosen to collect data. The online survey will be distributed conveniently. This type of sample will have to be analysed critically due to the possibility of bias [25]. The data will be processed and adequately analysed, considering the parameters of each theme. The created database may be used for one or more phases. The analysis and interpretation of data aim to present possible responses to research questions.

4 Results

As mentioned in the previous chapter, appropriate software will analyse the results.

WOS will be used for research (namely, in a systematic literature review). This prestigious database contains 1.9 billion cited references, 85.9 million records dating back to 1900, more than twenty-one thousand peer-reviewed journals, 17.2 million open-access records with full-text links, and 254 categories of subjects in the sciences, social sciences, arts, and humanities [26]. Scopus database will also be used.

Online means will be used to obtain data by doing the questionnaires or disseminating it on various platforms (e-mail, social networks, and others) to obtain as many answers as possible.

Specific software will be used to process data, whether textual or numerical. VOSviewer, Scispace, and IBM Statics - IPSS are planned at this stage. However, like Mendeley, other software and complements may be used.

5 Discussion

The study's results are expected to be adequately analysed considering the objectives.

The defined goals are: contextualise the existing literature on consumer behaviour in the face of AI; perceive how to gain consumer trust in the face of AI; identify better ways of interacting with the consumer using AI; understand how AI will be a service in the market; apprehend how Marketing can use AI to manage consumer behaviour toward brands; realise how to create engagement and maintain brand loyalty using AI; and, at least, clarify the future of brands in an AI era and how to maintain purchasing intentions.

The information obtained to respond to the first six objectives will contribute to responding to the last objective. In short, knowing consumer behaviour in the face of

AI, the AI market, and the use of AI in marketing is a complementary way of predicting the future of brands and purchasing. After answering the seven objectives, we will have a broader knowledge of consumer behaviour in this more technological era, and it will be possible to draw meaningful conclusions for the future of marketing.

6 Conclusion

In this digital era, AI is increasingly part of our lives, and the topic is increasingly being debated.

With new studies being published and journalists worldwide trying to explain to the public what IA is, speculation is an ally of ignorance. In this way, scientific studies, despite racing against time, in a process intended to be careful but on a topic that does not wait, is challenging but also necessary.

Scientific knowledge can and should use AI tools to its advantage, judiciously and without forgetting the critical capacity of the researcher, which gives him a character that is not only human and empathetic but also with the perspective that matters, the human perspective.

Since the project extends over several months and considers the speed of innovation in current software, tool choices may change during the project.

At the current time, the project is already in the systematic literature review phase, and it has already been possible to verify that in the space of about two months between an initial search and a final search for articles, the level of citations has increased, which causes a different bibliographic coupling result. Consequently, the inputs taken and the interests of the topics to be addressed may differ throughout the project, and new intentions for future studies may emerge. This living wheel of information in a machine that is constantly being fed produces innovation and knowledge.

The conclusions of the Ph.D project presented here will be as broad as possible and certainly as different from current reality as the time they take to be published. The challenge of the study is more than presenting conclusions. It presents them promptly when they still make sense and can contribute to marketing professionals.

References

1. Chintalapati, S., Pandey, S.K.: Artificial intelligence in marketing: a systematic literature review. Int. J. Mark. Res. **64**, 38–68 (2022)
2. Abrardi, L., Cambini, C., Rondi, L.: Artificial intelligence, firms and consumer behavior: a survey. J. Econ. Surv. **36**, 969–991 (2022)
3. Goanta, C., van Dijck, G., Spanakis, G.: Back to the future: waves of legal scholarship on artificial intelligence. In: Time, Law Chang, vol. 23529, pp. 1–45. Hart Publishing, Oxford (2019). Forthcom. Sofia Ranchordás Yaniv Roznai
4. Atwal, G., Bryson, D.: Antecedents of intention to adopt artificial intelligence services by consumers in personal financial investing. Strateg. Chang. **30**, 293–298 (2021)
5. Hasan, R., Shams, R., Rahman, M.: Consumer trust and perceived risk for voice-controlled artificial intelligence: the case of Siri. J. Bus. Res. **131**, 591–597 (2021)
6. Lalicic, L., Weismayer, C.: Consumers' reasons and perceived value co-creation of using artificial intelligence-enabled travel service agents. J. Bus. Res. **129**, 891–901 (2021)

7. Rasheed, H.M.W., He, Y.Q., Khizar, H.M.U., Abbas, H.S.M.: Exploring Consumer-Robot interaction in the hospitality sector: unpacking the reasons for adoption (or resistance) to artificial intelligence. Technol. Forecast. Soc. Change. **192** (2023)
8. Taghikhah, F., Voinov, A., Shukla, N., Filatova, T.: Shifts in consumer behavior towards organic products: theory-driven data analytics. J. Retail. Consum. Serv. **61** (2021)
9. Ameen, N., Tarhini, A., Reppel, A., Anand, A.: Customer experiences in the age of artificial intelligence. Comput. Hum. Behav. **114**, 106548 (2021)
10. Huang, B., Philp, M.: When AI-based services fail: examining the effect of the self-AI connection on willingness to share negative word-of-mouth after service failures. Serv. Ind. J. **41**, 877–899 (2021)
11. Minton, E.A., Kaplan, B., Cabano, F.G.: The influence of religiosity on consumers' evaluations of brands using artificial intelligence. Psychol. Mark. **39**, 2055–2071 (2022)
12. Hermann, E.: Anthropomorphized artificial intelligence, attachment, and consumer behavior. Mark. Lett. **33**, 157–162 (2022)
13. Kim, T.W., Lee, H., Kim, M.Y., Kim, S.A., Duhachek, A.: AI increases unethical consumer behavior due to reduced anticipatory guilt. J. Acad. Mark. Sci. **51**, 785–801 (2023)
14. Kumar, V., Rajan, B., Venkatesan, R., Lecinski, J.: Understanding the role of artificial intelligence in personalized engagement marketing. Calif. Manag. Rev. **61**, 135–155 (2019)
15. Poushneh, A.: Humanizing voice assistant: The impact of voice assistant personality on consumers' attitudes and behaviors. J. Retail. Consum. Serv. **58**, 102283 (2021)
16. Chen, H., Chan-Olmsted, S., Kim, J., Mayor Sanabria, I.: Consumers' perception on artificial intelligence applications in marketing communication. Qual. Mark. Res. **25**, 125–142 (2022)
17. Van Esch, P., Cui, Y., Jain, S.P.: Self-efficacy and callousness in consumer judgments of AI-enabled checkouts. Psychol. Mark. **38**, 1081–1100 (2021)
18. Davenport, T., Guha, A., Grewal, D., Bressgott, T.: How artificial intelligence will change the future of marketing. J. Acad. Mark. Sci. **48**, 24–42 (2020)
19. Perez-Vega, R., Kaartemo, V., Lages, C.R., Borghei Razavi, N., Männistö, J.: Reshaping the contexts of online customer engagement behavior via artificial intelligence: a conceptual framework. J. Bus. Res. **129**, 902–910 (2021)
20. Currim, I.S., Schneider, L.G.: A taxonomy of consumer purchase strategies in a promotion intensive environment. Mark. Sci. **10**, 91–110 (1991)
21. Jain, S., Gandhi, A.V.: Impact of artificial intelligence on impulse buying behaviour of Indian shoppers in fashion retail outlets. Int. J. Innov. Sci. **13**, 193–204 (2021)
22. Rethlefsen, M.L., Page, M.J.: PRISMA 2020 and PRISMA-S: common questions on tracking records and the flow diagram. J. Med. Libr. Assoc. **110**, 253–257 (2022)
23. Haddaway, N.R., Page, M.J., Pritchard, C.C., Mcguinness, L.A.: PRISMA2020: an R package and Shiny app for producing PRISMA 2020 - compliant flow diagrams, with interactivity for optimised digital transparency and Open Synthesis. Campbell Syst. Rev. 1–12 (2022)
24. Reis, A.T., Lay, M.C.D.: Análise quantitativa na área de estudos ambiente-comportamento. Ambient. Construído. **5**, 21–36 (2005)
25. Guimarães, P.R.B.: Material de Aula_Métodos Quantitativos Estatísticos. Não Aplicável. 252 (2012)
26. Clarivate: Web of Science Core Collection. https://clarivate.com/products/scientific-and-academic-research/research-discovery-and-workflow-solutions/webofscience-platform/web-of-science-core-collection/. Accessed 17 Feb 2024

Evaluation of the Effect of Side Information on LLM Rankers for Recommender Systems

Adrián Valera Román[1]($^{\boxtimes}$), Álvaro Lozano Murciego[2], and María N. Moreno-García[3]

[1] Department of Computer Science and Automatics, Faculty of Sciences, Universidad de Salamanca, Plaza de los Caídos s/n, 37008 Salamanca, Spain
adrianvalrom.usal@usal.es
[2] Expert Systems and Applications Lab, Department of Computer Science and Automatics, Faculty of Sciences, Universidad de Salamanca, Plaza de los Caídos s/n, 37008 Salamanca, Spain
loza@usal.es
[3] Data Mining Research Group, Department of Computer Science and Automatics, Faculty of Sciences, Universidad de Salamanca, Plaza de los Caídos s/n, 37008 Salamanca, Spain
mmg@usal.es

Abstract. Recently, Large Language Models (LLMs) have begun to be used in a wide variety of domains, marking their entry into the domain of recommender systems. Several approaches have been proposed to integrate LLMs into the recommendation process. One particularly promising approach is a two-stage system where a traditional recommender is first used to generate a candidate set, followed by an LLM acting as a ranker to refine the order of the top-k suggestions. This paper investigates how the results of training such an LLM ranker can be influenced by incorporating different side information about the candidates. The implications of these modifications are explored through a case study, with the results pointing to a consistent improvement of the retriever's performance due to the action of the LLM ranker.

Keywords: large language models · recommender system · llm ranker · item side information

1 Introduction

The advancements in natural language processing, the emergence of new neural network architectures like Transformers [9] and the development of Large Language Models (LLMs) have led to other fields within artificial intelligence starting to explore and incorporate the potential benefits offered by these models trained on large datasets. This spans from computer vision [3] to recommender systems (RS) [7].

In the field of recommendation, studies have begun on how to complement conventional recommender systems with LLMs [5] to address existing issues or propose new approaches to enhance their efficiency in specific contexts.

Nonetheless, the integration of these models and the new proposals are still very recent, and there is still a need to explore new ways to apply these LLMs in recommender systems [4]. Moreover, it is crucial to verify that they indeed lead to better results and under what circumstances.

Some preliminary studies [1] begin to explore the impact of this, generally focusing on addressing the shortcomings of recommender systems such as cold start and the initial scarcity of user data when relying on the LLM's knowledge. However, these models perform well in cold start scenarios (where users have little or no prior interactions) but do not outperform conventional recommender systems in a warm start scenario (where users have many previous interactions with items of the system).

Hence, new approaches have emerged, such as LlamaRec [11]. This work proposes a two-stage hybrid model that combines a retriever, which can be any conventional recommender system, with a ranker that reorders the results obtained in the first stage using LLMs like LLama 2 [8] trained for this ranker task. The aim of this work is to evaluate its performance against state-of-the-art recommender algorithms, showing promising results.

However, it is necessary to continue exploring such approaches under different conditions. For these reasons, in this article, we review the work conducted in [11] and the proposed approach. We reproduce some of its results and present the development of a case study to explore the inclusion of complementary information to find new ways of improving rankers.

The article is structured as follows: Sect. 2 shows the recent approaches to integrating LLM into recommender systems; Sect. 3 shows the characteristics of the rankers in a two-stage approach such as LLamaRec [11] and explains the case study and the methodology used to evaluate modifications in the training data of the rankers; Sect. 4 describes the results obtained after including this data; and finally, Sect. 5 discusses the main findings of this study and the conclusions drawn from this research and the future research lines to explore.

2 Background

2.1 Large Language Models for Recommendation

In the field of recommender systems, numerous studies are emerging aiming to utilize LLMs to enhance these systems. Such has been the interest that several literature review papers have been presented [2,7,10,12] with various approaches to classify the existing methods to date.

The work of Lin et al. [7] explores two main questions: where to employ LLMs for RS within the entire process and how to do so with two orthogonal taxonomy criteria: whether to freeze the parameters of the LLM and whether the process should utilize conventional recommender models (CRMs) as support.

This work is very interesting as it shows how it is now possible to apply LLMs in different stages of the recommendation pipeline: (1) feature engineering, (2) feature encoder, (3) scoring/ranking function, (4) user interaction, and (5) pipeline controller. Our work primarily focuses on applying LLMs in the scoring/ranking stage, as it is the core part of a recommender system.

On the other hand, various studies are highlighted that either utilize CRM in the recommendation process or not. Additionally, it presents works that use LLMs with frozen weights and others that opt for fine-tuning them, as seen in the TALLRec study of Bao et al. [1], one of the recent works yielding good results in this field.

2.2 Incorporating CRM to the LLM Approach

Some of the results emerging from the recent literature on LLMs applied in recommender systems [2,7,10,12] suggest that one promising approach is to employ CRM alongside LLMs to achieve better results. This can be realized through a hybrid approach. Within the studies following this approach, two notable works are LLamaRec [11] and A-LLMREC [6].

Based on sequential recommendation, LlamaRec proposes a two-stages recommendation framework integrating both retrieval and ranking stages, implementing a id-based sequential recommender as retriever and designing a LLM-based ranker algorithm. Experimental results on benchmark datasets show that LlamaRec consistently outperforms baseline methods and also the proposed verbalizer is able to improve the time and memory usage when a LLM is employed in the recommendation pipeline.

In the case of A-LLMREC an all-round LLM-based recommender system is proposed, laveraging collaborative knowledge from pre-trained collaborative filtering recommender CF-RecSys systems and textual knowledge for item descriptions, mainly on cold start, few-shot or cross-domain scenarios. In order to do this work proposes a two-stage approach: in the first stage, the system align the items from a frozen CF-RecSys, which is model agnostic and allows any system to be integrated, with the text embeddings obtained from pre-trained Sentence-BERT model, in the second stage the user representation and the textual knowledge onto the token space of the LLM, allowing the LLM to directly utilize the collaborative knowledge for the recommendation tasks.

Both of these systems are made up of two stages in the recommendation process. The first is the use of CRM systems and the second is the implementation of the LLM in a recommendation task.

3 Case Study

In this section we will explain the LLamaRec approach and our proposed modification of this approach as a case study.

3.1 LLamaRec: Two Stage Recommender System

LlamaRec [11] is a two-stage framework based on LLM for ranking-based recommendation and its main operation is described in the Fig. 1.

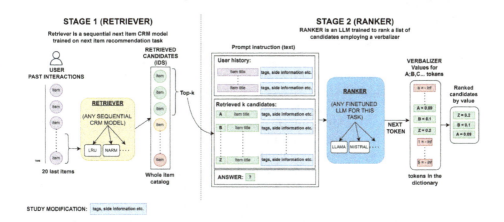

Fig. 1. Main diagram of the two steps LLamaRec approach

The first stage (left side of the Fig. 1) involves retrieving a series of candidates from the user's history using a sequential CRM (LRU, NARM, among others) previously trained. This provides an initial retrieval of good candidates that will subsequently be reordered in a second stage: the ranker stage. In the second stage, the user's interaction history and the candidates retrieved in the first stage are converted into a prompt instruction text, which will be provided as input to an LLM (e.g. LLama 2) acting as a ranker. This LLM is trained to provide a response with a single token corresponding to the candidates shown in the prompt (A, B, C, etc.). However, to function as a ranker, a component called a verbalizer is used, which only retrieves the probabilities for each of the valid tokens (A, B, C, etc., depending on the case), and these probabilities establish a new order for the provided list (as shown on the right side of Fig. 1).

In the second stage, when converting items to text, it is possible to decide what information to include, with the title and year being chosen in the original work of LLamaRec for the Movielens dataset. In this case study, we aim to exploit the possibility of including more information that could allow the ranker to establish a better order of the final candidates. At this point, certain tradeoffs need to be considered since the input context window of these models is limited and it is not possible to include all item information, thus requiring only the most relevant to be included. In the original work, in order to optimize the input length, the authors reduce the maximum amount of retrieved items to 20. This is performed for both, history and candidates items.

3.2 Proposed Modification

After reviewing this approach, it is proposed to answer the following research question:

- **RQ:** Including side information of items we obtain different results in the recommendation task?

The main hypothesis is that the richer the description provided in the prompt, the easier it will be for the LLM to identify the best candidates. Therefore, it is proposed to introduce item information in the training and inference process of the ranker. This information is usually available in the items and depends on the domain; therefore, for each of the datasets used in the original work (ml100k, beauty, and games), the item information has been obtained and added in the prompt construction. Since item information can be extensive (e.g., many tags for a movie) and the context length is limited, it is necessary to select only a limited amount of data from the item (in the proposed case study, 3 more descriptive tags of the available items). The LRU algorithm proposed in the original work has been trained as a retriever for each dataset with the best hyperparameters after a grid search (dropout = 0.5 and decay = 0), and the ranker has been trained including the additional item information mentioned for each dataset.

4 Results

Both the training results and all the hyperparameters used are documented in the following Weight&Biases[1] record of the experiments. Table 1 shows the results for each dataset and for each algorithm, presenting both the results reported in the original work and those obtained in the experiments conducted in this case study. The metrics are MRR (M), NDCG (N) and Recall (R) at different k values.

Table 1. Results reported on LLamaRec research paper and the results obtained on the case study adding the side information

	ML-100k				Beauty				Games			
	LRU	LamaRec	LRU*	LLamaRec*	LRU	LamaRec	LRU*	LLamaRec*	LRU	LamaRec	LRU*	LLamaRec*
M@5	0.0390	0.0440↑	0.0427	0.0439↑	0.0376	0.0385↑	0.0399	0.0394↓	0.0533	0.0600↑	0.0496	0.0564↑
N@5	0.0468	0.0543↑	0.0523	0.0529↑	0.0435	0.0450↑	0.0462	0.0460↓	0.0650	0.0714↑	0.0599	0.0679↑
R@5	0.0705	0.0852↑	0.0819	0.0803↓	0.0614	0.0648↑	0.0657	0.0660↑	0.0966	0.1061↑	0.0915	0.1032↑
M@10	0.0491	0.0529↑	0.0483	0.0509↑	0.0417	0.0428↑	0.0435	0.0432↓	0.0598	0.0671↑	0.0564	0.0632↑
N@10	0.0705	0.0759↑	0.0658	0.0701↑	0.0533	0.0554↑	0.0551	0.0552↑	0.0800	0.0887↑	0.0766	0.0846↑
R@10	0.1426	0.1524↑	0.1229	0.1344↑	0.0916	0.0971↑	0.0930	0.0947↑	0.1463	0.1599↑	0.1432	0.1551↑

(*) case study results with side information in the ranker stage.

The table shows the results reported in the original LLamaRec paper in the first two columns for each dataset and on the right, in the third and fourth

---

[1] Report: https://short-link.me/Eh9t.

columns for each dataset, the experiments performed in this work are shown. It has been marked with an up or down arrow if the LLamaRec ranker improves LRU results.

Regarding the LLamaRec approach, according to the data reported in the original paper, this approach consistently improves retriever results.

In our approach using retriever with the aforementioned hyperparameters, we consistently achieve improvements in the ML-100k and Games datasets, but not in beauty. It is possible that the labels in that dataset have added noise and do not add useful information for the ranker to perform better.

From the results we can see that the ranker is a good way to slightly improve the results offered by a CRM. As for adding extra information, no better results have been achieved than those reported in the original paper. Finally, it is also possible to see that the final results are very much conditioned to the retriever's performance.

5 Conclusions

This paper has reviewed the main ways to apply LLMs to recommender systems, specifically focusing on studies that combine CRM and LLMs. LLamaRec's work has been highlighted as a highly promising approach, and a case study has been proposed to examine how the inclusion of item information affects the ranker stage.

After experiments it has been proven that the LLamaRec approach consistently improves retriever performance even when side information is used. On the other hand, no better results than the original paper were achieved by including side information, although there seems to be a strong conditioning on the retriever results, and more research and experiments are needed in this direction.

As future work, various lines are proposed, such as: using LLama to generate this information if it is not available or is very long, as was the case with article or videogame descriptions. On the other hand, it is possible to use new models (e.g. Llama 3) with a larger context window.

Acknowledgments. We would like to thank the authors of LLamaRec for their work and for openly sharing the code of their experiments, which made this study possible.

References

1. Bao, K., Zhang, J., Zhang, Y., Wang, W., Feng, F., He, X.: TALLRec: an effective and efficient tuning framework to align large language model with recommendation. In: Proceedings of the 17th ACM Conference on Recommender Systems, RecSys 2023, pp. 1007–1014 (2023). https://doi.org/10.1145/3604915.3608857. http://arxiv.org/abs/2305.00447
2. Bao, K., Zhang, J., Zhang, Y., Wenjie, W., Feng, F., He, X.: Large language models for recommendation: progresses and future directions. In: SIGIR-AP 2023 - Annual International ACM SIGIR Conference on Research and Development in Information Retrieval in the Asia Pacific Region, pp. 306–309 (2023). https://doi.org/10.1145/3624918.3629550. https://dl.acm.org/doi/10.1145/3624918.3629550

3. Dosovitskiy, A., et al.: An image is worth 16 × 16 words: transformers for image recognition at scale. In: ICLR 2021 - 9th International Conference on Learning Representations (2020). https://arxiv.org/abs/2010.11929v2
4. Hua, W., Li, L., Xu, S., Chen, L., Zhang, Y.: Tutorial on large language models for recommendation. In: Proceedings of the 17th ACM Conference on Recommender Systems, RecSys 2023, pp. 1281–1283 (2023). https://doi.org/10.1145/3604915.3609494
5. Huang, C., Yu, T., Xie, K., Zhang, S., Yao, L., McAuley, J.: Foundation models for recommender systems: a survey and new perspectives (2024). http://arxiv.org/abs/2402.11143
6. Kim, S., Kang, H., Choi, S., Kim, D., Yang, M., Park, C.: Large language models meet collaborative filtering: an efficient all-round LLM-based recommender system (2024). https://arxiv.org/abs/2404.11343v1
7. Lin, J., et al.: How can recommender systems benefit from large language models: a survey (2023). https://arxiv.org/abs/2306.05817v5
8. Touvron, H., et al.: Llama 2: open foundation and fine-tuned chat models (2023). https://arxiv.org/abs/2307.09288v2
9. Vaswani, A., et al.: Attention is all you need. In: Advances in Neural Information Processing Systems, vol. 2017-December, pp. 5999–6009 (2017). https://arxiv.org/abs/1706.03762v7
10. Wu, L., et al.: A survey on large language models for recommendation (2023). https://arxiv.org/abs/2305.19860v4
11. Yue, Z., et al.: LlamaRec: two-stage recommendation using large language models for ranking. In: Woodstock 2018: ACM Symposium on Neural Gaze Detection, Woodstock, NY, 03–05 June 2018, vol. 1 (2023). https://arxiv.org/abs/2311.02089v1
12. Zhao, Z., et al.: Recommender systems in the era of large language models (LLMs). IEEE Trans. Knowl. Data Eng. 1 (2024)

A Monitoring Framework to Assess Air Quality on Car Parks

Tiago Almeida[1], Pedro Monteiro[1], João Henriques[1,2(✉)], Filipe Caldeira[1,2], and Cristina Wanzeller[1,2]

[1] Polytechnic of Viseu, Viseu, Portugal
{pv19896,pv20272}@alunos.estgv.ipv.pt,
{joaohenriques,caldeira,cwanzeller}@estgv.ipv.pt
[2] CISeD - Research Centre in Digital Services, Polytechnic of Viseu, Viseu, Portugal

Abstract. Looking for free parking places in cities is an increasing concern impacting drivers in terms of time, costs and health and also environment in terms of gas emissions and air quality.

To contribute to improve this scenario, it is proposed framework to collect data from sensors in order to report the available parking places and air pollution information. An application was developed as a Proof-of-Concept (PoC) collecting data from sensors to for air quality analysis. This application will report the parking lot locations and their respective air quality data by taking into account the respiratory conditions of each individual.

This work can contribute to the air quality in parks and respiratory health of the population. It's also important to highlight improvements in mental health, often affected by the stress and frustration associated with the inability to find parking spaces.

Keywords: parking · air quality · pollution · IoT

1 Introduction

When looking for free parking places in a city, drivers often experience long waiting queues, uncertainty about the availability of parking spaces, concerns about indoor air quality, and, in a new town, a need for more knowledge about the locations of potential parking cars. This work presents an innovative app designed to address and resolve these issues.

To overcome the issues above, this work explores the use of IoT to provide users with real-time information. It aims to report parking cars available in various cities in Portugal, along with the number of available places. This will enable users to make more informed decisions, thus saving them time and reducing the stress associated with waiting in queues.

Once parking spaces are available, drivers need to know the path to them. Hence, it is interesting to show the location of each parking lot. This is expected to improve the air quality and individuals' mental health. This way, it is possible to avoid unnecessary trips in search of parking cars, which often prove unsuccessful, thereby reducing the constant emission of pollutants along these routes and avoiding significant stress and frustration.

Considering that respiratory diseases are quite dangerous and require special care, it is of greater interest to take care of public health. Therefore, by using sensors that measure the levels of gases in the air, better management of ventilation systems can be achieved to reduce these quantities, thereby improving air quality, benefiting energy consumption, and reducing the overload associated with ventilation systems.

This paper is organized into five sections. Section 2 presents the state-of-the-art review. Section 3 presents the used architecture. Section 4 shows the data and how everything was done. Section 5 shows the obtained results. Finally, in Sect. 6, conclusions are drawn, followed by the introduction of future work guidelines.

2 Related Work

This section reviews the literature on air monitoring and parking solutions and compares them with the built application. Knowing that constant exposure to air pollutants can be hazardous to public health, especially to the respiratory system [1] and cardiovascular [2], the main goal of the present work was to install or make use of existing sensors in the parking cars to gather information about the pollutants existing in each park and, once the data was collected, it was essential to know if those values were acceptable or not to public health. [3] presents an air pollution monitoring solution supported by IoT. It describes how data gathering and analysis could be done using that architecture, making it possible to input accurate and reliable data into the database.

3 Architecture

The proposed application architecture is committed to modular design, scalability, and operational efficiency. Utilizing a microservices approach, distinct functionalities such as data simulation, MQTT protocol-based message handling, and data storage are encapsulated within Docker containers. This design ensures independent operation of each component, streamlining development, deployment, and scalability processes.

The adoption of the MQTT protocol for data transmission is driven by its lightweight features and efficiency, pushing it well-suited for simulating sensor data in a parking lot scenario. Simulated sensor data is transmitted through this protocol, facilitating rapid and reliable communication between sensors and the receiver. Additionally, the RabbitMQ message broker is integrated into the architecture to enhance the system's reliability and asynchronous messaging capabilities. RabbitMQ acts as a central hub, efficiently managing the communication flow between the sender (simulated sensors) and the receiver (data storage), further contributing to the robustness of the overall architecture.

Docker containers are instrumental in achieving encapsulation, portability, and scalability objectives. Each component, such as data simulation, RabbitMQ, and data storage, resides within its Docker container. This approach simplifies deployment and ensures a consistent runtime environment across diverse deployments, mitigating potential compatibility issues. Docker volumes are employed to store sensor data in an Excel file for persistent data storage. This strategic choice leverages Docker's persistent volumes alongside Excel's familiar and user-friendly format, facilitating straightforward and reliable data management.

To further increase the data analysis effectiveness and visualization, a Proof of Concept (PoC) of the framework was implemented to read stored sensor data and generate graphical representations. This separation of concerns allows for the utilization of specialized tools or libraries in creating charts or graphs, contributing to a robust solution for data analysis.

In conclusion, the proposed architecture combines the flexibility of microservices, the efficiency of the MQTT protocol for data transmission, the reliability of RabbitMQ as a message broker, and the encapsulation benefits of Docker containers. Each architectural choice serves a specific purpose in achieving the overall objectives of the application, making it a well-rounded system for simulating, storing, and visualizing sensor data from a parking lot.

4 Experimental Work

The implementation of a PoC of the framework required reviewing the literature about the state of the art. Moreover, it is essential to understand how to operate them. Thus, after evaluating the existing methodologies and technologies it was possible to implement the proposed framework. However, since no sensors were available for testing, it was necessary to define algorithms simulating the operation of the sensors. Therefore, these implementations were implemented using the Python programming language [4] and some of its libraries, such as the Numpy library [5].

With the data acquisition completed, creating a new application capable of analyzing the received data was necessary. Thus, once again, using Python, an algorithm was designed to meet these needs. However, for this algorithm to receive the data, it was necessary to implement a message exchange system between applications, and it was decided to use microservices.

The evaluation of Docker [6] and RabbitMQ [7] made it possible to set up both the algorithms and the protocol, thus achieving the expected result: real-time data acquisition and analysis through designated sensors.

To complete the framework, another algorithm was defined. However, in this case, it required using an Excel file provided by the data receiver. Using that file, the data was utilized to produce graphs to facilitate and enhance data visualization, making it more accessible and organized.

5 Results and Analysis

After the implementation of the PoC, experiments were run, and results will be presented and discussed.

First of all, the main goal of this application was to gather information about air quality using sensors. As a result, the expected results were to receive and store that information and check if it was correct. As mentioned, using the implemented architecture, data gathering was successful, and, using an algorithm, according to the data depicted in Fig. 1.

1	Timestamp	Sensor Id	CO2	NO2	Humidity	Temperature	Last Read Date
2	2024-01-18 23:53:55	12	359,8	45,6	62,3	8,1	18/01/2024 23:53:55
3	2024-01-18 23:53:56	8	369,2	34	76,1	15,5	18/01/2024 23:53:55
4	2024-01-18 23:54:03	0	410,4	45,1	73,1	21,6	18/01/2024 23:54:03
5	2024-01-18 23:54:05	12	278	23,9	74	20,5	18/01/2024 23:54:05
6	2024-01-18 23:54:05	8	454,5	59,7	74,3	16,4	18/01/2024 23:54:05
7	2024-01-18 23:54:13	0	340,1	46,6	66,4	13,2	18/01/2024 23:54:13
8	2024-01-18 23:54:15	12	365,1	35,8	75,7	20	18/01/2024 23:54:15
9	2024-01-18 23:54:15	8	420,2	57,8	74,3	13,6	18/01/2024 23:54:15
10	2024-01-18 23:54:23	0	278,3	74,2	66	10,5	18/01/2024 23:54:23
11	2024-01-18 23:54:25	12	278	37,7	72,1	16	18/01/2024 23:54:25
12	2024-01-18 23:54:25	8	338,9	48,1	90,6	17,4	18/01/2024 23:54:25
13	2024-01-18 23:54:33	0	395,3	38,7	65,8	6,5	18/01/2024 23:54:33
14	2024-01-18 23:54:35	12	375	57,6	72,4	6,4	18/01/2024 23:54:35
15	2024-01-18 23:54:35	8	369	42,1	54,2	15,5	18/01/2024 23:54:35
16	2024-01-18 23:54:43	0	389,2	63,7	74,7	12,2	18/01/2024 23:54:43
17	2024-01-18 23:54:45	12	333,8	47,3	73,4	17,7	18/01/2024 23:54:45
18	2024-01-18 23:54:45	8	394,7	62,8	70,6	14,3	18/01/2024 23:54:45
19	2024-01-18 23:54:53	0	412,6	67,2	69,2	16,7	18/01/2024 23:54:53
20	2024-01-18 23:54:55	12	335,5	46,4	80,3	15,7	18/01/2024 23:54:55
21	2024-01-18 23:54:55	8	406,3	51,9	59,5	20,4	18/01/2024 23:54:55
22	2024-01-21 21:44:12	25	412,6	56,6	67,1	13,8	21/01/2024 21:44:12
23	2024-01-21 21:44:12	8	323,4	28,9	67,6	19,7	21/01/2024 21:44:12
24	2024-01-21 21:44:13	15	387,5	34,1	70,9	16,1	21/01/2024 21:44:13
25	2024-01-21 21:44:16	12	347,8	42,6	72,7	15,3	21/01/2024 21:44:16
26	2024-01-21 21:44:21	0	432	59,4	56,1	15,1	21/01/2024 21:44:21
27	2024-01-21 21:44:22	25	439,2	33,7	99,7	11,9	21/01/2024 21:44:22
28	2024-01-21 21:44:22	8	334,5	58,1	64,2	16	21/01/2024 21:44:22
29	2024-01-21 21:44:23	15	278	50,4	83,1	12,7	21/01/2024 21:44:23
30	2024-01-21 21:44:26	12	340,6	32,2	78	17,9	21/01/2024 21:44:26
31	2024-01-21 21:44:31	0	338,7	53,5	60,8	17	21/01/2024 21:44:31
32	2024-01-21 21:44:32	25	482	52,5	71,2	20,2	21/01/2024 21:44:32
33	2024-01-21 21:44:32	8	382	51,5	82,2	17	21/01/2024 21:44:32
34	2024-01-21 21:44:33	15	301,2	39,5	67,3	17,8	21/01/2024 21:44:33
35	2024-01-21 21:44:36	12	318	52	62,4	12,8	21/01/2024 21:44:36
36	2024-01-21 21:44:41	0	417,3	47,1	74,1	14,1	21/01/2024 21:44:41
37	2024-01-21 21:44:42	25	362,2	64,3	62,3	13	21/01/2024 21:44:42
38	2024-01-21 21:44:42	8	280,7	52,2	60,2	14,4	21/01/2024 21:44:42
39	2024-01-21 21:44:43	15	278	48	83,2	10,7	21/01/2024 21:44:43
40	2024-01-21 21:44:46	12	426,3	61,5	66,8	28,3	21/01/2024 21:44:46
41	2024-01-21 21:44:51	0	370	32,9	73,2	13,2	21/01/2024 21:44:51
42	2024-01-21 21:44:52	25	308,7	51,3	68,6	23	21/01/2024 21:44:52
43	2024-01-21 21:44:52	8	278,9	70,4	56	21,5	21/01/2024 21:44:52
44	2024-01-21 21:44:53	15	278	37,7	70,5	18,8	21/01/2024 21:44:53
45	2024-01-21 21:44:56	12	370,2	48,8	73,5	18,9	21/01/2024 21:44:56
46	2024-01-21 21:45:01	0	278	64,5	84,2	6,4	21/01/2024 21:45:01
47	2024-01-21 21:45:02	25	367,7	62,4	59,6	19,2	21/01/2024 21:45:02
48	2024-01-21 21:45:02	8	365,1	55,3	46,9	27,1	21/01/2024 21:45:02
49	2024-01-21 21:45:03	15	341,4	38,5	67,1	12,4	21/01/2024 21:45:03
50	2024-01-21 21:45:06	12	358,4	66,2	86,5	10,7	21/01/2024 21:45:06
51	2024-01-21 21:45:11	0	315,6	43,3	55,2	6,2	21/01/2024 21:45:11
52	2024-01-21 21:45:12	25	296,3	57,7	65,4	22,3	21/01/2024 21:45:12
53	2024-01-21 21:45:12	8	346,7	56,3	79,7	17,9	21/01/2024 21:45:12
54	2024-01-21 21:45:13	15	362,6	48,5	62,2	20,9	21/01/2024 21:45:13
55	2024-01-21 21:45:16	12	461,9	44,9	82,8	11,3	21/01/2024 21:45:16

Fig. 1. Simulated data

After gathering the data, implementation, and evaluation, the next stage was to analyze data and produce graphs to streamline visualization aspects. A new algorithm was implemented, and the results were obtained successfully, as presented in Figs. 2 and 3.

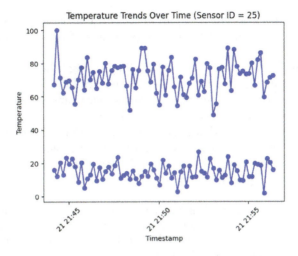

Fig. 2. Temperature graph

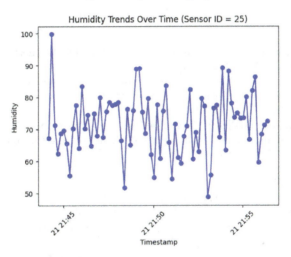

Fig. 3. Humidity graph

6 Conclusions and Future Work

The gas emissions in parking cars and the associated health risks raise significant concerns, highlighting the importance of innovative solutions to monitor and mitigate these impacts. The implementation supported by technologies such as Docker proves crucial, leveraging its encapsulation, portability, and scalability properties. Docker scalability is vital for streamlining development, deployment, and consistent maintenance across diverse scenarios, providing a robust foundation for complex systems. Furthermore, incorporating IoT devices to enhance the lives of societies underscores the pursuit of technological solutions that address immediate issues, such as efficient parking management, and contribute to the overall quality of life. This highlights the relevance of technological progress in promoting community well-being.

Future paths for this work may include integrating other sensors on parking area availability, such as cameras and IoT [8] devices for live data. For that, it is also essential to expand the capability of the services and adapt the algorithms to access more data and analyze it without errors.

Additionally, integrating this work with a mobile application should be considered to allow society to access air quality data and parking cars availability on their favorite parking cars. Cameras would help the park administrators have better control of their parks.

Acknowledgements. This work is funded by National Funds through the FCT - Foundation for Science and Technology, I.P., within the scope of the project Ref UIDB/05583/2020. Furthermore, we thank the Research Centre in Digital Services (CISeD), the Polytechnic of Viseu, for their support.

References

1. Kurt, O.K., Zhang, J., Pinkerton, K.E.: Pulmonary health effects of air pollution. Curr. Opin. Pulm. Med. **22**(2), 138 (2016)
2. Mirowsky, J.E., et al.: Ozone exposure is associated with acute changes in inflammation, fibrinolysis, and endothelial cell function in coronary artery disease patients. Environ. Health **16**(1), 1–10 (2017)
3. Chen, X., Liu, X., and Xu, P.: IoT-based air pollution monitoring and forecasting system. In: 2015 International Conference on Computer and Computational Sciences (ICCCS), pp. 257–260. IEEE (2015)
4. Python. Python (2024)
5. Numphy. Numphy (2024)
6. Anderson, C.: Docker [software engineering]. IEEE Softw. **32**(3), 102-c3 (2015)
7. Ionescu, V.M.: The analysis of the performance of RabbitMQ and ActiveMQ. In: 2015 14th RoEduNet International Conference-Networking in Education and Research (RoEduNet NER), pp. 132–137. IEEE (2015)
8. Madakam, S., Lake, V., Lake, V., Lake, V., et al.: Internet of things (IoT): a literature review. J. Comput. Commun. **3**(05), 164 (2015)

Disruptive Technologies Applied to Digital Education

Gamification, PBL Methodology and Neural Network to Boost and Improve Academic Performance

Filipe William C. Almeida[1]((⊠)), André Fabiano de Moraes[1], Rafael de Moura Speroni[1], and Luis Augusto Silva[2]

[1] IT - Information Systems, IFC, Camboriú, SC, Brazil
lipe.will18@gmail.com, {andre.moraes,rafael.speroni}@ifc.edu.br
[2] Departamento de Informática y Automática, USAL, Salamanca, Spain
luisaugustos@usal.es

Abstract. The article proposes to develop a computational architecture using gamification and project-based learning as active teaching methodologies to encourage and motivate students against their lack of interest and engagement in their education. The intention is to make them the main actors in the learning process, getting them involved in this environment and, consequently, improving their academic performance. To do this, teachers, with the help of an AI, will prepare tasks for students that will be answered on a playful, inclusive, collaborative and immersive platform. This platform allows the group, as a whole, to interact, participate and cooperate due to the progression of their knowledge. With data from student responses, the architecture also uses neural networks to identify patterns in the data and return the metrics to teachers, supporting teacher's decision-making in creating study strategies adapted to their needs. Students in their difficulties.

Keywords: Artificial Intelligence · Gamification · Education

1 Introduction

The contemporary educational scenario demands new approaches to captivate and motivate students. The traditional teaching model, characterized by expository and passive classes, does not keep up with the demands of the current generation, resulting in a lack of interest and lack of engagement [1]. Studies prove that motivation is fundamental to successful learning. Vieira [2] states that "motivation is the decisive element in the learning process" and that "the teacher will not achieve effective learning if the student is not willing to make efforts to learn voluntarily". Faced with this reality, this work proposes an innovative solution that integrates gamification, project-based learning (PBL) and artificial neural networks (AI) to assist in strategic and pedagogical decision-making.

The main objective is to transform the learning experience into something playful, collaborative and participatory through a computational architecture, taking advantage of the benefits of AI to stimulate students and contribute to improving their academic performance. To achieve this objective, data was collected from 39 students from the 4th to the 9th year of elementary school of the investigated educational institution. Data analysis reveals two worrying patterns: the drop in academic performance over the years and the inertia of performance, highlighting the need for new strategies to motivate students. The article is divided into five chapters: the first contextualizes the problem and presents the objective of the article; the second chapter presents the literary review that is the theoretical basis for the development; the third is about the development of the computational architecture itself; the fourth, presents results achieved by development; and, finally, in the fifth chapter, the conclusions.

2 Study Review

2.1 Gamification

To deal with the general lack of interest in some daily activities, including learning, institutions are looking for playful strategies to increase engagement and student performance [3]. Gamification is one of these strategies, defined as applying game mechanics in scenarios outside of games. In education, gamification allows the creation of "learning spaces mediated by challenge, pleasure and entertainment" [4, 5]. Gamification differs from games by taking rules into account and using elements outside the context of the game as a whole. Among these elements are scoring systems, levels, badges, feedback and ranking among players. When used well, this learning methodology appears as a good alternative to the current teaching model [6].

When participating in gamified activities, students are not merely passive recipients of knowledge but active protagonists of their own learning. They are encouraged to make decisions, solve problems and apply knowledge in situations that facilitate the retention and transfer of knowledge to different contexts. Gamification also allows students to receive feedback that contributes to self-regulation and continuous improvement in learning. Furthermore, through reward systems, scores and rankings, students can understand their progress over time and identify improvement areas. Gamification in education seeks to create a playful, collaborative and participatory learning environment, which contributes to increasing student motivation and engagement [7].

2.2 PBL Methodology

To understand the relationship between learning efficiency and study methods, psychiatrist William Glasser concluded that the best learning methods were those that involved the student in active participation in their process, leaving aside passive models that focus only on maintaining the content learned is "memorized" without using it in a practice in which the student becomes the main actor [8]. Glasser's study then produced the well-known "Learning Pyramid", which indicates the percentage of efficiency a learning method has on the content retained by the student. The pyramid (see Fig. 1) is divided into two main learning groups: passive, which has a low degree of effectiveness in its learning methods, and active, which has a high degree of effectiveness in retaining content.

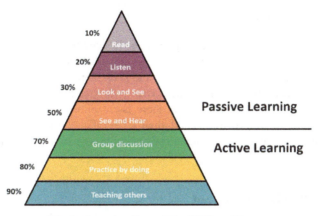

Fig. 1. Learning Pyramid by William Glasser.

Thus, as a proposal to increase learning efficiency, active learning methodologies emerged. One of these methodologies is Project-Based Learning (PBL) [9], which complements gamification by placing the student as the protagonist of the learning process. Specifically, the PBL methodology focuses on relevant and collaborative projects, which stimulate autonomy, contribute to developing skills and knowledge relevant to the job market, improve collective work collaboration [10], and increase creativity and development of students' critical thinking [11].

2.3 Neural Network

Since the discovery of the cellular structure of the nervous system at the end of the 19th century, the brain has attracted a lot of attention around the world and, thanks to this, in 1960, the term "neuroscience" emerged to represent a specific discipline studied by scientists. From different areas of knowledge, such as biology, psychology, medicine and physics, all of them collaborating in a more integrated way to understand the functioning of this highly complex, non-linear and parallel "computer" [12].

These studies discovered a constituent structure of the brain, neurons, which are organized to carry out certain processes more quickly than current computers. As examples of these processes, the brain takes around 100 to 400 ms to recognise familiar faces and, in circumstances of intense heat stimulation, such as placing the hand on fire, the brain receives this information and begins the movement of removing the hand in less time. 100 ms. Since then, Artificial Neural Networks (AI), inspired by the functioning of the human brain [13], have been widely used in the field of pattern recognition. Thus, they have demonstrated great effectiveness in tasks that involve identifying and classifying patterns in data and are suitable for dealing with problems in which the relationship between input data and output is complex or nonlinear. This is due to its ability to learn hierarchical and abstract representations of data, which has evolved since its first presentation in massive amounts of different types of learning. This is to allow the identification of discriminative characteristics in different contexts and needs.

2.4 Preliminary Aspects

Following these principles and observing the benefits they bring, combining them with a system proposal is a step towards achieving the defined objectives of the work. That is, combining gamification to motivate students and keep them engaged by understanding their evolution, together with the PBL methodology to give the student a collaborative, cooperative and active experience so that there is a feeling of belonging and mutual advancement within the same project; and, finally, neural networks, which help teachers identify patterns and analyses for better decisions in the educational direction of a group of students [14].

3 Architecture Development

Following these principles and observing the benefits they bring, combining them with a system proposal architecture (see Fig. 2) is a step towards achieving the defined objectives of the work. That is, combining gamification to motivate students and keep them engaged by understanding their evolution, together with the PBL methodology to give the student a collaborative, cooperative and active experience so that there is a feeling of belonging and mutual advancement within the same project; and, finally, neural networks, which help teachers identify patterns and analyses for better decisions in the educational direction of a group of students.

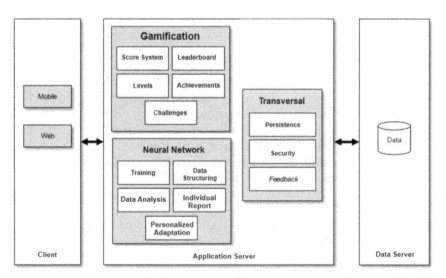

Fig. 2. Final system architecture.

In addition to user interfaces, there is a need for communication with third-party resources (for example, GPT Chat API). Communication occurs through an API developed in NodeJS located on the main server for this interoperability between systems. The choice to use an API in this way was precisely due to the need for resources to be "spread" across several "connection points" that use different technologies and, therefore, different interfaces, but both with the ability to make HTTPS (Hypertext Transfer Protocol Secure) requests (see Fig. 3).

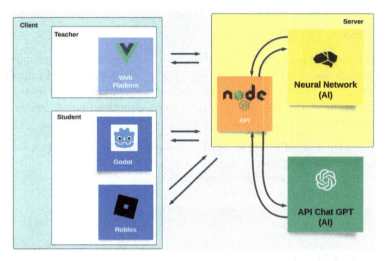

Fig. 3. Representation of the project's interoperational communication.

The communication with the GPT-3.5 Chat API is through an HTTPS request with a fixed structure: "Please generate a question with the theme 'theme chosen' returning the question in JSON format. The question will be returned as 'title'; its difficulty from 0 to 2 as 'level'; 4 alternatives to answer the question as 'alternative1', 'alternative2'; and the index (starting at 1 and not 0) of the correct alternative as 'correctAlternative'" and, in this way, in addition to allowing the use of other "ends" other than just the GPT-3.5 Chat (changing only the route and request authentication) will already be in the correct structure to be saved in the database (see Fig. 4).

```
{
  "title": "What is the main component of the central nervous system?",
  "level": 1,
  "alternative1": "Blood cells",
  "alternative2": "Neurons",
  "alternative3": "Bone cells",
  "alternative4": "Muscle cells",
  "correctAlternative": 2
}
```

Fig. 4. ChatGPT API JSON returns with the correct structure of question.

As said, the expected answer is the fixed structure above for saving in the database, however it is simple to scale this answer to larger content such as additional texts for the questions. Also, there is the possibility of adapting the number of responses per request in order to keep the ChatGPT API price as low as possible.

The other access interfaces developed in Vue (web), Godot and Roblox (native) communicate by making requests to the server that "orchestrates" interoperability.

3.1 Use Cases

The system is divided between 3 types of users: administrators, students and teachers. Although the administrator user type exists, it will not be described in detail in this chapter, and its actions are restricted to the creation of institutions and users. Students and teachers are the main actors and have other more specific actions related to the work proposal, which are:

- View individual and collective analysis of students in their subjects (teacher);
- Register new tasks (teacher);
- View student performance in their subjects (teacher).
- Carry out proposed tasks (student);
- View performance history in subjects (student);
- Redeem rewards (student).

3.2 Implemented Neural Networks

Our system utilizes two supervised learning neural network models. The first, developed in TypeScript with Brain.js, analyzes student performance data (response time, question difficulty, answer correctness) to identify performance trends and generate personalized reports. This basic Perceptron (neural network model) with a single hidden layer uses the Backpropagation algorithm and Stochastic Gradient Descent (SGD) optimizer for training (Fig. 5).

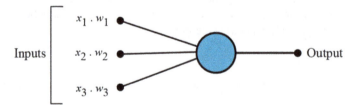

Fig. 5. Example of Perceptron neural network model with one hidden layer and three inputs.

The second model, developed in Python using PyTorch, classifies the difficulty of teacher-generated questions based on a predefined set. This model also leverages Backpropagation for training, but employs the AdamW optimizer.

Both are based on the Perceptron model that mimics the behavior of biological neurons. It receives input signals, processes them using a weighted sum, and generates an output signal based on a threshold activation function. Both neural network models are available in the project's GitHub repository: (Filipe William, André Moraes, Igor Ribeiro, 2023): (https://github.com/FilipeWilliam/TCC).

4 Results Achieved

The results achieved were obtained through four different experiments. Each of them implemented a new functionality in the system to improve the proposal and improve previous experiments. To begin the experiments, ten records were separated from the 39 available notes of the investigated institution. These ten records simulate the grade of ten students in an activity with ten questions. The lowest grade is 0, and the maximum is 10. The data was treated before being used because they were true notes and, therefore, were not made up of only whole numbers, so any decimal note greater than ".5" was rounded to the nearest largest positive whole number. In contrast, the smaller ones were rounded to the nearest positive integer (see Fig. 6).

Student 10001 Average	8,9	9	6,8	6,2	7,5	9,5
Student 10001 Utilized Average	9	9	7	6	8	10

Fig. 6. Rounding of averages used in student tasks.

After processing the students' averages, the average time spent on each question and their difficulty were randomly generated to generate student metrics (see Fig. 7).

Student 10001	6	8	8	7	7	7
Student 10002	9	9	7	6	8	10
Student 10003	8	9	8	8	8	9
Student 10004	10	10	9	10	9	10
Student 10005	8	9	6	5	9	3
Student 10006	9	9	8	9	7	9
Student 10007	9	9	9	8	8	9
Student 10008	8	6	5	5	6	8
Student 10009	7	8	7	7	8	6
Student 10010	9	10	10	9	8	8

Fig. 7. Averages used in student tasks.

4.1 First Experiment

In the first experiment, an interface was implemented that allowed teachers to register tasks consisting of questions and answers, using OpenAI's ChatGPT-3.5 API to generate questions. The teacher started the process by registering a task and being able to choose between registering it manually or generating questions based on a provided topic. In the case of automatic generation, the teacher interacts with the AI, providing the desired topic. Due to financial limitations, the API was adapted using the free OpenAI service to generate many questions with 4 alternatives and difficulty levels. These questions were stored in a database, and upon receiving the topic from the teacher, they were retrieved and displayed randomly.

Before completing the task registration, the teacher could modify the generated question, adjusting the question, alternatives, correct alternatives, and difficulty. Once registration was completed, the task was displayed on the to-do list for students enrolled in the course. After answering a question, the student received feedback. If the answer was incorrect, the wrong alternatives were visually deactivated, highlighting only the correct one. An encouraging message was displayed in a pop-up if the answer was correct. Furthermore, in cases of consecutive correct answers, additional messages were presented in the same pop-up, encouraging the student to stay focused (see Fig. 8).

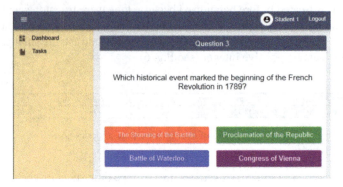

Fig. 8. Task response screen.

4.2 Second Experiment

In the second experiment the focus was on optimizing the process of teachers creating tasks and understanding the patterns behind difficulty attribution. Instead of relying on ChatGPT-3.5's suggestions, a single-neuron neural network was implemented to identify patterns in questions and categorize their difficulty levels.

To bridge the gap between human language and computer processing, the Google BERT natural language processing (NLP) algorithm was employed. BERT essentially acts as a translator, converting questions from human language (English in this case) into a format that the neural network can understand. This allows the network to analyze the content of each question, including the words and their relationships.

However, a significant challenge arises with increasing reliance on data: the neural network can become biased towards frequently occurring terms in questions, leading to inaccurate difficulty categorizations. To mitigate this, a supervised learning approach was adopted, where the network is trained using pre-categorized questions by teachers of a specific subject.

The amount of data significantly impacts the accuracy of categorizations. For example, commonly used terms like "man landing on the moon" might be wrongly grouped under the same difficulty level, neglecting important nuances. Similarly, ambiguous terms like "author" can also lead to misinterpretations.

Therefore, to create a truly helpful tool for teachers, a substantial database is crucial. This allows the AI system to analyze the content and context of each question more effectively, minimizing bias towards frequent terms and enabling it to identify potential relationships between difficulty levels and the words within questions. This approach aims to mimic ChatGPT-3.5's ability to understand the nuances of human language but through a data-driven learning process.

4.3 Third Experiment

In the third experiment, an improvement was developed to evaluate student performance collectively and individually to provide teachers with a deeper understanding of students' difficulties and facilitate the planning of more effective teaching strategies.

Four neural networks were created to identify student performance params and predict results based on their task histories, generating metrics and reports. Each neural network was trained for a specific analysis: "Probability of getting it right when going 1 time," "Average time per difficulty," "Average time of correct answer per difficulty," and "Probability of getting it right." (see Fig. 9).

The first analysis predicts the probability of success considering the possibility of the student losing focus once during the attempt, indicating potential problems with engagement or lack of concentration. For this analysis, the formula is defined by:

$$Probability = g(W * quantityLossFocus + b') \qquad (1)$$

where g is the activation function; W is the final layer's weight of the number of times the focus was lost.

The second metric reveals the average time it takes students to answer questions of different difficulties, providing insights into assessment time and identifying time patterns between difficulties. For this analysis, the formula is defined by:

$$AverageTime = g(W * difficulty + b') \qquad (2)$$

where g is the activation function, W is the difficulty weight.

The third analysis, similar to the previous one, only considers correct answers, allowing the teacher to understand whether more difficult questions require more time for reflection. For this analysis, the formula is defined by:

$$AverageTime = g(W_{difficulty} * difficulty + W_{hits} + hits * b') \qquad (3)$$

where g is the activation function; $W_{difficulty}$ is the weight of the difficulty; W_{hits} is the hit weight.

The latter analysis indicates the probability of correcting any question at different difficulty levels, offering an overview of students' understanding of the material. For this analysis, the formula is defined by

$$Probability = g(W * difficulty + b') \qquad (4)$$

where g is the activation function, W is the difficulty weight.

For all formulas, the term b (bias) ensures that the network adjusts and learns patterns even when the inputs are null. Adjusting these values during training is fundamental to a flexible neural network learning process. It is also crucial to highlight that these

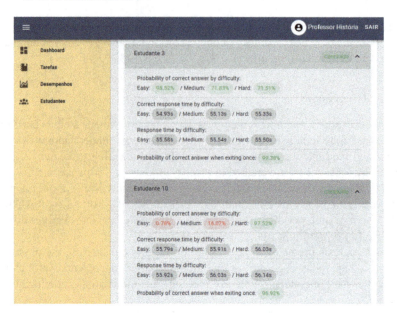

Fig. 9. Individual performance analysis.

metrics must be analyzed together, as a single isolated metric can lead to erroneous conclusions. Continuous data collection and joint metrics analysis help identify more accurate patterns, facilitating more informed and effective decisions in the teaching process.

These performance params are valuable tools for teachers helping to comprehensively assess student progress and identify areas for improvement, empowering them to gain deeper insights into their students' strengths, weaknesses, and learning patterns to make informed decisions that enhance student learning and foster academic growth of each one.

4.4 Fourth Experiment

In the fourth experiment, the focus was on interaction with the student, providing a more playful experience through developing a game using the "Godot" game engine. The objective was to cover different levels of education, capture students' attention and ensure inclusion, especially for those with visual impairments.

The game, implemented following the PBL (Project-Based Learning) methodology, offers an interactive and collaborative interface, promoting participation in collective activities. Students can share experiences with colleagues from similar or different disciplines, encouraging academic engagement. Furthermore, features similar to the platform in the first version are included, allowing students to view and carry out pending tasks. For better accessibility, a mode for people with visual problems was incorporated, enabling the use of the platform using voice commands and dynamic screen readers.

When starting the session, students are directed to a character editing module, allowing them to create a unique identity. Interaction with tasks occurs as the character sails on a boat to islands, representing the students' pending tasks. Navigation and confirmation of tasks are carried out playfully (see Fig. 10).

Question feedback occurs in real time, with the project's API centralizing information across platforms. After completing a task, the API updates the student's score, exchanging points for improvements to the boat or new clothing for the character, encouraging and rewarding students' efforts. The system displays student performance data for teachers, including the number of correct answers each student has.

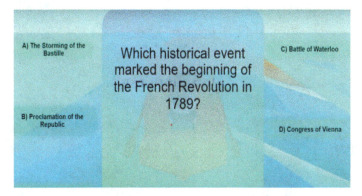

Fig. 10. Student interaction on task inside the game.

Furthermore, a student ranking was added to the web platform that indicates the student's performance relationship with the group and their progression throughout the year, boosting them to improve (see Fig. 11).

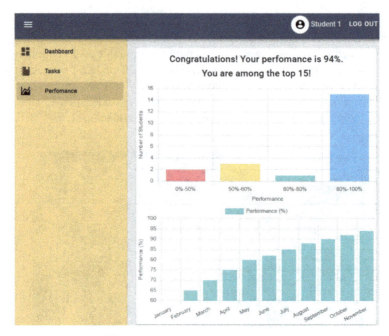

Fig. 11. Student ranking on platform.

5 Conclusion

This project provides a deep understanding of trends and changes in student performance over time, enabling the definition of specific goals and the comparison between actual performance and established goals. In addition, graphics were generated to track their own progress and adjust their study strategies as needed promoting a sense of responsibility and self-management in students. By encouraging students and providing teachers with tools to focus on strategic levels of teaching aims to boost student's. These features combined with the possibilities offered by AI, promotes a promising teaching model, resulting in improved academic performance and signaling the beginning of a more efficient and engaging educational approach.

References

1. Deci, E.L., Ryan, R.M.: The "what" and "why" of goal pursuits: human needs and the self-determination of behavior. Psychol. Inq. **11**, 227–268 (2000)
2. Vieira, F., Silva, G., Peres, J., Alves, E.: Causas do desinteresse e desmotivação dos alunos nas aulas de Biologia (2010). https://www.publicacoesacademicas.uniceub.br/universitashumanas/article/download/1061/1238. Accessed 11 Nov 2023
3. Silva J., Sales G., Castro J.: Gamificação como estratégia de aprendizagem ativa no ensino de Física (2019). https://www.scielo.br/j/rbef/a/Tx3KQcf5G9PvcgQB4vswPbq/. Accessed 03 June 2023

4. Alves, L., Minho, M., Diniz, M. Gamificação: diálogos com a educação. In Fadel. L, Ulbricht, V., Batista, C., Vanzin, T. (Org.). Gamificação na educação, pp. 74–97. Pimenta Cultural, São Paulo (2014)
5. Deterding, S., Dixon, D., Khaled, R., Nacke, L.: From game design elements to gamefulness: defining "gamification". ACM (2011)
6. Silva, M., Toassi, P., Harvey, M.: Metodologias Ativas e Ensino de Língua Estrangeira: Objetos de Aprendizagem como recurso didático no contexto da Gamificação. Revista Linguagem em Foco **12**(2), 227–247 (2020)
7. Filatro, A., Cavalcanti, C.: Metodologias inov-ativas: Na educação presencial, a distância e corporativa. Saraiva, São Paulo (2018)
8. Glasser, W.C.T.: A New Psychology of Personal Freedom. HarperCollins, NewYork (1999)
9. Thomas, J.W.: A review of research on project-based learning (2000). https://tecfa.unige.ch/proj/eteach-net/Thomas_researchreview_PBL.pdf. Accessed 11 Oct 2023
10. Bender, W.N.: Aprendizagem Baseada em Projetos: Educação Diferenciada para o Século XXI. Penso, Porto Alegre (2012)
11. Venturelli, J.: Educación Médica: nuevos enfoques, metas y métodos (1997). https://pesquisa.bvsalud.org/portal/resource/pt/lil-228796. Accessed 15 July 2023
12. Haykin, S.: Redes Neurais: Princípios e Prática - 2ª Edição. Bookman, Porto Alegre (2000)
13. Rauber, T.: Redes Neurais Artificiais. https://www.researchgate.net/profile/Thomas-Rauber-2/publication/228686464_Redes_neurais_artificiais/links/02e7e521381602f2bd000000/Redes-neurais-artificiais.pdf. Accessed 10 Nov 2023
14. Luckin, R., Holmes, W., Griffiths, M., Forceir, L.: Intelligence Unleashed: An Argument for AI in Education. Pearson (2016)

Wrap Your Mind Around Education: Applying Hugging Face to a Chatbot with AI

Dafne Itzel Rojas González[✉], Josue Aaron Soriano Rivero, and Jatziri Hernandez Hernandez

Universidad Tecnológica de Tecámac, 55740 Tecamac, Estado de Mexico, Mexico
daffi.rojas@gmail.com

Abstract. The search for information in educational settings is often slow, confusing, and bound by conventional methods. To address this issue, the development of a software that leverages a chatbot for information retrieval was proposed, aiming to expedite the consultation process.

As a result, a chatbot trained with specific, delimited information has been created, in addition to being adaptable and accessible, all this is possible due to the implementation of AI tools.

Furthermore, this approach enables easy access to information on any subject, achieving a chatbot that is adaptable, available, accessible and that contributes to the digital transformation, also supporting the education of new generations.

In conclusion, it is essential to begin integrating AI into education to facilitate the search and obtaining of information in the classrooms, starting a contribution to the digital transformation and new ways of learning that optimize time and further nurture students' skills.

Keywords: Artificial intelligence (AI) · Education · Chatbot · Large Language Model (LLM) · Digital Transformation

1 Introduction

Are we ready to trust Artificial Intelligence with the education of the next generations?

Clearly, every technological advancement poses a direct challenge to the traditional methods of education. Since the democratization of internet access, there has been a significant shift in information retrieval practices: the conventional visit to the library has been supplanted by the convenience of mobile devices. It is likely that many of today's teenagers have never set foot in a library nor understand the process of borrowing a book. Given this context, what can people expect with the integration of AI tools in education?

Additionally, it is increasingly common to encounter videos on social media demonstrating tools that assist students with a variety of tasks. These tools range from automatically generating APA-style references for citing articles to creating diagrams, maps, and even comprehensive research projects on demand.

Therefore, this advancement presents a formidable prospect for sceptics of formal education, as AI could assimilate knowledge within minutes that would traditionally require years to acquire.

Hence, this paper aims to investigate the symbiotic relationship between AI and education. To achieve this objective, the article presents the development of a chatbot tailored to enhance comprehension in specific subjects or academic courses. Essentially, a chatbot simulates a dialogue akin to conversing with ChatGPT, albeit with an educational emphasis. The paper delineates the tools utilized, the development methodology, and the challenges encountered during the process, with a particular focus on a designated chatbot. Additionally, the article delves into diverse prospective applications of AI in educational support.

2 Overview of Education and AI

Certainly, the concept of education is widely understood, as it encompasses the acquisition of skills, knowledge, and values through pedagogy, a fundamental aspect of human development. Similarly, Artificial Intelligence (AI), a discipline within computer science, denotes a technological domain enabling machines to emulate human behaviour and execute sophisticated tasks, including problem-solving, recommendation algorithms, and beyond [4, 11].

2.1 Importance of AI in Education

Firstly, AI has revolutionized the way people obtain information, making it easier to obtain resources. Presently, individuals equipped with internet connectivity can leverage the AI infrastructure readily available across networks.

Consequently, the amalgamation of AI and education presents a formidable synergy, facilitating enhanced accessibility to information for both students and educators. This symbiotic relationship empowers them to accomplish more within shorter timeframes, while also serving as a catalyst for skill acquisition and development [22].

Above all, the true importance is that education can present a plus if it is used appropriately as a complement to support learning and obtain information in a better way, without harming the development of the skills of students and teachers.

2.2 Students not Afraid to Ask Their Questions and Concerns

Thus, through the integration of an AI-powered chatbot, students can foster a heightened sense of confidence in their ability to articulate and address inquiries effectively. This technological resource facilitates the consolidation of information discussed in class through synthesis and query resolution. Consequently, the utilization of such a tool serves to fortify students' understanding and efficiency in managing their academic endeavours.

Further, recognizing the pivotal role of student confidence and comfort in soliciting questions within the classroom environment is imperative for their continuous progression and developmental advancement.

2.3 Approaches to Artificial Intelligence in Education and Content Customization

Also, the application of AI extends beyond digital education, encompassing the challenge of integrating it within traditional face-to-face educational settings. Leveraging machine learning, chatbots can be tailored to focus on specific subjects, courses, or domains, thereby enhancing student engagement and concentration within targeted areas of study.

Notably, a chatbot has the advantage of being fed only with useful information or that is of interest to the public to which it is directed, this precision significantly streamlines the chatbot's focus, minimizing extraneous content and optimizing resource allocation. Consequently, the implementation of customized solutions proves invaluable, offering a multitude of benefits across various educational contexts [21].

2.4 Automatic Content Generation

Especially, the chatbot will swiftly furnish responses to inquiries posed by students, drawing upon a repository of pre-existing content. By leveraging the information it has been provided, the chatbot will autonomously seek out and provide pertinent answers to students' queries on the specified subject matter.

Furthermore, the chatbot may enhance its response capabilities through the application of AI, amalgamating the data it has been fed to furnish more comprehensive and refined answers to inquiries.

3 Introduction to Hugging Face

Notably, the concept of open source can be elucidated through a comparison with proprietary software such as Windows 11, developed by a private entity whose underlying code remains inaccessible and necessitates licensing for use. In contrast, open-source software endeavours to circumvent singular control, affording a community the opportunity to access, replicate, employ, and collectively refine its features.

Thus, the open-source analogue to Windows finds expression in Linux, an open and freely available operating system that has engendered numerous specialized variants. Extending this analogy to the realm of artificial intelligence, Hugging Face assumes a role akin to Linux, providing a collaborative platform for the development and utilization of AI technologies.

3.1 Understanding Hugging Face

Initially, Hugging Face began as a website featuring an AI-driven chatbot for engaging in conversations. However, its evolution has seen the establishment of strategic partnerships with prominent cloud computing entities like Amazon Web Services (AWS) and Azure, facilitating the seamless integration of models into its platform.

Because this platform has garnered trust from leading AI enterprises such as OpenAI, Meta, and Google, who leverage it to publicly unveil novel models.

So, the transformer architecture represents a paradigm in model development, consolidating knowledge and computational prowess. Initially conceived for Natural Language Processing (NLP) applications, this architecture has since been extended to diverse domains including video, audio, imagery, and reinforcement learning.

3.2 Best Features of Hugging Face

To grasp the essence of Hugging Face's offerings, it is essential to commence with its repository of models, housing over 600,000 AI models, readily accessible with a single click. Analogous to GitHub but specialized in AI, this repository empowers users to explore datasets and even craft their own models should the need arise.

Upon model creation, users can proceed to develop applications and seamlessly publish them on the platform utilizing designated spaces. In the event that a model gains significant relevance and requires deployment on alternative platforms, the Application Programming Interface (API) service's inference functionality facilitates this process effortlessly, generating an endpoint to receive requests with a single click. Furthermore, for individuals inclined towards a non-programmatic approach, an intuitive no-code service named Auto Train is available.

3.3 Benefits of Using Hugging Face for Chatbots

Undoubtedly, the platform operates under an open-source framework, affording users the liberty to utilize and customize the available models. This not only economizes time and resources but also expedites development processes by harnessing pre-trained models that can be tailored to specific NLP tasks.

Furthermore, Hugging Face's libraries and tools is meticulously crafted to ensure user-friendliness, bolstered by comprehensive documentation elucidating intricacies. This concerted effort alleviates concerns surrounding the complexity of NLP model implementation, facilitating seamless integration into projects for developers and data scientists alike.

In addition, the platform boasts an engaged community of developers and users, fostering an environment conducive to support, guidance, and collaborative endeavours in the realm of NLP projects.

3.4 Transparency Issues with Hugging Face Models

While leveraging pre-trained models offers numerous advantages, such as time and resource efficiencies, democratization of AI, and enhanced accessibility, developers must navigate certain considerations when incorporating them into their projects.

Undoubtedly, it is crucial to review how they are documented and whether it is possible to access the datasets used for training, identify, and address possible biases present in the models, and finally, it is important to understand the licensing of the models and their compatibility with software licenses.

According to research conducted by collaborators from several universities, only 14% of the models analyzed report datasets as labels in the model files. Out of a total of 389 manually analyzed models, only 72 (18%) describe their possible bias.

Furthermore, 31% of the Hugging Face models declare a license, most of which (91%) have a permissive license. Machine Learning (ML) specific licenses are also used, which restricts a "responsible" use of the models [21].

4 Building a Chatbot

Presently, the construction of a chatbot mandates proficiency in the programming language selected for its development. Predominantly, Python, JavaScript, and TypeScript stand as the prevailing choices [4]. However, it is foreseeable that additional programming languages will emerge as viable alternatives in the near future [23].

Surely, with the use of a Large Language Model (LLM), the chatbot will have better capabilities, since the LLM intervenes in the characteristics it may have, as well as in the responses and content it can generate [3, 7, 16].

Likewise, a server is necessary to execute the code and a platform for its use and management. Of course, the information necessary to feed the chatbot, the area to be delimited and the target audience are also essential to start building a chatbot.

4.1 Understanding Chatbot Functionality

Of course, the functionality of a chatbot focuses on the area in which it will be used, each developer will define the area in which it will be applied and therefore the main functionality or functionalities of the bot will also be defined, so the functionalities are totally dependent on the field to which it will be used that is directed [8, 16, 17].

4.2 Choosing the Right AI Framework

And then, regarding the choice of the AI framework, it is necessary to choose it according to the one that best suits user's needs, in this case, LangChain will be chosen, because is an open-source framework for building applications based on LLMs [2, 14, 18] and will provide tools to connect to LLMs and thus simplify processes [9, 10].

4.3 Collecting and Preparing Data

Therefore, for this process, a prior analysis of the information to be used is necessary, to later collect the information and select which one best suits the audience to which it is directed and will be most useful for its use [1].

Additionally, it is important to avoid biases, for this the analysis stage is essential and utmost attention is required in this process, to have a better result.

Besides, the data collection and preparation phase are one of the first critical steps for the success of chatbots in education. It is important to explore methods to obtain relevant data, such as lecture notes, study materials, and frequently asked questions from students. In addition, the steps of preparing this data, including text cleansing, authorization, and removal of confidential information, should also be considered to ensure the quality and consistency of the training data set [13].

To facilitate the provision of requisite information for each subject, it is imperative to feed the Lechuzobot through the following steps: delineate specific subjects along with their respective topics, meticulously prepare the information by cleansing the data and converting it to PDF format, securely store the resultant PDFs on a server for accessibility, and ultimately integrate the information into the bot's source code. This integration enables the bot to furnish tailored responses in accordance with the data contained within the PDFs.

4.4 Integrating Hugging Face into the Chatbot

First, Gradio is a Python library that allows you to quickly create customizable web apps for ML models [26] and Gradio apps can be deployed on Hugging Face Spaces [26], this is how Hugging Face is integrated into the chatbot.

Due to this, Gradio library allows Hugging Face to be integrated to develop and customize the functionalities that are required and that the model allows.

To deploy Lechuzobot in Hugging Face Spaces, you need to set up an environment using Gradio. This environment must include all necessary dependencies in a "requirements.txt" file, and the environment variables must be correctly configured within the Space. Next, host the language model and the vector database in the available Space. Finally, launch the Gradio interface to enable interaction with the chatbot.

5 Prompting vs Fine-Tuning for Training the Chatbot

The process of training a chatbot represents a critical juncture, wherein decisions regarding model adaptation to the unique requirements of a given project are paramount. It is imperative to discern between two primary methodologies: initial stimulation and refinement of pre-existing models [12, 16, 20].

In an educational context, comprehending the strategies employed for training chatbots utilizing Hugging Face assumes significance. Initial stimulation entails the gathering and preprocessing of data from inception, while refinement of pre-trained models involves tailoring existing models to align with specific training requirements. Both approaches wield substantial influence over the efficacy and efficiency of chatbots, particularly in their role of augmenting the learning process [15].

In the scope of Lechuzobot resources, it is essential to first establish the OpenAI API, enabling the chatbot to utilize OpenAI's language models. Subsequently, various documents in PDF format are downloaded and processed to extract their content. This content is then divided into manageable fragments, from which embeddings are generated for each fragment and stored in the vector database. This setup allows the chatbot to efficiently search for and retrieve relevant information from the PDF documents.

5.1 Fine-Tuning the Pre-Trained Model

Moreover, pre-trained models are highly customizable in Hugging Face, as their use and deployment provide a unique opportunity to tailor the chatbot to specific training needs. At this stage, it is extremely important to examine the refinement processes, including the selection of hyperparameters, the selection of model architecture, and the optimization of the model weight, to improve the ability of the chatbot to understand and generate answers relevant to educational contexts [20].

5.2 Evaluating the Chatbot's Performance

Additionally, evaluating the performance of a chatbot in an educational setting is a fundamental aspect to ensure its usefulness and effectiveness. In this regard, describing the assessment criteria, such as the accuracy of the answers, the coherence of the conversation, and the ability to provide appropriate educational support, will serve as metrics to determine whether the chatbot meets the main needs correctly.

Likewise, an analysis of the assessment methods that will be conducted should also be carried out, which may include student questionnaires, automated assessments, and qualitative comments to identify areas for improvement and refinement of the chatbot and its responses [20].

5.3 Adding Natural Language Processing Features

Furthermore, following the approach of education, the application of advanced NLP skills is an indispensable aspect to improve interactions between students and chatbots. At this stage, NLP techniques such as semantic analysis and user question recognition should be explored and implemented to enhance the chatbot's ability to understand complex questions, adapt to learners' natural language, and provide contextualized, objective, and useful answers [16, 19, 25].

The utilization of various NLP techniques in Lechuzobot ensures a more efficient and precise interaction. By employing the OpenAI text-embedding-ada-002 model, embeddings are generated as numerical representations of the text and stored in the vector database. This facilitates efficient and rapid similarity-based searches. Additionally, the 'ConversationBufferMemory' technique aids in maintaining the context of conversations with users. This NLP technique enables the chatbot to recall previous interactions within the current session, thereby enhancing the coherence and relevance of responses. This method is essential for retrieving pertinent fragments from PDFs when users submit queries.

5.4 Incorporate Response Suggestions

Along, the integration of response suggestions becomes a valuable component to facilitate the learning process and interaction with the chatbot. Implementing Hugging Face helps to explore opportunities to generate and present suggestions for answers based on the context of the conversation and the student's needs. These suggestions serve not only as a guide for the student, but also to encourage the exploration and deepening of the topics covered [24–25].

6 Deploying the Chatbot

Clearly, the effective implementation of the chatbot within a real educational environment is crucial to its practical usefulness. Consequently, this process involves carefully selecting the deployment platform that best suits educational and technological needs. For this, it is important to consider accessibility for students, scalability to handle multiple users, and integration with existing Learning Management Systems (LMS), although this will depend on the educational environment to which it is directed [25].

In the landscape of digital learning tools, Lechuzobot presently responds to inquiries from students utilizing the information it has been fed. While internal testing has been conducted solely with the development team and a limited audience, the project has not yet undergone full deployment. Nonetheless, it represents an ambitious and scalable endeavour poised for potential adaptation for classroom utilization.

6.1 Choosing the Deployment Platform

In this regard, the available options of the platform are evaluated considering factors such as accessibility, scalability, and integration with other education systems. Hugging Face facilitates the deployment of the chatbot through a dynamic and flexible interface. It also takes into account ethical and privacy considerations to ensure a safe and respectful learning environment [25].

Upon selecting the appropriate platform, the chatbot undergoes setup, integration, and comprehensive testing using diverse question sets to ascertain its optimal performance prior to full deployment within an educational setting. The successful integration of the chatbot represents a potent augmentation to the learning journey, thereby facilitating enhanced access to a myriad of high-quality educational resources, thereby enriching the overall learning experience for students.

7 Results: Overcoming Limitations and Technical Challenges

Although natural language models are adept at generating text, they can be uncontrollable and unstable. One of the common problems is when they do not follow the instructions given to them or generate incorrect answers, this phenomenon is referred to as "hallucinations".

This is because AI works on probabilities, so in the case of not having an answer, LLMs will often not indicate "I don't know" but will provide the most likely answer and be convinced that it is the absolute truth. Large models such as ChatGPT have an estimated hallucination rate of 15% to 20% [5, 25].

The LLM used has a limit of 16,385 tokens per request, which means that providing it with context must be optimized using a vector database - this is an advanced technical issue that may hinder its accessibility to the public.

In addition, the execution space within Hugging Face is limited by the current demand for that model means that the time it will take for the model to execute a request will depend on how many people are using it at the time.

To speed up processing, users can clone the space in their private Hugging Face repository, which offers 2vCPU and 16 GB of RAM for free, however, if a user needs more processing power, he will need to pay for it.

Similarly, another aspect to consider is that users depend on the capacity and limitations of each LLM, so they must adjust to model capabilities or find ways to optimize its performance, which becomes even more complicated when using a model that depends on other models for different types of processing.

7.1 Advantages and Disadvantages of the Lechuzobot

More importantly, within a conventional classroom, the chatbot offers significant advantages, such as 24/7 availability, allowing it to accommodate the unique study schedules of each student. Additionally, the bot can tailor challenges and questions to the individual comprehension levels of students, ensuring that information is presented in the most suitable format for each learner.

However, a primary disadvantage is the potential for AI-generated content to supplant human-generated content. For instance, a teacher lacking ideas for a class might resort to using artificial intelligence to generate research assignments for students. Subsequently, a student might use a large language model (LLM) to complete the research assignment.

The following day, the teacher might employ AI to verify the accuracy of the assignments, revealing a situation where only AI models are interacting with each other. This scenario compromises the quality of information, as it lacks the necessary human oversight to ensure the accuracy and relevance of the content.

Furthermore, it was determined that the bot shall be christened "Lechuzobot" as a deliberate homage to the esteemed community of students and professors at the Universidad Tecnológica de Tecámac. This decision is anchored in the university's emblematic mascot, the owl, symbolizing wisdom and knowledge. Hence, the chosen name aligns seamlessly with the institution's identity, fostering a sense of unity and recognition among its members.

7.2 Challenges and Future Scope

Although the best way to predict the future is to create it, it is impossible to know what might happen. Most of people did not expect the creation of ChatGPT. According to Karl Popper, 'predicting a new invention requires defining what that new invention is, which would only happen in the future' [6].

It is deduced that Lechuzobot possesses the capability to respond to inquiries based on the information it has been fed, meticulously searching for the most pertinent answers within its data repository. In the event of unavailability, the bot will suggest referring to the information contained within the PDFs.

Lechuzobot features a Gradio-powered interface comprising a dialogue box, a welcoming message, query suggestions for the bot, and three buttons: one to remove the preceding answer, another to clear the screen and the submit button for sending the query, as shown in Fig. 1.

7.3 Ethical Considerations in AI Education and Privacy Issues

Eventually, it is imperative to recognize that the data amassed pertains to individuals, thus necessitating the unwavering adherence to foundational principles such as respect, fairness, accountability, and human rights.

As well, safeguarding protection and privacy constitutes paramount considerations, necessitating the maintenance of essential security properties such as confidentiality, integrity, and availability of information. This mandates that access and modification rights are restricted to authorized users, ensuring data availability as required.

Furthermore, it is incumbent upon developers to inform and empower users, elucidating the rationale behind data collection, its methodologies, and the intended purposes

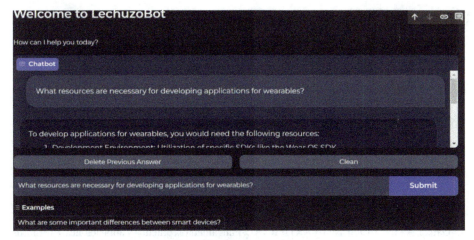

Fig. 1. Lechuzobot interface.

thereof. AI models must be engineered with the overarching goal of societal benefit, underpinned by principles of accountability and the capacity for scrutiny and rectification of erroneous decisions.

8 Conclusion

First, the integration of AI in education through chatbots marks a significant advancement in teaching and learning methodologies. Today, its importance in education is indisputable, as it provides simplified access to information and personalized content tailored to specific needs. AI-powered chatbots, trained through machine learning and equipped with relevant data, can create a learning environment where students feel comfortable and confident in expressing and resolving their doubts and questions.

Furthermore, implementing these chatbots makes obtaining information on any topic as effortless as conversing with a friend. This results in an adaptable, easy-to-implement tool that accelerates information dissemination, optimizes students' time, and resolves doubts quickly. A standout feature is their 24/7 availability, enhancing the efficiency of the educational process.

Consequently, this advancement represents a leap toward a future where learning is more accessible, personalized, and efficient. The application of chatbots in both face-to-face and digital education promises an enriched educational landscape that facilitates intellectual development and skill acquisition while promoting greater effectiveness in the educational process. It is essential to recognize that while this technology offers numerous advantages, it does not replace the irreplaceable value of human interaction in teaching.

In conclusion, it is crucial to continue exploring and developing AI-powered chatbots to ensure they complement and enrich the educational experience for both students and teachers. The goal is to enhance education, as the potential applications are vast and, if implemented correctly, could radically transform the educational field for future generations.

References

1. Aguinis, H., Hill, N.S., Bailey, J.R.: Best practices in data collection and preparation. Organ. Res. Methods **24**(4), 680–689 (2021)
2. Amazon. https://aws.amazon.com/what-is/data-preparation/. Accessed 14 Apr 2024
3. Amazon. https://aws.amazon.com/what-is/langchain/. Accessed 17 Apr 2024
4. Amazon. https://aws.amazon.com/es/what-is/large-language-model/. Accessed 16 Apr 2024
5. Bilan, M.: https://masterofcode.com/blog/hallucinations-in-llms-what-you-need-to-know-before-integration. Accessed 15 Apr 2024
6. Course Hero. https://www.coursehero.com/file/p3qfnfsj/Qu%C3%A9-dice-el-argumento-de-Karl-Popper-sobre-la-imposibilidad-de-predecir-nuevas/. Accessed 16 Apr 2024
7. IBM Watson Assistant. https://www.ibm.com/mx-es/products/watsonx-assistant/resources/how-to-build-a-chatbot. Accessed 17 Apr 2024
8. Feine, J., Morana, S., Maedche, A.: A chatbot response generation system. In: Proceedings of Mensch und Computer (2020)
9. García, A.: INTELIGENCIA ARTIFICIAL. Fundamentos, práctica y aplicaciones. RC Libros (2012)
10. Google Cloud. https://cloud.google.com/learn/what-is-artificial-intelligence. Accessed 14 Apr 2024
11. Gradio. https://www.gradio.app/guides/using-hugging-face-integrations. Accessed 16 Apr 2024
12. Haugeland, J.: La inteligencia artificial, 4th edn. (2001)
13. Harris, Y.: https://powell-software.com/resources/blog/how-to-train-chatbot-on-your-own-data/. Accessed 15 Apr 2024
14. IBM. https://www.ibm.com/topics/artificial-intelligence. Accessed 16 Apr 2024
15. Martinsanz, G.P., Peñas, M.S.: Inteligencia artificial e ingeniería del conocimiento. Ra-Ma (2005)
16. Mingondo, J., López, J., Spontón, H., Migoya, M., Englebienne, G.: Embracing the Power of AI. A Gentle CXO Guide. Roundtree press, Petaluma (2018)
17. Munir, H., Vogel, B., Jacobsson, A.: Artificial intelligence and machine learning approaches in digital education: a systematic revision. Inf. **13**(4), 2–8 (2022)
18. Nam, D., Macvean, A., Hellendoorn, V., Vasilescu, B., Myers, B.: Using an LLM to help with code understanding. In: IEEE/ACM 46th International Conference on Software Engineering (ICSE), EEE Computer Society (2024)
19. Nilsson, N. J.: INTELIGENCIA ARTIFICIAL. Una nueva síntesis. McGraw-Hill/Interamericana de España, S.A.U. (2001)
20. Notomoro. https://webisoft.com/articles/ai-based-chatbot/. Accessed 16 Apr 2024
21. Pepe, F.: How do hugging face models document datasets, bias, and licenses? An empirical study. In: 32nd IEEE/ACM International Conference on Program Comprehension (ICPC 2024), Lisbon, Portugal (2024). https://doi.org/10.1145/3643916.3644412
22. Rajendran, R. M.: Importance of using generative AI in education: dawn of a new era. J. Sci. Technol. **4**, 35–44 (2023)
23. Rahman, A.M., Al Mamun, A., Islam, A.: Programming challenges of chatbot: current and future prospective. In: 2017 IEEE Region 10 Humanitarian Technology Conference (R10-HTC), pp. 75–78. IEEE (2017)
24. Serrano, A.G.: Inteligencia artificial: fundamentos, práctica y aplicaciones. ALFAOMEGA GRUPO EDITOR, S.A de C. V. (2017)
25. Spinak, E.: https://blog.scielo.org/es/2023/12/20/es-que-la-inteligencia-artificial-tiene-alucinaciones/. Accessed 18 Apr 2024
26. Turing. https://www.turing.com/resources/finetuning-large-language-models. Accessed 15 Apr 2024

An Explainable Clustering Methodology for the Categorization of Teachers in Digital Learning Platforms Based on Their Performance

Emma Pérez García(✉)[iD], María Rosa Hortelano Díaz, and Eva Martín Rodríguez

Computer Science Faculty, Pontifical University of Salamanca, 37002 Salamanca, Spain
{eperezga.inf,mrhortelanodi.inf,emartinro.inf}@upsa.es

Abstract. Categorizing teachers using artificial intelligence (AI) on digital platforms opens up new ways to optimize teacher performance. By applying unsupervised clustering techniques together with supervised methods such as Random Forest and decision trees, significant patterns in teacher behavior and performance can be identified. These patterns are crucial for designing personalized educational interventions and improving the allocation of training resources, which significantly increases the effectiveness of teacher training programs. The results of these analyses highlight the potential of AI to provide valuable insights into teacher effectiveness, thus promoting more adaptive and equitable education. This methodological approach makes it possible to develop a deeper understanding of educational needs and to adjust pedagogical strategies effectively. The implementation of these technologies not only enriches the educational experience for teachers but also establishes a robust foundation for the expansion of AI use in a variety of educational settings. Furthermore, the study underlines the importance of maintaining a balance between technological innovation and fundamental pedagogical needs. Ensuring that the adoption of advanced technologies in education is beneficially inclusive and accessible to all stakeholders is essential to fostering a positive and lasting impact on the educational field.

Keywords: Artificial Intelligence · Machine learning · Teaching performance

1 Introduction

In the era of digital transformation, education systems around the world face the need to adapt to rapid and profound changes in technology and pedagogical methodology. AI (Artificial intelligence) emerges as an essential tool in this context, not only as an opportunity for innovation but also as a requirement for the reformulation of teaching and teaching practice. Within this framework, the ProFuturo Fundación[1], a joint initiative of Fundación Telefónica and Fundación "la Caixa", is dedicated to improving the quality of education in vulnerable regions of Latin America, the Caribbean, Africa,

---

[1] Further details regarding Fundación ProFuturo can be found at https://profuturo.education/quienes-somos/.

and Asia. Its mission is to reduce the educational gap not only by offering quality education to children in disadvantaged environments but also by strengthening this education through the continuous training of teachers in pedagogical methodologies and educational technologies.

As noted in previous studies [1], the identification and categorization of teachers according to their performance and acquired competencies are crucial for providing personalized training and targeting those teachers with less optimal performance. Although some studies have explored this issue, there is still a significant gap in the in-depth identification of patterns that characterize different types of Teachers [2]. A detailed understanding of the groups and characteristics of teachers will enable the design of personalized training strategies that focus specifically on their strengths and weaknesses.

Traditionally, assigning a teacher to a group based on their competencies, knowledge, and skills is a process that requires the intervention of human experts. However, in platforms such as ProFuturo, where the number of teachers and students is considerably high, carrying out this process manually would result in a significant economic cost. Fortunately, since the training that teachers receive is carried out through the Moodle digital platform, all relevant information for their evaluation is digitally recorded, thus facilitating its systematic analysis.

The use of advanced artificial intelligence techniques, such as clustering models, offers a promising solution to automatically identify different groups of teachers without the need for external supervision [3]. This not only reduces the associated costs but also ensures the scalability of the model. Although student clustering based on Moodle indicators has already been the subject of research, for ProFuturo it is essential to deepen the knowledge of the patterns that define each group of teachers. This task, traditionally less explored in unsupervised clustering techniques, can greatly benefit from the development of explanatory artificial intelligence models, such as those used in Deep Learning.

In this article, we propose a framework for clustering teachers according to their performance, using advanced clustering techniques that facilitate the interpretation and practical application of the results obtained. This approach not only improves the accuracy of educational interventions but also optimizes the allocation of educational resources, enabling more adaptive and user-centered education.

Using unsupervised clustering algorithms and supervised methods such as Random Forest and decision trees, this study seeks to identify meaningful patterns in the data that enable effective and strategic personalization of teacher training. We expect that the implementation of this system will contribute significantly to the optimization of educational resources, improve the sustainability of the education system, and respond effectively to the challenges of educational inequality.

This document is structured as follows: Sect. 2 reviews current methods and the state of the art. Section 3 details the proposed methods. Section 4 presents the results obtained and discusses their implications. Finally, Sect. 5 provides a summary of the conclusions and suggests future lines of research that could expand and deepen the initial findings of this study.

2 Background

The global education system faces unprecedented challenges in the era of digital transformation, exacerbating existing disparities in regions with limited access to advanced technology. In this context, the mission is to provide quality education to children in vulnerable environments by empowering teachers with new pedagogical and technological methodologies essential for sustainable educational development.

Teachers in these regions face numerous barriers that impede their professional development and the adoption of innovative educational practices. The ability to effectively identify and differentiate teacher profiles through advanced technologies is crucial, facilitating the personalization of teacher training and the optimization of the allocation of educational resources, which significantly increases the effectiveness of training programs.

This study introduces an innovative AI-based learning framework that combines unsupervised clustering techniques and supervised methods such as decision trees and Random Forest to analyze meaningful patterns in teaching data. Through a study on the Moodle platform, we have observed how performance-based clustering can differentiate between teachers who require additional development and those who are effectively optimizing their pedagogical practices. This approach not only uncovers these patterns but also allows for a precise segmentation of teacher profiles, thus facilitating highly personalized educational interventions aligned to individual learner needs. These preliminary findings suggest that data-driven personalization can significantly improve educational intervention.

AI systems analyze large volumes of data on teacher performance and behavior on digital platforms. They also offer unprecedented opportunities for evaluation and improvement of teaching. Big data-driven educational assessment is making remarkable progress, with various algorithmic models enabling more refined and personalized assessment in different educational contexts, from higher education to online learning environments [4].

One of the most illustrative examples of the application of learning analytics in real-world settings is presented by Herodotou et al. (2019). They described the implementation of the "OU Analyse" system at the Open University in the UK. This system uses machine learning methods for the early identification of students at risk of not submitting their assigned homework. The study found that teachers' access to interactive dashboards, which predict student risk every week, facilitated more informed and timely interventions. This type of predictive analytics not only helped teachers improve student performance but also enabled more effective personalization of educational intervention, demonstrating a clear benefit of integrating AI into teaching practice [5].

The transformative influence of AI on teacher education highlights how AI-enhanced methods, such as personalized learning environments, are reshaping educational landscapes. The integration of AI in teacher training, while promising, presents challenges related to ethical considerations, cultural adaptability, and technical issues. Such as the correct interpretation of performance data or adaptability, as solutions need to be sensitive to pedagogical norms and contexts without losing accuracy. Addressing these challenges is critical to ensure that the implementation of advanced technologies reinforces equitable and high-quality education [6] and does not replace essential human interactions in education. In this way, AI does not replace but expands and enriches human interactions in education.

The framework is evaluated through rigorous testing and data analysis, including assessments of pedagogy, digital citizenship, and professional development competencies. The analysis of this data helps to better understand the needs and progress of teachers, contributing to the optimization of educational resources and promoting more inclusive and equitable education.

This study seeks not only to improve the quality of education provision but also to proactively address the challenges posed by educational inequalities, providing a replicable model for future research and the application of advanced technologies in other educational contexts. This holistic approach ensures that the framework is not only a temporary solution, but also lays the foundation for ongoing innovations in education, adapting to the changing needs of society and technology, and highlighting the importance of considering the ethical and practical dimensions of technology in educational settings. The proposed framework is designed to adapt and scale with technological advances and changes in the education sector.

3 Methods

3.1 Data Collections

Data collection for this study is carried out through two platforms managed by Fundación ProFuturo. All data collected were fully anonymized to ensure privacy and compliance with data protection regulations. In addition, security measures were adopted, storing the information on personal computers protected with passwords.

Digital Skills Self-assessment Platform. Fundación ProFuturo offers a self-assessment platform[2] where teachers assess their competencies in three key areas: Pedagogy, Digital Citizenship, and Professional Development. Each area assesses the following four specific competencies:

- Pedagogy: Pedagogical Practice, Curation and Creation, Evaluation, Personalization.
- Digital Citizenship: Responsible Use, Safe Use, Critical Use, Inclusion.
- Professional Development: Self-Development, Self-Assessment, Sharing, Communication.

The data collected includes user identifiers, dates of assessments, and scores on a scale of 1 to 5 for each competency. This data allows for the analysis of teachers' progress and professional development.

Educational Training Platform. This platform, implemented on the Moodle learning management system, provides data on teachers' performance, participation, and interaction in learning activities. The information collected includes:

---

[2] Further information regarding the Fundación Profuturo digital skills self-assessment platform can be found at https://competencyassessment.profuturo.education/.

- Demographics: User ID, country, city, and gender.
- Educational Activity: Data on the number of courses completed, and progression through the courses, including course identifiers and dates of completion of the first and last module.

This data collection process is carried out from the Moodle database, using the following tables:

- Context: Provides the organizational structure and location of courses within the system.
- Course Categories: Details the categories under which the courses offered are grouped.
- Course Completions: Records the courses completed by each user.
- Course Modules and Course Modules Completion: Provides information on course modules and their completion by users.
- Enrolments: Lists user enrolments in the various courses.
- User Data: Contains demographic data and other additional records provided by users.

3.2 Population and Samples

The study sample includes teachers active on the ProFuturo training platform who have completed at least two self-assessments, ensuring that the data reflect continuous development and progress in their competencies. The sample size selected is 228 teachers (see Table 1), which is considered adequate for statistical analysis and the application of ML algorithms.

Table 1. Sample size distribution by gender and country.

Country	Men	Women	I prefer not to say	NaN	Total
Ecuador	32	71	0	6	109
México	33	60	0	7	100
Other	4	9	1	5	19
Total	69	140	1	18	228

Note: Countries included in the "Other" category: Chile, El Salvador, Guatemala, Panama, Spain and Uruguay.

3.3 Procedures

For the study of teacher profiles, we adopted a systematic methodology integrating advanced Machine Learning techniques for the clustering and detailed analysis of the data. The specific procedures employed are described below:

Initially, data were collected from multiple platforms dedicated to teacher evaluation and training. To preserve the confidentiality and integrity of the data, we implemented anonymization procedures and undertook a meticulous cleaning process to standardize data formats, which was essential to facilitate subsequent analyses.

Extensive exploratory analysis was carried out to identify patterns and correlations between variables. This crucial step uncovered the inherent structure of the data, guiding the informed selection of the most appropriate Machine Learning techniques to address the research hypotheses.

Significant or critical variables were transformed, calculating averages and differences that reflected the development and competencies of educators over time. This approach reduced the dimensionality of the data and focused the analysis on the most impactful attributes of teacher effectiveness.

Finally, the dataset was carefully prepared for clustering by selecting only the most significant variables. This structuring facilitated the use of advanced clustering techniques to analyze the data, optimizing the detection of significant groupings among teacher profiles.

Unsupervised Learning. The KMeans algorithm was chosen to segment the data due to its efficiency in handling large volumes of data and the interpretive clarity it offers. The optimal number of clusters was determined by evaluating the silhouette score [7] which measures the cohesion and separation of clusters, thus indicating how appropriately the data is grouped. The goal of KMeans was to mini-measure inertia (Eq. (1)), formally defined as:

$$\text{Inertia} = \sum_{i=1}^{n} \min_{\mu_j \in C} \|x_i - \mu_i\|^2 \qquad (1)$$

where $\|x_i - \mu_i\|^2$ represents the squared distance between a point x_i and the centroid of its cluster μ_i y C is the set of all centroids. A lower inertia indicates more compact and homogeneous clusters, contributing directly to the internal cohesion of the cluster.

Despite its advantages, KMeans presents challenges such as sensitivity to the scale of the data and the initial selection of centroids, problems that can be mitigated by normalizing the data.

Supervised Learning. For the supervised models, Random Forest and Decision Trees classifiers were used. The hyperparameters of each model were manually adjusted, focusing the evaluation on accuracy (Eq. (2).) and F1 score (Eq. (3)), calculated by cross-validation.

The Random Forest model was employed to identify the most significant variables in cluster prediction, providing a deep understanding of the factors impacting the clustering of the data. On the other hand, the Decision Tree was instrumental in deriving clear and understandable classification rules, which facilitates the interpretation of the models and their justification in applied contexts, underlining the importance of explainability in Machine Learning models[3].

$$\text{Precision} = \frac{TP}{TP + FP} \qquad (2)$$

$$F1 = 2 \cdot \frac{\text{Precision} \cdot \text{Sensibility}}{\text{Precision} + \text{Sensibility}} \qquad (3)$$

This focus on explainability not only improves confidence in the model results but also ensures that decisions based on the clustering results are transparent and justifiable, a crucial aspect when these models are applied in educational and other settings where decisions have significant impacts. All these procedures were performed using the Scikit-learn library in Python, which ensures reproducibility and efficiency in the implementation of the models.

All ML procedures were performed using the Scikit-learn Python library [8].

3.4 Interpretability

The results are encapsulated in a structured JSON file that synthesizes the information derived from the clustering and supervised classification analysis. Each object within the array represents a unique cluster identified in the study, providing a detailed view of the analyzed teacher segments. Key elements within each cluster object include:

- Cluster: Unique numerical identifier for each cluster.
- Number of users: Reflects the number of teachers associated with each cluster, allowing an appreciation of the magnitude of each group.
- Importance of characteristics: Details that assign a numerical value to the relevance of different characteristics, reflecting their weight in shaping the cluster and their influence on teaching behavior patterns.
- Rules: A set of logical criteria obtained from the decision tree model, which specify the conditions under which a teacher is assigned to a cluster.
- Categorical variables: List of variables relevant to the cluster along with their possible values, which highlight key differences between groups.
- Numerical variables: Includes descriptive statistics such as mean and standard deviation for the relevant numerical variables, providing a quantitative summary of the behavior of these variables within the cluster, which facilitates comparisons and analysis.

In addition, to complement the interpretation of the clusters and to provide an intuitive visualization of the differences and similarities between the identified groups, a principal component analysis (PCA) was applied. This method reduces the dimensionality of the data while maintaining as much variation as possible.

---

[3] Note: In the equations above, TP (True Positives) represents the number of positive cases correctly identified by the model, FP (False Positives) is the number of negative cases incorrectly classified as positive, and Sensitivity (also known as the true positive or recall rate) measures the proportion of true positives that were correctly identified (Sensitivity = TP/TP + FN).

4 Experimental Results and Discussion

A Through the implementation of the techniques described in Sect. 3, three clusters were identified that reflect different levels of progression in the competencies analyzed (see Table 2),

Table 2. Sample size distribution by cluster.

Cluster	% of Users	Number of Records
Cluster 0	56.58%	129
Cluster 1	18.86%	43
Cluster 2	24.56%	56

In our analysis, three clusters representative of different levels of teaching competence were identified within the sample. Cluster 0, which constitutes 56.58% of the participants, includes teachers with high performance, highlighting their effectiveness in the integration of technologies and advanced pedagogical methodologies. Cluster 1, with 18.86% of the sample, groups teachers with intermediate performance who still have room for improvement in key competencies. Finally, Cluster 2, comprising 24.56%, consists of teachers in the early stages of adopting new practices, showing the need for additional training support.

One of the main limitations of the study includes the reliance on self-assessment in the ratings, which may introduce bias. Teachers may overestimate or underestimate their abilities, which affects the accuracy of the data collected. To address these limitations in future research, we recommend incorporating multiple data sources, including peer evaluations and analysis of direct classroom interaction, to gain a more complete and balanced understanding of teacher performance.

5 Conclusion and Future Works

The integration of artificial intelligence technologies in teacher training within digital learning platforms has a significant high impact. By employing a combined approach of unsupervised clustering techniques and supervised methods, such as Random Forest and decision trees, a detailed understanding and effective categorization of teacher performance has been achieved. This methodological approach has enabled the development of highly personalized educational interventions, optimizing the allocation of resources and consequently improving the quality of education provided.

The findings of this study highlight the ability of AI to provide deep and analytical insights into teacher effectiveness, leading to more adaptive and equitable education. The implementation of these technologies not only enriches the educational experience for teachers but also establishes a robust foundation for the expanded use of AI in diverse educational settings. Furthermore, the paper confronts and discusses the ethical and adaptive challenges inherent in the integration of AI systems in education, highlighting the importance of a careful balance between technological innovation and fundamental pedagogical needs.

The field of educational data mining continues to expand into the use of complex models to discover applicable knowledge [3]. Our work on teacher categorization through advanced clustering answers this call, opening new avenues for future research in large-scale educational personalization.

It is essential to continue to evaluate the ethical implications and ensure that the adoption of advanced technologies in education is beneficially inclusive and accessible to all stakeholders. The results not only validate the utility of AI in improving educational programs but also raise recommendations for future research, promoting an ongoing dialogue on best practices for the implementation of advanced technologies in education. This study serves as a call to action for educational researchers and practitioners to further explore the possibilities that AI can offer, ensuring that its implementation contributes effectively and ethically to the educational field. In addition, this paper addresses the ethical and adaptive challenges of integrating AI systems in education, highlighting the importance of maintaining a balance between technological innovation and essential pedagogical needs. Ferguson's [9] review of learning analytics highlights how these technologies can be implemented to improve both learning and the environment in which it occurs, suggesting that optimization of these processes must be done with careful consideration of ethical implications and accessibility, ensuring that the benefits of AI in education are inclusive and widely accessible [4].

References

1. Green, T., Brown, M.: Machine learning in higher education: a case study on building predictive models for students retention. Artif. Intell. Soc. 257–269 (2020)
2. Koedinger, K., McLaughlin, E., Heffernan, N.: A quasi-experimental evaluation of an on-line formative assessment and tutoring system. J. Educ. Comput. Res. **43**, 489–510 (2010)
3. Baker R., Yacef, K.: The state of educational data mining in 2009: a review and future visions. J. Educ. Data Min. **1**, 3–17 (2009)
4. Lin, L., Zhou, D., Wang, J., Wang, Y.: A systematic review of big data driven education evaluation. SAGE Open (2024)
5. Herodotou, C., Hlosta, M., Boroowa, A., Rienties, B., Zdrahal, Z., Mangafa, C.: Empowering online teachers through predictive learning analytics. Br. J. Edu. Technol. **50**(6), 3064–3079 (2019)
6. Singh, V., Ram, S.: Impact of artificial intelligence on teacher education. Shodh Sari-Int. Multidiscip. J. 243–266 (2024)
7. Shutaywi, M., Kachouie, N.: Silhouette analysis for performance evaluation in machine learning with applications to clustering. Entropy **23**, 759 (2021)
8. Pedregosa, F., et al.: Scikit-learn: machine learning in Python. J. Mach. Learn. Res. 2825–2830 (2011)
9. Ferguson, R.: Learning analytics: drivers, developments and challenges. Int. J. Technol. Enhanc. Learn. **4**, 304–317 (2013)

Educational Platform for Inclusive Learning with Deep Camera Integration and Serious Games

Héctor Sánchez San Blas[1], Rocío Galache Iglesias[2], Enrique Maya-Cámara[2], Blanca García-Riaza[3], Ana Paula Couceiro Figueira[4], Josué Prieto-Prieto[5], and André Sales Mendes[1(✉)]

[1] Expert Systems and Applications Lab - ESALAB, Faculty of Science, University of Salamanca, Plaza de los Caídos s/n, 37008 Salamanca, Spain
{hectorsanchezsanblas,andremendes}@usal.es

[2] University of Salamanca, Patio de Escuelas s/n, 37008 Salamanca, Spain
{rociogalache,enriquemayacamara}@usal.es

[3] Department of English Philology, University of Salamanca, 37008 Salamanca, Spain
bgr@usal.es

[4] Faculty of Psychology and Educational Sciences, University of Coimbra, Coimbra, Portugal
apouceiro@fpce.uc.pt

[5] Institute of Education Sciences (IUCE), Department of Didactics of Musical, Plastic and Body Expression, University of Salamanca, 37008 Salamanca, Spain
josueprieto@usal.es

Abstract. This research presents the development and implementation of an inclusive learning platform that uses serious games and interactive projection technology to facilitate educational access for people with disabilities, the elderly, and other vulnerable groups. The platform transforms any flat surface into an interactive space by integrating a projector and a depth camera to detect and respond to user interactions in real-time. This technology aims to promote a more inclusive and adaptive educational environment that adjusts to each user's needs and abilities. The serious games developed for this platform focus on improving motor, cognitive, and social skills, and educators can customize them to cover various subjects and difficulty levels. This approach aims to increase user engagement and motivation and facilitate more effective knowledge acquisition. This study contributes to educational technology by demonstrating how the combination of serious games and interactive technology can be a powerful tool for inclusion and adaptive education.

Keywords: inclusive education · serious games · interactive projection technology · user interaction

1 Introduction and Background

Games have been an integral part of human development since antiquity, serving as a means of entertainment and pedagogical tools. For instance, Medrano points out that board games have been used since Roman times, from which the term "ludo" is derived, signaling their playful nature and purpose beyond mere entertainment [1]. Braghirolli et al. emphasize the duality of the concept of games, which includes both commercially oriented entertainment games and those specifically designed for educational purposes [2].

In the educational environment, games have been categorized in various ways to reflect their diversity and specificity, encompassing terms such as digital learning games, game-based learning, educational entertainment games, persuasive games, epistemic games, instructional games, serious games, and purposeful games [3]. The latter category, purposeful games, refers to those designed with a deliberate and explicit educational objective beyond mere entertainment to promote specific and measurable learning outcomes [4,5].

Although often interchangeable, serious games and game-based learning have distinct applications and audiences. For example, serious games have been employed in contexts ranging from business training to formal education, aiming to teach and induce behavioral changes in diverse areas such as business, health, and education [6].

Studies such as those by Kebritchi and Hirumi argue that educational games are particularly effective for teaching complex procedures and skills because they facilitate active learning, motivate students, and provide immediate feedback, all within contexts that simulate real-life situations interactively [7]. Modern learning theories support this view, indicating that learning is most effective when it is active, experiential, situated, and problem-based [8].

As game development technologies advance, these pedagogical tools gain increasing relevance in the teaching-learning processes due to their ability to adapt to multiple learning styles and their potential to effectively include vulnerable groups whom more traditional teaching methods may have historically marginalized. Thus, educational games emerge as crucial components in the evolution of contemporary pedagogical practices, demonstrating their value as engagement tools and powerful facilitators of comprehensive and accessible learning [9].

Gestural iteration in educational games has become crucial for facilitating a more immersive and natural learning experience. Image processing and gesture detection allow the games to respond to users' physical actions in real-time, providing an interactive platform that captures the player's attention but also promotes kinesthetic learning [10]. These technologies use depth cameras to analyze body movements and translate them into actions within the game, enabling students to interact directly with the educational content through physical movements.

Advancements in computer vision techniques have enabled the development of games that can accurately interpret complex and varied gestures. This achievement is made possible by integrating context-based architectures that detect standard movements and learn and adapt to the individual peculiarities of each player's gestures [11]. This customization ensures the game is accessible to many physical abilities and learning styles. Also, it enhances the educational experience by allowing dynamic adjustments that align with the specific learning needs of each user.

Furthermore, integrating image processing techniques in educational games facilitates the automatic assessment of learning progress. For example, the accuracy and speed with which students perform the required gestures can indicate their understanding and proficiency in the addressed subjects [12]. This approach enhances the games' ability to teach practical skills and concepts and allows educators to monitor and evaluate student performance more efficiently and objectively.

These technological innovations in gesture detection and image processing transform educational games from simple teaching tools into complex learning platforms that offer deeply interactive and personalized experiences [13]. This development reflects a significant shift in how digital education is conceptualized and underscores the importance of emerging technologies in the evolution of teaching methods that are both inclusive and effective.

Analyzing the state of the art, the role of educational games has significantly expanded due to technological advances in areas such as image processing and gesture detection. These innovations have enriched the learning experience, making it more interactive and personalized, and have also transformed educational games into powerful tools for teaching and assessing complex skills. As these technologies evolve, we will see even greater integration of intelligent tools in educational environments, promising to revolutionize how skills are taught and learned. These advancements highlight the importance of continuing to explore and develop new applications for technology in education, ensuring that all students, regardless of their physical abilities or learning styles, have access to rich and engaging learning opportunities.

2 Proposed System

The inclusive learning process for people with disabilities, older adults, and other vulnerable groups can vary significantly based on the specific needs of the individual and their educational context. Typically, inclusive education involves a combination of adaptive pedagogical methods and support technologies. Although traditional methods are effective, they are often costly, and their implementation can be limited by the availability of resources, making them less accessible.

In response to these limitations, this work proposes the development of a playful learning platform that uses interactive projection technology and serious games to provide an accessible and motivating educational environment. The platform's main goal is to enhance the inclusion and personalization of the educational process through gamification and computer vision.

The proposed exercises are carried out using different types of games adapted to each user's characteristics and capabilities. For example, the games may require the user to perform specific gestures, movements, or displacements that assist in treating and improving cognitive and motor skills.

To detect the actions performed by the user who will interact with the platform, an image processing system uses a depth camera that captures the movements of the user's arms and estimates the position of the hands and arms with high precision. This information obtained through the analysis of camera images is processed by the platform, allowing the user to interact with the virtual environment. All data generated by the platform are analyzed to adapt the user's learning process according to their progress in performing the proposed exercises.

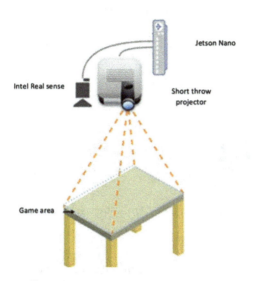

Fig. 1. Diagram of the hardware elements that are imperative in the platform.

The platform uses low-cost and widely available components such as the Jetson Nano microcomputer, RealSense depth camera, and a projector commonly found in most modern classrooms. The specific arrangement of these elements can be observed in Fig. 1, facilitating their implementation and accessibility in educational environments. The components that make up the system are described in detail below:

- **Projector:** A short-throw projector positioned vertically from the ceiling projects images and interactive games onto any flat surface, such as a table, a carpet, or a floor, transforming ordinary spaces into dynamic learning areas.
- **Depth Camera:** The RealSense depth camera captures three-dimensional data of the environment and user movements, generating a point cloud to precisely detect the position of hands and arms, facilitating immersive interaction with the virtual environment.
- **Microcomputer:** Nvidia's Jetson Nano microcomputer is optimized for image processing, allowing real-time processing of the information collected by the camera and analyzing the point cloud to provide input to the games.

3 Tracking System

Access to the system is allowed through a procedure that detects the user's interactions with the play area. This procedure (Fig. 2) starts with an image captured by the RealSense camera. It involves several phases with distinctly separated functionalities that, once completed, generate input for the game.

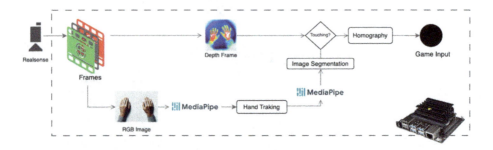

Fig. 2. Iteration detection flowchart of user's hands.

- *Fase 1: Frame Capture:* The RealSense camera captures two types of frames simultaneously: RGB images and depth frames. The RGB images are used for visual display. At the same time, the depth frames, representing the three-dimensional information of the captured space, are crucial for determining the position and movement of the user's hands.
- *Fase 2:Hand Detection:* Once the frames are captured, the RGB frame and MediaPipe are used to track hand movements. MediaPipe processes the RGB images to detect and track the hands in real-time. This method accurately identifies the position and movements of the hands, which are crucial for subsequent interaction with the game.
- *Fase 3:Image Segmentation and Depth Analysis:* Once the hands are located within the image, the system uses image segmentation to determine the region where the hand is present. With the region identified, information from the depth frame is used to ascertain the hand's distance from the table. This helps evaluate whether the hand is truly "touching" the table or a game object, thus dismissing invalid interactions or potential movements significantly distant from the table. Figure 3 shows two examples of hand detection, either touching or not touching the table with just a 5 cm height difference from the table.
- *Fase 4:Homography:* A homographic transformation is performed to align and adapt the coordinates of the detected hands to the coordinate system of the game environment. In Fig. 4, an example of the transformation of the hand coordinate from the left window to the transformation of the window on the

(a)   (b)

Fig. 3. Detection of iterations with the game using the half-pipe and point cloud to determine the distance to the surface. A) Table touching. B) Not touching the table

right, which is the direct input to the game, can be seen. This homography is essential to ensure that physical movements are accurately translated into the virtual space, allowing for precise and consistent interaction.

Fig. 4. Example of homographic transformation applied to the hand coordinates showing the conversion of the left window to the right window as a direct input to the game.

- *Fase 5:Interaction with the Game:* Finally, all processed and validated interactions are translated into inputs for the game in the form of coordinates that contain information about the type of hand, the gesture made, whether or not it is touching the table and the distance from the table. This information is sent via WebSockets to an application developed in PhaserJS, responsible for generating the game logic and projecting the information onto the corresponding projector window. This type of communication allows for easy adaptation of the input system to any language or game development environment. An example of the JSON structure used for data transmission can be seen in Listing 1.

Listing 1.1: JSON Data Example

```
{
  "hand_type": "right",
  "gesture": "pointing",
  "touching_table": true,
  "distance_from_table": "15",
  "coordinates": {
    "x": 102,
    "y": 348
  }
}
```

4 Complete System

This section aims to explore the proposed platform's potential. Through simulation and analysis of its design, we discuss the system's capabilities in an ideal educational environment.

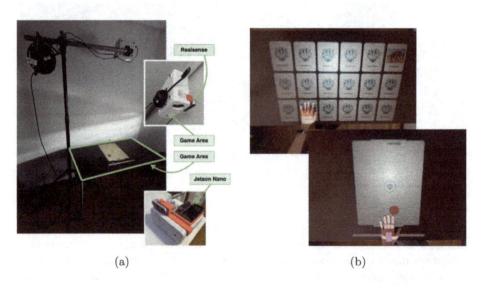

(a)   (b)

Fig. 5. a) System component's location. b) User's interaction with the system.

As shown in Fig. 5a, the system consists of the components described in Sect. 2 arranged over a play area on a table. This setup could easily be transferred to a lower surface, such as the floor or a carpet, particularly when working with children who prefer these surfaces. This arrangement enables precise gestural interaction, essential for manipulating the interactive elements of the proposed games (Fig. 5b).

The platform includes a variety of games created using the PhaserJS framework. These games have been designed to enrich cognitive and motor skills, employing interactive technology to create an accessible and engaging educational environment. Figure 6 displays the three different games developed for the platform.

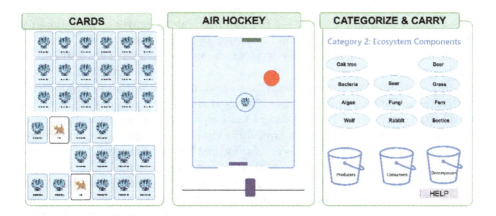

Fig. 6. Schedule about developed games

One of the games is "*CARDS*" as shown in Fig. 6, based on the classic memory game where players must find pairs of cards laid face down. Through gestures captured by the depth camera, users virtually "flip" the cards to reveal the images and find matches. This game is entertaining and enhances memory and concentration, fundamental skills for ongoing learning and everyday life.

Another game integrated into the platform is "*AIR HOCKEY*", which follows the traditional rules of the physical game. Users interact with the game through pushing gestures to control a virtual mallet and strike a puck to score to the opponent. This game requires quick reflexes and precision, promoting hand-eye coordination and improving motor skills while providing a fun and dynamic environment for light exercise.

The third game, "*CATEGORIZE and CARRY*", focuses on education by classifying elements representing topic-specific items. Players must drag these elements to the appropriate containers for the corresponding category, utilizing gestures captured by the camera in various contexts.

These games have been carefully selected to maximize the interactive capabilities of the platform, providing users with an educational experience that is both informative and engaging. Combining interactive projection technology with serious games offers a powerful educational tool tailored to enhance various skills and promote meaningful learning applicable in real-world contexts. The implementation of such technologies not only makes learning more accessible and appealing to a broad audience, including those with special needs or from vulnerable groups.

Compared to traditional interactive educational systems, which often rely on tactile interfaces or keyboards, our system innovates by employing computer vision technology to create a far more dynamic and accessible interface. This approach eliminates physical barriers associated with conventional devices, allowing users to interact with the system through natural movements and gestures. Such advancements facilitate physical interaction, particularly for those with mobility limitations or motor difficulties, and promote greater inclusion.

The algorithm designed for the system enables it to detect and respond to a wide range of physical movements, which is particularly beneficial for users who might struggle with more restrictive input methods such as keyboards or touch screens. This flexibility ensures that individuals of all ages and abilities can actively participate, making education a more equitable and adaptive experience.

Moreover, this technology can transform virtually any space into an interactive learning environment. This is crucial for institutions that cater to a diverse population, including those with physical or cognitive disabilities, as it adapts learning to individual needs and encourages greater autonomy and participation in the educational process. Ultimately, integrating platforms like the one designed into educational systems represents a step towards creating truly inclusive and accessible learning environments for all.

5 Conclusions

Following the evaluation of the project's results, it is evident that the developed system effectively promotes educational interaction through games that utilize gesture recognition techniques. This system is not only innovative but also economically accessible for educational institutions, as it allows for the reuse of existing devices or the acquisition of new ones at reduced costs.

Implementing image recognition mechanisms alongside depth cameras has been crucial for the system's full functionality. The tests conducted demonstrated the platform's correct functioning. However, these results are preliminary, and it is crucial to extend the testing to gather more data that confirm and expand our understanding of the system's efficacy within educational environments.

For the next stages of this project, there will be an emphasis on conducting comprehensive tests in educational settings to certify the system's functionality and correct approach. There is an anticipation of integrating this technology with established educational platforms, such as Moodle, to maximize its reach and effectiveness. Additionally, features such as self-assessment systems that promote greater autonomy in student learning will be explored. The scope of the study will also be expanded by conducting tests with a broad sample of students and teachers, which will allow for a robust and diversified evaluation of the educational impact of the system. This multidisciplinary approach aims to strengthen the foundations of interactive and adaptive learning, proposing continuous improvements that meet emerging educational needs and significantly contribute to the evolution of pedagogical practices.

Acknowledgments. Héctor Sánchez San Blas's research was supported by the Spanish Ministry of Universities (FPU Fellowship under Grant FPU20/03014). This project results from a research stay at the University of Coimbra in 2023. Special thanks are extended for their hospitality and support, which were instrumental in developing this work.

References

1. Medrano, N.: El gran libro de los juegos de mesa. Ediciones Andremeda, Buenos Aires (2005)
2. Braghirolli, L.F., Ribeiro, J.L.D., Weise, A.D., Pizzolato, M.: Benefits of educational games as an introductory activity in industrial engineering education. Comput. Hum. Behav. **58**, 315–324 (2016). https://doi.org/10.1016/j.chb.2015.12.063
3. Djaouti, D., Alvarez, J., Jessel, JP., Rampnoux, O.: Origins of serious games. In: Ma, M., Oikonomou, A., Jain, L. (eds.) Serious Games and Edutainment Applications, pp. 25–43. Springer, London (2011). https://doi.org/10.1007/978-1-4471-2161-9_3
4. Abt, C.C.: Serious games. University Press of America (1970)
5. Riemer, V., Schrader, C.: Learning with quizzes, simulations, and adventures: students' attitudes, perceptions, and intentions to learn with different types of serious games. Comput. Educ. **88**, 160–168 (2015). https://doi.org/10.1016/j.compedu.2015.05.003
6. Connolly, T.M., Boyle, E.A., MacArthur, E., Hainey, T., Boyle, J.M.: A systematic literature review of empirical evidence on computer games and serious games. Comput. Educ. **59**(2), 661–686 (2012). https://doi.org/10.1016/j.compedu.2012.03.004
7. Kebritchi, M., Hirumi, A.: Examining the pedagogical foundations of modern educational computer games. Comput. Educ. **51**(4), 1729–1743 (2008). https://doi.org/10.1016/j.compedu.2008.05.004
8. Arnab, S., Berta, R., Earp, J., De Freitas, S., Popescu, M.: Framing the adoption of serious games in formal education. Electron. J. e-Learn. **5**(2), 159–171 (2012)
9. Yıldız, D., et al.: Development and evaluation of an image processing-based kinesthetic learning system. Appl. Sci. **14**, 2186 (2024). https://doi.org/10.3390/app14052186
10. Sánchez San Blas, H., Sales Mendes, A., de la Iglesia, D.H., Silva, L.A., Villarrubia González, G.: A multiagent platform for promoting physical activity and learning through interactive educational games using the depth camera recognition system. Entertain. Comput. **49**, 100629 (2024). https://doi.org/10.1016/j.entcom.2023.100629
11. Sales Mendes, A.F., Sánchez San Blas, H., Pérez Robledo, F., et al.: A novel multiagent system for cervical motor control evaluation and individualized therapy: integrating gamification and portable solutions. Multimed. Syst. **30**, 131 (2024). https://doi.org/10.1007/s00530-024-01328-6

12. Serrano, Á., Marchiori, E.J., del Blanco, Á., Torrente, J., Fernández-Manjón, B.: A framework to improve evaluation in educational games. In: Proceedings of the 2012 IEEE Global Engineering Education Conference (EDUCON), Marrakech, Morocco, pp. 1–8 (2012). https://doi.org/10.1109/EDUCON.2012.6201154
13. Tobias, J.L., Di Mitri, D.: Using accessible motion capture in educational games for sign language learning. In: Viberg, O., Jivet, I., Muñoz-Merino, P., Perifanou, M., Papathoma, T. (eds.) Responsive and Sustainable Educational Futures. EC-TEL 2023. LNCS, vol. 14200, pp. 762–767. Springer, Cham (2023). https://doi.org/10.1007/978-3-031-42682-7_74

Author Index

A

Abbasi, Maryam 3, 123, 363
Abomhara, Mohamed 97
Abrantes, José Luís 148, 171, 410
Aidos, Manuel 389
Alejano, Fernando Lobato 372
Almeida, Filipe William C. 431
Almeida, Tiago 423
Álvarez-Sánchez, Arturo 62
Alves, Victor 319
Artman, Henrik 40
Ashaduzzaman, Md. 73

B

Bano, Taranum 85
Bastos, André 380
Berry, Ann L. Anderson 85
Blas, Héctor Sánchez San 231, 464
Bom Jesus, Vagner 244

C

Caetano, Filipe 194, 219
Caldeira, Filipe 133, 159, 180, 256, 281, 380, 389, 423
Cardoso, Filipe 3, 363
Carneiro, Davide 319
Carreto, Carlos 244
Cláudio, Ricardo 180
Corbacho, Carlos Chinchilla 372
Correia, Luciano 133, 281
Cristovam, Rodrigo 389
Cunha, Carlos A. 28

D

Daniel, José 194
de la Iglesia, Alejandro H. 372
de la Iglesia, Daniel H. 148, 171
de Lima, José Donizetti 15
de Moraes, André Fabiano 53, 431
de Moura Speroni, Rafael 431
de Paz Santana, Juan Francisco 15
Deon, Samara 15
Dexe, Jacob 40
Díaz, María Rosa Hortelano 455
Dib, Jorge Zakour 372
Diego, Belen Curto 231
dos Anjos, Julio Cesar Santos 15
Dranka, Geremi Gilson 15
Duarte, Rui P. 28
Duri, Taha 355

F

Fahad, Muhammad 97
Fernandim, Laura 219
Ferraz, João 363
Ferreira, Rui 269
Figueira, Ana Paula Couceiro 464
Figueiredo, Diana M. 28
Franke, Ulrik 40

G

García, Emma Pérez 455
García-Riaza, Blanca 464
Gardenier, Anne Marte 399
Gaudêncio, Bernardo 363
Gonçalves, Celestino 194, 219, 269
Gonçalves, Vicente 194
González, Gabriel Villarrubia 231
González, Sergio García 231
Gouveia, Eduardo 148
Gouveia, Sónia 148, 171

H

Helgesson Hallström, Celine 40
Henriques, João 133, 159, 180, 256, 281, 380, 389, 423
Hernandez Hernandez, Jatziri 444

I

Iglesias, Rocío Galache 464

J
Jiménez-Bravo, Diego M. 62
Jorge, Henrique 389

K
Karki, Bishwa 85

L
Leitão, Jorge 180
Leithardt, Valderi Reis Quietinho 15
Lopes, Manuel 133, 281
López-Rivero, Alfonso 109, 207
Lozano Murciego, Álvaro 62

M
Magalhães, Jhony Reinheimer 53
Marques, Rafael 269
Martín-Gómez, Lucía 331
Martins, Pedro 3, 109, 123, 207, 363
Maya-Cámara, Enrique 464
Mendes, André Filipe Sales 231
Mobeen, Noor E. 97
Monteiro, Pedro 423
Moreno-García, María N. 416
Murciego, Álvaro Lozano 416

N
Nguyen, Thi 73
Nweke, Livinus Obiora 97

P
Pais, Luís 380
Palumbo, Guilherme 319
Parra, Daniel Blanco 331
Pina, Eduardo 256
Pinto, João 123, 159
Prieto-Prieto, Josué 464

R
Ramos, José 256
Raperger, Gabriel 207
Ribeiro, Ana 410
Ribeiro, Matheus Henrique Dal Molin 15

Rivero, Alfonso J. López 148, 171
Rivero, Alfonso 410
Rodilla, Vidal Moreno 231
Rodríguez, Eva Martín 455
Rodríguez, Francisco Javier Blanco 231
Rojas González, Dafne Itzel 444
Román, Adrián Valera 416
Rosado, José 3
Royakkers, Lambèr 295

S
Sales Mendes, André 62, 464
Santos, Vasco 148
Shaikh, Sarang 97
Silva, Carlos 380
Silva, Diogo 123, 159
Silva, Elisabete 148
Silva, José 3, 109, 123, 207, 363
Silva, Luís Augusto 53, 62, 431
Silva, Marco 3
Silveira, Clara 194, 219, 244
Slobodenyuk, Nadiya 346
Soares, Francisco 180
Soriano Rivero, Josue Aaron 444
Sousa, David 109

T
Thoene, Melissa K. 85
Tsai, Chun-Hua 73, 85

V
Vadapalli, Jagadeesh 85
van der Puil, Roxanne 307
VanOrmer, Matt 85
Varanda, José 123, 159
Váz, Paulo 3, 109, 123, 207, 363

W
Wanzeller, Cristina 159, 180, 256, 380, 389, 423

Y
Yayilgan, Sule Yildirim 97

Printed in the USA
CPSIA information can be obtained
at www.ICGtesting.com
CBHW070329230924
14771CB00003B/35

9 783031 666346